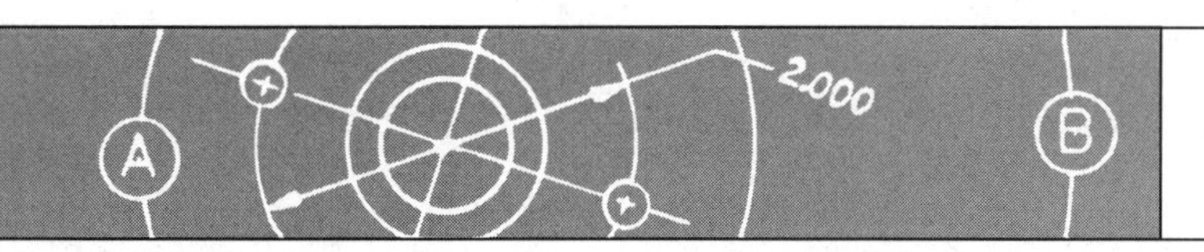

Blueprint Reading for the Machine Trades

4th EDITION

RUSS SCHULTZ
Hawkeye Institute of Technology, Ret.
Waterloo, Iowa

LARRY SMITH
St. Clair College of Applied Arts
and Technology
Windsor, Ontario

Upper Saddle River, New Jersey
Columbus, Ohio

Library of Congress Cataloging-in-Publication Data

Schultz, Russ.

Blueprint reading for the machine trades / Russ Schultz, Larry Smith. — 4th ed.

p. cm.

ISBN 0-13-084677-5

1. Blueprints. 2. Machinery — Drawings. I. Smith, Larry (Loran Walter) II. Title.

T379 .S29 2001

604.2′5—dc21 00-030736

Vice President and Publisher: Dave Garza
Editor in Chief: Stephen Helba
Executive Editor: Ed Francis
Production Editor: Christine M. Buckendahl
Production Coordination: Megan Hill, Lithokraft II
Design Coordinator: Robin G. Chukes
Cover Designer: Linda Fares
Cover photo: FPG
Production Manager: Brian Fox
Marketing Manager: Jamie Van Voorhis

This book was set in Zapf Calligraphy by Lithokraft II and was printed and bound by Banta Book Group. The cover was printed by Phoenix Color Corp.

10 9 8 7 6 5 4 3 2
ISBN: 0-13-084677-5

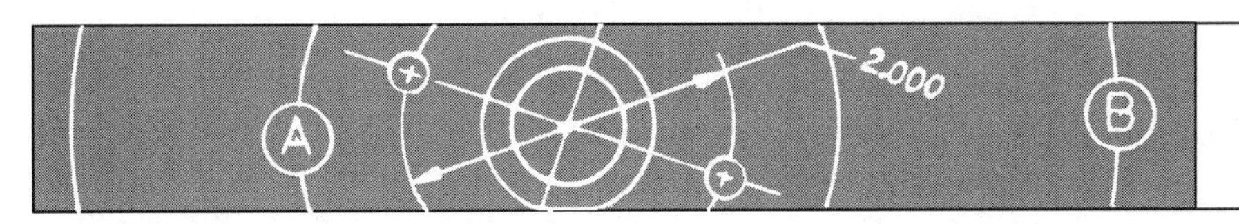

CONTENTS

UNIT 3

UNIT 4

UNIT 5

UNIT 6

UNIT 7

UNIT 8

UNIT 9

UNIT 10

UNIT 11

UNIT 15

UNIT 16

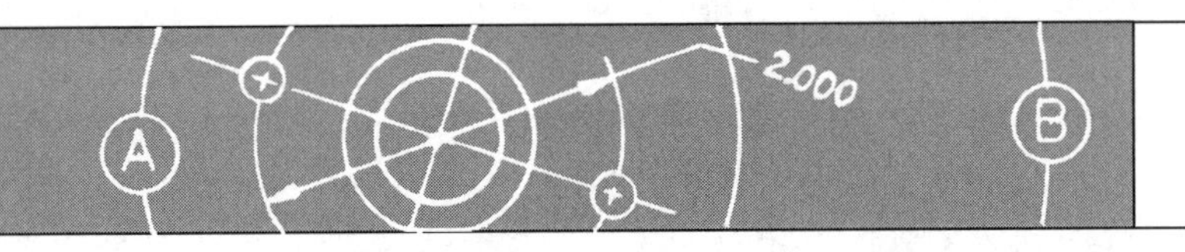

PREFACE

Being able to read and correctly interpret a blueprint is a necessary skill in the industrial world today. *Blueprint Reading for the Machine Trades* is written as a workbook that requires your responses. By writing your answers and solving the problems, you can more easily retain important information.

No prior knowledge of blueprint reading is required, and no additional materials are required other than a pencil and eraser. The first of sixteen units begins with the very basics of print reading, progressing to visualization, and then to multi-view drawings. Upon completion of this book, you should be well prepared to enter the industrial field of work.

The authors use the successfully proven teaching technique of introducing new material followed by an immediate application of it through assignments. As new material continues to be introduced, earlier information must also be retained. Therefore, questions pertaining to previously introduced material will continue to appear throughout the book. A variety of questions have been used to stimulate your interest. These include short answer, multiple choice, true/false, and sketching. The book is written in a style that lends itself to classroom or industrial settings, and it also has been successfully used for self-teaching.

This fourth edition of *Blueprint Reading for the Machine Trades* has been expanded to include numerous optional math exercises and a mathematics appendix with formulas for right angle trigonometry, measurement of dovetails, measurement of vees, determining chords, and measurement over pins. Units 13 and 14, which focus on Geometric Dimensioning and Tolerancing (GD&T), have been expanded to include many more illustrations and examples. All drawings in this edition now include the latest standards established by the American National Standards Institute (ANSI) and the Canadian Standards Association (CSA). Both US Customary Units (inches) and the International System of Units (metric) are used throughout the book. The ASME Y14.5M-1994 Dimensioning and Tolerancing standard and the CAN3-B78.1-M83 (reaffirmed 1990) Technical Drawings—General Principles standard have been followed. These are both current standards as of this writing.

A Solutions Manual with the correct answers and solutions is available from the publisher.

We thank the following reviewers for their helpful suggestions and comments: Pat Carney II, Jackson Area Career Center; Jeffrey B. Hellwig, Alfred State College; Ed Janecek, Waukesha County Technical College; and Ben Lee Thompsson, Alfred State College.

We also wish to thank the following organizations and corporations for granting permission to use their graphic or printed materials: The American Society of Mechanical Engineers, The American Welding Society, The Industrial Press L.S. Starrett Co., International Business Machines Corp., Houston Instrument, Computervision Corp., Great Dane Trailers, Deere and Co., Harnischfeger Corp., and Viking Pump, a unit of Idex Corp.

Finally, we thank our wives for their assistance, patience, and understanding during the many hours of time devoted to this edition.

Russ Schultz
Cedar Falls, IA

Larry Smith
St. Clair Beach, ONT

UNIT 1

DICTIONARY OF TERMS

It is important for you to become acquainted with terminology commonly used in the machine trades area to effectively interpret blueprints. Following is a partial list of terms that should be helpful to you. Briefly review them before continuing, then test your knowledge with the quizzes that follow.

ABRASIVE: A material used to cut other materials softer than itself.

ALLOWANCE: The minimum clearance (or maximum interference) between two mating parts, such as a shaft and a hole (illus. on p. 40).

ALLOY: A mixture of two or more elements, at least one of which is metallic, melted together to form a new metal.

ANGULAR DIMENSION: A dimension measured in degrees, minutes, and seconds.

ANNEAL: The process of slowly cooling hot iron-base metals to remove stresses and reduce the hardness.

ARBOR: A shaft upon which a cutting tool is mounted, or a spindle for holding the workpiece.

ASSEMBLY DRAWING: A drawing that shows the relationship of the various components as they fit together.

AUXILIARY VIEW: An orthographic view projected angularly, used to show features appearing on inclined or oblique surfaces (illus. on p. 276).

AXIS: A central line about which parts are symmetrically arranged that may or may not revolve upon it.

BASELINE DIMENSIONING: A system of dimensioning used to locate features of a part from a common set of datums. (illus. on p. 79)

BASIC DIMENSION: An exact value used to describe the size, shape, or location of a feature without tolerance.

BASIC SIZE: The exact theoretical size from which limits of size are derived by the application of allowances and tolerances.

BEVEL: A flat, slanted surface between two other surfaces at right angles to one another.

BILATERAL: Since "bi" means two, it is used to indicate that two sides are involved.

BLIND HOLE: A hole that does not pass all the way through (illus. on p. 127).

BOLT CIRCLE: A circular centerline upon which two or more hole centers are located (illus. on p. 37).

BORE: To enlarge a hole to an accurate size. Held to close tolerances, it is usually specified with limit dimensions.

BOSS: A raised, cylindrical projection used to provide extra metal around a hole in a casting or forging. The boss and the pad are similar, except that the boss is always circular, and the pad may be any other shape (illus. on p. 153).

BROACH: To produce a desired shape inside a cylindrical hole, such as a keyway or hexagon. The broaching tool has a series of teeth, gradually increasing in size, which do the cutting as the tool is pushed or pulled through the hole.

BURNISH: To finish or polish metal by rolling or sliding a tool over the surface under pressure.

BURR: A jagged edge of metal produced by working the metal. A common note on a blueprint is to "remove all burrs."

BUSHING: A replaceable insert to provide metal with better wearing quality.

CAD DRAWING: A drawing created by computer-aided drafting methods.

CAM: A mechanical device used to change rotary motion to some other motion, such as reciprocating or sliding.

CARBURIZE: Heating low-carbon steel to a temperature below its melting point in carbonaceous solids, liquids, or gases to raise the level of carbon in the exterior, then cooling slowly in preparation for heat treating. Abbreviated CARB.

CASE HARDENING: To harden ferrous alloy so that the surface layer is harder than the interior core.

CASTING: A metal object made by pouring molten metal into a mold.

CHAIN DIMENSIONING: Successive dimensions that extend from one feature to another, rather than each originating at a datum. Tolerances accumulate with chain dimensions (illus. on p. 79).

CHAMFER: A corner that has been removed from the end of a cylindrical surface, at an angle to the face. Used to facilitate assembly (illus. on p. 127).

CLEARANCE HOLE: A hole slightly larger than the bolt or fastener that is intended to pass through it.

COAXIALITY: Coaxiality of cylindrical features exists when two or more cylindrical features have a common axis.

COLLAR: A projecting ring around a shaft.

CONCENTRICITY: The condition where the median points of a feature's diametrically opposed elements are located within a cylindrical zone that is equally centered about a datum axis.

CORE: The part of a mold that shapes the interior of a casting.

COUNTERBORE: To enlarge a hole cylindrically at the end. A drawing specification will include its diameter and depth (illus. on p. 134).

COUNTERSINK: To enlarge a hole conically at the end to accept a flatheaded fastener. A drawing specification will include its diameter and included angle (illus. on p. 134).

DATUM: A point, line, surface, or plane from which the location of other features is established.

DEGREE: A unit of angular measurement. (There are 360 degrees in a circle.)

DETAIL DRAWING: A drawing of a single object, complete with dimensions and all other information necessary to produce the part.

DIE: A tool used to cut external threads. Also a tool used to shape, mold, stamp, or cut metal.

DIE CASTING: A metal object produced by injecting molten material into metal dies under pressure.

DRAFT: The angle provided on a pattern or mold that enables the pattern or molded part to be easily withdrawn from the mold.

ECCENTRIC: The term means off center. Unlike concentric, eccentric diameters do not share a common axis.

EXTRUSION: A metal object made by forcing hot or cold material through dies of the desired shape.

FACE: To machine a flat surface perpendicular to the axis, such as the end of a shaft.

FASTENER: A connector used to secure two or more parts. Bolts, nuts, screws, and rivets are types of fasteners.

FERROUS: Metals that contain iron as their base material, such as steel.

FILLET: A concave surface (interior radius) at the intersection of two surfaces of an object (illus. on p. 139).

FIN: A thin projecting edge on cast or molded parts.

FINISH MARKS: Symbols appearing on the edge view of surfaces to be machined. Usually confined to drawings of castings and forging (illus. on p. 138).

FIXTURE: A device designed to position and hold a part in a machine tool. It does not guide the cutting tool.

FLAME HARDENING: A process of hardening steel by heating with oxyacetylene torch and quenching.

FLANGE: A projecting rim or collar on an object. It may include mounting holes to secure it in place.

FORGE: To force hot metal into a desired shape by hammering or squeezing.

GAUGE: The thickness of sheet metal or the diameter of wire designated by a number rather than dimension.

GEOMETRIC TOLERANCING: Tolerances on drawings with emphasis on the actual function or relationship of part features where interchangeability is critical (illus. on p. 292).

GUSSET: An angular piece of metal fastened in the angle of a metal frame to give strength or stiffness.

HARDNESS TEST: A test that measures the degree of hardness of metals.

HEAT TREATMENT: The application of heat to metals to produce desired properties.

HORIZONTAL: Parallel to the horizon.

HUB: The central part of a wheel, such as the part into which the spokes are inserted.

INCLINED SURFACE: A flat surface slanting in one direction. It will appear as an edge in one principal view, distorted in others (illus. on p. 276).

INCLUDED ANGLE: The angle formed between one side and another, always less than 180° (illus. on p. 107).

ISOMETRIC DRAWING: A three-dimensional pictorial drawing that has its horizontal surfaces drawn on 30° axes from horizontal (illus. on p. 61).

JIG: A device designed to hold a part to be machined. It also positions and guides the cutting tool.

KEY: A metal bar or wedge used to secure gears or pulleys to a shaft and prevent rotary motion between the two pieces.

KEYSEAT: An axially located groove in a shaft that positions the key (illus. on p. 129).

KEYWAY: An axially located groove in a hub that positions the key (illus. on p. 129).

LAP: To produce a surface finish by use of a soft metal impregnated with a fine abrasive powder.

LEAST MATERIAL CONDITION: When a feature contains the minimum amount of material (maximum hole diameter and minimum shaft diameter). Abbreviated LMC (illus. on p. 184).

LIMITS: The maximum and minimum permissible dimensions. Arrived at by applying tolerances to basic dimension.

LINEAR DIMENSION: A dimension measured in a straight line.

MAXIMUM MATERIAL CONDITION: When a feature contains the maximum amount of material (minimum hole diameter and maximum shaft diameter). Abbreviated MMC (illus. on p. 40).

NECK: An external groove cut in a cylindrical piece at a change in diameter, usually where another part is to fit against a shoulder (illus. on p. 161).

NOMINAL SIZE: A term used for the purpose of general identification (3/4-in. plate, 1-in, pipe, etc.).

NONFERROUS: Metals that do not contain iron, such as brass, bronze, and aluminum.

NORMALIZING: The process of heating steel, then cooling in still air to room temperature to restore uniform grain structure and relieve internal stresses.

OBLIQUE SURFACE: A flat surface slanting in two directions. It will appear distorted in all principal views (illus. on p. 276).

ORTHOGRAPHIC PROJECTION: The process of projecting the essential views of a three-dimensional object onto a flat plane, such as a piece of paper. This process is used for machinist's blueprints (illus. on p. 46).

PAD: A raised projection on a casting. Used to provide extra metal around a series of holes or a slotted hole. *See* BOSS for comparison (illus. on p. 153).

PARALLEL: Extending in the same direction, such as two lines, everywhere equidistant, and not meeting.

PATTERN: A model, made of wood, metal, or other material, used to form a cavity in the sand for pouring castings.

PERPENDICULAR: Being at a right angle (90°) to a given line or plane.

PICKLE: To remove stains or oxide scale from parts by immersion in an acid solution.

PINION: The smaller of two mating gears.

PITCH: The distance from a point on a thread to a corresponding point on the next adjacent thread measured parallel to the axis.

PROFILE VIEW: The view that most clearly illustrates the shape of a feature is considered the profile view of that feature. With a hole it would be the view which shows that it is round.

QUENCHING: Cooling metals rapidly by immersing them in water or oil.

RADIUS: The distance from the center of a circle or an arc to its circumference. Equal to one-half a diameter. The plural of radius is "radii."

REAM: A machine operation consisting of enlarging a hole slightly with a rotating fluted tool to provide greater accuracy, and better finish.

RECIPROCATION: A straight-line, back-and-forth motion. To move in alternate directions.

REFERENCE DIMENSION: Used only for informational purposes. Does not receive the standard print tolerance and should not be used for production or inspection (illus. on p. 127).

REGARDLESS OF FEATURE SIZE: The condition where tolerance of position or form must be met irrespective of where the feature lies within its size tolerance. Abbreviated RFS.

RELIEF: The amount one plane surface of a piece is set below another plane, usually for clearance.

RIB: A relatively thin flat member acting as a brace or support (illus. on p. 189).

ROUND: A convex surface (exterior radius) at the intersection of two surfaces of a casting or forging (illus. on p. 139).

SCALE: Refers to the relative size of the drawing and the size of the part. The first number represents the size of the drawing and second number the size of the part. Dimensions always represent the size of the part, not the print.

SECTION: An interior view of an object, drawn to expose features not otherwise visible (illus. on p. 177).

SERRATED: A surface or edge having notches or sharp teeth is said to be serrated.

SHIM: A thin piece of metal or other material placed between two parts to adjust the fit.

SHOULDER: A plane surface on a shaft, normal to the axis and formed by a difference in diameter.

SPLINE: A raised area on a shaft (external) or hub (internal) parallel to the axis, and designed to fit into a recessed area of a mating part (illus. on p. 355).

SPOTFACE: To machine a round spot on a rough surface, usually around a hole, to give a good seat to a nut or bolthead. Abbreviated SF or SFACE (illus. on p. 185).

SPRINGS: Basically springs fall into two categories: helical or flat. The three most common helical springs are listed below.

COMPRESSION: Tends to supply force when compressed, either in a hole or around a rod.

EXTENSION: Tends to supply force when it is extended (pulled from end to end). Contains hooks on each end for connections.

TORSION: Designed to resist twisting or rotary motion. Contains hooks or projections for connections.

STUD: A round piece of metal threaded on both ends.

SURFACE ROUGHNESS: Roughness refers to the finely spaced irregularities produced in fabricating a part. The height of the irregularities is rated in microinches or micrometers (illus. on p. 144).

SYMMETRICAL: The same on each side. Equal halves. A centerline usually indicates symmetry.

TABULAR DIMENSIONING: A type of dimensioning where letters are substituted for dimensions that are listed in a table on the drawing. This permits more than one part to appear on the same drawing (illus. on p. 226).

TANGENT: A line drawn to the surface of an arc or circle so that it contacts the arc or circle at only one point.

TAP: A tool for cutting threads in a hole drilled to a particular size for the threading operation.

TAPER: Conical shape that permits a shaft or a hole to become gradually smaller at one end.

TEMPER: To change the physical characteristics of hardened steel by reheating to an intermediate temperature range for the purpose of toughening the steel. Also has a softening effect.

TEMPLATE: A thin sheet used as a guide for the form and shape of the piece to be made.

TOLERANCE: Permissible variation from a given dimension.

TRIANGLE: A three-sided figure. (There are 180° in a triangle.)

TYPICAL: A term used with drawing dimensions to indicate that all similar undimensioned features are intended to have the same dimension as the one marked TYP (abbreviation).

UNDERCUT: A groove cut on the inside of another cut (a hole, for example) (illus. on p. 181).

UNILATERAL: Having only one, or one side only.

UPSET: To strike a piece of metal until it becomes greater in its cross section. To hammer metal into a different shape.

VELLUM: A translucent paper. Blueprints are made from drawings created on this material.

VERTICAL: Perpendicular to the horizon, or 90° from the horizontal plane.

WOODRUFF KEY: A flat key, semicircular in shape. The two basic dimensions (diameter and width) are coded into the key number. The last two digits designate the diameter in 1/8ths of an inch. The preceding digits designate the width in 1/32nds of an inch. *Example:* Woodruff key No. 608 (8/8 = 1 in. diameter; 6/32 = 3/16 in. wide).

Terminology Quiz 1

INSTRUCTIONS: The following definitions were selected from the Dictionary of Terms. Enter the proper term for each description.

1. A machining operation that produces a round, flat spot around a hole. Abbreviated SF. — 1. ______________
2. A machining operation that uses a toothed cutting tool to produce shapes inside of a hole, such as a keyway. — 2. ______________
3. The plural of "radius." — 3. ______________
4. The term that means two sides are involved. — 4. ______________
5. A tool used for cutting internal threads in a hole. — 5. ______________
6. Two lines equidistant from each other, extending in the same direction. — 6. ______________
7. The term that describes two lines or planes that are at right angles to one another. — 7. ______________
8. A line drawn to an arc that contacts the arc at one point only, without crossing. — 8. ______________
9. The thickness of sheet metal by a number rather than a dimension. — 9. ______________
10. A shaft or spindle for holding the cutting tool or workpiece. — 10. ______________
11. A thin piece of metal inserted between two parts to adjust the fit. — 11. ______________
12. The smaller of two mating gears. — 12. ______________
13. A hole slightly larger than the fastener intended to pass through it. — 13. ______________
14. A projecting rim or collar used to attach to another object. — 14. ______________
15. A flat surface perpendicular to the axis, such as the end of a shaft. — 15. ______________
16. A point, line, plane, or surface from where other features are located. — 16. ______________
17. A spring designed to supply force when pulled end to end. — 17. ______________
18. The cooling of metals rapidly by immersing them in a liquid. — 18. ______________
19. A dimension used only for information, not for production or inspection. — 19. ______________
20. A drawing made by computer-aided drafting methods. — 20. ______________

Terminology Quiz 2

INSTRUCTIONS: The following definitions were selected from the Dictionary of Terms. Enter the proper term for each description.

1. A machining operation that enlarges the end of a hole cylindrically to a specified diameter and depth. 1. ____________
2. A machining operation that produces a cone-shaped end to a hole, usually to accept a flat-head fastener. 2. ____________
3. The permissible variation from a specified dimension. 3. ____________
4. The maximum and minimum permissible dimensions. 4. ____________
5. A tool used for cutting external threads. 5. ____________
6. Cylindrical surfaces that are equally centered about a common axis. 6. ____________
7. Equal halves. The same shape on both sides of a common centerline. 7. ____________
8. A circular centerline upon which two or more hole centers are located. 8. ____________
9. An object made by pouring molten metal into a mold. 9. ____________
10. A model of the part to be cast. Used to create a cavity in the sand. 10. ____________
11. An interior radius (concave) between intersecting surfaces of an object. 11. ____________
12. A circular, raised portion around a hole in a casting or a forging. 12. ____________
13. Metals that contain iron as their base material, such as steel. 13. ____________
14. Metals without iron content, such as brass, copper, or aluminum. 14. ____________
15. Conical shape that permits a shaft or a hole to become gradually smaller from one end to the other. 15. ____________
16. To remove a small amount of material from the end of a shaft or hole to facilitate assembly. 16. ____________
17. A groove in a shaft to position a key. 17. ____________
18. An external groove at a change in diameter of a shaft, usually for another part to fit against its shoulder. 18. ____________
19. The term used to define the angle formed between one side and another. 19. ____________
20. The minimum clearance (or maximum interference) between two mating parts, such as a shaft and a hole. 20. ____________

Terminology Crossword Puzzle

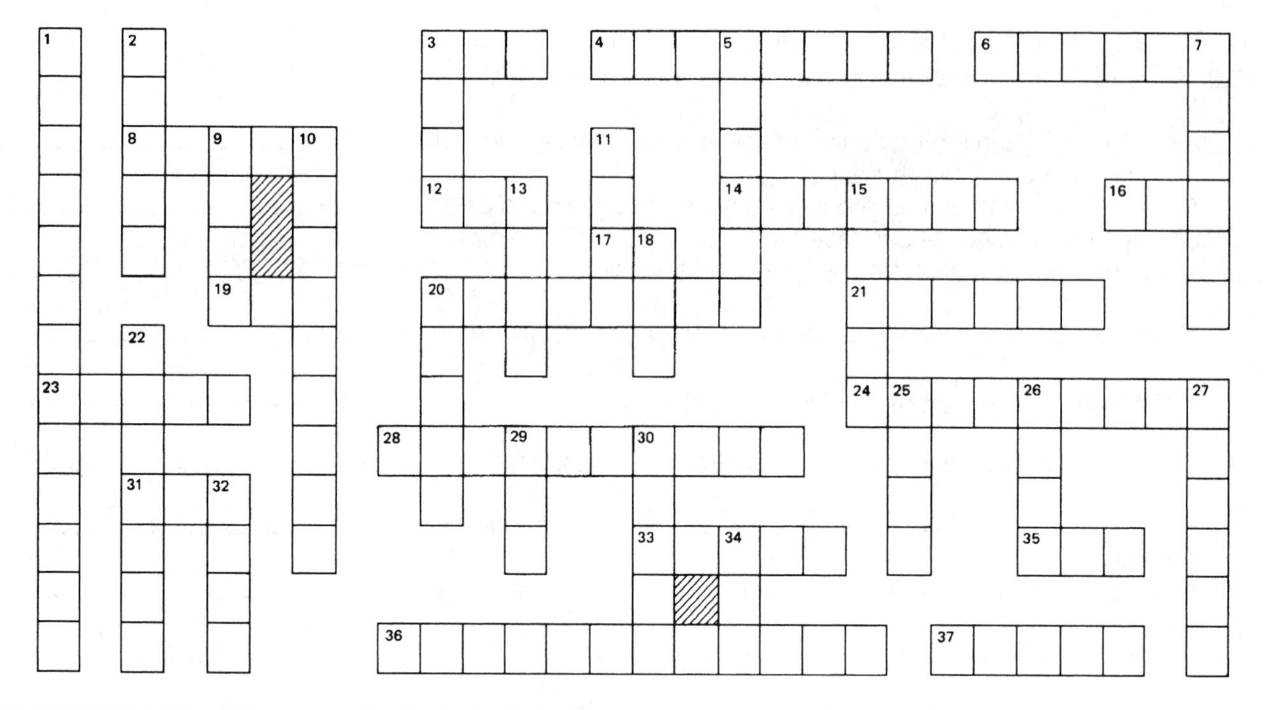

INSTRUCTIONS: Many words from the Dictionary of Terms are included in this puzzle. Enter the terms or their abbreviations from the definitions that follow.

HORIZONTAL

3. Abbreviation for "regardless of feature size."
4. Thin plate used as a guide for form or shape.
6. Angular piece of metal used to add strength.
8. Strike a piece of metal on its end to increase its cross section.
12. Abbreviation for "maximum material condition."
14. Groove that positions the key in a shaft.
16. Tool for cutting threads in a drilled hole.
17. Abbreviation for "spotface."
19. Tool for cutting external threads by hand.
20. Ceramics used to wear away softer materials.
21. Cool hot metal slowly to remove stresses.
23. Exterior radius between cast surfaces.
24. Off center. Also cylindrical surfaces with uncommon axes.
28. Cylindrical surfaces with a common axis.
31. Central part of a wheel.
33. Distance between common points on adjacent threads.
35. Metal piece used to prevent rotary movement between hub and shaft.
36. Type of projection used on machinist's blueprints.
37. Shapes hot metal by hammering or squeezing.

VERTICAL

1. Type of drawing which includes all information necessary to produce the object.
2. Machine a hand grip on a cylindrical surface.
3. Slightly enlarge a drilled hole.
5. Remove oxide scale with an acid solution.
7. Reheat hardened steel.
9. Cylindrical piece threaded on both ends.
10. Permissible variation from a given dimension.
11. Circular, raised portion around a hole in a casting.
13. Mold part that shapes the interior of a casting.
15. Proportion between two sets of dimensions: those of a drawing and the actual object.
18. Thin projecting edge on a casting.
20. Mixture of two or more metals to form a new metal.
22. Insert that provides metal with a better wearing quality.
25. Abbreviation for "carburize."
26. Groove cut into a cylindrical surface.
27. Cylindrical ring around a shaft.
29. Device that changes rotary motion to reciprocating motion.
30. Constant change in diameter over the length of a cylinder.
32. Unwanted projecting edge of metal.
34. Abbreviation for "typical."

Terminology Loop-A-Word

INSTRUCTIONS: Most of the words from the Dictionary of Terms (pages 1-5) are hidden in this puzzle. They appear either horizontal (left to right) or vertical (top to bottom) only. How many can you find?

```
R E F E R E N C E D I M E N S I O N N A A L L O W A N C E D
E O R L E A S T M A T E R I A L C O N D I T I O N U O A B I
G S Y M M E T R I C A L K E Y S E A T H R S N G X X N D O E
A B B Y D I X R C B R O A C H F E R R O U S C A L I F D I C
R F I X T U R E F A S T E N E R T D A R S C L U G L E R F A
D Y R C O S T A S S E M B L Y D R A W I N G I G N I R A S S
L E S P O T F A C E C C E N T R I C C Z O O N E I A R W N T
E B I L A T E R A L D D X C N W F E L O G K E Y S R O I O I
S P L I N E A R D I M E N S I O N T E N D I D S O Y U N R N
S E R R A T E D S N A A T R L F C O A T M U S R M V S G M G
O S E C T I O N V E R T I C A L O R R A S T U D E I L L A P
F O B P E R P E N D I C U L A R U S A L B E R S T E D U L T
F T A B U L A R D I M E N S I O N I N G C D F O R W L N I D
E Y S L I P Y T E M P L A T E A T O C A L B A S I C S I Z E
A P I O B L I Q U E S U R F A C E N E N A G C H C O U L I T
T I C W P I N I O N V N A T I M R S H G R S E O D M R A N A
U C D B I K L B E S E D C E A A B P O U B R X U R P F T G I
R A I A C G R U T I L E H M B X O R L L O T T L A R A E P L
E L M C K N U R L O L R A P R I R I E A R U R D W E C R A D
S I E K L S N R N N U C M E A M E N T R T E U E I S E A R R
I A N N E A L S P I M U F R S U R G O D H D S R N S R L A A
Z X S C O O R D I N A T E D I M E N S I O N I N G I O L L W
E I I C A S T I N G G R R V V M S H I M G B O S S O U O L I
E S O M E D I T B O L S M E E A T O L E R A N C E N G Y E N
F E N O M I N A L S I Z E P A T T E R N A T C L L S H U L G
E X T E N S I O N S P R I N G E A D P S P I T C H P N E I U
E F F I N I S H M A R K S B R R N A T I H U B R I R E A M S
F I L L E T O F D I B E S D C I B T I O I B U R N I S H I S
I C A S E H A R D E N I N G P A L U M N C D R I P N S M T E
C A M M L D R Z B O N D D R R L X M B S P A D B E G A T S T
O D E F I K B O R C A A X S O C A R B U R I Z E F L A N G E
N U H A R D N E S S T E S T F O R G E C O U N T E R S I N K
C H A I N D I M E N S I O N I N G B R N J I G B U S H I N G
E R R O U N D U S T I F C B L D M M Q U E N C H I N G R S T
N E D R A F T P A X C L O E E I B O L T C I R C L E B R T X
T L E K M C O S C A L E L V V T N H E A T T R E A T M E N T
R I N C L U D E D A N G L E I I E R E C I P R O C A T I O N
I E I O B O T T A P E R A L E O C B W O O D R U F F K E Y T
C F N R A D I U S D X B R V W N K D C F N X A A T O D N E S
A T G E O M E T R I C T O L E R A N C I N G F T A N G E N T
O D E G R E E S T R I A N G L E E F A B L I N D H O L E G E
```

STANDARD ABBREVIATIONS

Most blueprints contain many abbreviated words. This is a common practice used to conserve valuable drawing space. Therefore, it is important for you to become familiar with standard abbreviations. Briefly review the list below, then test your knowledge by attempting the abbreviation quizzes that follow. Note the absence of periods except when the abbreviation spells a different word, such as IN. for inch. Also observe the absence of space between letters when the abbreviation consists of more than one word, such as AWG for American Wire Gauge.

ACCESS.	Accessory
ADJ	Adjustable, Adjust
ADPT	Adapter
ADV	Advance
AL	Aluminum
ALLOW.	Allowance
ALT	Alternate
ALY	Alloy
AMT	Amount
ANL	Anneal
ANSI	Amer Natl Stds Institute
APPROX	Approximate
ASME	Amer Society of Mech Engrs
ASSEM	Assemble
ASSY	Assembly
AUTH	Authorized
AUTO.	Automatic
AUX	Auxiliary
AVG	Average
AWG	American Wire Gauge
BC	Bolt Circle
BET.	Between
BEV	Bevel
BHN	Brinell Hardness Number
BLK	Blank, Block
B/M	Bill of Material
BOT	Bottom
BP or B/P	Blueprint
BRG	Bearing
BRK	Break
BRKT	Bracket
BRO	Broach
BRS	Brass
BRZ	Bronze
B&S	Brown & Sharpe
BSC	Basic
BUSH.	Bushing
BWG	Birmingham Wire Gauge
C TO C	Center-to-Center
CAD	Computer-Aided Drafting
CAM	Computer-Aided Mfg
CAP.	Capacity
CAP SCR	Cap Screw
CARB	Carburize
CBORE	Counterbore
CCW	Counter Clockwise
CDRILL	Counterdrill
CDS	Cold-Drawn Steel
CFS	Cold-Finished Steel
CH	Case Harden
CHAM	Chamfer
CHAN	Channel
CHG	Change
CHK	Check
CI	Cast Iron
CIR	Circle, Circular
CIRC	Circumference
CL	Centerline
CLP	Clamp
CNC	Computer Numerical Control
COMB.	Combination
COML	Commercial
CONC	Concentric
CONN	Connect, Connector
COV	Cover
CPLG	Coupling
CQ	Commercial Quality
CRS	Cold-Rolled Steel
CRT	Cathode Ray Tube
CS	Cast Steel
CSA	Canadian Stds Association
CSK	Countersink
CSTG	Casting
CTR	Center
CU	Cubic
CW	Clockwise
CYL	Cylinder, Cylindrical
DBL	Double
DEC	Decimal
DEG	Degree
DET	Detail
DEV	Develop
DFT	Draft
DIA	Diameter
DIM.	Dimension
DIST	Distance
DN	Down
DP	Deep, Diametral Pitch
DR	Drill, Drill Rod
DSGN	Design
DVTL	Dovetail
DWG	Drawing
DWL	Dowel
DWN	Drawn
EA	Each
ECC	Eccentric
EFF	Effective
ENCL	Enclose, Enclosure
ENG	Engine
ENGR	Engineer
ENGRG	Engineering
EQL SP	Equally Spaced
EQUIV	Equivalent
EST	Estimate
EX	Extra
EXH	Exhaust
EXP	Experimental
EXT	Extension, External
FAB	Fabricate
FAO	Finish All Over
FDRY	Foundry
FIG.	Figure
FIL	Fillet, Fillister
FIM	Full Indicator Movement
FIN.	Finish
FIX.	Fixture
FL	Floor, Fluid, Flush
FLEX.	Flexible
FLG	Flange
FORG	Forging
FR	Frame, Front
FTG	Fitting
FURN	Furnish
FWD	Forward
GA	Gage, Gauge
GALV	Galvanized
GR	Grade
GRD	Grind
GRV	Groove
GSKT	Gasket
H&G	Harden and Grind
HD	Head
HDL	Handle
HDLS	Headless
HDN	Harden
HDW	Hardware
HEX	Hexagon
HGR	Hanger
HGT	Height
HOR	Horizontal
HRS	Hot-Rolled Steel
HSG	Housing
HT TR	Heat Treat

Abbreviation	Meaning
HVY	Heavy
HYD	Hydraulic
ID	Inside Diameter
IDENT	Identification
ILLUS	Illustration
IN.	Inch
INCL	Include, Including
INCR	Increase
INFO	Information
INSP	Inspect
INSTL	Install
INST	Instruct, Instrument
INT	Interior, Internal, Intersect
IR	Inside Radius
ISO	Internatl Stds Organization
JCT	Junction
JNL	Journal
JT	Joint
K	Key
KNRL	Knurl
KST	Keyseat
KWY	Keyway
LB	Pound
LBL	Label
LG	Length, Long
LH	Left Hand
LMC	Least Material Condition
LOC	Locate
LT	Light
LTR	Letter
LUB	Lubricate
MACH	Machine
MAINT	Maintenance
MATL	Material
MAX	Maximum
MECH	Mechanical, Mechanism
MED	Medium
MFG	Manufacturing
MI	Malleable Iron
MIN	Minimum, Minute
MISC	Miscellaneous
mm	Millimeter
MMC	Max Material Condition
MS	Machine Steel
MTG	Mounting
MULT	Multiple
MWG	Music Wire Gauge
NA	Not Applicable
NATL	National
NC	Numerical Control
NEG	Negative
NO.	Number
NOM	Nominal
NPSM	Natl Pipe Straight Mech
NPT	Natl Pipe Tapered
NTS	Not to Scale
OA	Over All
OBS	Obsolete
OC	On Center
OD	Outside Diameter
OPP	Opposite
OPTL	Optional
OR	Outside Radius
ORIG	Original
PAT.	Patent
PATT	Pattern
PC	Piece, Pitch Circle
PCH	Punch
PD	Pitch Diameter
PERF	Perforate
PERM	Permanent
PERP	Perpendicular
PFD	Preferred
PKG	Package, Packing
PL	Parting Line, Places, Plate
PNEU	Pneumatic
PNL	Panel
POL	Polish
POS	Position, Positive
PR	Pair
PRI	Primary
PROC	Process
PROD	Product, Production
PSI	Pounds per Square Inch
PT	Part, Point
QTR	Quarter
QTY	Quantity
QUAL	Quality
R	Radius
RA	Rockwell Hardness, A-Scale
RB	Rockwell Hardness, B-Scale
RC	Rockwell Hardness, C-Scale
RECD	Received
RECT	Rectangle
REF	Reference
REINF	Reinforce
REL	Release, Relief
REM	Remove
REQD	Required
RET.	Retainer, Return
REV	Reverse, Revision, Revolution
RFS	Regardless of Feature Size
RGH	Rough
RH	Right Hand
RIV	Rivet
RM	Ream
RND	Round
RPM	Revolutions per Minute
RPW	Resistance Projection Weld
SAE	Society of Automotive Engrs
SCH	Schedule
SCR	Screw
SEC	Second
SECT	Section
SEP	Separate
SEQ	Sequence
SER	Serial, Series
SERR	Serrate
SF	Spotface
SFT	Shaft
SGL	Single
SH	Sheet
SI	Intl System of Units
SL	Slide
SLV	Sleeve
SOC	Socket
SP	Space, Spaced, Spare
SPL	Special
SPEC	Specification
SPG	Spring
SPHER	Spherical
SPRKT	Sprocket
SQ	Square
SST	Stainless Steel
STD	Standard
STK	Stock
STL	Steel
STR	Straight, Strip
SUB	Substitute
SUP.	Supply, Support
SURF	Surface
SYM	Symmetrical
SYS	System
T	Teeth, Tooth
TECH	Technical
TEMP	Template, Temporary
THD	Thread
THK	Thick
TOL	Tolerance
TOT	Total
TPF	Taper per Foot
TPI	Taper per In., Threads per Inch
TPR	Taper
TS	Tool Steel
TYP	Typical
UNC	Unified Natl Coarse
UNEF	Unified Natl Extra Fine
UNF	Unified Natl Fine
UNIV	Universal
VAR	Variable
VERT	Vertical
VOL	Volume
VS	Versus
W	Wide, Width
WASH.	Washer
WDF	Woodruff
WI	Wrought Iron
WT	Weight

Abbreviation Quiz 1

INSTRUCTIONS: Enter the word or words that represent the following standard abbreviations found on drawings.

1. CI	1. ______________
2. CRS	2. ______________
3. CBORE	3. ______________
4. CSK	4. ______________
5. DIA	5. ______________
6. DR	6. ______________
7. FIL	7. ______________
8. FAO	8. ______________
9. GA	9. ______________
10. HRS	10. ______________
11. ID	11. ______________
12. LH	12. ______________
13. MATL	13. ______________
14. MAX	14. ______________
15. OD	15. ______________
16. SECT	16. ______________
17. STL	17. ______________
18. THK	18. ______________
19. THD	19. ______________
20. TOL	20. ______________

Abbreviation Quiz 2

INSTRUCTIONS: Enter the word or words that represent the following standard abbreviations found on drawings.

1. BC	1. ______________
2. BHN	2. ______________
3. CHAM	3. ______________
4. CDS	4. ______________
5. DIM.	5. ______________
6. FIN.	6. ______________
7. GRD	7. ______________
8. HT TR	8. ______________
9. LG	9. ______________
10. MI	10. ______________
11. MIN	11. ______________
12. NTS	12. ______________
13. PC	13. ______________
14. RM	14. ______________
15. REF	15. ______________
16. RND	16. ______________
17. SCR	17. ______________
18. SPEC	18. ______________
19. SYM	19. ______________
20. TYP	20. ______________

Abbreviation Loop-A-Word

INSTRUCTIONS: Listed below are a number of words that have their abbreviations hidden in the puzzle. They appear either horizontally (left to right) or vertically (top to bottom). Can you find them?

ACCESSORY
ALLOWANCE
APPROXIMATE
ASSEMBLE
ASSEMBLY
AUTOMATIC
AUXILIARY
BEARING
BRACKET
CASTING
CHAMFER
COLD-ROLLED STEEL
COUNTERBORE
COUNTERCLOCKWISE
COUNTERSINK
CYLINDER
DIAMETER
DRAWING
EQUIVALENT
FINISH ALL OVER
FOUNDRY
GASKET
GRIND
HARDEN
HEXAGON
HOT-ROLLED STEEL
HOUSING
INTERIOR
JUNCTION
KEYSEAT
LUBRICATE
MACHINE
MATERIAL
MAXIMUM
MINIMUM
MISCELLANEOUS
NATIONAL
NOT TO SCALE
PIECE
QUANTITY
REAM
REFERENCE
SPECIFICATION
STAINLESS STEEL
STEEL

```
B C C W Z J G K A U T O B H R S N I L Y D
M E I K D O D W G F U J D Q K U G N F H I
J C T B R C V P L N G I M R E F O T D T A
F Q E Y M S O N A T L C A V Z H J X K O B
B R K T I K Y D H Z P V X A B S E M I N F
J X D Z G X Q V Z X U T V Q Z G M G C I H
E H O B Z P L N C Y L F R Z X P H Y Q L U
Q Y A S S E M Z Y R K O M Y J A P P R O X
U K N V J I N D M Z A J Y V N F Z B P U S
I G T X S S T J A U F Y A G Z K R G Y X P
V Q S E B M L Y C Z M Z L V L U B H B E E
I J N H X K A K H F I B L F Z D L V R V C
D A C C E S S Q P D S Z O Y H U N P G F O
O K U M V J S B N X C O W M E J C Z R H C
K I Q T Y Q Y F Y N N J I K X T A L P A B
S V P L O G A D Z Q K P M H E I U X N U O
T H E F D R Y X P F I C B X F Q X E G V R
C C J U F I J R K A H O M V T R G Y R J E
M S V S Y K P M Q O B P H D N D U P D I L
J T O T D C R S U G F J E L L Q A Q E Q N
D G I L H E F B L D C H A M I B G S K T G
```

ALPHABET OF LINES

Various types of lines are found on blueprints, and each has a definite purpose. It is important that you learn to recognize and to understand the meaning of these lines. Study the lines shown below, then test your knowledge by attempting to solve the Alphabet of Lines quizzes that follow.

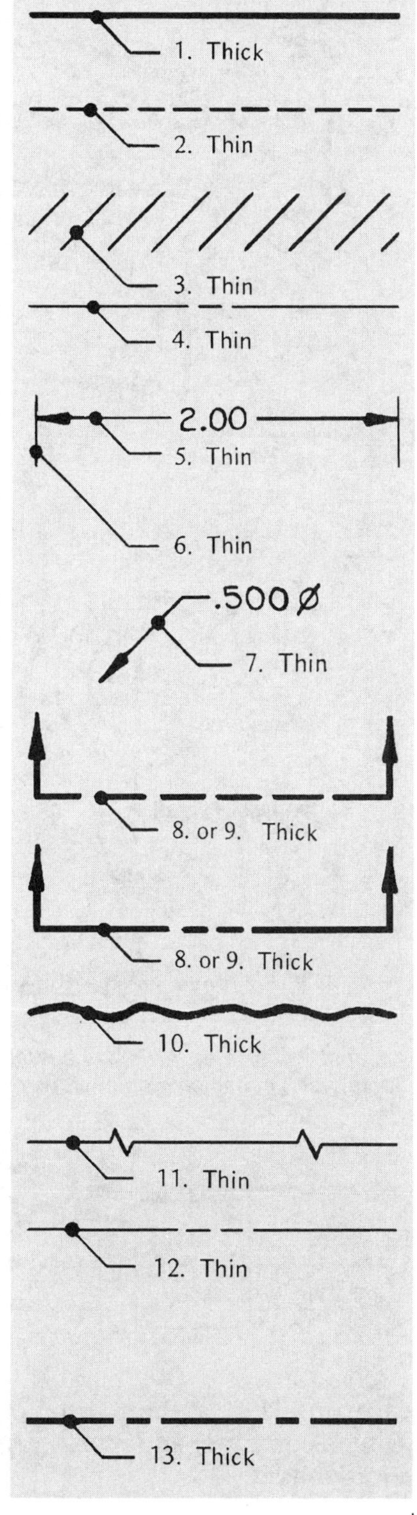

1. VISIBLE LINE: Used to represent visible edges and outlines of the object being drawn.

2. HIDDEN LINE: Represents edges and outlines of an object that are not visible in a given view.

3. SECTION LINES: Appear only on a sectional view where the surface has been cut. Normally drawn diagonally.

4. CENTERLINE: Indicates symmetry, center points, or axes. Consists of alternate long and short dashes.

5. DIMENSION LINE: Denotes the extent of the dimension when used in conjunction with arrowheads.

6. EXTENSION LINE: Extends the surface or point away from the view for the purpose of dimensioning.

7. LEADER: Drawn diagonally, it directs a dimension or note to the area where it applies. Normally terminates with an arrowhead, but may terminate with a dot.

8. CUTTING-PLANE LINE: Shows where the imaginary cutting takes place to create a sectional view. May be a series of long dashes, or alternately two short dashes between single long dashes. Arrowheads indicate the direction of sight.

or

9. VIEWING-PLANE LINE: Used in conjunction with removed views to show where the view would normally appear. Arrowheads indicate the direction of sight. (Illustrated the same as the cutting-plane line.)

10. SHORT BREAK LINE: Terminates a view to conserve space and avoid congestion. Also separates internal and external features with broken-out sections.

11. LONG BREAK LINE: Allows removal of a long central portion of an object to shorten a view. Normally used in pairs.

12. PHANTOM LINE: Represents the outline of an adjacent part; shows alternate position of a given part; or replaces repetitive detail, such as spring coils, gear teeth, threads, etc. Drawn with two short dashes between single long dashes, but thinner than the cutting/viewing-plane line.

13. CHAIN LINE: Used to indicate an area or portion of a surface that is to receive special treatment.

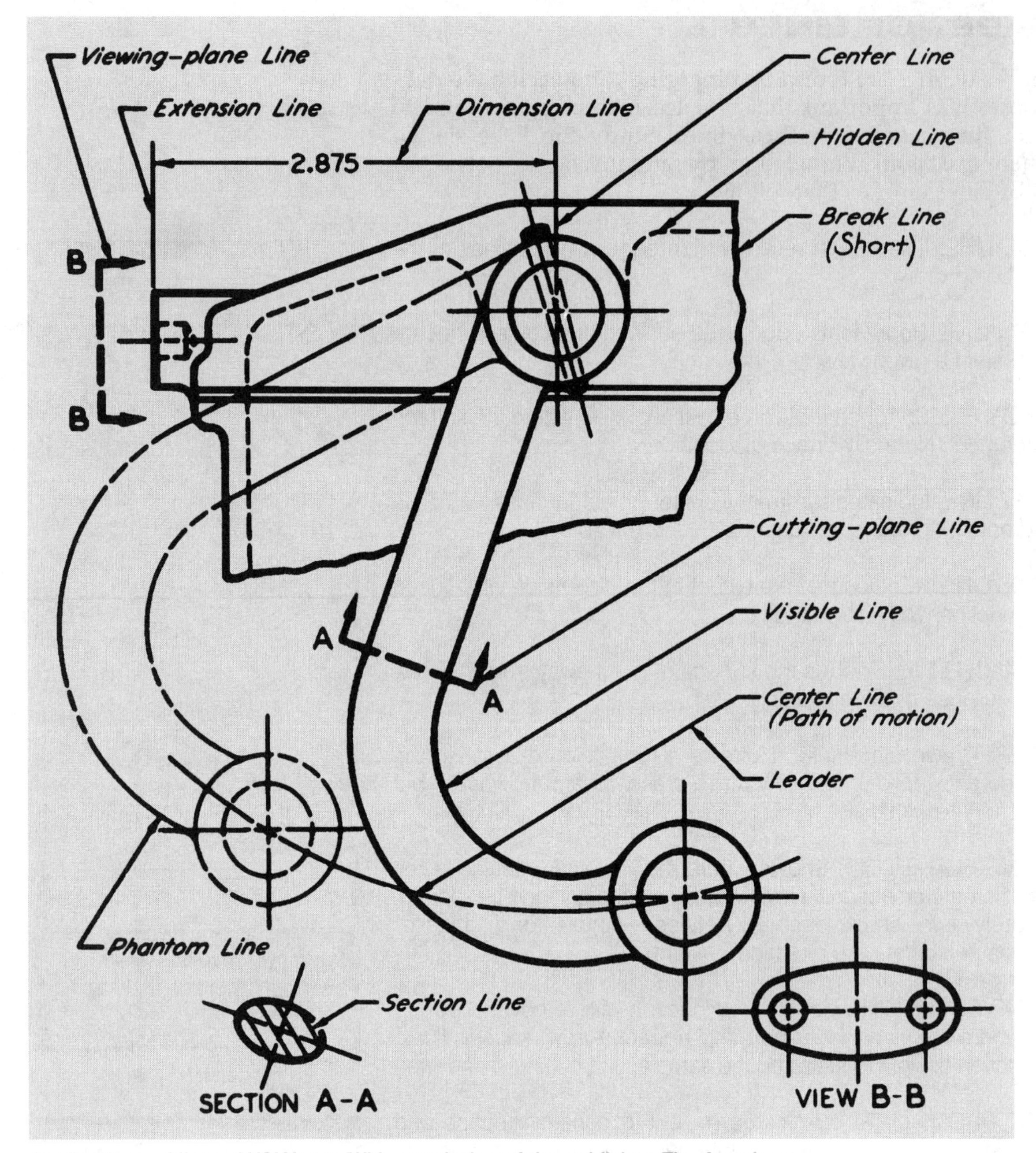

Applications of lines. ANSI Y14.2 (With permission of the publisher, The American Society of Mechanical Engineers)

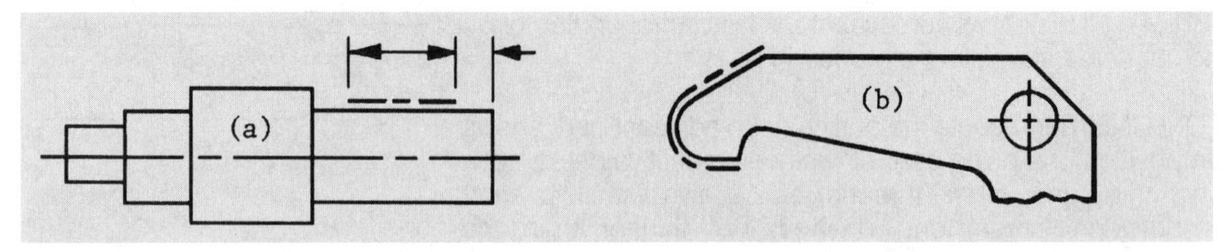

A chain line is drawn parallel to the surface profile at a short distance from it. Dimensions are added for length and location (Fig. a).

If the chain line clearly indicates the location and extent of the surface area, dimensions may be omitted (Fig. b).

Alphabet of Lines Quiz 1

INSTRUCTIONS: Answer the following questions by referring to the illustrations on the preceding page. In questions 1 through 11, enter the name of the line that does the following:

1. Shows the outline of the object. 1. ____________
2. Indicates the place from where the removed view is viewed. 2. ____________
3. Points diagonally to an area or a feature. 3. ____________
4. Represents a surface not visible in the view drawn. 4. ____________
5. Shows an alternate position of the movable arm. 5. ____________
6. Terminates with arrowheads and encloses a dimension figure. 6. ____________
7. Extends the visible line for the purpose of dimensioning to it. 7. ____________
8. Indicates the place where the section is cut. 8. ____________
9. Shows the axis of symmetrical parts and the arm's path of motion. 9. ____________
10. Permits the use of a partial view to conserve space and avoid congestion. 10. ____________
11. Represents the exposed surface of a sectioned feature. 11. ____________
12. List the five lines that are drawn thick. 12. ____________

____

____

____

____

13. List the six lines that include short dashes. 13. ____________

____

____

____

____

____

14. List the four lines that include arrowheads. 14. ____________

____

____

____

15. How does the cutting-plane line differ from the viewing-plane line (in application)? 15. ____________

____

Alphabet of Lines Quiz 2

INSTRUCTIONS: Enter the proper names of the various types of lines used in the illustration below.

①	______________	⑦	______________
②	______________	⑧	______________
③	______________	⑨	______________
④	______________	⑩	______________
⑤	______________	⑪	______________
⑥	______________	⑫	______________

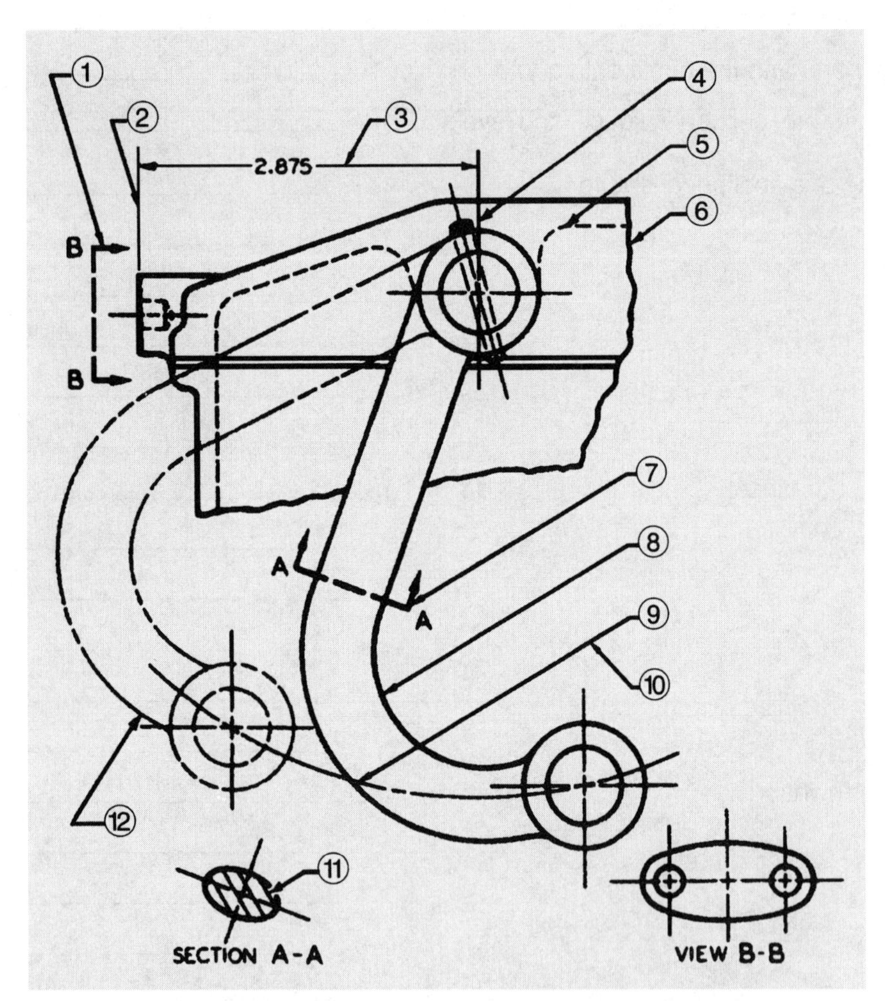

Adapted from ANSI Y14.2 (With permission of the publisher, the American Society of Mechanical Engineers)

Alphabet of Lines Quiz 3

INSTRUCTIONS: Match the lines used in the drawing with their correct names from the list below.

Centerline	Hidden Line	Section Line
Chain Line	Leader	Short Break Line
Cutting-plane Line	Long Break Line	Viewing-plane Line
Dimension Line	Phantom Line	Visible Line
Extension Line		

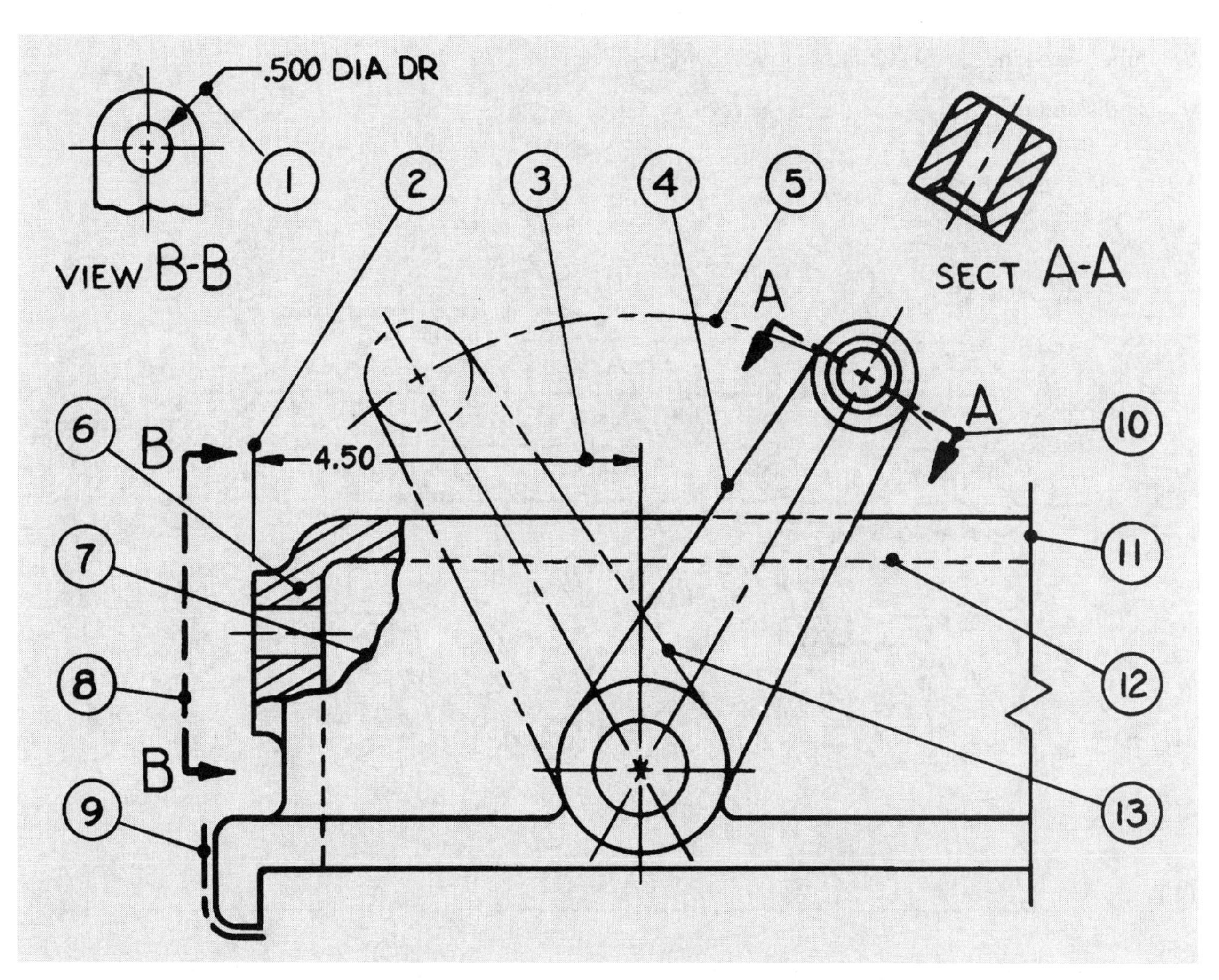

① ____________________ ⑤ ____________________ ⑨ ____________________

② ____________________ ⑥ ____________________ ⑩ ____________________

③ ____________________ ⑦ ____________________ ⑪ ____________________

④ ____________________ ⑧ ____________________ ⑫ ____________________

⑬ ____________________

Alphabet of Lines Quiz 4

INSTRUCTIONS: Listed below are the various lines that comprise the Alphabet of Lines. Match the letters from the list with the numbers in the illustration.

Ⓐ Centerline
Ⓑ Chain Line
Ⓒ Cutting-plane Line
Ⓓ Dimension Line
Ⓔ Extension Line
Ⓕ Hidden Line
Ⓖ Leader
Ⓗ Long Break Line
Ⓘ Phantom Line
Ⓙ Section Line
Ⓚ Short Break Line
Ⓛ Viewing-plane Line
Ⓜ Visible Line

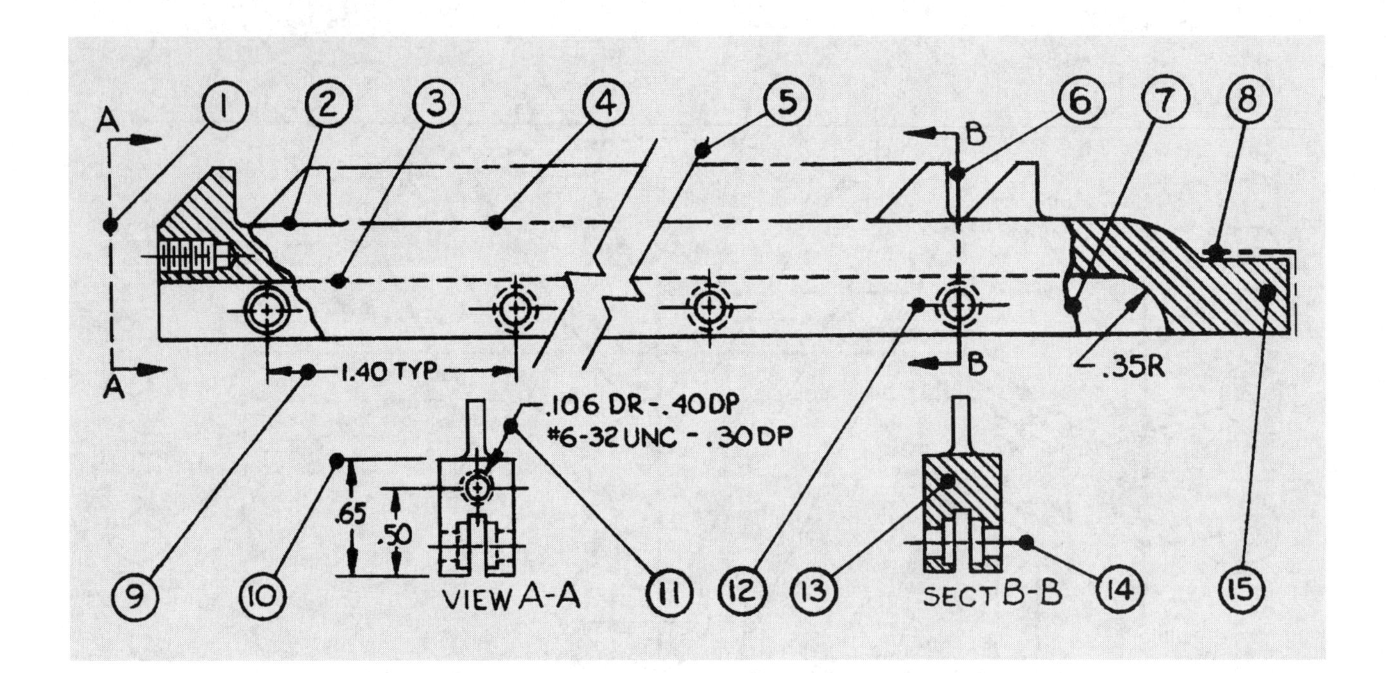

① ____________
② ____________
③ ____________
④ ____________
⑤ ____________
⑥ ____________
⑦ ____________
⑧ ____________
⑨ ____________
⑩ ____________
⑪ ____________
⑫ ____________
⑬ ____________
⑭ ____________
⑮ ____________

UNIT 2

DIMENSIONING SYSTEMS

Shown below are examples of the four different systems of dimensioning being used in the U.S. and Canada today. Fractional inch dimensions (Fig. 1) have been replaced with decimal inch dimensions (Fig. 2) by most industries in recent years. There are several advantages to the decimal system, such as: (1) it is easier to add and subtract; (2) closer tolerances can be applied; and (3) it is a natural transition to metric dimensioning. Many large U.S. industries, particularly those involved in world export, have already begun to use metric dimensions on their drawings. Dual dimensioning (Fig. 3) is a system that includes the metric equivalent (millimeters) in brackets adjacent to the decimal inch dimension. The SI metric system (Fig. 4) uses metric dimensions only, but may include a decimal inch conversion table elsewhere on the drawing.

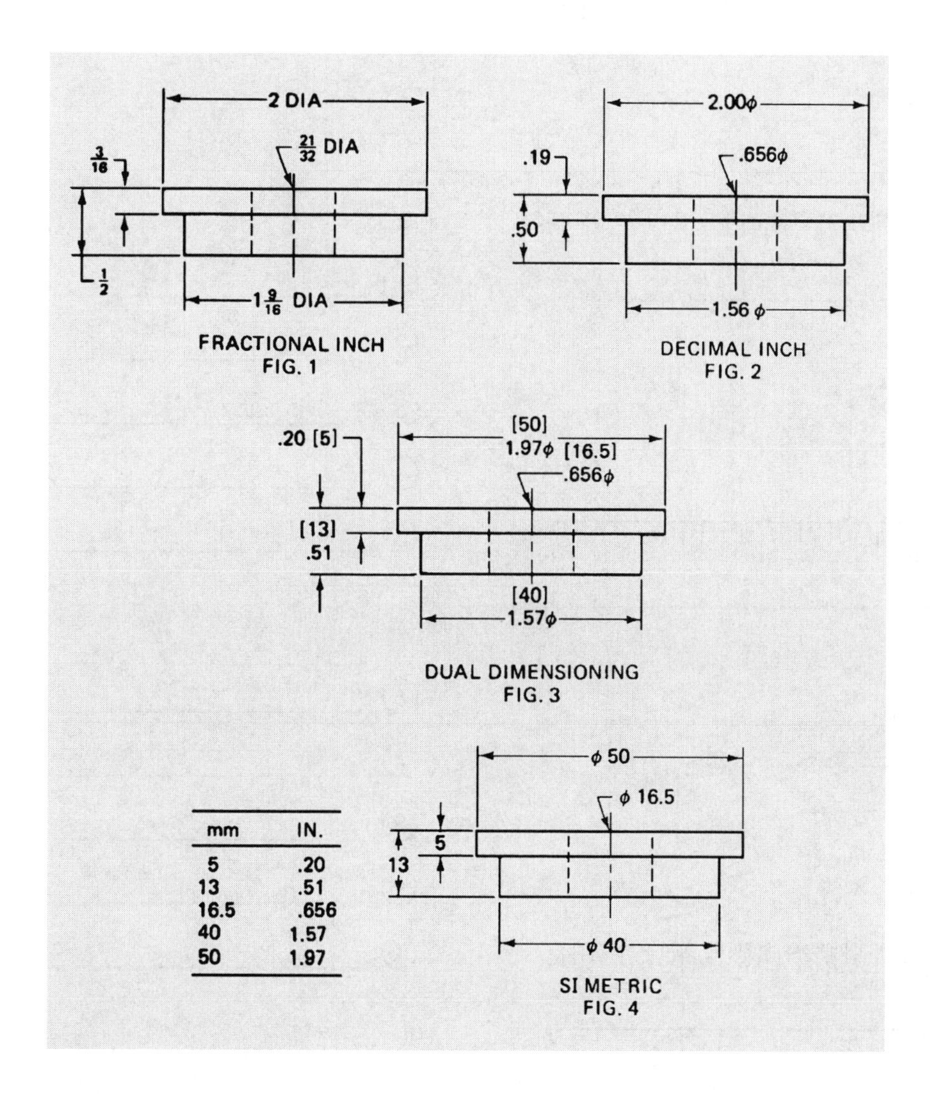

mm	IN.
5	.20
13	.51
16.5	.656
40	1.57
50	1.97

FRACTIONAL INCH
FIG. 1

DECIMAL INCH
FIG. 2

DUAL DIMENSIONING
FIG. 3

SI METRIC
FIG. 4

Fractional Dimensioning

Although the majority of prints in this workbook will be decimally dimensioned, there are still a great number of industrial blueprints that are fractionally dimensioned. A machine operator should become skilled at reading the fractional scale and metric scale as well as the decimal scale.

Most fractional-type steel rules (scales) are subdivided into units of 1/16, 1/32, or 1/64 in. Major graduations along the rule are accented by longer lines. Some have small numbers printed adjacent to the lines that correspond with the number of graduations. *Example:* The number 16 on a scale graduated into 32's would represent 16/32, which would reduce to 1/2 in. Always reduce a fraction to its lowest terms. If the numerator is an even number, you know that it can be further reduced. It is often quicker to begin your count from a major graduation close to the reading, rather than to begin from a full inch. *Example:* A reading of $2\,{}^{7}/_{16}$ could be read faster by starting at $2\,{}^{1}/_{2}''$ instead of 2″. (See no. ③ below.)

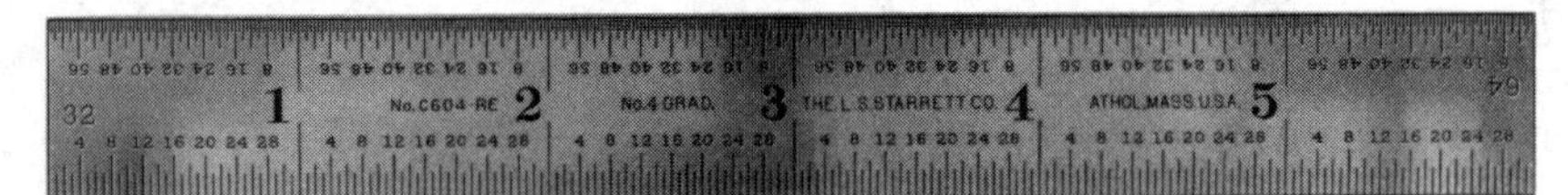

(Courtesy The L. S. Starrett Co.)

Scale Reading Quiz 1

INSTRUCTIONS: Determine the readings from the illustrations shown below and enter your answers in the appropriate spaces.

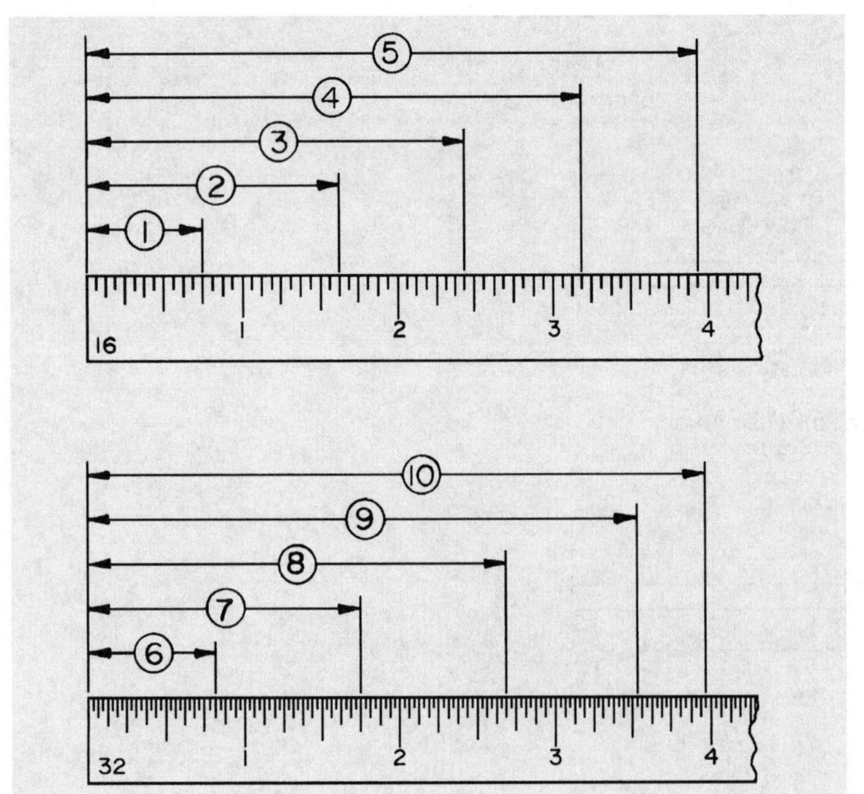

① ______________

② ______________

③ ______________

④ ______________

⑤ ______________

⑥ ______________

⑦ ______________

⑧ ______________

⑨ ______________

⑩ ______________

Decimal Scales

Most decimal-type scales (steel rules) are subdivided into units of 1/10, 1/20, 1/50, or 1/100 in. The 1/50 will provide sufficiently accurate readings for most applications. This scale can usually be recognized by the number 50 stamped near the left edge. With 50 increments to the inch, each increment then represents .02 in. By reading between these increments, the machinist is capable of determining the closest .01 measurement (ten thousandths of an inch). Closer measurements require the use of other instruments, such as the micrometer, vernier calipers, etc.

Major graduations along the scale are accented by longer lines, usually every 1/10 in. (.10). In addition, they may include numbers as shown in the illustration below. A popular graduation style has the .04 and .06 increment lines slightly longer than the .02 and .08 increment lines. This allows for quicker reading of the commonly used .05 increment.

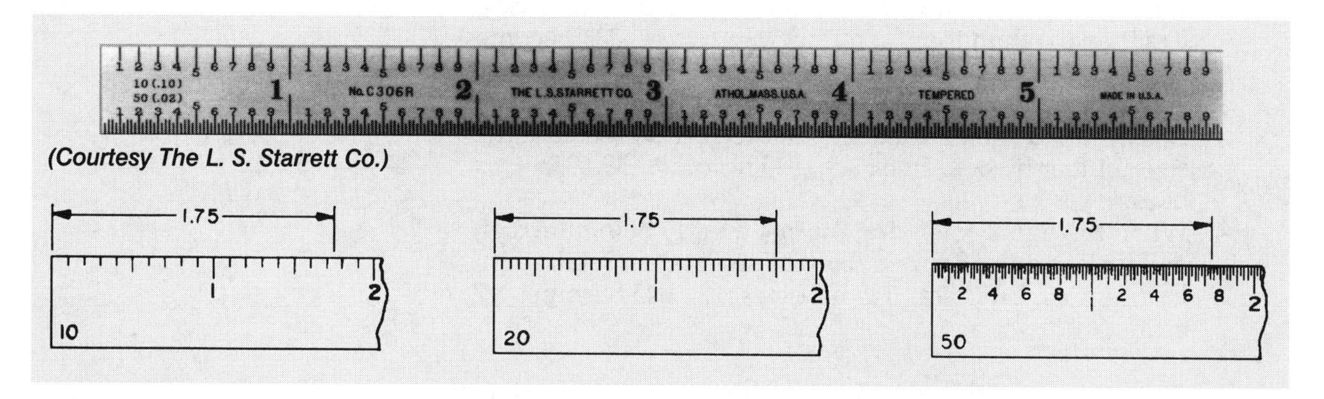

(Courtesy The L. S. Starrett Co.)

Scale Reading Quiz 2

INSTRUCTIONS: Determine the readings from the illustration shown below, and enter your answers to the closest .01 in.

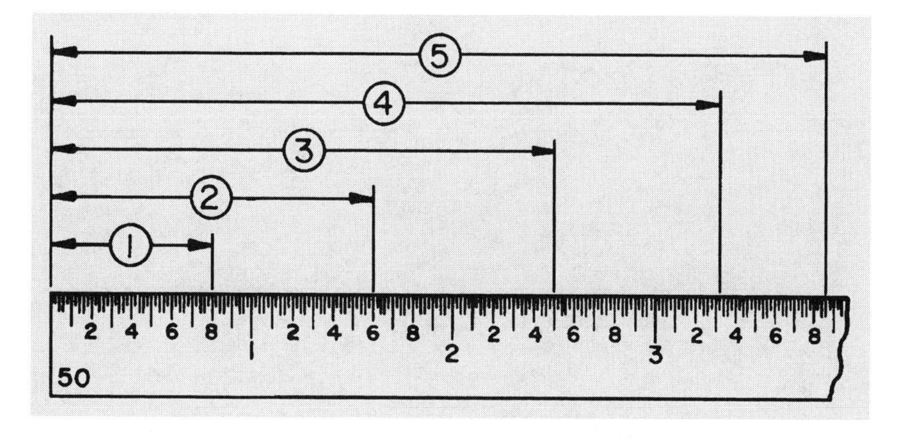

① ______________

② ______________

③ ______________

④ ______________

⑤ ______________

Decimal Dimensioning

A drawing may be dimensioned using two- or three-place decimals. The reason for this difference may be that the amount of tolerance allowed for a given dimension is determined by the number of decimal places that the dimension contains. *Example:* .50 may have a tolerance of ± .01, .500 may have a tolerance of ± .001, and .5000 may have a tolerance of ± .0001 because a dimension and its tolerance contain the same number of decimal places.

When a fraction is converted to a decimal dimension, the conversion often results in displaying too many decimal places. To round off a decimal value to the number of decimal places desired, the following rules should apply:

1. When the figure following the last digit to be retained is greater than 5, increase the last number by one. *Examples:* .016 becomes .02; .047 becomes .05; .078 becomes .08; etc.
2. When the figure following the last digit to be retained is less than 5, retain the last number. *Examples:* .031 becomes .03; .062 becomes .06; .093 becomes .09; etc.
3. When the figure following the last digit to be retained is exactly 5 and the figure to be retained is *odd,* increase the last number by one. *Examples:* .375 becomes .38; .875 becomes .88; etc.
4. When the figure following the last digit to be retained is exactly 5 and the figure to be retained is *even,* retain the last number. *Examples:* .125 becomes .12; .625 becomes .62; etc.

Whereas rules 1 and 2 are probably the same ones that you learned years ago, rules 3 and 4 may be contrary to what you have been taught. They apply only when the figure following the last digit to be retained is exactly 5. These two rules were established in an effort to retain even numbers as often as possible. Even numbers are preferred so that when a dimension is divided by 2, such as diameters to find radii, the results will contain the same number of decimal places.

Decimal Rounding Quiz

INSTRUCTIONS: Round the following three-place decimals to two-place decimals. Follow the rules given on page 24.

1. .109 =	1. ________	11. 1.056 =	11. ________
2. .125 =	2. ________	12. 1.672 =	12. ________
3. .156 =	3. ________	13. 2.205 =	13. ________
4. .234 =	4. ________	14. 2.454 =	14. ________
5. .438 =	5. ________	15. 3.335 =	15. ________
6. .547 =	6. ________	16. 3.767 =	16. ________
7. .562 =	7. ________	17. 5.555 =	17. ________
8. .641 =	8. ________	18. 6.665 =	18. ________
9. .797 =	9. ________	19. 7.045 =	19. ________
10. .875 =	10. ________	20. 8.885 =	20. ________

Decimal and Metric Equivalents of Fractions

Fraction	*Four-Place Decimal*	*Three-Place Decimal*	*Two-Place Decimal*	*Milli-meters*	*Fraction*	*Four-Place Decimal*	*Three-Place Decimal*	*Two-Place Decimal*	*Milli-meters*
1/64	.0156	.016	.02	.397	33/64	.5156	.516	.52	13.097
1/32	.0312	.031	.03	.794	17/32	.5312	.531	.53	13.494
3/64	.0469	.047	.05	1.191	35/64	.5469	.547	.55	13.891
1/16	.0625	.062	.06	1.588	9/16	.5625	.562	.56	14.288
5/64	.0781	.078	.08	1.984	37/64	.5781	.578	.58	14.684
3/32	.0938	.094	.09	2.381	19/32	.5938	.594	.59	15.081
7/64	.1094	.109	.11	2.778	39/64	.6094	.609	.61	15.478
1/8	.1250	.125	.12	3.175	5/8	.6250	.625	.62	15.875
9/64	.1406	.141	.14	3.572	41/64	.6406	.641	.64	16.272
5/32	.1562	.156	.16	3.969	21/32	.6562	.656	.66	16.669
11/64	.1719	.172	.17	4.366	43/64	.6719	.672	.67	17.066
3/16	.1875	.188	.19	4.762	11/16	.6875	.688	.69	17.462
13/64	.2031	.203	.20	5.159	45/64	.7031	.703	.70	17.859
7/32	.2188	.219	.22	5.556	23/32	.7188	.719	.72	18.256
15/64	.2344	.234	.23	5.953	47/64	.7344	.734	.73	18.653
1/4	.2500	.250	.25	6.350	3/4	.7500	.750	.75	19.050
17/64	.2656	.266	.27	6.747	49/64	.7656	.766	.77	19.447
9/32	.2812	.281	.28	7.144	25/32	.7812	.781	.78	19.844
19/64	.2969	.297	.30	7.541	51/64	.7969	.797	.80	20.241
5/16	.3125	.312	.31	7.938	13/16	.8125	.812	.81	20.638
21/64	.3281	.328	.33	8.334	53/64	.8281	.828	.83	21.034
11/32	.3438	.344	.34	8.731	27/32	.8438	.844	.84	21.431
23/64	.3594	.359	.36	9.128	55/64	.8594	.859	.86	21.828
3/8	.3750	.375	.38	9.525	7/8	.8750	.875	.88	22.225
25/64	.3906	.391	.39	9.922	57/64	.8906	.891	.89	22.622
13/32	.4062	.406	.41	10.319	29/32	.9062	.906	.91	23.019
27/64	.4219	.422	.42	10.716	59/64	.9219	.922	.92	23.416
7/16	.4375	.438	.44	11.112	15/16	.9375	.938	.94	23.812
29/64	.4531	.453	.45	11.509	61/64	.9531	.953	.95	24.209
15/32	.4688	.469	.47	11.906	31/32	.9688	.969	.97	24.606
31/64	.4844	.484	.48	12.303	63/64	.9844	.984	.98	25.003
1/2	.5000	.500	.50	12.700	1"	1.0000	1.000	1.00	25.400

Metric Dimensioning

The increment of measurement on metric drawings for the machine trade is the millimeter, unless otherwise specified. A note to this effect is normally located in the area of the title block on a drawing. Many industries in the U.S. and Canada that use the metric system of dimensioning also include the inch equivalent in decimal form. The dual dimensioning system and the SI metric system are illustrated on page 21, "Dimensioning Systems."

If you encounter a drawing dimensioned only in millimeters, you may convert each dimension to inches by multiplying it times .03937, or you may use a conversion chart. However, metric scales are available to eliminate the need for conversion. These scales have graduations every millimeter or half-millimeter, and may be recognized by the designation "mm" or "1/2 mm" stamped near the left edge. Normally, every fifth millimeter is accented by a longer line and every tenth millimeter is identified by number, as shown below.

(Courtesy The L. S. Starrett Co.)

Scale Reading Quiz 3

INSTRUCTIONS: Determine the readings from the illustration shown below, and enter your answers to the closest 0.5 mm.

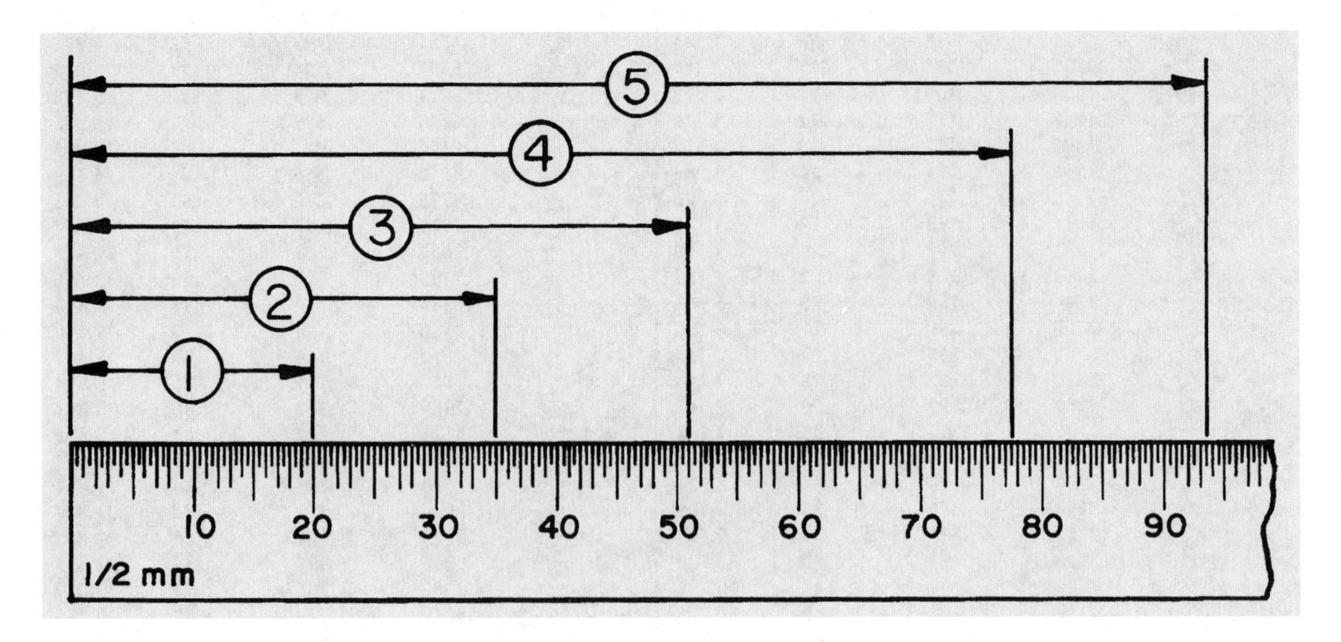

① ____________________ ③ ____________________ ⑤ ____________________

② ____________________ ④ ____________________

BLUEPRINTS VS. WHITEPRINTS

Most of us continue to use the term "blueprint" when referring to prints of engineering drawings, although today the majority of prints produced are actually "whiteprints." The blueprint process is essentially a photographic process in which the original drawing is the negative. The true blueprint contains white lines on a blue background.

A more efficient method of producing prints today is the diazo process, which is based on sensitivity to light and ammonia vapors. Prints produced by this process usually contain blue lines on a white background, and may be referred to as whiteprints, or bluelines. Black or red lines are also possible, although they are not as commonly used.

The diazo process requires the original drawing to be made on a translucent material that allows light to pass through. One of the most commonly used materials is vellum, an oil-treated tracing paper. Either ink or pencil may be used on vellum to create the opaque lines necessary for producing a clear image.

The diazo process uses a special print paper that contains a dye coating capable of decomposing when exposed to light and forming an azo dye when exposed to ammonia vapors. The exposure and development stages are illustrated below.

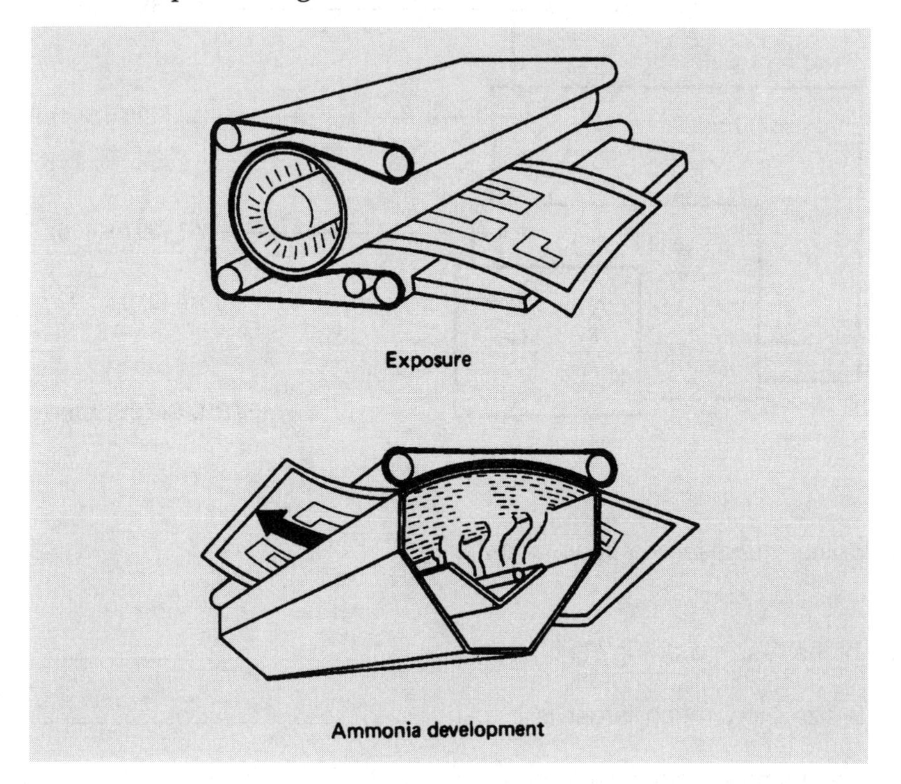

ENGINEERING DRAWINGS

The original drawings made by a drafter or engineer are referred to as engineering drawings. Prints made from these drawings may be referred to as blueprints, whiteprints, or blue-lines. By the diazo process, prints are reproduced to the same size as the original drawing. These sizes have been standardized, and are shown on the illustration below. The $8\frac{1}{2} \times 11$ inch A-size is the same size as typing paper, thereby fitting into mailing envelopes, file drawers, and other standard office equipment. B-size through E-size are all in multiples of $8\frac{1}{2} \times 11$ so that they too will fit when folded. Larger drawings may be drawn on rolls of vellum. Metric drawing sizes have been standardized on the width-to-length ratio of 1 to $\sqrt{2}$ and are shown at the right.

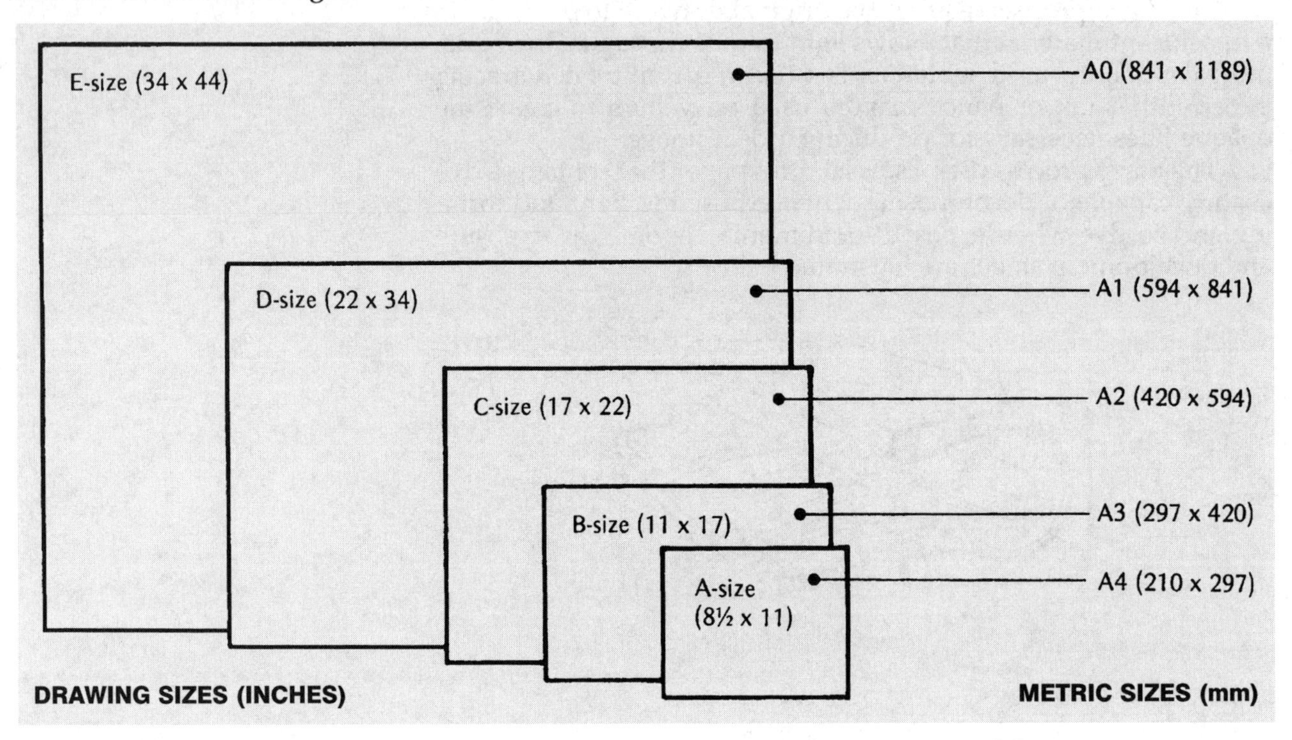

Drawing Size Quiz

INSTRUCTIONS: Answer the following questions about drawing sizes.

1. What are the basic dimensions of the A-size drawing? (Inches) 1. ______
2. If C-size is twice as large as B-size, how much larger is it than A-size? 2. ______
3. Will an E-size print fold to fit in the same envelope with an A-size print? 3. ______
4. What metric size is twice as large as the A3 size? 4. ______
5. Are the A4 metric size and the A inch size exactly the same? (Use 1 in. = 25.4 mm, or 1 mm = .03937 in. to compare.) 5. ______

TITLE BLOCKS

Each sheet of vellum is normally preprinted with a border and title block for the inclusion of information pertinent to the drawing. The most prominent lettering in the title block is the drawing number (part number), since it acts as a filing system. The title (part name) is also lettered larger to make it stand out. An important entry in the title block for you, the blueprint reader, is entitled "scale." This information will inform you whether or not the illustration appears actual size.

If an item is too large to comfortably fit onto a standard size sheet, it may be drawn to a smaller scale (size). Similarly, if it is too small to adequately show all details or dimensions, it may be drawn to a larger scale. However, you must understand that only the size of the illustration is affected, not the dimensions that appear adjacent to the illustration. For example, if a 6-in. cube were drawn to half-scale, the cube would be drawn 3 in., but the dimension alongside it would still be written 6″. A 1-in. sphere drawn four times scale would actually measure 4 in., but its dimension would be written 1″. The dimension figure is always the actual size, regardless of scale.

Mechanical engineering drawings may be reduced to half-size, quarter-size, or one-eighth size. If they are enlarged, it is customary to use double size, four times size, or ten times size. Several methods of designating the scale may be used. For example, half-scale may appear as 1/2, 1:2, or 1/2″ = 1″, whereas twice scale would appear as 2/1, 2:1, 2–1, 2″ = 1″, or 2X. Note that the arrangement of numbers (1:2 vs. 2:1) indicates the relationship between the drawing and the actual part.

INSTRUCTIONS: Circle the scale specified in each of the sample title blocks shown below and on the next page.

UNLESS OTHERWISE SPECIFIED DIMENSIONS ARE IN INCHES			APP JR	REL 12-7-00	GREAT DANE TRAILERS, INC. SAVANNAH, GA.	
TOLERANCES ON DECIMALS	2 PLACE	3 PLACE	SCALE 1/2			
0 TO 6 DIM. INCL.	±.02	±.005				
OVER 6 TO 24 DIM. INCL.	±.03	±.010	DWN SCOTT S.		TITLE	
OVER 24	±.06	±.015				
ANGLES ± 1/2°			DATE 12-2-00		DOUBLER PLATE	
MAT 5052-H34 ALUM			USED ON		RESTRICTED NOSE BOX	
WEIGHT 2.11 LBS					DWG NO 040-47315	REV A
SHEET 1 OF 1						

(Courtesy Great Dane Trailers, Inc.)

THIS DRAWING AND THE INFORMATION HEREON ARE OUR PROPERTY AND MAY BE USED BY OTHERS ONLY AS AUTHORIZED BY US. UNPUBLISHED-ALL RIGHTS RESERVED UNDER THE COPYRIGHT LAWS ©	DWN R. GONZALES 9-4-00		DEERE & COMPANY MOLINE, ILLINOIS 61265				
	CHK K. KADER 9-11-00						
REFERENCES:	ENGR J. COOPER 9-19-00		NAME				
JDS-G25 GENERAL TOLERANCES	CHK DWM		BRACKET, GUSSET				
JDS-G113 DRAFTING							
	APVD JCM	DESIGN CONTROL RE	SIZE E	FSCM NO. 1	PART NUMBER RE 65234	VERSION RE40	REV B
	C/C GLA	TERM CODE 0900 01	SCALE 1:1	METRIC	⌖◁	SHEET 1 OF 1	

(Courtesy Deere & Co.)

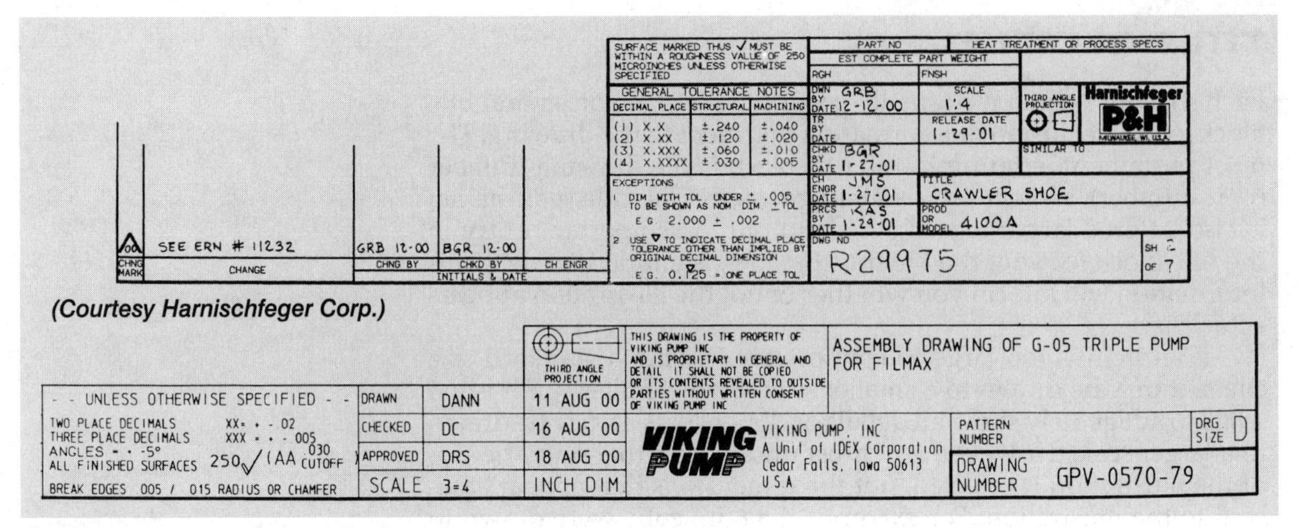

(Courtesy Harnischfeger Corp.)

(Courtesy Viking Pump, Unit of Idex Corp.)

DRAWING NOTES

Notes are classified as local notes when they apply only to specific items or areas, and as general notes when they apply to the entire drawing or product. Local notes use leaders to direct the notes to where they apply. General notes contain information pertaining to machining, finishing, heat treatment, material, tolerances, etc., and will normally appear just above the drawing title block. Abbreviations are commonly used in notes to keep them brief. Typical examples of drawing notes you might encounter are shown below.

(GENERAL)	(LOCAL)
UNLESS OTHERWISE SPECIFIED:	
1. METRIC UNITS ARE mm	↖ 5 HOLES EQL SP ON 6.000 BC
2. FILS & RNDS .12R	↖ .344 DR, .500 CBORE, .30 DP
3. PART SYM ABOUT CL	↖ .281 DR, 82° CSK, .53 DIA

Drawing Notes Quiz

INSTRUCTIONS: Answer the following questions pertaining to the notes above.

1. What type of note applies to an entire drawing? 1. ____________
2. What type of line is used with local notes? 2. ____________
3. Where do general notes usually appear on a drawing? 3. ____________
4. Interpret the abbreviations SYM and CL used in the third general note. 4. ____________
5. Interpret the abbreviations DR and CSK used in the third local note. 5. ____________

SINGLE-VIEW DRAWINGS

Often, a single view of an object is adequate to describe it completely. Such is the case with drawing 11A001, page 32. An additional view would serve no other purpose than to provide for a thickness dimension, which ordinarily is already listed in the material space of the title block. Therefore, thin, flat objects will normally be drawn as single-view drawings.

Some geometric shapes, such as the sphere or the cube, would also appear as single-view drawings. Whenever a shape can be determined from a note, abbreviation, or symbol, it is possible to describe the object with a single view. The abbreviation SQ or DIA following the dimension is a sufficient description. Nearly all shafts, bushings, bolts, screws, and similar parts are represented by single-view drawings in this manner.

DETAIL DRAWINGS

The terms "production drawing" and "working drawing" are general terms that include all types of engineering drawings. The type that we will concentrate our interest on throughout this book is called a detail drawing. A detail drawing contains a sufficient number of views, dimensions, notes, and other pertinent information necessary to produce a specific part. Thus, a detail may be complete with one view, or it may be necessary to include an unlimited number of views to adequately describe a complex shape.

CENTERLINES

Observe from the illustration below how intersecting centerlines are drawn through each hole. When used for location dimensioning, centerlines are extended outside the view and dimension lines are attached. The same is true for radii that also have a definite location. The .38R and .62R share the same center locations as the holes, which is a common design practice.

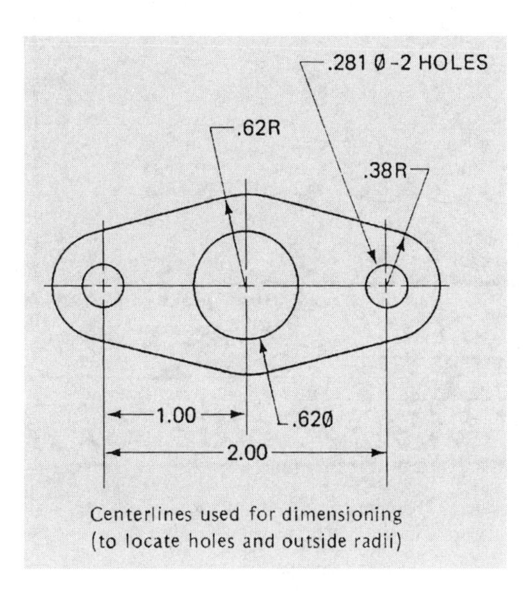

Centerlines used for dimensioning
(to locate holes and outside radii)

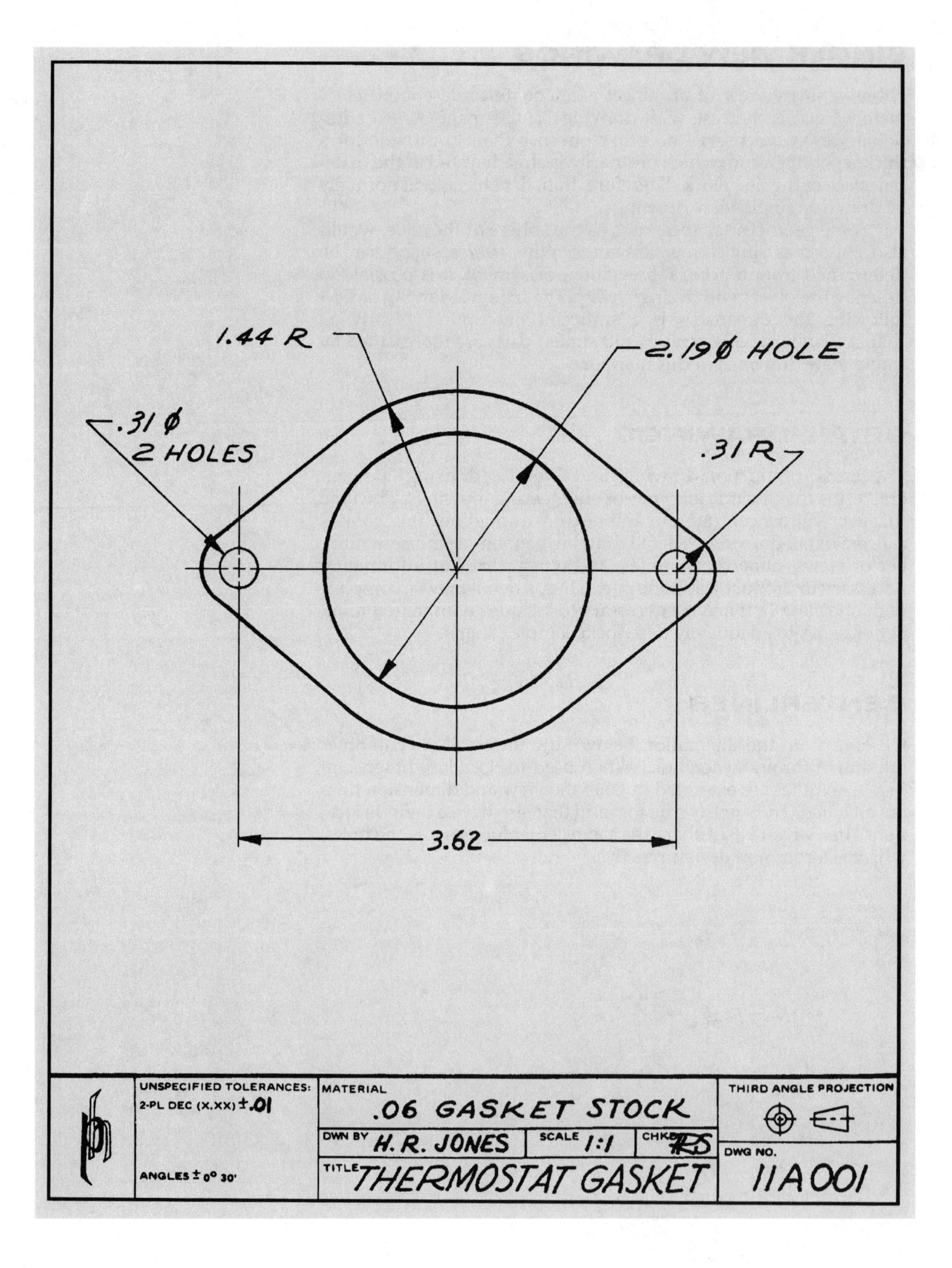
1.44 R
2.19 ⌀ HOLE
.31 ⌀
2 HOLES
.31 R
3.62
UNSPECIFIED TOLERANCES:
2-PL DEC (X.XX) ±.01
ANGLES ± 0° 30'
MATERIAL
.06 GASKET STOCK
DWN BY H. R. JONES
SCALE 1:1
CHKD
TITLE THERMOSTAT GASKET
THIRD ANGLE PROJECTION
DWG NO.
11A001

DIAMETERS AND RADII

Observe the symbol ∅ that follows the dimensions for the holes in the gasket in drawing 11A001. This is an internationally recognized symbol for diameter, and is used in place of the abbreviation DIA. When more than one hole of the same size is specified, common practice uses only one leader and includes the quantity in the local note. The letter R is used with radial dimensions, but unlike dimensioning holes, the quantity is usually not included.

Note that the 3.62 dimension appearing in the gasket drawing is used for locating the holes and also for locating the radii that surround them. Round holes will always be located from their centers, *not* by their sides.

Shapes such as the one illustrated in drawing 11A001 do not ordinarily include overall dimensions, but you can calculate them by adding the radii to their locations. Material remaining around a hole can be determined by either of the concentric calculation methods illustrated at the right.

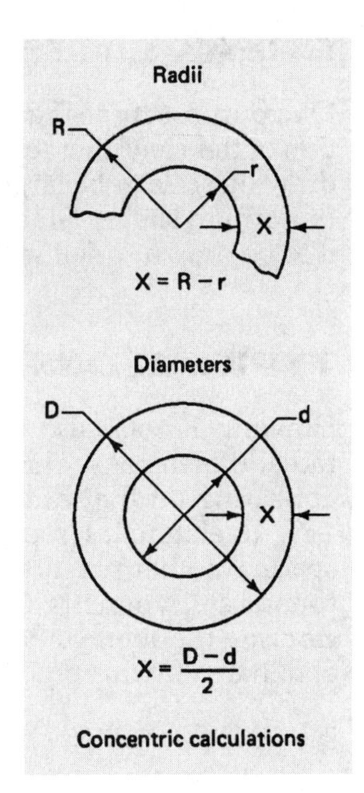

Concentric calculations

INSTRUCTIONS: Study the title block and dimensions in drawing 11A001 before answering the following questions.

1. What is the part number? 1. ______
2. What is the name of the part? 2. ______
3. What is the drawing size? (A, B, or C) 3. ______
4. What scale is the drawing? 4. ______
5. Use words to explain the answer to question 4. 5. ______
6. How thick is the part? 6. ______
7. Is the part symmetrical? 7. ______
8. Which system of dimensioning was used? (Refer to page 21.) 8. ______
9. What is the fractional equivalent of the center hole *size*? (Refer to the equivalency chart, page 25.) 9. ______
10. What is the fractional equivalent of the mounting hole *spacing*? (Refer to the equivalency chart, page 25.) 10. ______
11. Calculate the longest overall dimension of the part. 11. ______
12. Calculate the other overall dimension of the part (height). 12. ______
13. Calculate the material remaining between the large hole and the nearest outside edge. 13. ______
14. Calculate the material remaining between the small hole and the nearest outside edge. 14. ______
15. Calculate the material remaining between the large hole and the nearest small hole. 15. ______

BREAK LINES

One purpose for using break lines is to permit the removal of a portion of the view, thereby allowing the paper size to be smaller or the drawing scale to be larger. The portion removed, however, must be uniform in shape. Drawing 11A002 uses long break lines, allowing it to be drawn to full scale without requiring B-size vellum.

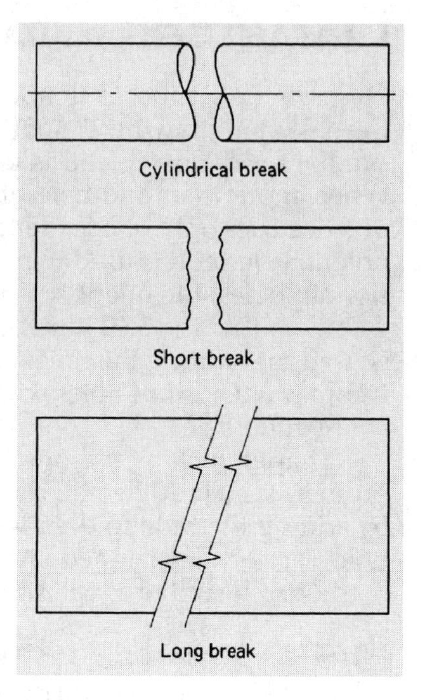

TYPICAL DIMENSIONS

Drawing 11A002 also uses the abbreviation TYP after one of the radial dimensions. This means that all other undimensioned radii appearing on that drawing are the same size as the one marked TYP. It eliminates repetition of dimensions and saves drawing space. Another practice used to reduce dimensions on drawings is to indicate symmetry either by note, or with the view's centerline carrying the abbreviation ℄. Dimension Ⓐ on drawing 11A002 was eliminated by this practice.

INSTRUCTIONS: Refer to drawing 11A002. Enter the dimensions for the following letters.

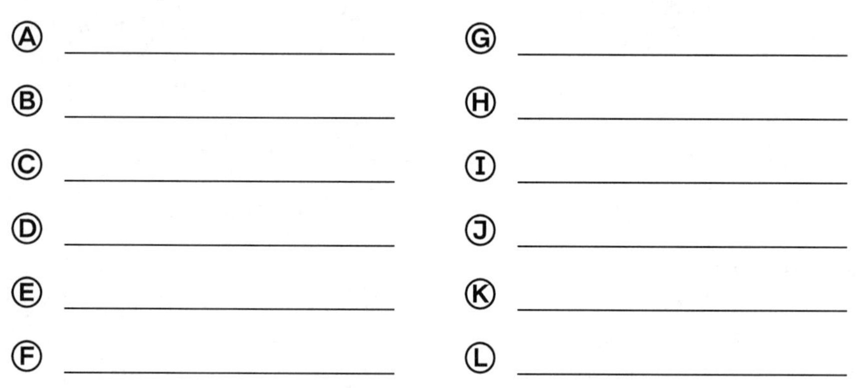

INSTRUCTIONS: Answer the following questions.

1. What type of line was drawn to indicate symmetry? 1. ____________
2. What word is abbreviated TYP? 2. ____________
3. What is the thickness of the plate? 3. ____________
4. How many round holes does the plate contain? 4. ____________
5. What is the fractional equivalent of the small hole diameter? (Refer to page 25.) 5. ____________
6. How much material remains between the .62 hole and the nearest outside edge? 6. ____________
7. How much material remains between a .28 hole and the nearest outside edge? 7. ____________
8. Calculate the longest overall dimension of the plate. 8. ____________

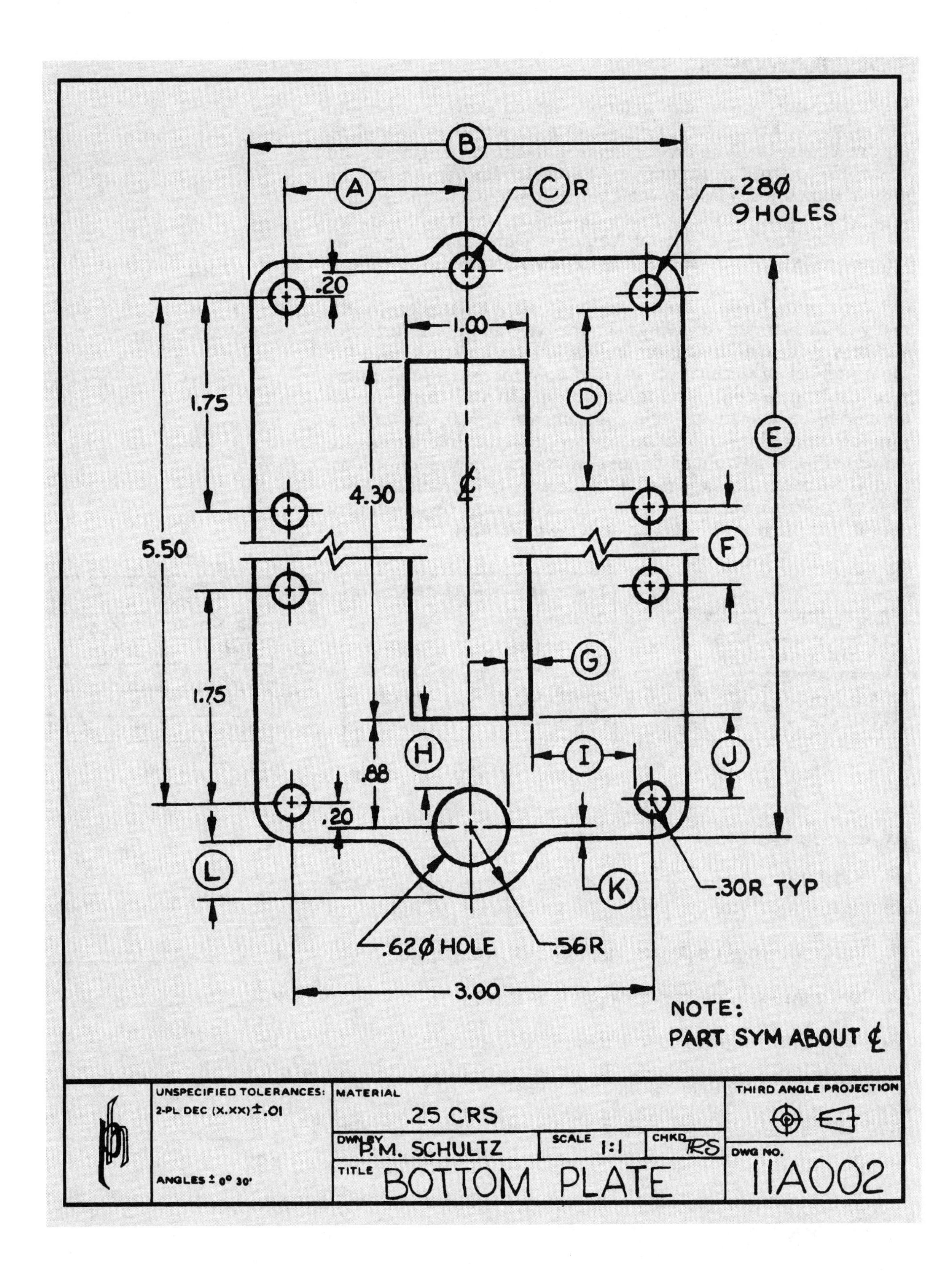

B
A
C R
.28Ø
9 HOLES
.20
1.00
1.75
D
E
4.30
℄
5.50
F
G
1.75
H
I
J
.88
.20
L
K
.30R TYP
.62Ø HOLE
.56R
3.00
NOTE:
PART SYM ABOUT ℄
UNSPECIFIED TOLERANCES:
2-PL DEC (X.XX) ±.01
ANGLES ± 0° 30'
MATERIAL
.25 CRS
DWN BY P.M. SCHULTZ
SCALE 1:1
CHKD RS
THIRD ANGLE PROJECTION
DWG NO.
11A002
TITLE BOTTOM PLATE

TOLERANCES

Detail drawings will have a tolerance assigned to every dimension that appears. Recognizing the fact that perfect sizes cannot be obtained consistently in production manufacturing, engineers and designers assign the maximum acceptable deviation from the desired dimension. This allowable variation is the tolerance. It may be shown individually alongside a dimension, or it may be shown in the title block as a general tolerance. Dimensions appearing without individual tolerances are automatically covered by general tolerances.

A common method used to assign general tolerances to decimally dimensioned drawings is the decimal-place method. Whereas a decimal dimension and its tolerance always have the same number of decimal places, it is easy for you to determine which tolerance applies. (The dimension .50 will carry a two-decimal-place tolerance, while the dimension .500 will carry a three-decimal-place tolerance.) Most general tolerances are expressed bilaterally, although not always equal in both directions. Such is the case with the drilled hole tolerance in Example A below. General tolerance values apply only to the drawing on which they appear, and often vary from one drawing to another.

UNSPECIFIED TOLERANCES:
2-PL DEC IN. (.XX) ± .03
3-PL DEC IN. (.XXX) ± .005
METRIC (MM) ± 0.25
DRILLED HOLES $^{+.010}_{-.002}$
ANGLES ± 0° 30'

Example A

Tolerances Unless Noted Otherwise	
One Place Decimal	± .1
Two Place Decimal	± .02
Three Place Decimal	± .004
Angular	± 0°30'
Concentricity	± .02 T.I.R.

Example B

UNLESS OTHERWISE SPECIFIED:	
FRACTIONAL	± 1/64
.000	±.003
.00	±.01
.0	±.1
ANGULAR	± 0°30'

Example C

Tolerance Quiz

INSTRUCTIONS: Answer the following questions pertaining to the examples above.

1. What is the two-place decimal inch tolerance in Example A? 1. ______________
2. What is the tolerance for drilled holes in Example A? 2. ______________
3. What is the three-place decimal tolerance in Example B? 3. ______________
4. What is the fractional tolerance in Example C? 4. ______________
5. What is the angular tolerance in Example C? 5. ______________

BOLT CIRCLES

When three or more holes are positioned around a common center (concentrically), they will be located angularly on a circular centerline, called a bolt circle. The dimension of the bolt circle may appear inside of a diagonal dimension line, or it may appear (with the abbreviation BC) after the hole data. See examples using both locations shown below.

If the holes are spaced equally on the bolt circle, the abbreviation EQL SP will follow the quantity in the local note. For you to determine the angular dimension of equally spaced holes, merely divide 360° by the number of holes. For example, 360°/3 holes = 120°, 360°/6 holes = 60°, etc. One of the holes will be drawn on the vertical or the horizontal centerline, or it will be angularly dimensioned to one of them.

Holes may also be located coordinately, using only vertical or horizontal dimensions. See the example shown below.

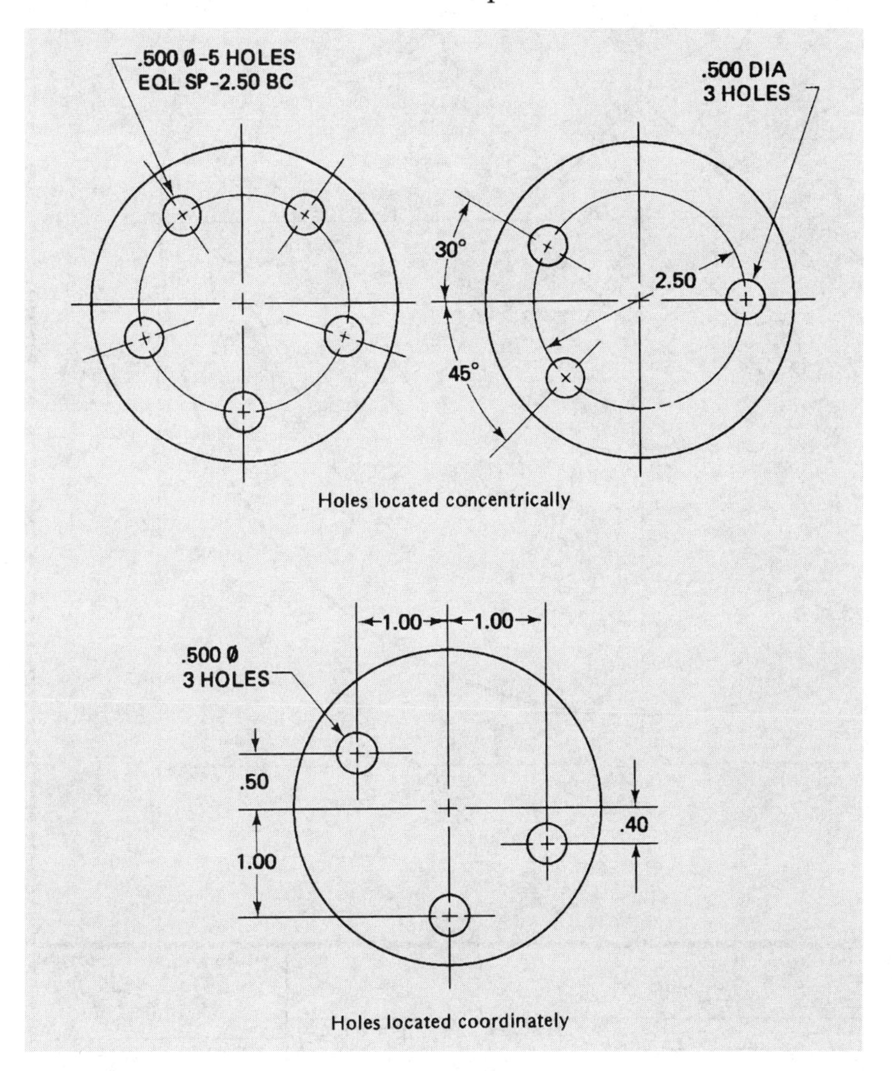

Holes located concentrically

Holes located coordinately

Student Note:

For converting concentrically located holes to coordinate locations, refer to the mathematics appendix beginning on page 373.

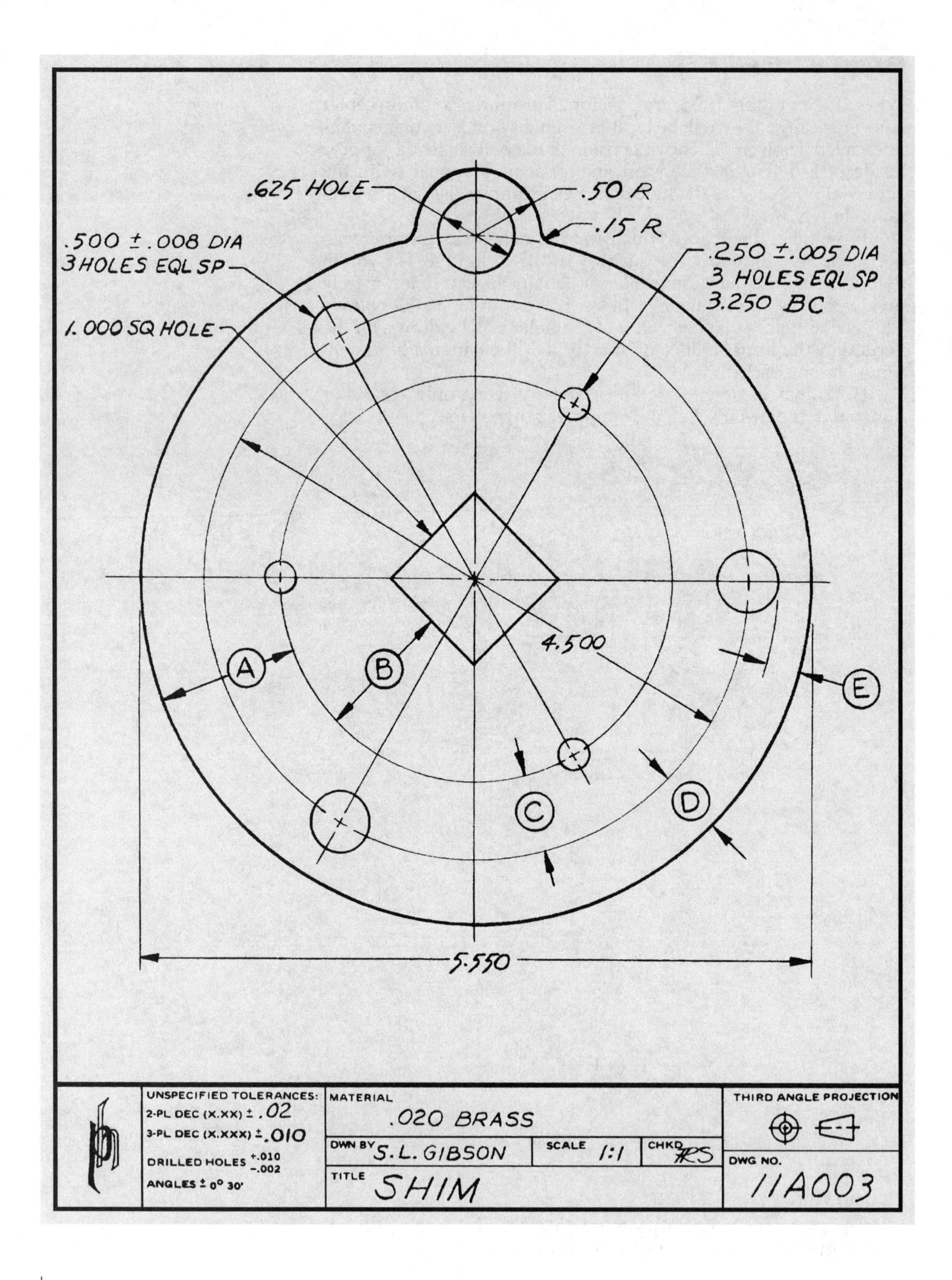
.625 HOLE
.50 R
.15 R
.500 ±.008 DIA
3 HOLES EQL SP
.250 ±.005 DIA
3 HOLES EQL SP
3.250 BC
1.000 SQ HOLE
4.500
A
B
C
D
E
5.550
UNSPECIFIED TOLERANCES:
2-PL DEC (X.XX) ± .02
3-PL DEC (X.XXX) ± .010
DRILLED HOLES +.010 −.002
ANGLES ± 0° 30'
MATERIAL
.020 BRASS
DWN BY S.L. GIBSON
SCALE 1:1
CHKD RS
TITLE SHIM
THIRD ANGLE PROJECTION
DWG NO.
11A003

INSTRUCTIONS: Answer the following questions pertaining to drawing 11A003.

1. How thick is the shim? 1. ____________
2. Is the shim symmetrical? 2. ____________
3. Are the bolt circles concentric? 3. ____________
4. How many holes are in the shim? 4. ____________
5. What is the tolerance on the .250∅ holes? 5. ____________
6. What is the tolerance on the .500∅ holes? 6. ____________
7. What is the tolerance on the 1.000 hole? 7. ____________
8. What is the tolerance on the bolt circles? 8. ____________
9. What is the tolerance on the radii? 9. ____________
10. How much material is between the 5/8″ hole and the nearest outside edge? 10. ____________
11. How far apart angularly are the three .500∅ holes? 11. ____________
12. How far apart angularly are a .250∅ hole and the *nearest* .500∅ hole? 12. ____________
13. How far apart angularly are a .250∅ hole and the *farthest* .500∅ hole? 13. ____________
14. How far apart angularly is the 5/8″ hole from the nearest 1/2″ hole? 14. ____________
15. What is the angular tolerance for the hole locations? 15. ____________

INSTRUCTIONS: Enter the dimensions for the following letters.

Ⓐ ____________

Ⓑ ____________

Ⓒ ____________

Ⓓ ____________

Ⓔ ____________

Optional Math Exercises:

1a. Using the trigonometry tables from the Math Appendix, calculate the Cartesian coordinates (X & Y) for the .500∅ and the .250∅ holes.

2a. Using the formula from the Math Appendix, determine the distance over pins of adjacent holes for both the .500∅ and the .250∅ holes.

MAXIMUM MATERIAL CONDITION

Maximum Material Condition (MMC) is the condition whereby a feature of size contains the maximum amount of material within assigned tolerances. For example, an external feature such as a shaft is at MMC when its positive (plus) tolerance has been *added* to its size dimension. Thus, in the illustration below, the MMC of the shaft is .76 (.75 size dimension, plus .01 tolerance).

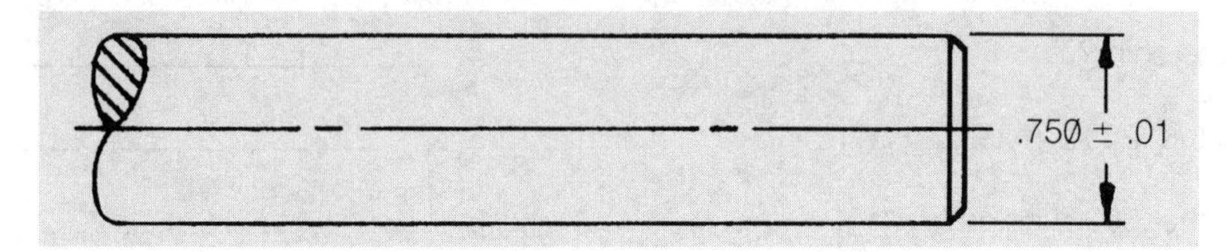

An internal feature such as a hole is at MMC when its negative (minus) tolerance has been *subtracted* from its size dimension. Thus, in the illustration below, the MMC of the hole is .74 (.75 size dimension, minus .01 tolerance). Visually compare the exaggerated .74∅ and the .76∅ examples below to verify which one leaves the maximum amount of material remaining.

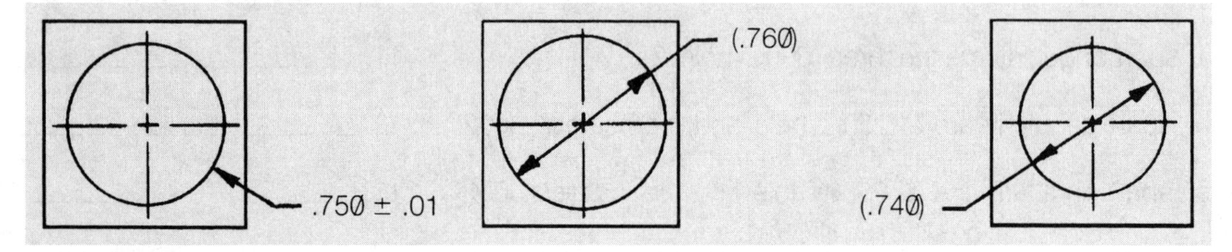

ALLOWANCE

Allowance may be defined as the minimum clearance (or maximum interference) between mating parts, such as a shaft and a hole. Allowance may be calculated by subtracting the MMC of the shaft from the MMC of the hole. If the shaft's MMC is *smaller* than the hole's MMC, it is considered a positive allowance and provides a clearance fit. If the shaft's MMC *is larger* than the hole's MMC, it is considered a negative allowance, which provides an interference fit. In either case (positive or negative), allowance is always the tightest possible fit between mating parts. It can only be calculated by using the MMC of each part. See examples below and on page 41.

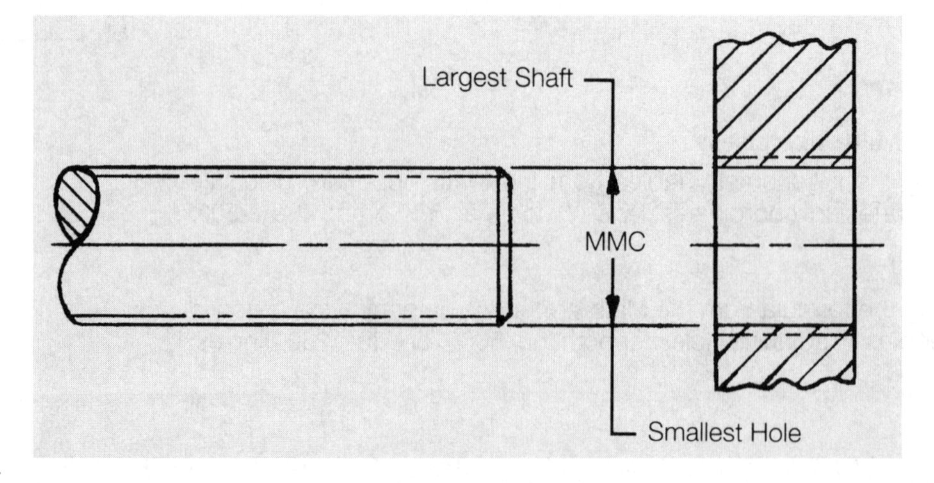

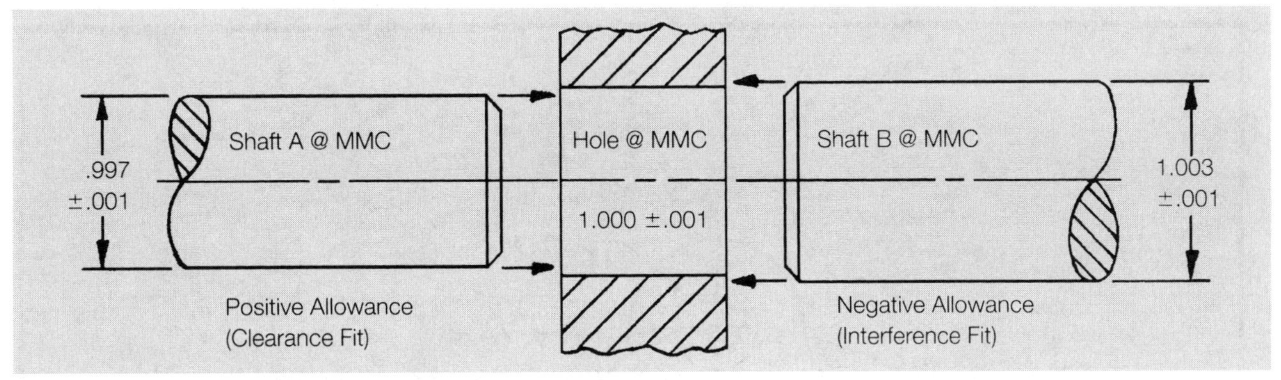

Use the following formula for calculating allowance.

MMC Hole
− MMC Shaft
Allowance

Refer to the example above as you verify the following figures:

MMC Hole	= .999	MMC Hole	= .999
− MMC Shaft A	= .998	− MMC Shaft B	= 1.004
Allowance	= .001 (positive)	Allowance	= −.005 (negative)

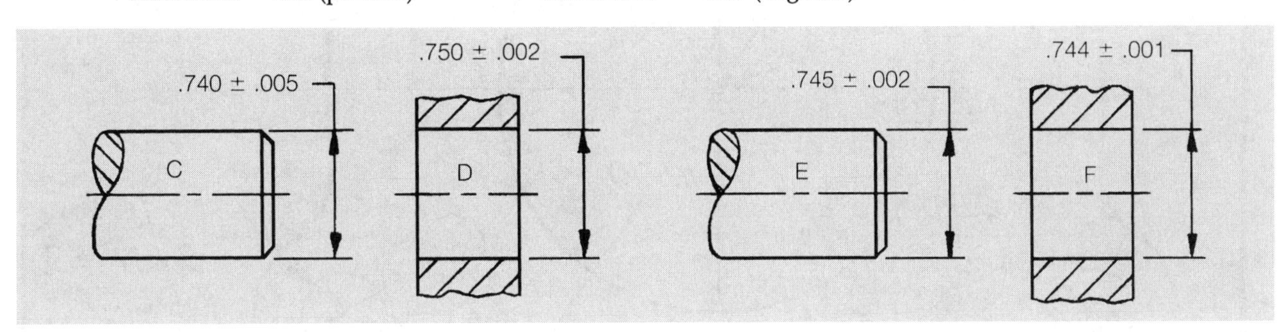

INSTRUCTIONS: Refer to the illustrations above to answer the following questions.

MMC/Allowance Quiz

1. What is the MMC of hole D? 1. ______
2. What is the MMC of shaft C? 2. ______
3. What is the allowance between C and D? 3. ______
4. Is the allowance positive or negative? 4. ______
5. Is it a clearance or interference fit? 5. ______
6. What is the MMC of hole F? 6. ______
7. What is the MMC of shaft E? 7. ______
8. What is the allowance between E and F? 8. ______
9. Is the allowance positive or negative? 9. ______
10. Is it a clearance or interference fit? 10. ______

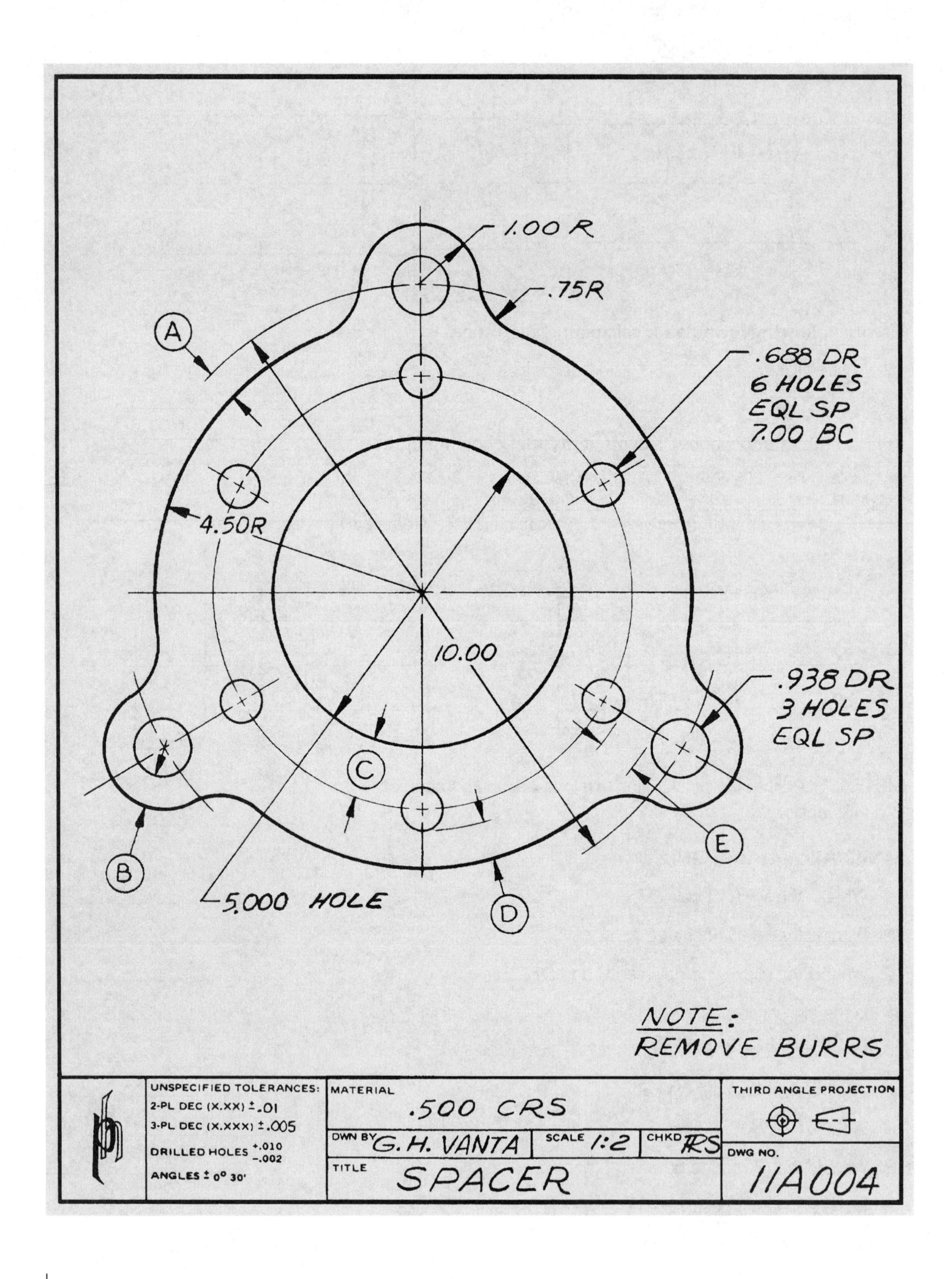
1.00 R
.75R
A
.688 DR
6 HOLES
EQL SP
7.00 BC
4.50R
10.00
.938 DR
3 HOLES
EQL SP
C
E
B
D
5.000 HOLE
NOTE:
REMOVE BURRS
UNSPECIFIED TOLERANCES:
2-PL DEC (X.XX) ±.01
3-PL DEC (X.XXX) ±.005
DRILLED HOLES +.010 −.002
ANGLES ± 0° 30'
MATERIAL
.500 CRS
DWN BY G.H. VANTA
SCALE 1:2
CHKD RS
TITLE
SPACER
THIRD ANGLE PROJECTION
DWG NO.
11A004

ARCS AND CIRCLES

When an arc is less than a complete circle, it is dimensioned by its radius, abbreviated R. A radius may have a definite location, such as the 1.00R & 4.50R in drawing 11A004 or it may be a "floating radius" that runs tangent with other arcs or lines, as with the .75R. A radial dimension is not normally used for circles or diameters, since it would double the tolerance. For example, A 1.00 diameter with a print tolerance of ± .01 would permit dimensions between .99 and 1.01, but a .50 radius ± .01 would permit diametric dimensions between .98 and 1.02.

INSTRUCTIONS: *Refer to drawing 11A004 to answer the following questions.*

1. Is the print drawn to (a) half-size or (b) twice-size? 1. ______
2. What is the thickness of the spacer? 2. ______
3. Interpret the abbreviation CRS used in the title block. 3. ______
4. How many holes does the spacer contain? 4. ______
5. Is the spacer symmetrical? 5. ______
6. What is the angular dimension between each small hole? 6. ______
7. What is the angular dimension between each large hole? 7. ______
8. What tolerance is assigned to the angular spacing of the drilled holes? 8. ______
9. What is the tolerance on the bolt circles? 9. ______
10. What is the tolerance on the center hole diameter? 10. ______
11. What is the MMC of the center hole? 11. ______
12. What is the tolerance on the drilled holes? 12. ______
13. What is the MMC of the .688 holes? 13. ______
14. What is the overall dimension along the horizontal CL? 14. ______
15. What is the overall dimension along the vertical CL? 15. ______

INSTRUCTIONS: *Calculate the dimensions for the following letters.*

Ⓐ ______

Ⓑ ______

Ⓒ ______

Ⓓ ______

Ⓔ ______

Optional Math Exercises:

1a. Using the trigonometry tables from the Math Appendix, calculate the Cartesian coordinates (X & Y) for the .688∅ and the .938∅ holes.

2a. Using the formula from the Math Appendix, determine the distance over pins of adjacent holes for both the .688∅ and the .938∅ holes.

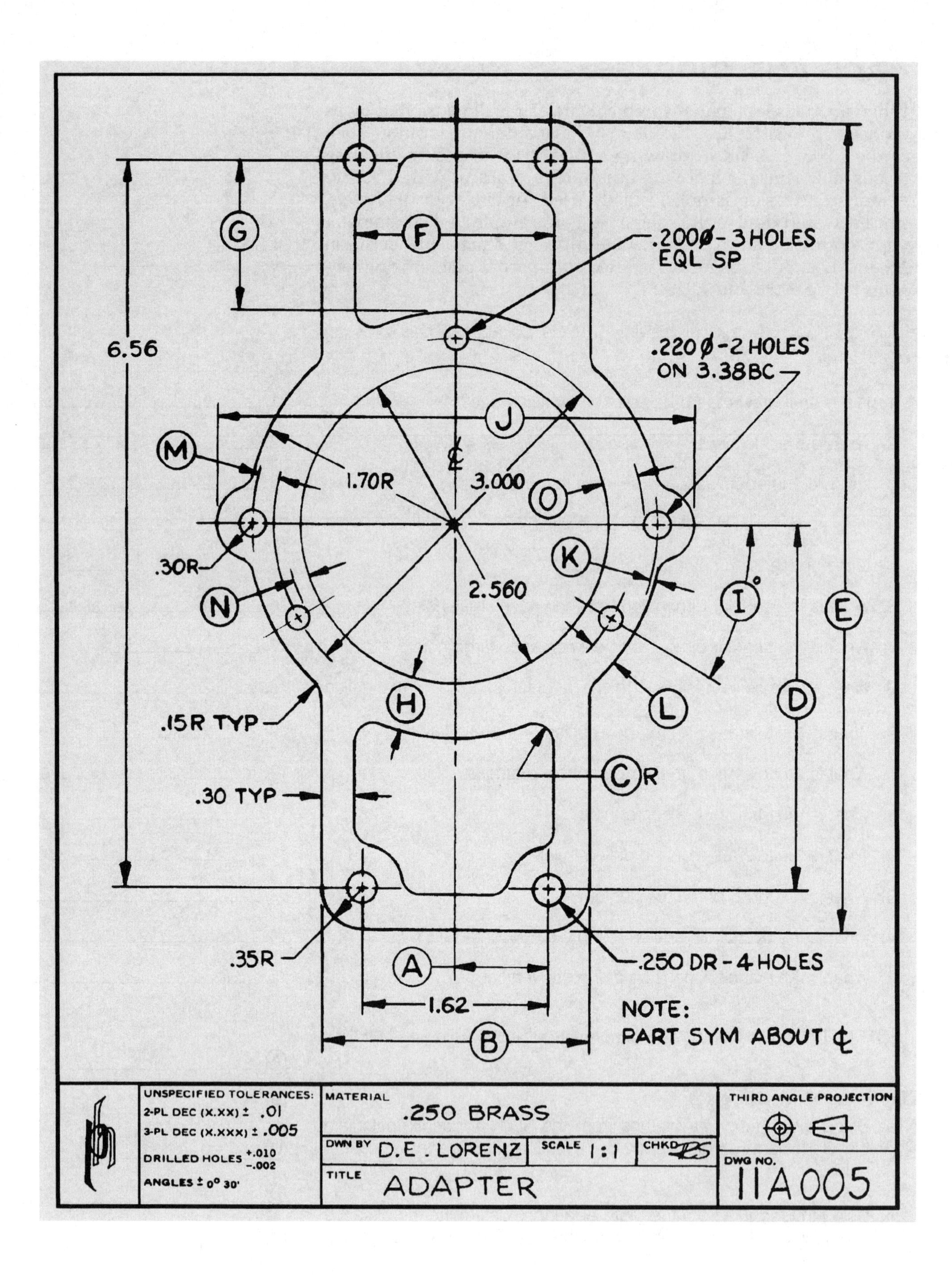

.200Ø - 3 HOLES
EQL SP
.220Ø - 2 HOLES
ON 3.38BC
6.56
1.70R
3.000
2.560
.30R
.15R TYP
.30 TYP
.35R
.250 DR - 4 HOLES
1.62
NOTE:
PART SYM ABOUT ℄
UNSPECIFIED TOLERANCES:
2-PL DEC (X.XX) ± .01
3-PL DEC (X.XXX) ± .005
DRILLED HOLES +.010 −.002
ANGLES ± 0° 30'
MATERIAL
.250 BRASS
DWN BY D.E. LORENZ
SCALE 1:1
CHKD
TITLE ADAPTER
THIRD ANGLE PROJECTION
DWG NO.
11A005

INSTRUCTIONS: Enter the dimensions for the following letters found on drawing 11A005.

Ⓐ ____________

Ⓑ ____________

Ⓒ ____________

Ⓓ ____________

Ⓔ ____________

Ⓕ ____________

Ⓖ ____________

Ⓗ ____________

Ⓘ ____________

Ⓙ ____________

Ⓚ ____________

Ⓛ ____________

Ⓜ ____________

Ⓝ ____________

Ⓞ ____________

INSTRUCTIONS: Refer to drawing 11A005 to answer the following questions.

1. How much material remains between a .250 hole and the outside edge?
2. How much material remains between a .220 hole and the outside edge?
3. Are the .200 holes located in the exact center between ID and OD? (calculate)
4. What is the MMC of the center hole?
5. What is the MMC of the .250 holes?

1. ____________

2. ____________

3. ____________

4. ____________

5. ____________

Optional Math Exercises:

1a. Using the trigonometry tables from the Math Appendix, calculate the Cartesian coordinates (X & Y) for the .220∅ holes.

2a. Using the formula from the Math Appendix, determine the distance over pins of adjacent holes for the .220∅ holes.

UNIT 3

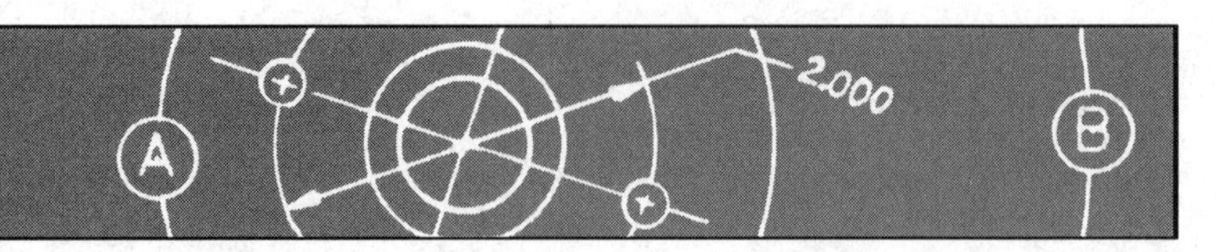

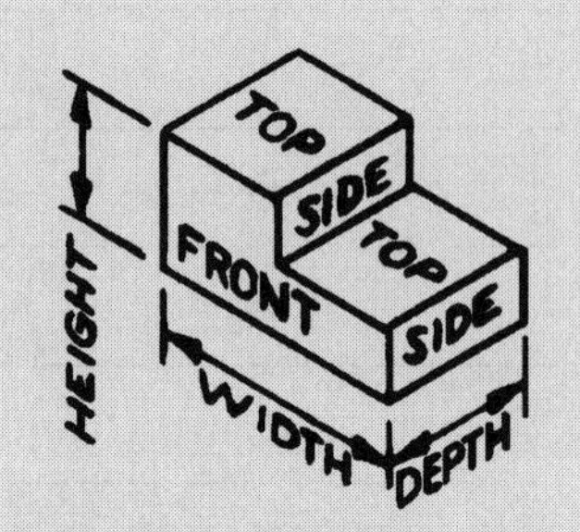

Pictorial Drawing

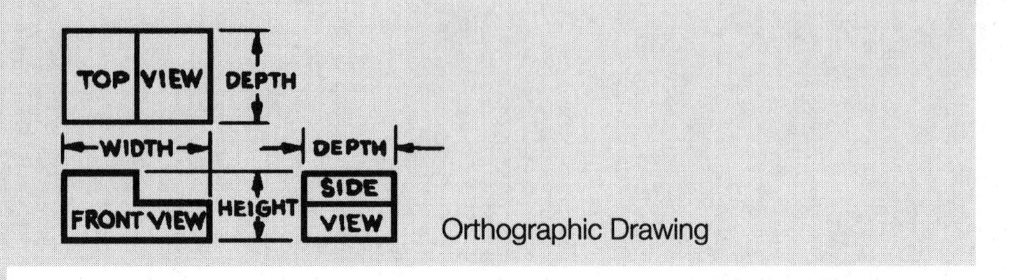

Orthographic Drawing

ORTHOGRAPHIC PROJECTION

Pictorial drawings show the three basic dimensions (height, width, depth) on a common view. This helps the viewer to visualize the shape of an object easily. However, with the exception of some very simple shapes it would be difficult to dimension this type of drawing for accurate interpretation. Therefore, detail drawings will be drawn by the method of orthographic projection. This method permits the drafter to draw all views to true shape. Although it limits each view to two dimensions, the third dimension can be found in an adjacent view. With practice, you will soon be able to visualize three-dimensional shapes from two-dimensional views drawn orthographically. Compare the two types of drawings illustrated above.

The views of an orthographic drawing are projected at right angles (90°) to each other and have a definite relationship. This can best be visualized by cutting and unfolding a cardboard box, as illustrated below. Compare the unfolded view with the parallel block drawing 11A006. The views of the drawing are separated, but always remain in direct projection from one another.

Some textbooks teach the principles of orthographic projection by having the student visualize the object suspended inside a glass cube. Then the image of the object is projected to each glass panel, and the hinged panels of the cube are unfolded. Since there are six sides to a cube, there are six principal views of any object. Proper arrangement of the six views is shown below. Observe that each adjacent view includes the third dimension. Also notice that every-other-view repeats its size.

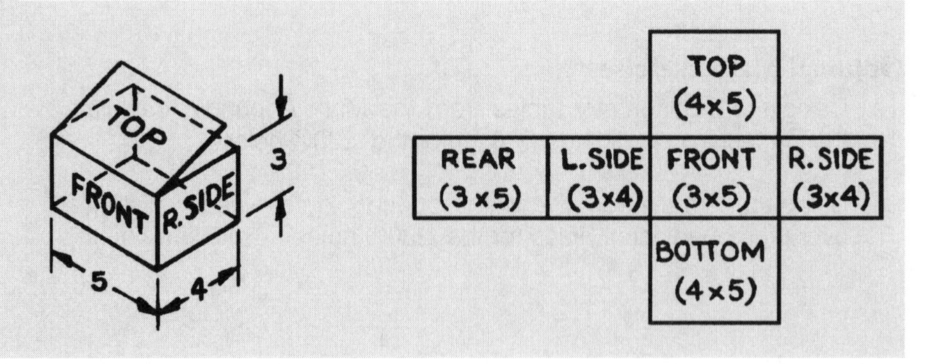

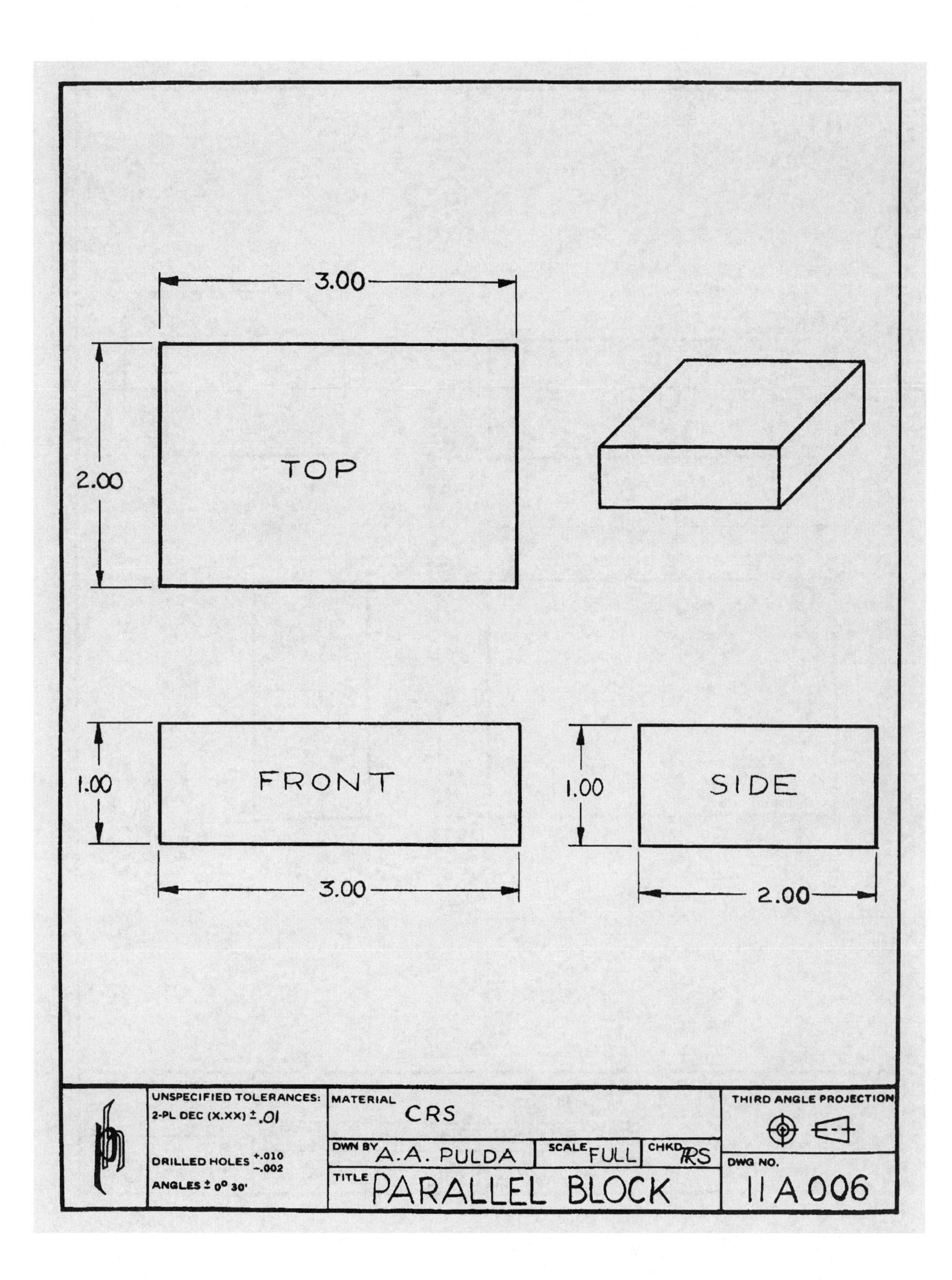
3.00
2.00
TOP
FRONT
1.00
3.00
1.00
SIDE
2.00
UNSPECIFIED TOLERANCES:
2-PL DEC (X.XX) ± .01
DRILLED HOLES +.010 −.002
ANGLES ± 0° 30'
MATERIAL CRS
DWN BY A.A. PULDA
SCALE FULL
CHKD RS
TITLE PARALLEL BLOCK
THIRD ANGLE PROJECTION
DWG NO.
11A006

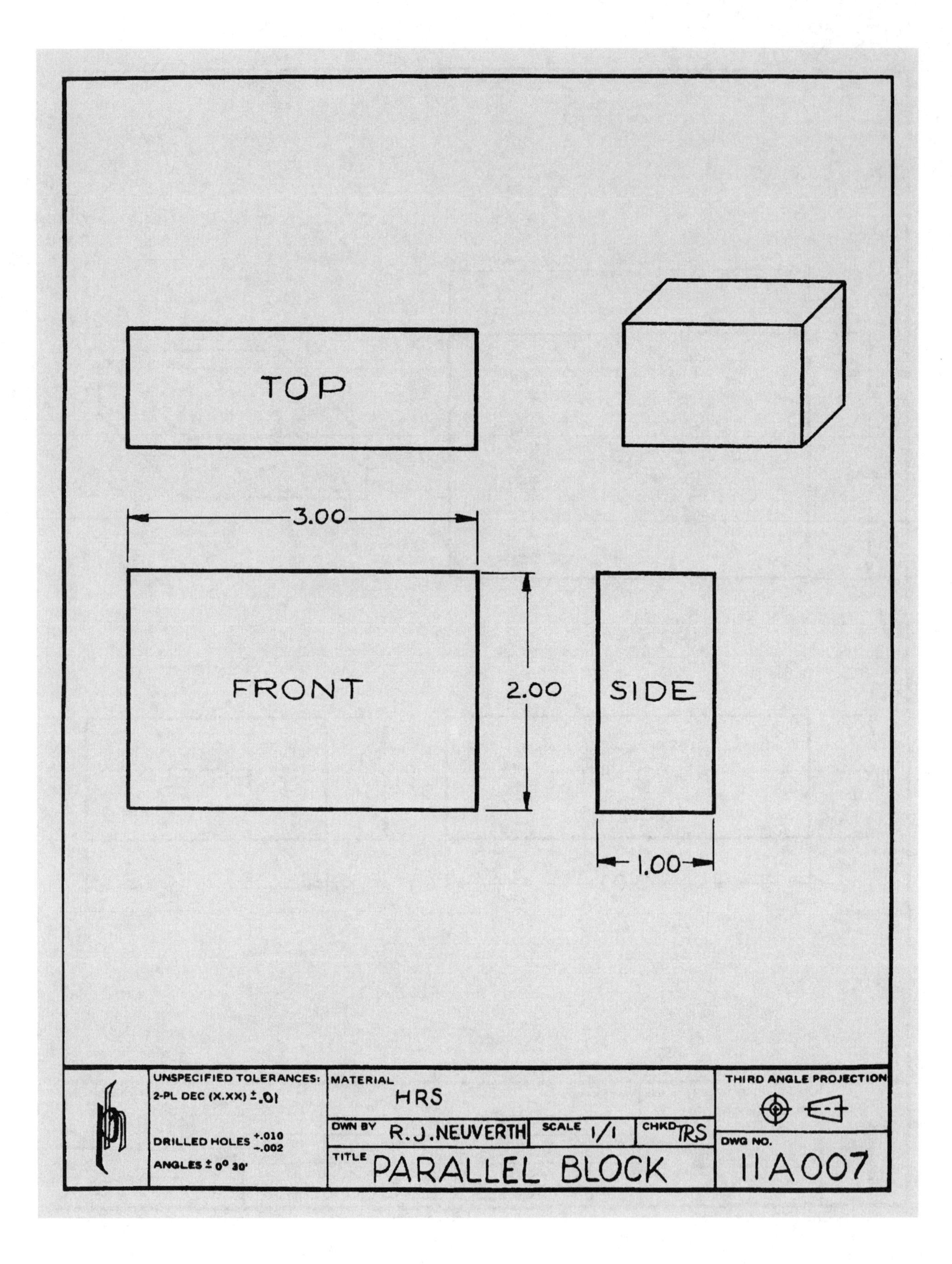
TOP
3.00
FRONT
2.00
SIDE
1.00
UNSPECIFIED TOLERANCES:
2-PL DEC (X.XX) ±.01
DRILLED HOLES +.010 -.002
ANGLES ± 0° 30'
MATERIAL
HRS
DWN BY R.J.NEUVERTH
SCALE 1/1
CHKD TRS
TITLE PARALLEL BLOCK
THIRD ANGLE PROJECTION
DWG NO.
11A007

Naming the Views

Compare the names assigned to views on drawings 11A006 and 11A007. The front view is usually the one drawn first, then by projecting horizontally and vertically the remaining views are constructed. Normally, the view that shows the most profile (contour) is selected for the front view. Therefore, the front view of a drawing may not be the front of an object as it is ordinarily used. For example, since the side of an automobile shows more distinguishing profile than the front, it would likely be selected as the front view on a drawing.

The relationship of views, however, is much more important than the names we assign to them. For example, a front view to one person may be called a top view by another as long as both people realize that adjacent views are always at right angles to each other, and always in direct projection from one another.

Observe from drawing 11A006 that each of the overall dimensions is visible on two views of a three-view drawing. Each will appear only once on a production drawing, as on 11A007. Whenever practical, dimensions are placed between the views.

When referring to dimensions on drawings, the preferred terms are height, width, and depth. The front view will show height and width, while any view surrounding it will show depth. Sometimes, when describing the object itself, the terms length and thickness may be used. When this is the case, length is normally the largest dimension and thickness is normally the smallest.

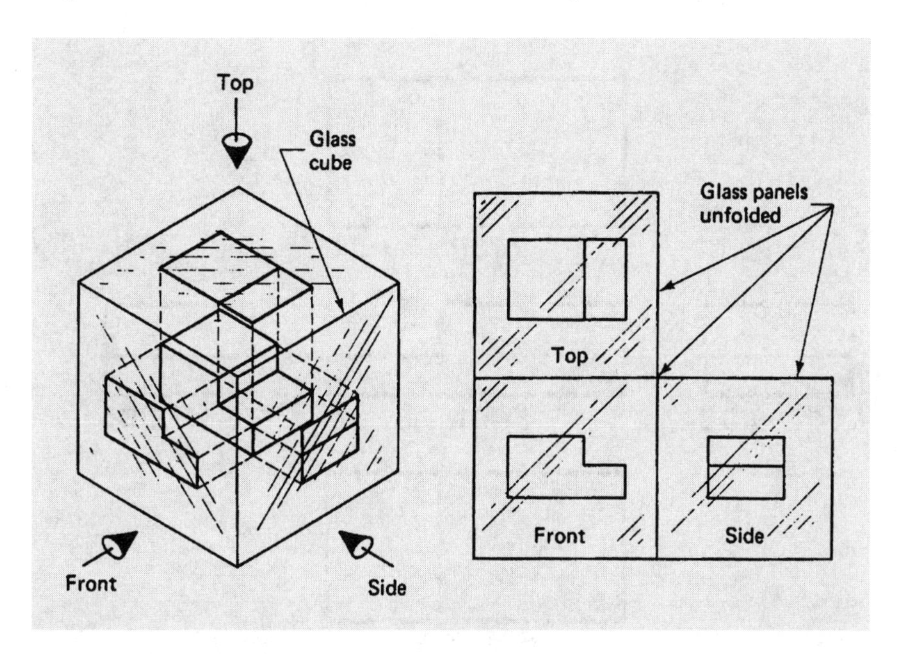

THIRD-ANGLE PROJECTION

When the vertical and horizontal planes of projection intersect one another, the quadrants create the first, second, third, and fourth angles (see illustration on opposite page). if the object to be viewed is placed below the horizontal plane and behind the vertical plane, it appears in the third-angle. If the object is placed above the horizontal plane and in front of the vertical plane, it appears in the first-angle. When the planes are unfolded, the views will arrange themselves in opposite positions around the front view. A comparison of the first-angle and third-angle view arrangements is shown on the opposite page.

The third-angle method of projection has been used exclusively in the United States and Canada since the early twentieth century. It was introduced as the most logical or natural positioning of orthographic views. However, the first-angle method of projection continues to be used in most all of the other industrial nations. A tapered plug symbol has been adopted by the International Standards Organization (ISO) to distinguish between first-angle and third-angle projections used on drawings. These symbols will appear in or near the title block.

Illustrated below are the six principal views arranged properly by the third-angle projection method. Notice that the right side view coincides with the view you would observe if you positioned yourself to the right of the object. This would also be true with each of the other positions: top, bottom, left side, and rear.

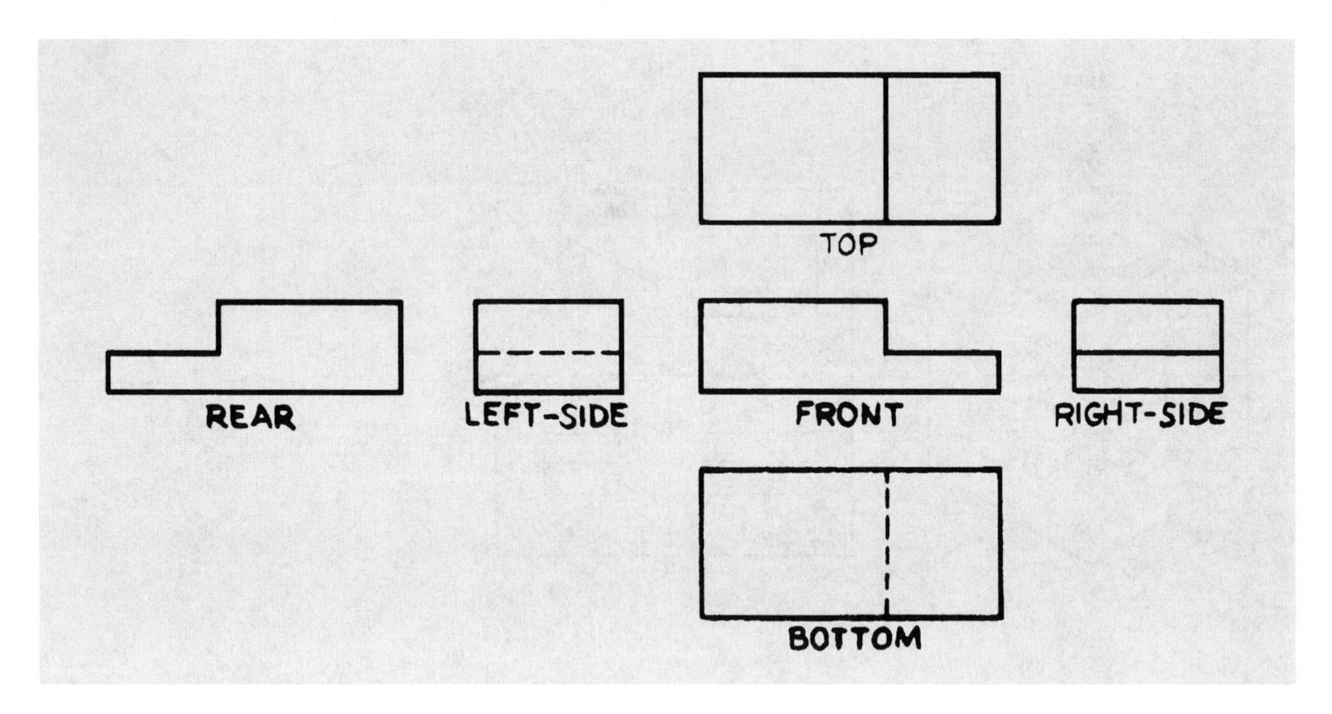

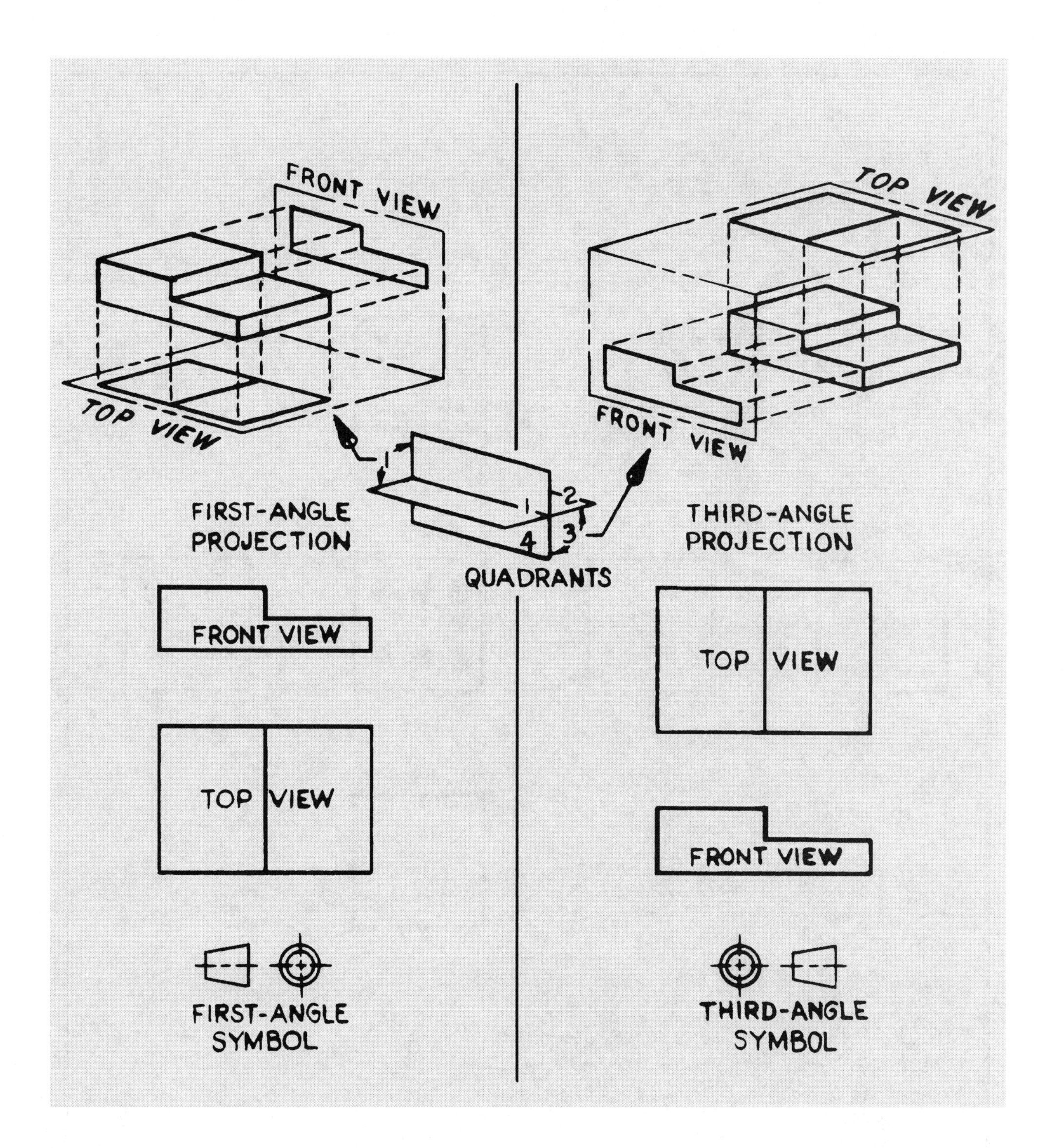
FRONT VIEW
TOP VIEW
TOP VIEW
FRONT VIEW
1
2
3
4
FIRST-ANGLE
PROJECTION
THIRD-ANGLE
PROJECTION
QUADRANTS
FRONT VIEW
TOP VIEW
TOP VIEW
FRONT VIEW
FIRST-ANGLE
SYMBOL
THIRD-ANGLE
SYMBOL

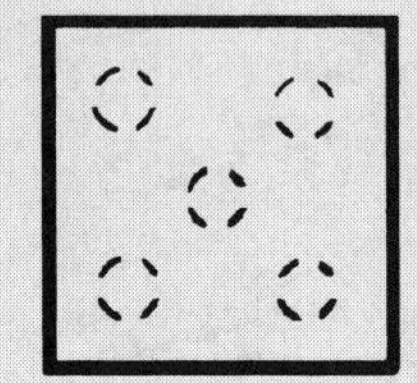

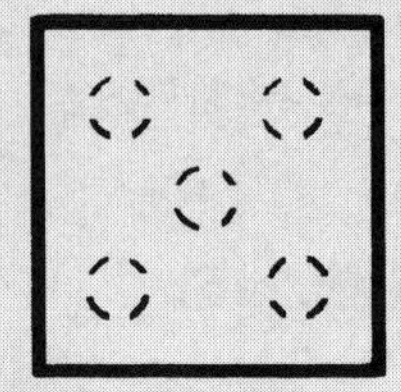

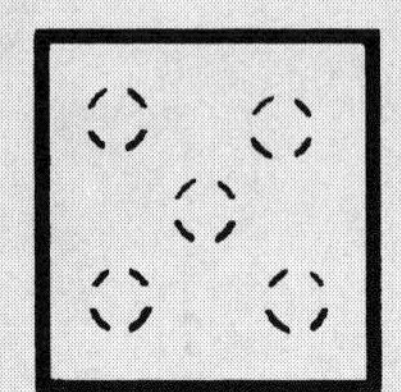

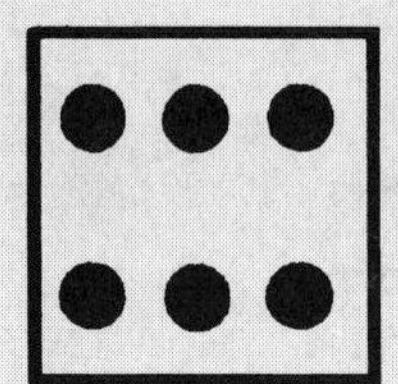

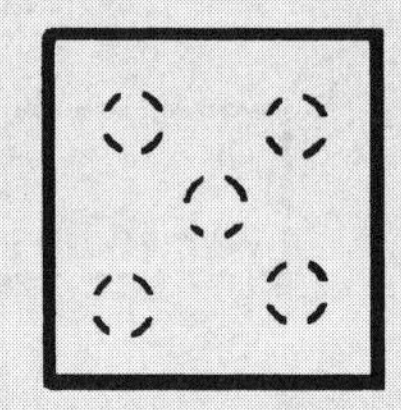

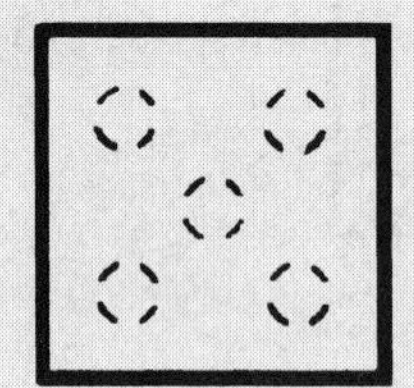

INSTRUCTIONS: Fill in the appropriate dots in the orthographic views of the die illustrated above. (Note: opposite sides of a die always equal seven.)

UNSPECIFIED TOLERANCES:	MATERIAL IVORY			THIRD ANGLE PROJECTION
2-PL DEC (X.XX) ±				
3-PL DEC (X.XXX) ±	DWN BY J. J. HEFFERON	SCALE ~	CHKD TRS	
DRILLED HOLES +.010 −.002	TITLE DIE			DWG NO. 11A008
ANGLES ± 0° 30'				

FIRST-ANGLE PROJECTION

The first-angle method of projection continues to be used in most all of the other industrial nations except the U.S. and Canada. With the amount of export/import of products taking place today, you will probably be exposed to drawings produced by the first-angle method sooner or later. Therefore, you should become acquainted with this arrangement of views. Compare the arrangement of the first-angle views shown below with the arrangement of the third-angle method illustrating the same object on page 50.

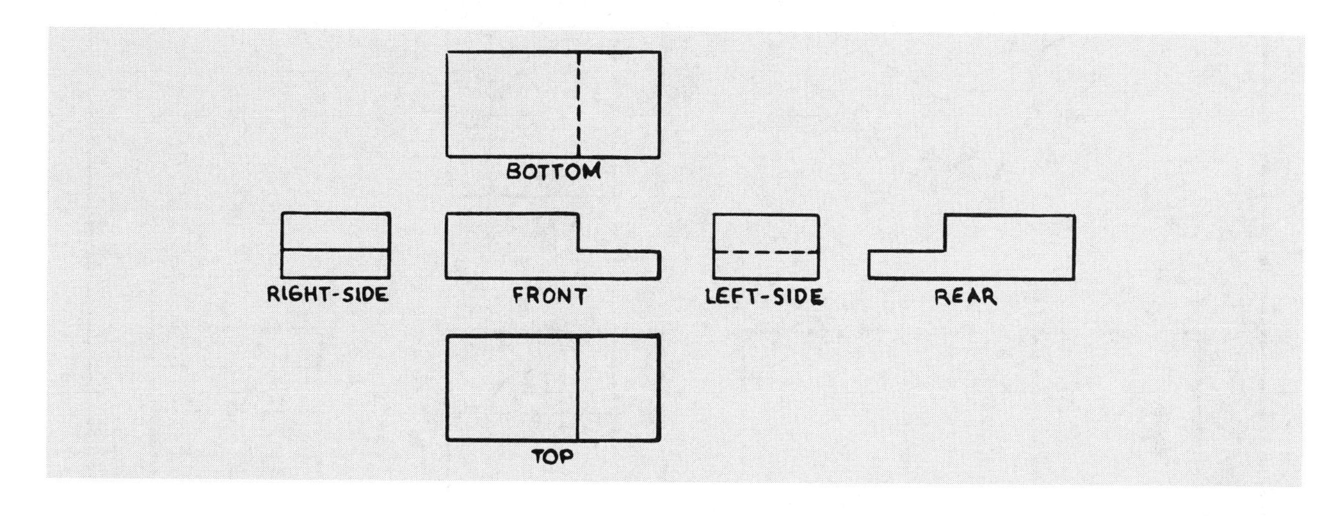

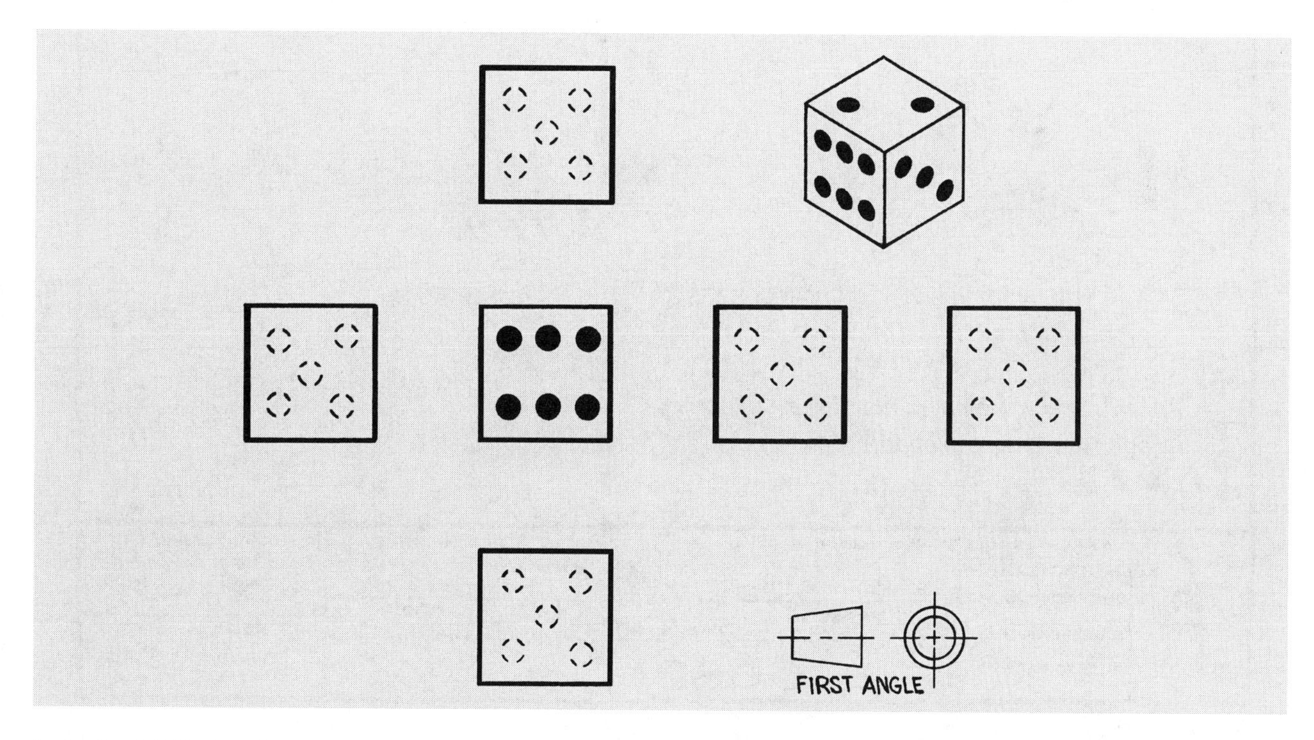

INSTRUCTIONS: Shade the appropriate dots in the orthographic views, as in 11A008, but follow the first-angle projection method.

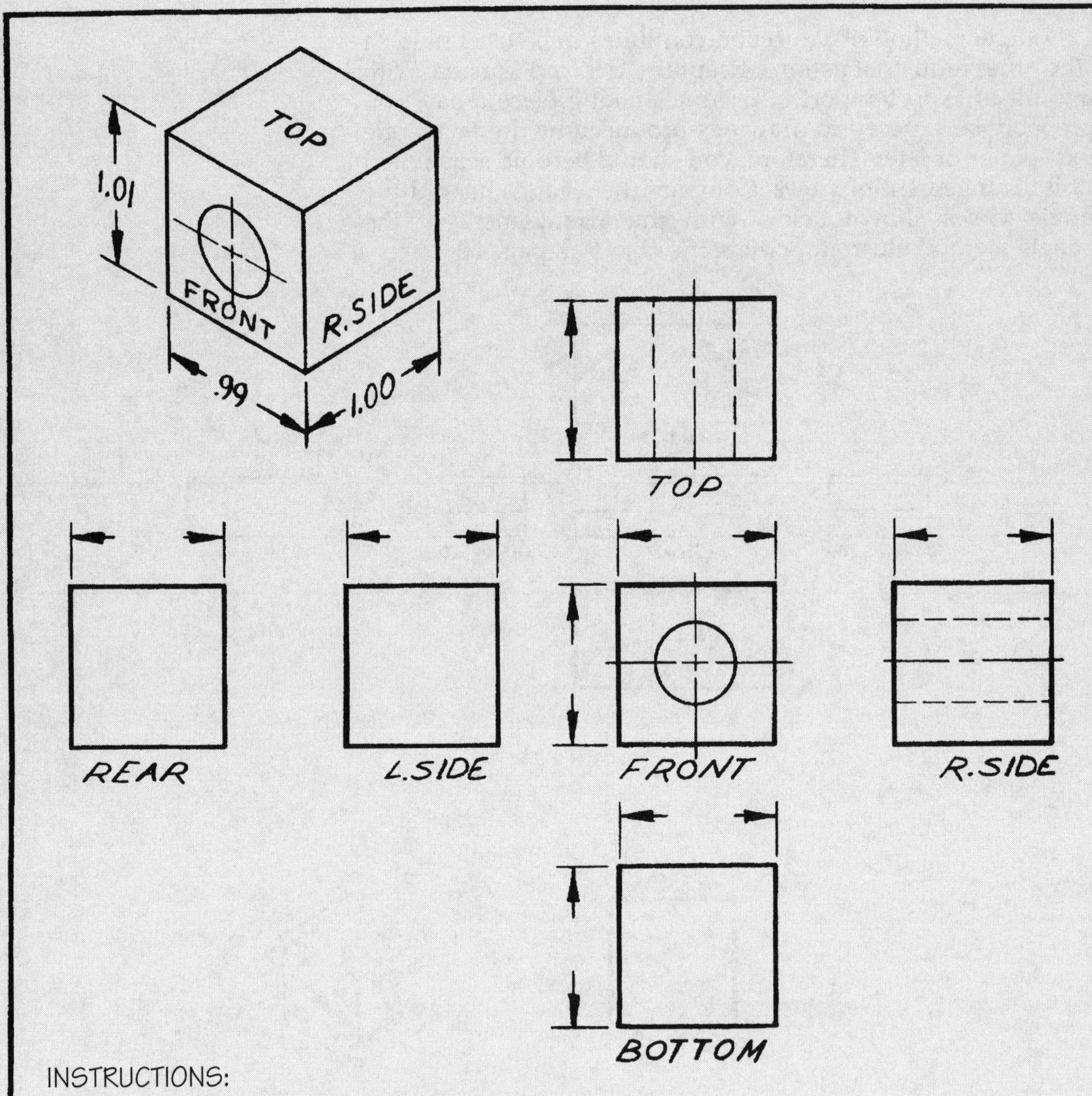

INSTRUCTIONS:

1. Add the missing lines to the incomplete orthographic views (rear, left side, and bottom).
2. Add the missing dimensions to correspond with the pictorial view.

UNSPECIFIED TOLERANCES:	MATERIAL			THIRD ANGLE PROJECTION
2-PL DEC (X.XX) ±	ALUMINUM			
3-PL DEC (X.XXX) ±	DWN BY D.L. SCANTLING	SCALE ~	CHKD RS	DWG NO.
DRILLED HOLES +.010 −.002	TITLE			
ANGLES ± 0° 30'	CUBE			11A009

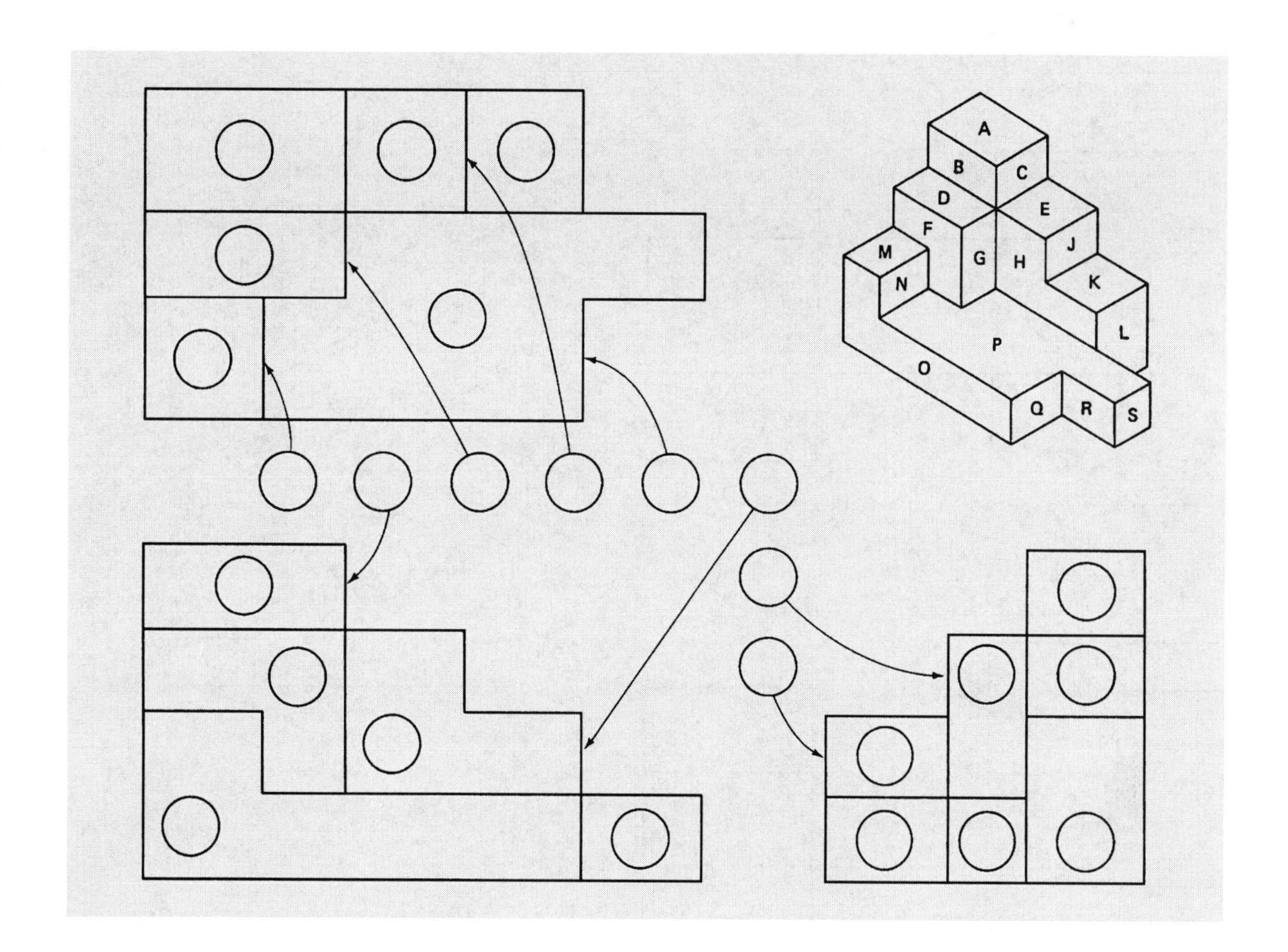

Surface Identification 1

INSTRUCTIONS: Enter the letters from the pictorial view into the corresponding balloons on the orthographic views.

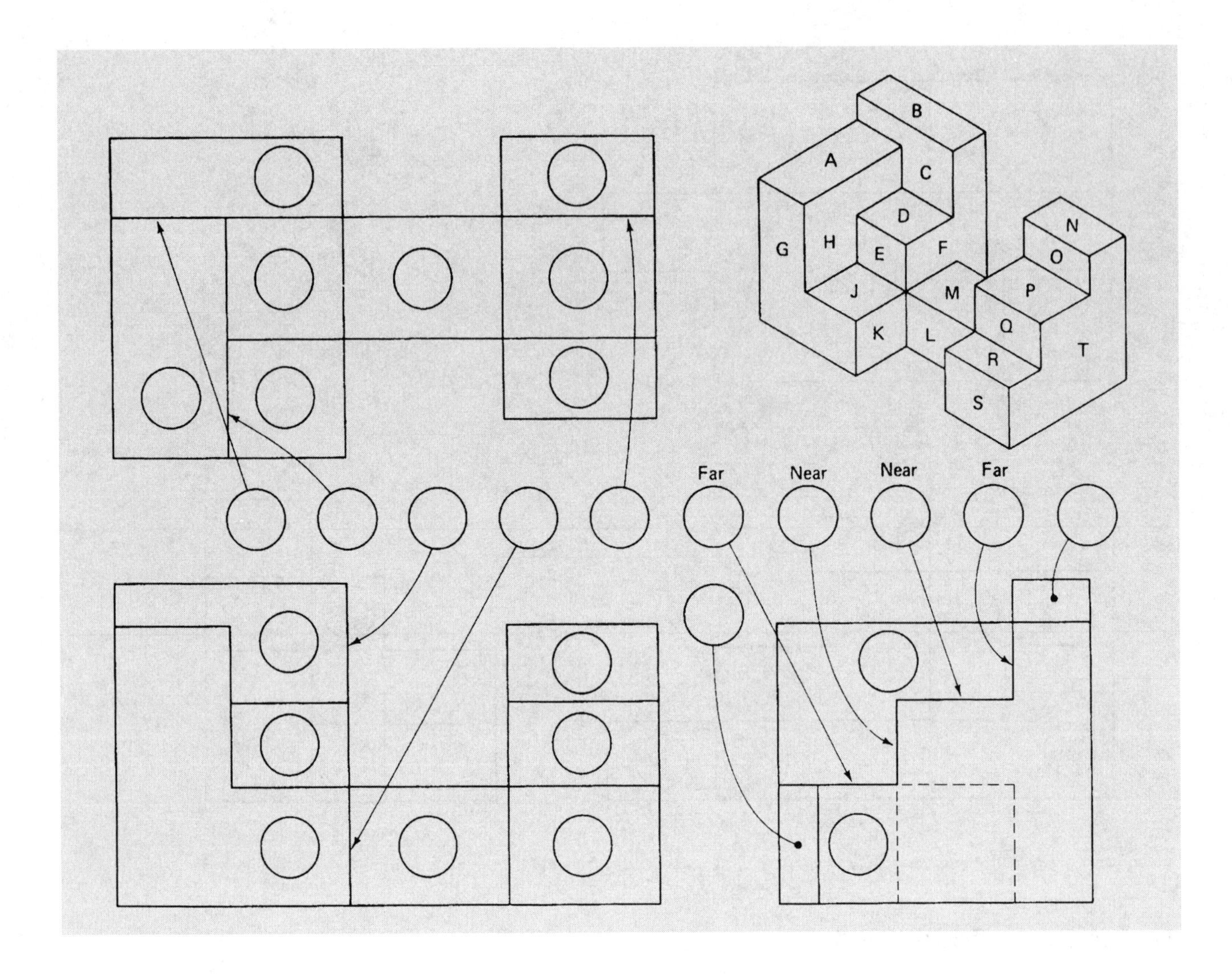

Surface Identification 2

INSTRUCTIONS: Enter the letters from the pictorial view into the corresponding balloons on the orthographic views.

NOTE: When two or more surfaces occupy the same position on an orthographic view, they are represented by one common line. Such is the case with surfaces CO, DP, EQ, and JR in the right-side view of the object drawn above. To distinguish between them, I have designated "near" or "far" to represent which surface is nearest or farthest from the person's eye in that particular view (using third-angle projection).

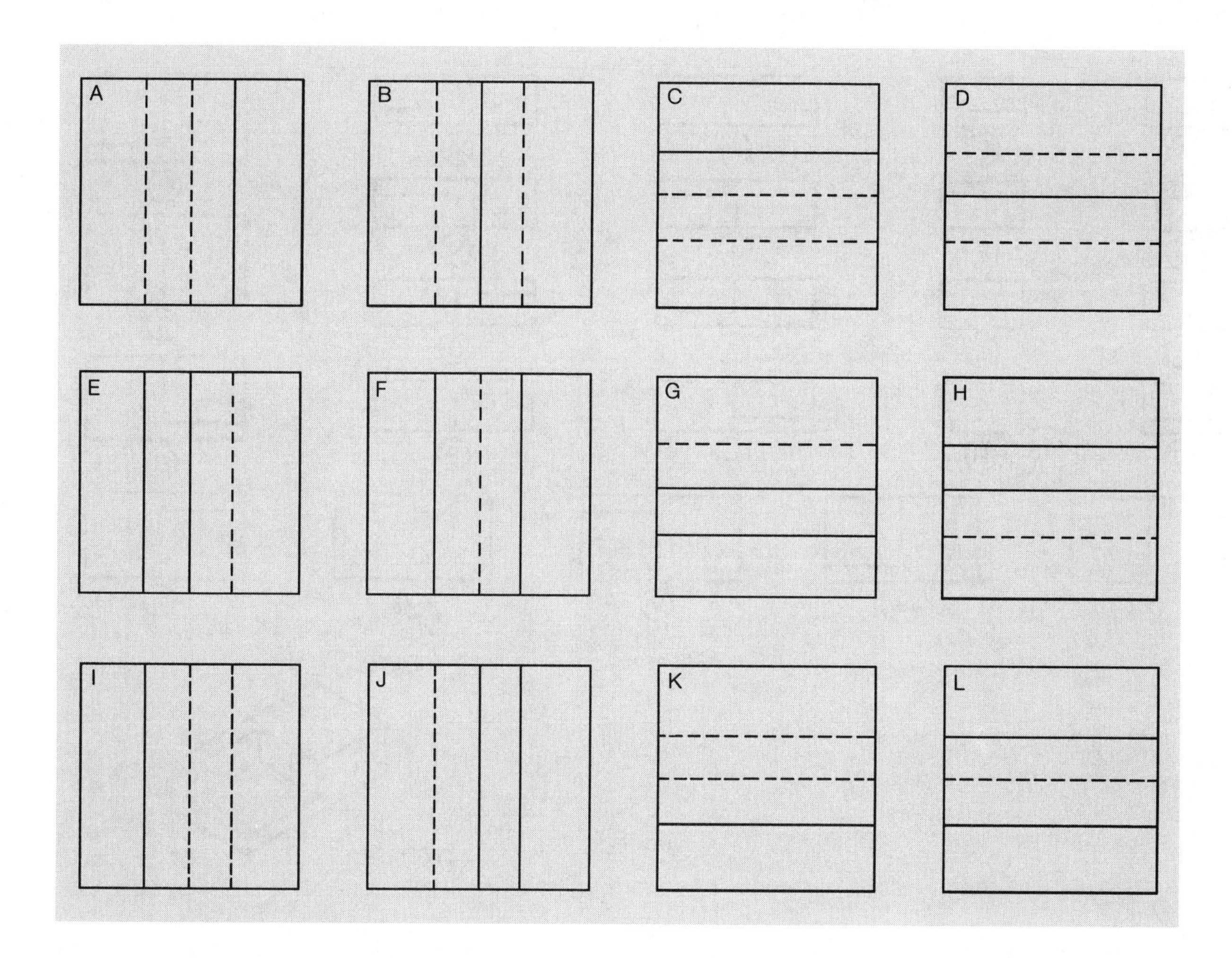

View Identification 1

INSTRUCTIONS: Circle the letter that corresponds with the correct view of the object shown at the right. (Views may not be turned from their positions above.)

1. Top view: A B C D E F G H I J K L
2. Right-side view: A B C D E F G H I J K L
3. Bottom view: A B C D E F G H I J K L
4. Left-side view: A B C D E F G H I J K L

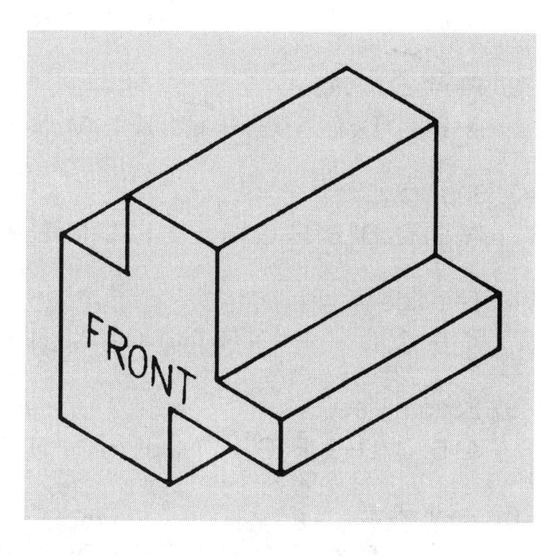

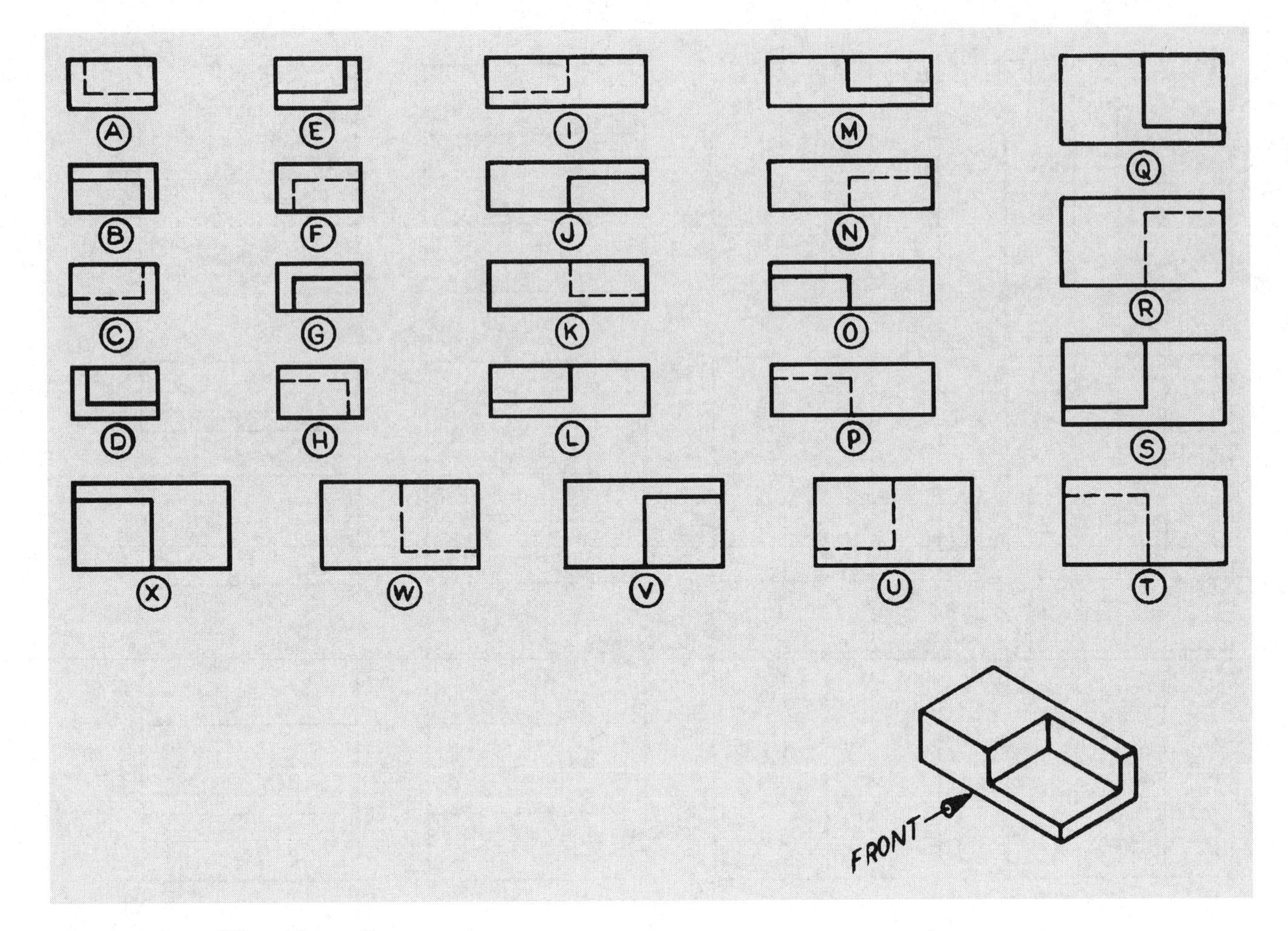

View Identification 2

INSTRUCTIONS: Circle the letter that corresponds with the correct view of the object shown at the right. (Views may not be turned from their positions above.)

1. Front view:
   A B C D E F G H I J K L M N O P Q R S T U V W X

2. Right-side view:
   A B C D E F G H I J K L M N O P Q R S T U V W X

3. Top view:
   A B C D E F G H I J K L M N O P Q R S T U V W X

4. Left-side view:
   A B C D E F G H I J K L M N O P Q R S T U V W X

5. Bottom view:
   A B C D E F G H I J K L M N O P Q R S T U V W X

6. Rear view:
   A B C D E F G H I J K L M N O P Q R S T U V W X

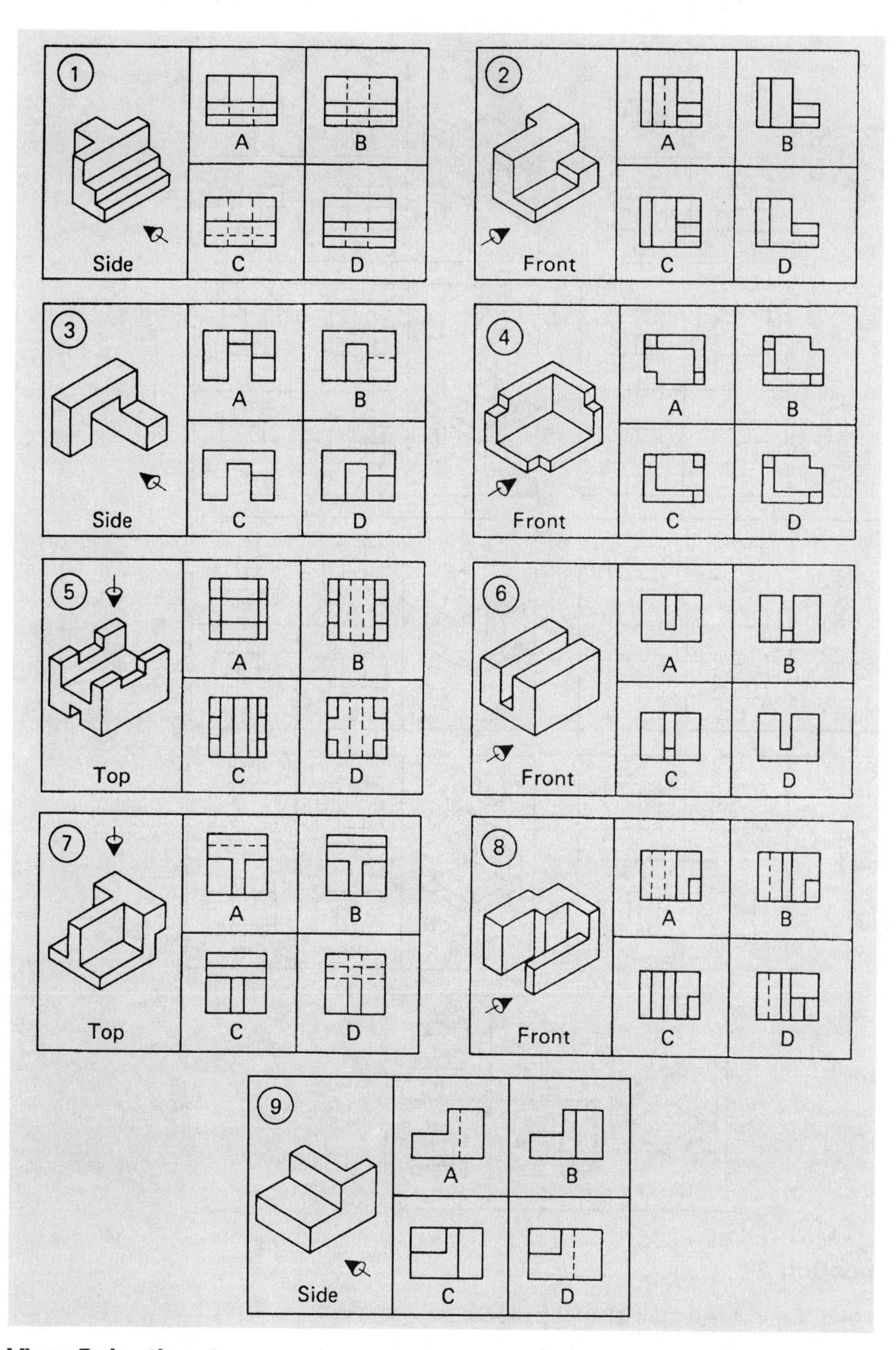

View Selection 1

INSTRUCTIONS: Circle the letter that represents the correct orthographic view for each of the pictorial views shown above.

View Selection 2

INSTRUCTIONS: Circle the letter that represents the correct orthographic view for each of the pictorial views shown above.

ISOMETRIC SKETCHING

It usually helps to learn to visualize shape if the blueprint reader can make a three-dimensional sketch of the object. One of the easiest pictorial drawings to sketch is the isometric. Isometric sketch paper contains a grid of light guidelines which are drawn along three axes, 30° above left horizontal, 30° above right horizontal, and vertical.

Observe the following from the example below:

1. The horizontal lines in the front view follow the 30° guidelines to the left.
2. The vertical lines in the front view follow the vertical guidelines.
3. The horizontal lines in the side view (vertical in the top view) follow the 30° guidelines to the right.

To produce a full-size sketch such as the example shown below, draw each line the same length (number of squares) as in the orthographic view.

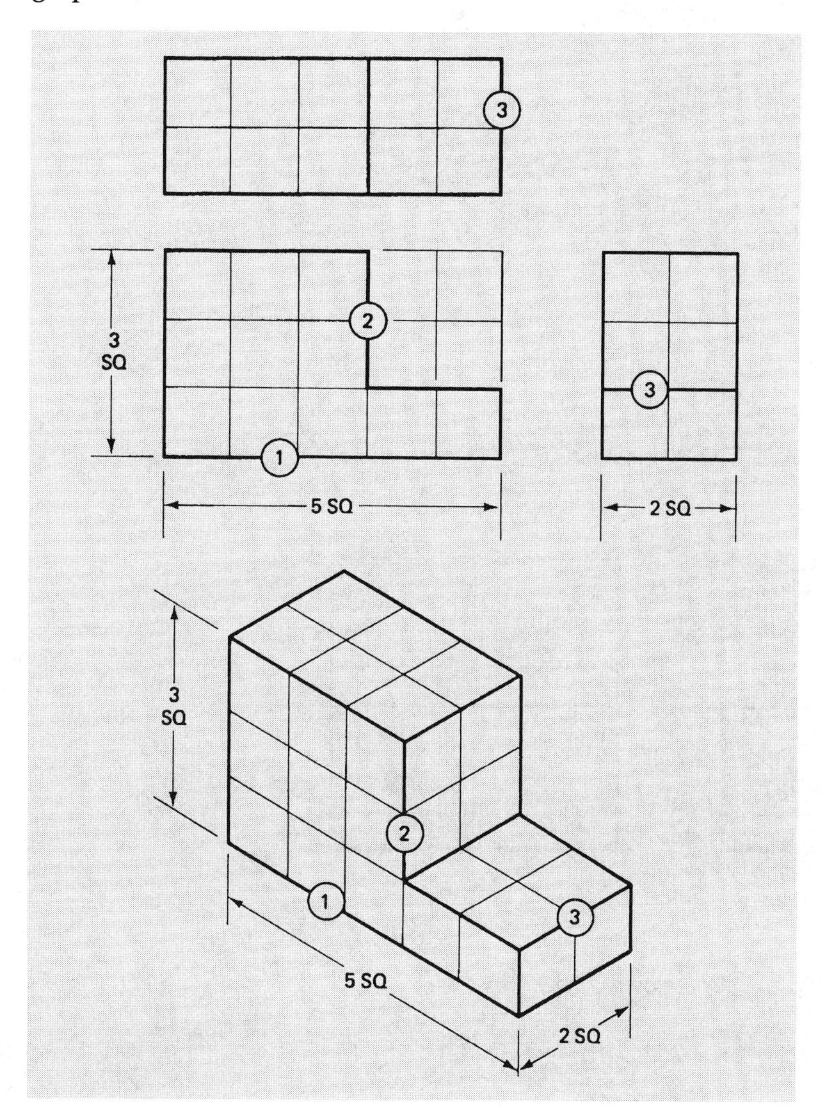

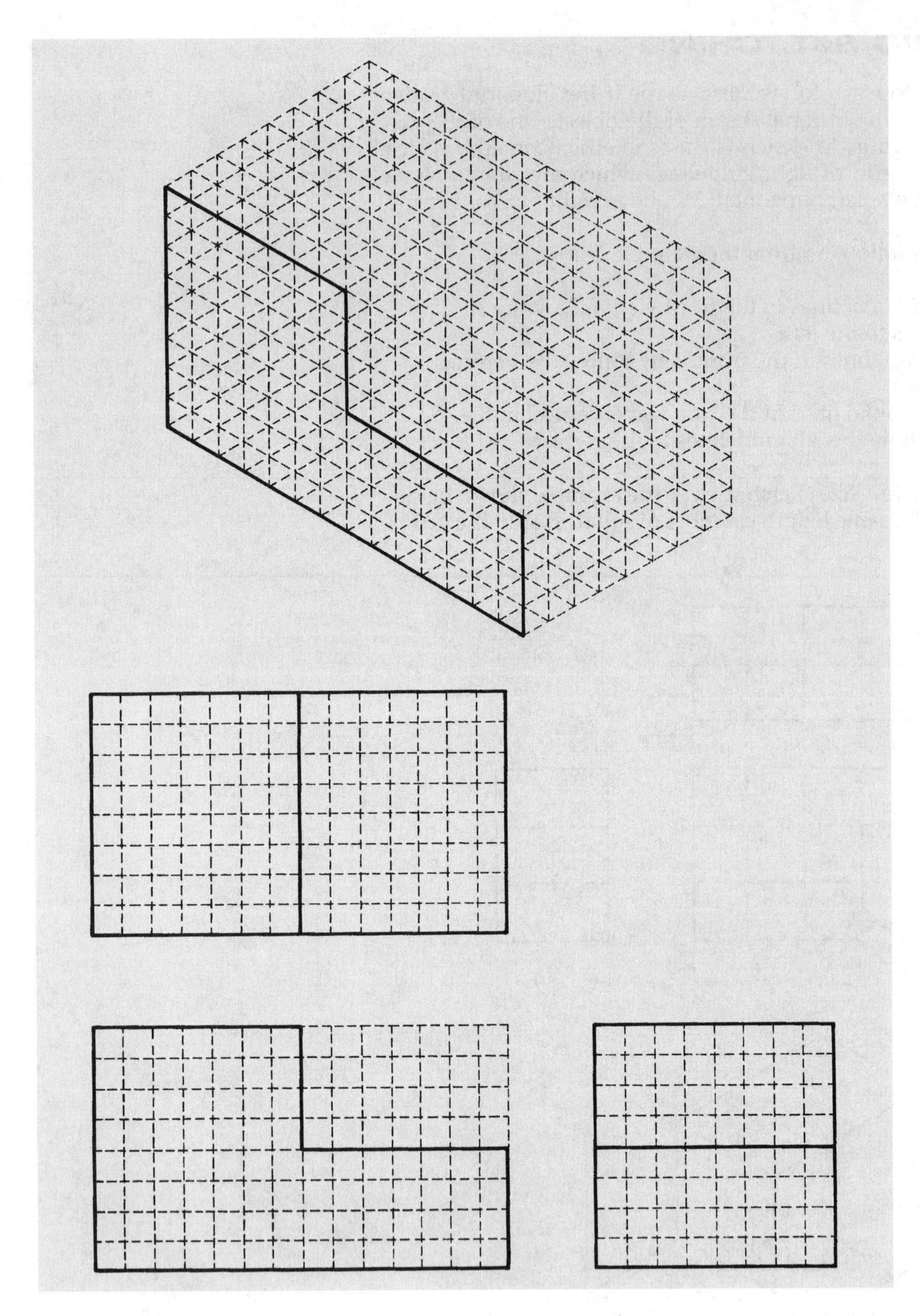

Isometric Sketch 1

INSTRUCTIONS: Complete the isometric sketch shown above.

OPTIONAL EXERCISE: Add dimensions to the orthographic views.
(Each square represents 1/4".)

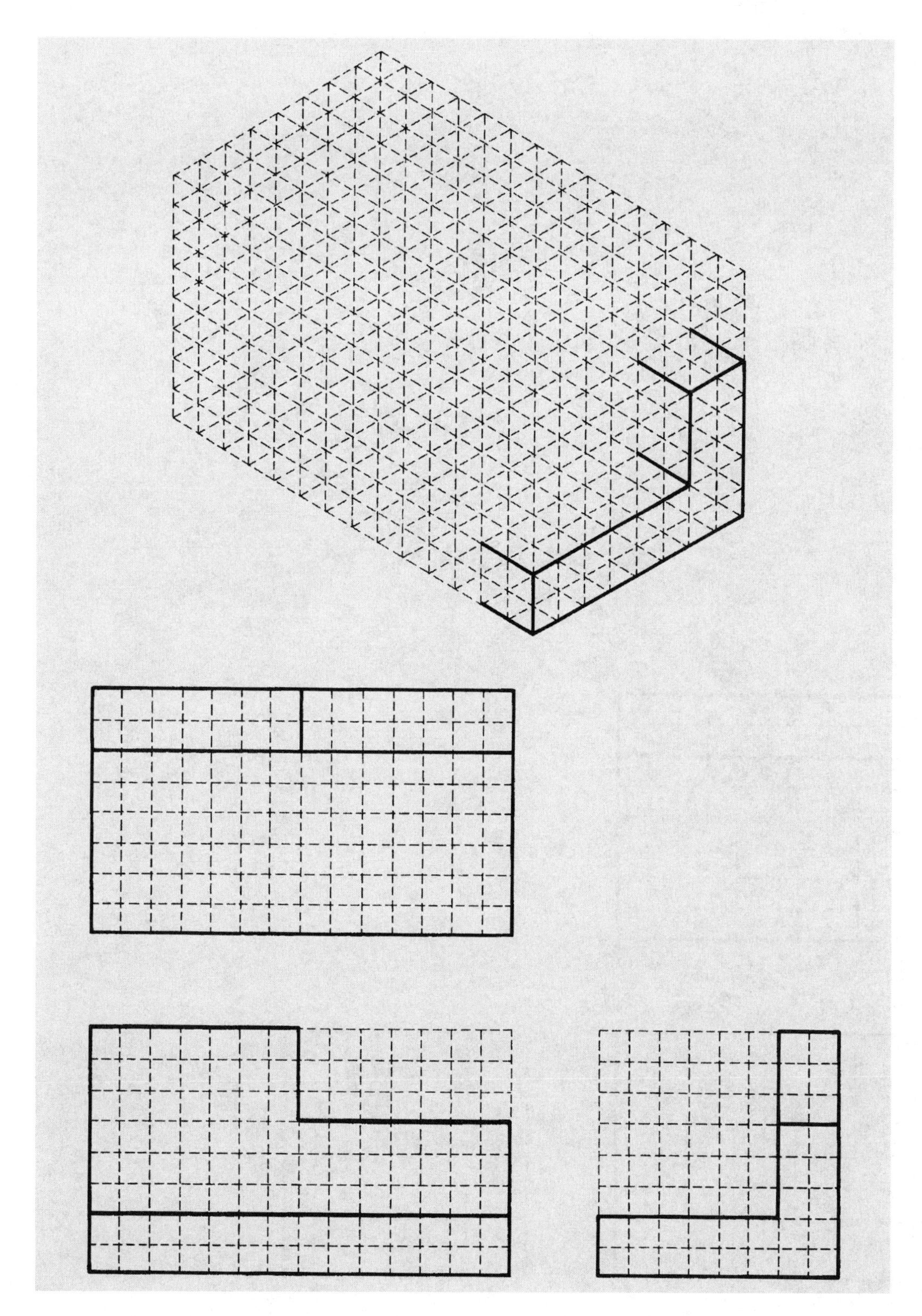

Isometric Sketch 2

INSTRUCTIONS: Complete the isometric sketch shown above.

OPTIONAL EXERCISE: Add dimensions to the orthographic views.
(Each square represents 1/4".)

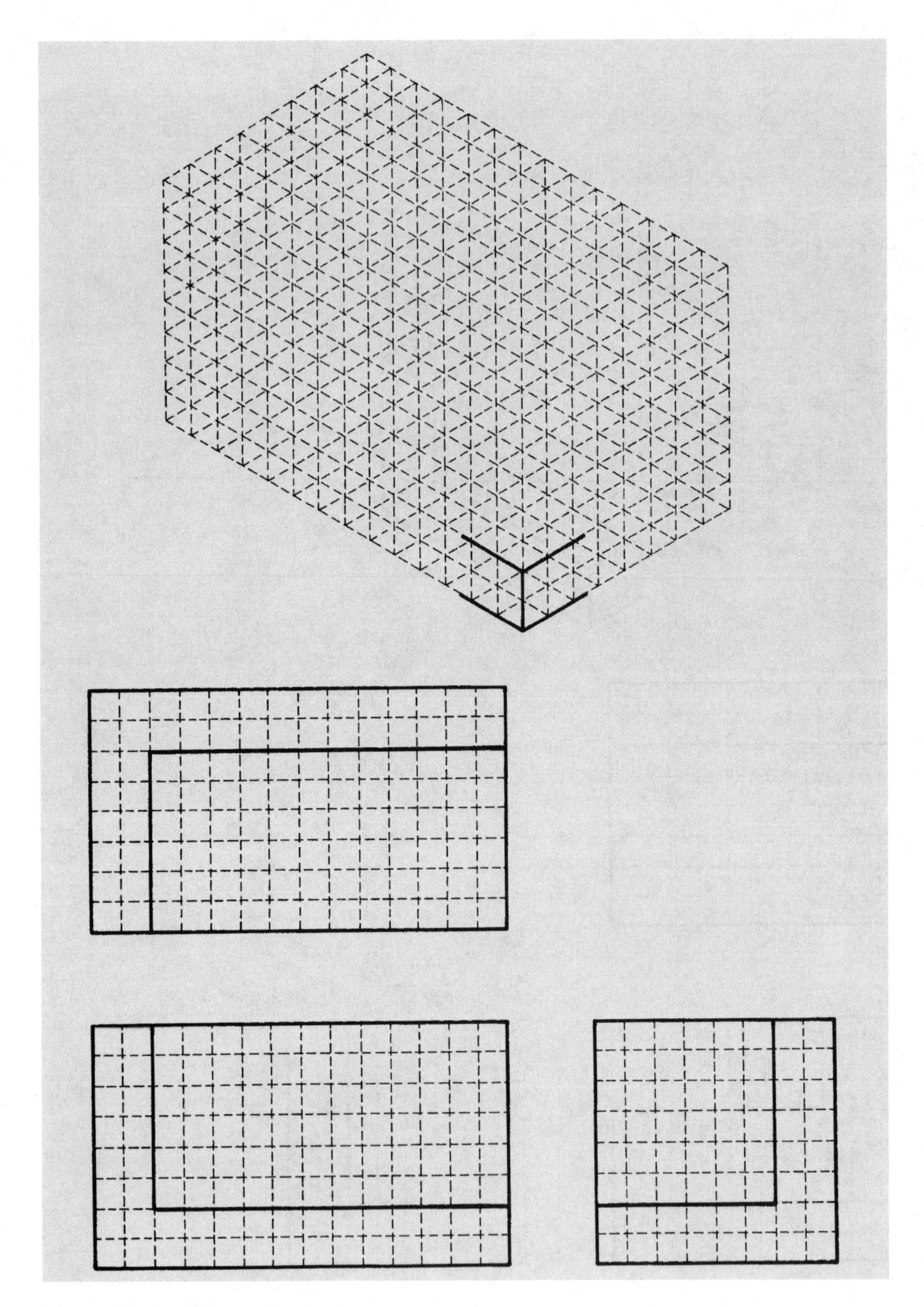

Isometric Sketch 3

INSTRUCTIONS: Complete the isometric sketch shown above.

OPTIONAL EXERCISE: Add dimensions to the orthographic views.
(Each square represents 1/4".)

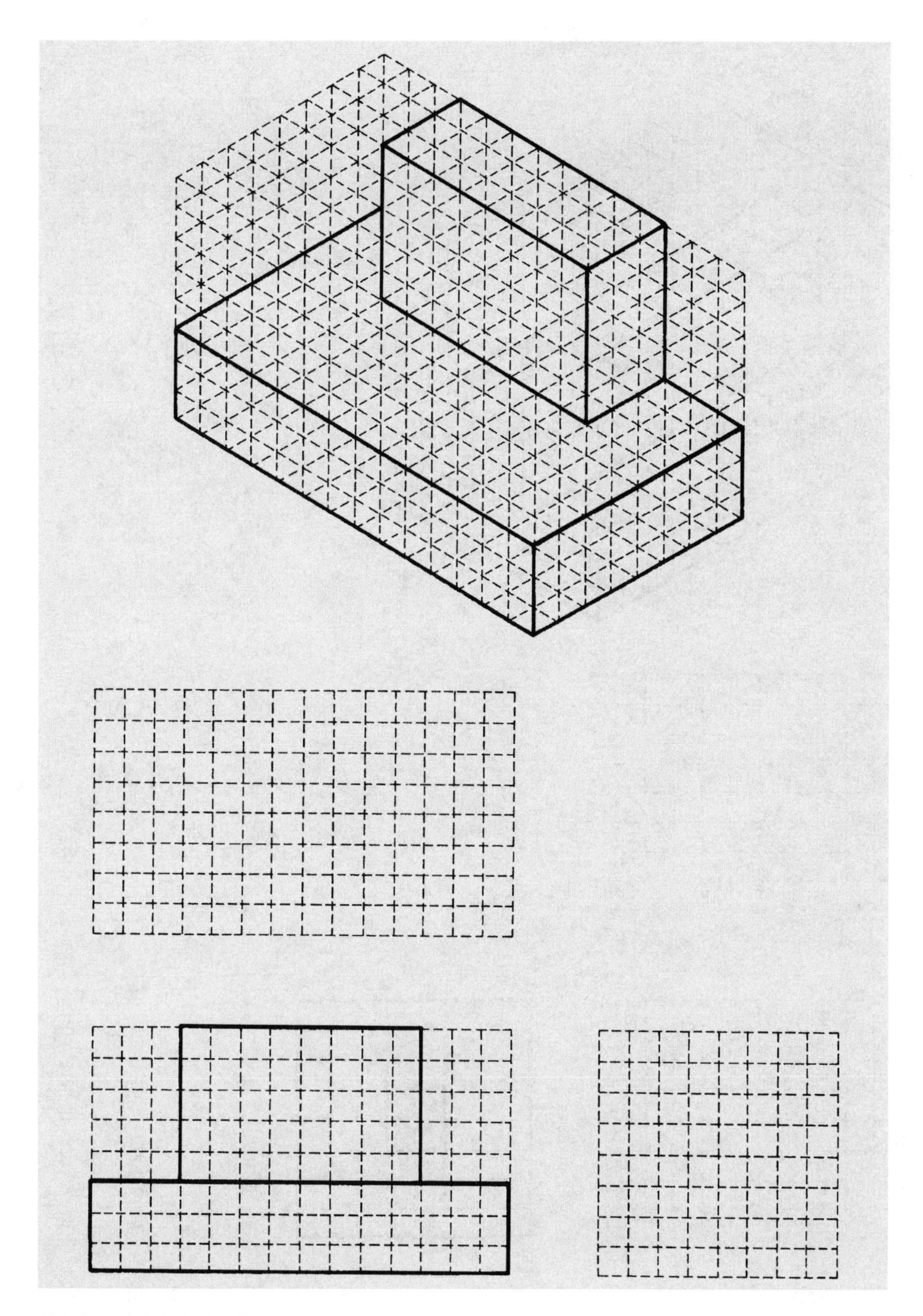

Orthographic Sketch 1

INSTRUCTIONS: Draw the two missing orthographic views of the object shown above.

OPTIONAL EXERCISE: Add dimensions to the orthographic views. (Each square represents 1/4".)

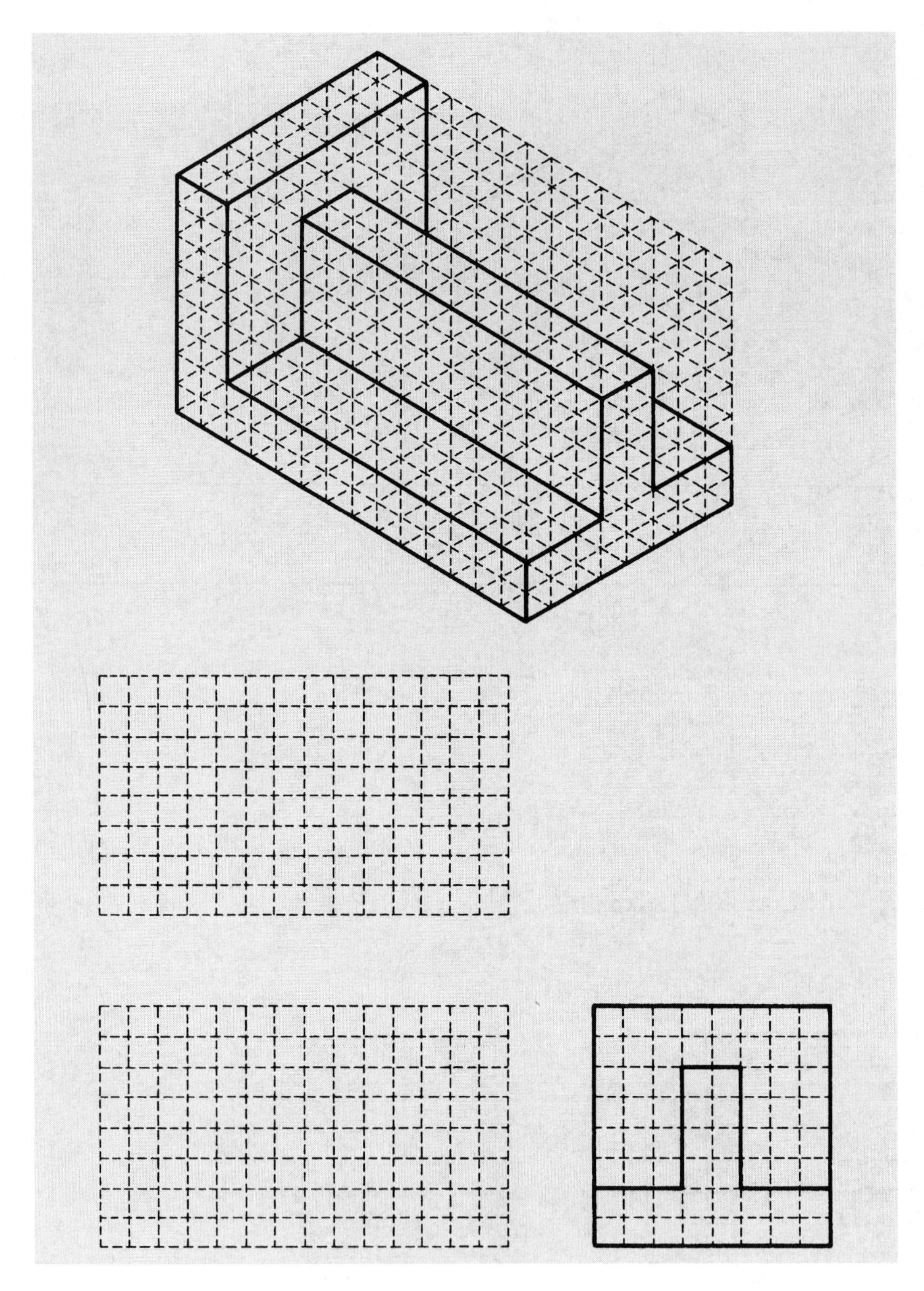

Orthographic Sketch 2

INSTRUCTIONS: Draw the two missing orthographic views of the object shown above.

OPTIONAL EXERCISE: Add dimensions to the orthographic views. (Each square represents 1/4".)

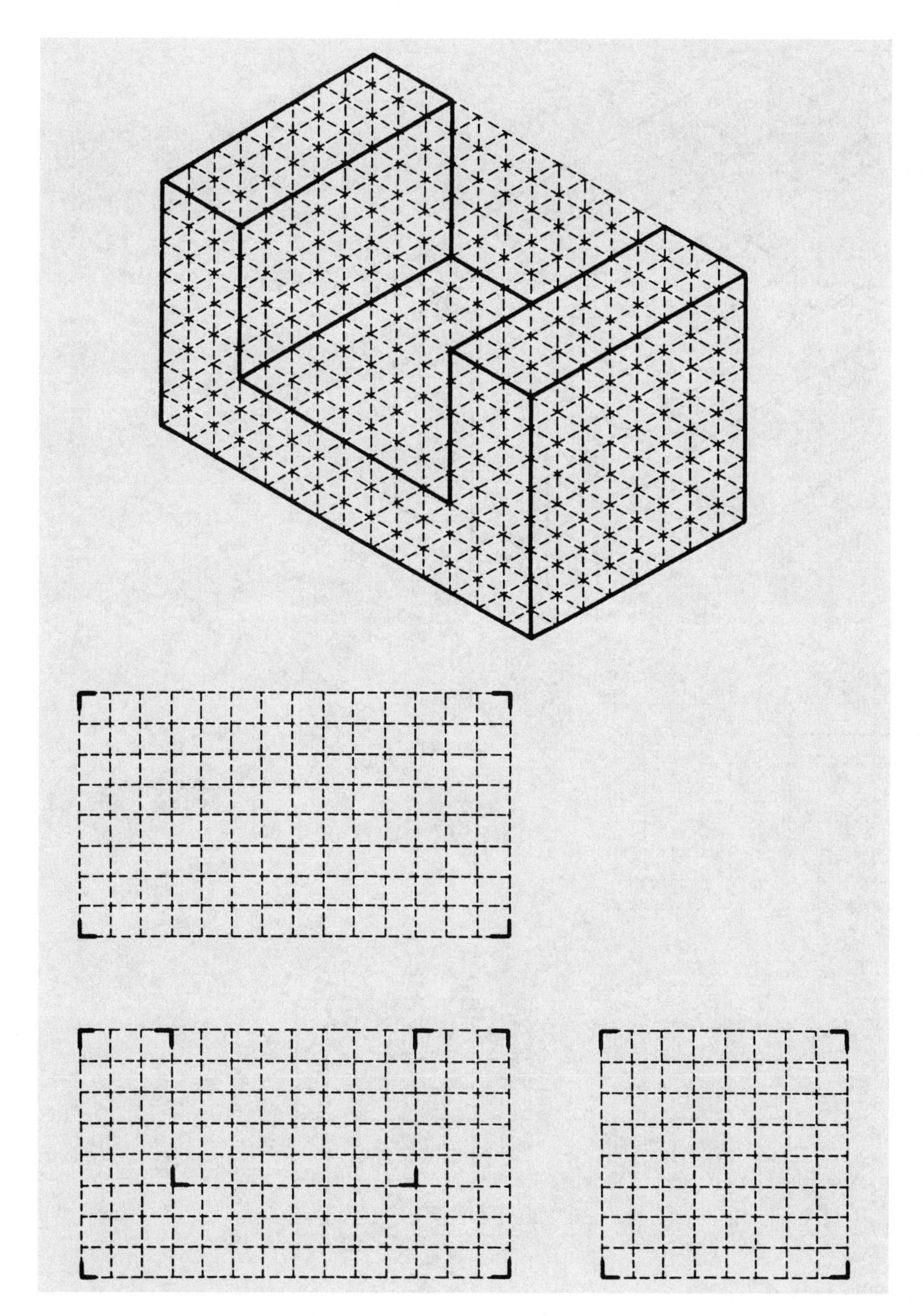

Orthographic Sketch 3

INSTRUCTIONS: Draw the three orthographic views of the object shown above. Be sure to include the hidden line in the side view.

OPTIONAL EXERCISE: Add dimensions to the orthographic views. (Each square represents 1/4".)

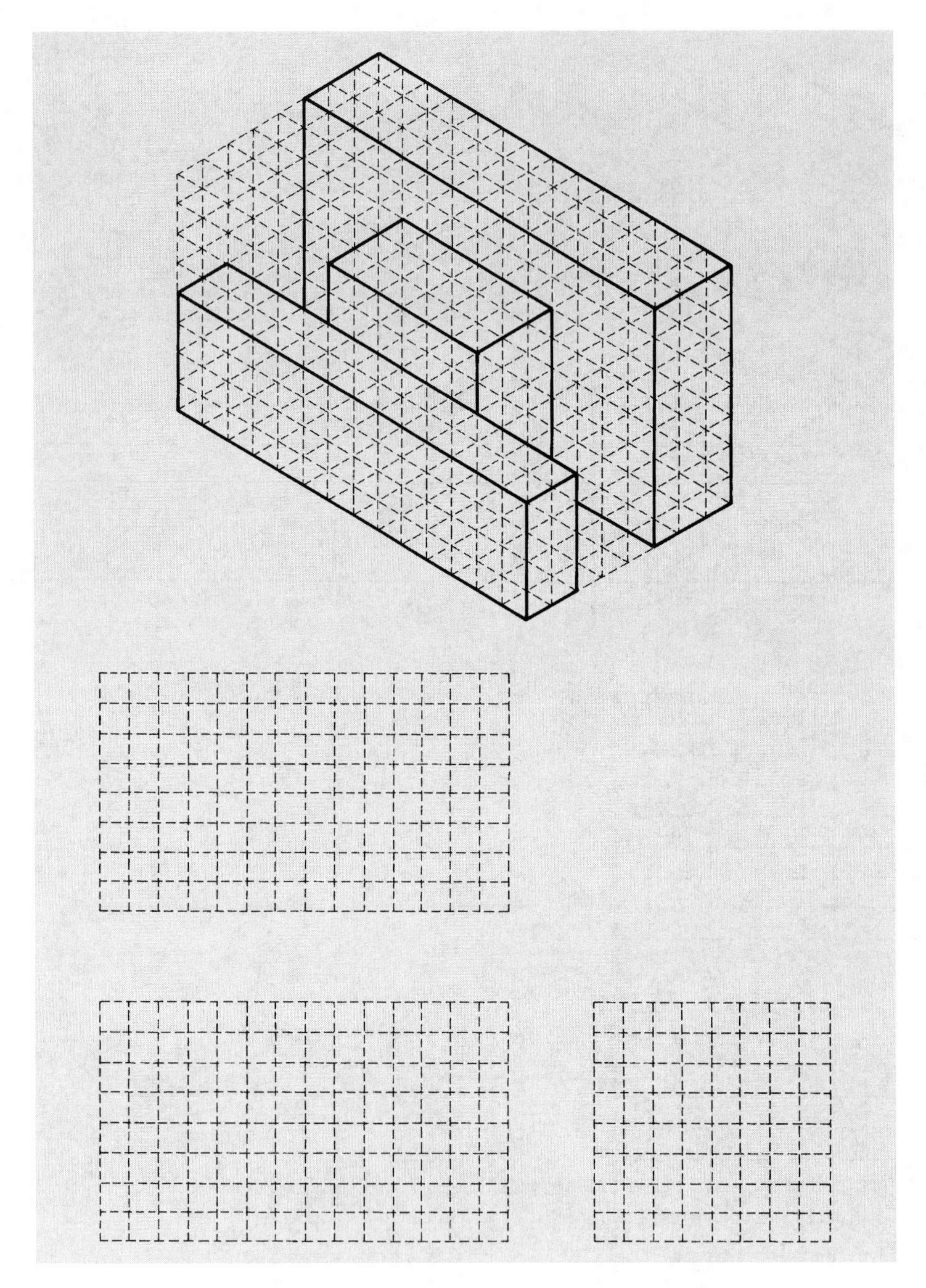

Orthographic Sketch 4

INSTRUCTIONS: Draw the orthographic views of the object shown above. Be sure to include the hidden lines in the front view.

OPTIONAL EXERCISE: Add dimensions to the orthographic views. (Each square represents 1/4".)

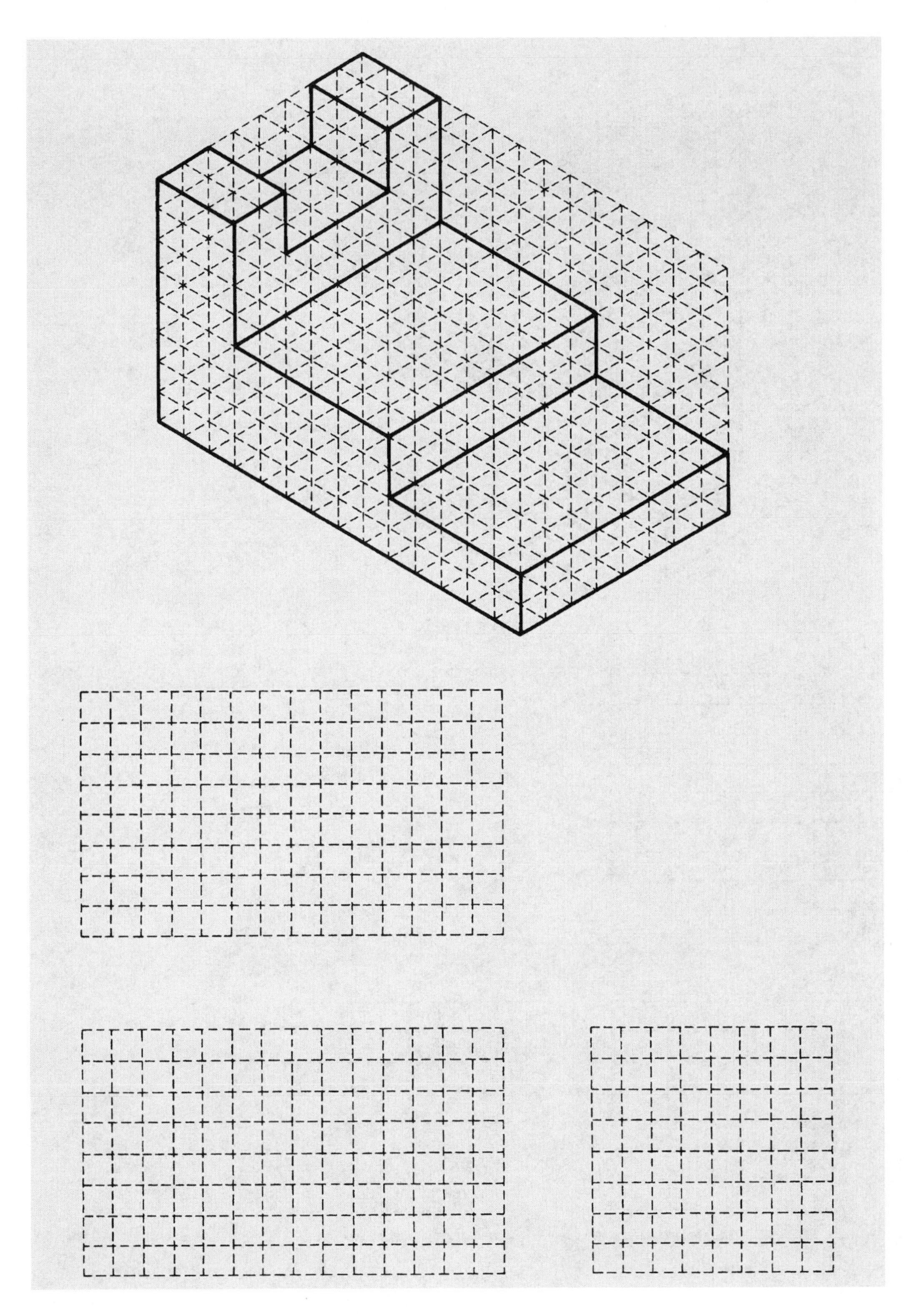

Orthographic Sketch 5

INSTRUCTIONS: Draw the orthographic views of the object shown above. Include the hidden line.

OPTIONAL EXERCISE: Add dimensions to the orthographic views. (Each square represents 1/4".)

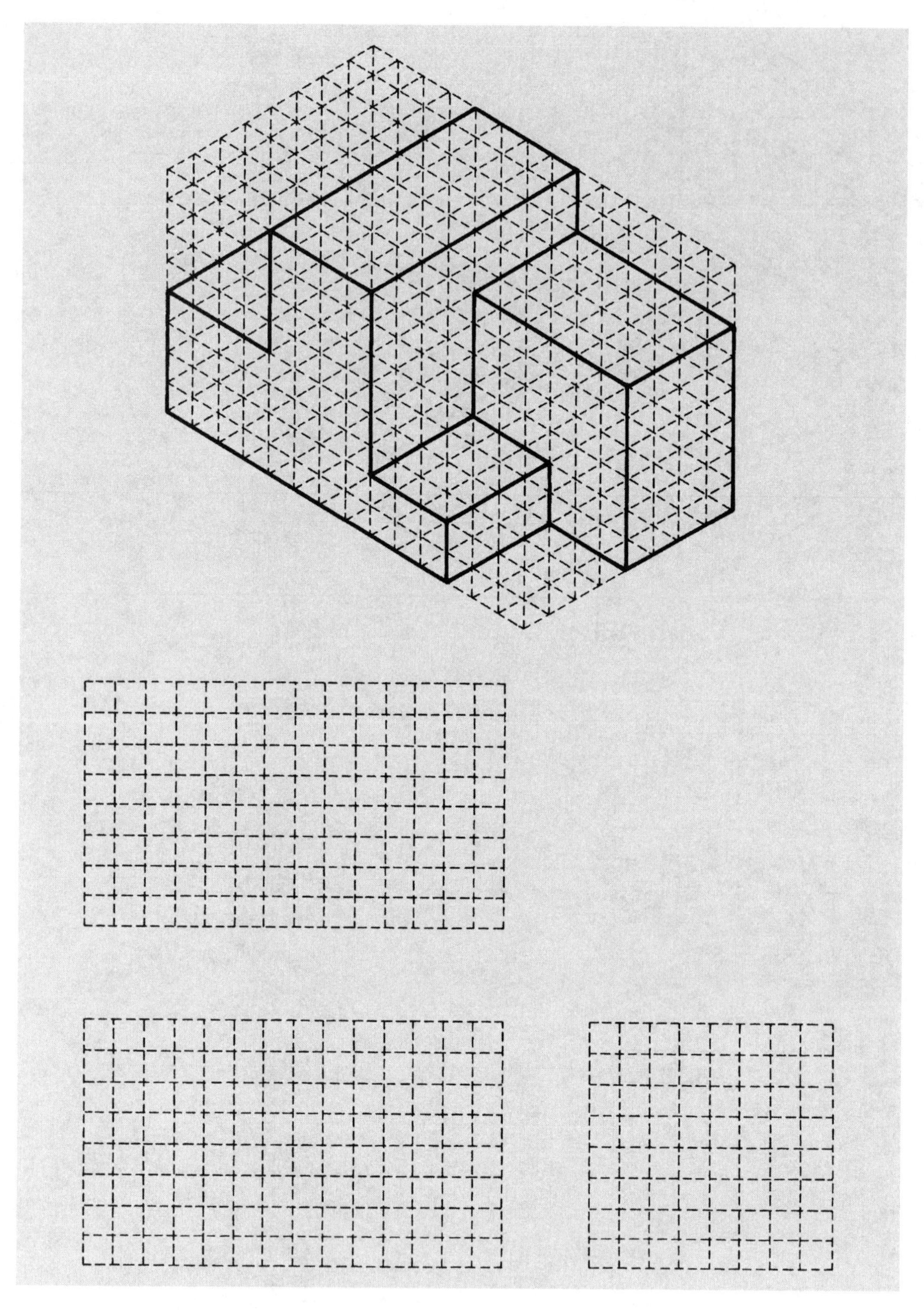

Orthographic Sketch 6

INSTRUCTIONS: Draw the orthographic views of the object shown above. Include the hidden line.

OPTIONAL EXERCISE: Add dimensions to the orthographic views. (Each square represents 1/4".)

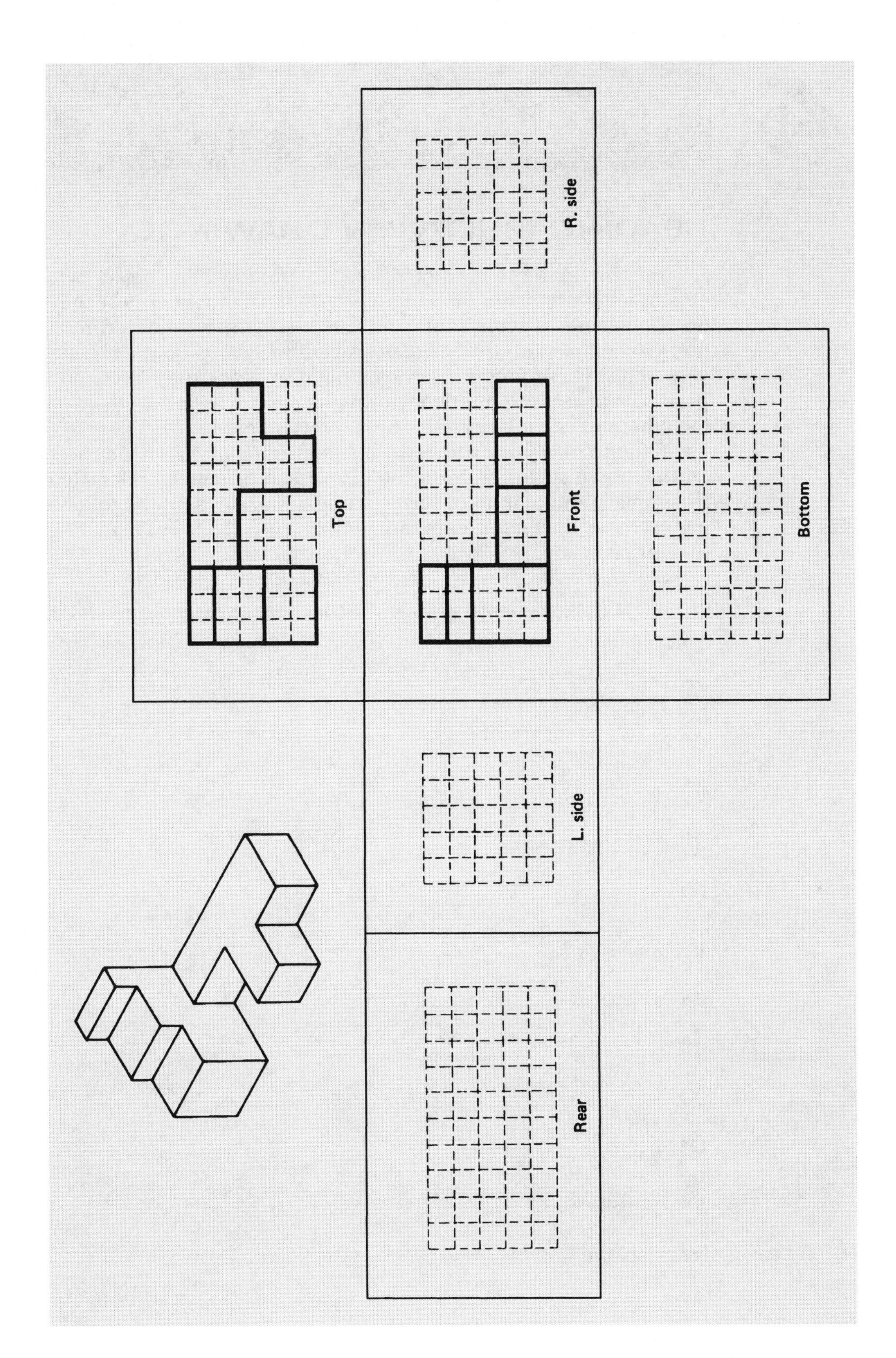

Missing Views 1

INSTRUCTIONS: Draw the four remaining principal views of the object shown above. Include all hidden lines.

UNIT 4

READING MULTIVIEW DRAWINGS

A sufficient amount of dimensions to produce the object will appear on every detail drawing. Each dimension will appear only once, however, to avoid confusion and possible error in interpretation. Sometimes the drafter may place dimensions on a different view than where you think they should appear, so by becoming proficient at reading multiview drawings you will know which of the other views could contain those dimensions.

Often, a machinist will desire a dimension that does not appear on the drawing. When this is the case, it can usually be calculated by simple manipulation of surrounding dimensions. In the following exercises, increased emphasis will be placed upon calculating dimensions.

INSTRUCTIONS: Refer to drawing 11A010 to calculate the following dimensions.

Ⓐ ____________________

Ⓑ ____________________

Ⓒ ____________________

Ⓓ ____________________

Ⓔ ____________________

Ⓕ ____________________

Ⓖ ____________________

Ⓗ ____________________

Ⓘ ____________________

Ⓙ ____________________

Ⓚ ____________________

Ⓛ ____________________

Ⓜ ____________________

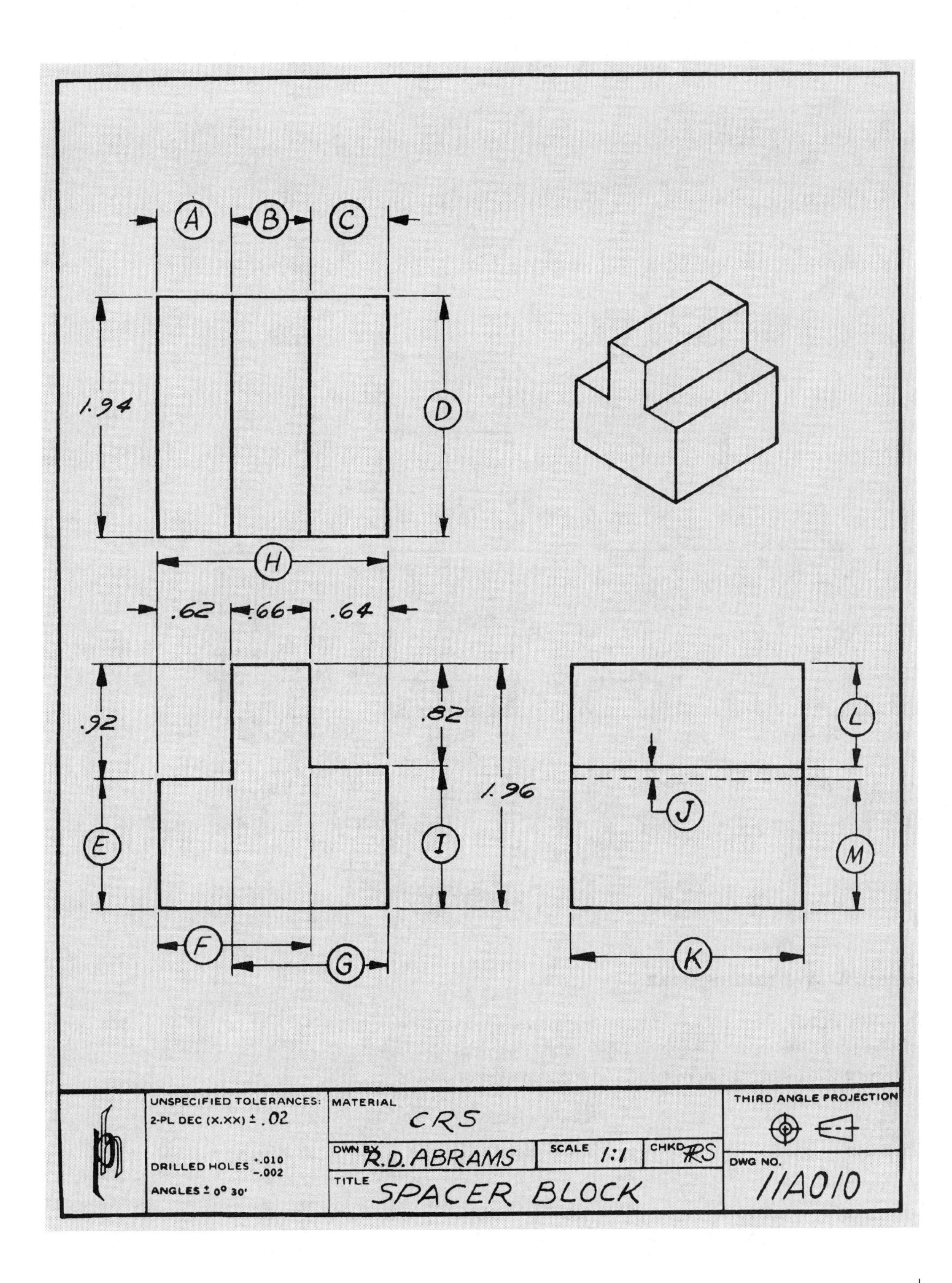
A
B
C
1.94
D
H
.62
.66
.64
.92
.82
1.96
E
I
F
G
L
J
M
K
UNSPECIFIED TOLERANCES:
2-PL DEC (X.XX) ± .02
DRILLED HOLES +.010 −.002
ANGLES ± 0° 30'
MATERIAL
CRS
DWN BY R.D. ABRAMS
SCALE 1:1
CHKD RS
TITLE SPACER BLOCK
THIRD ANGLE PROJECTION
DWG NO.
11A010

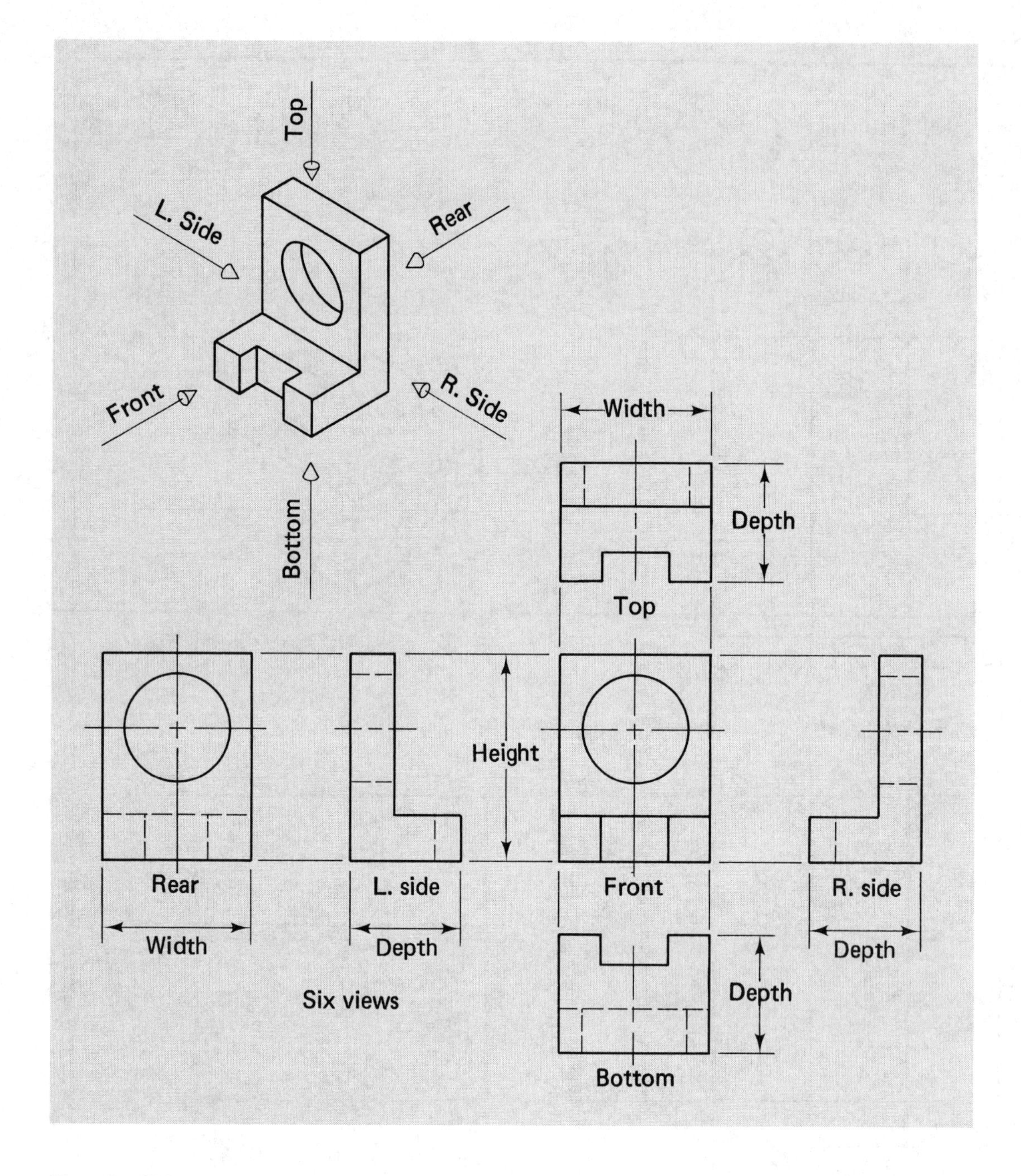

Basic Dimensions Quiz

INSTRUCTIONS: Each of the orthographic views will always show two of the three basic dimensions (height, width, or depth). Circle the corresponding letters (H, W, or D) for the views listed below.

Front view:	H	W	D	Bottom view:	H	W	D
Top view:	H	W	D	L.-side view:	H	W	D
R.-side view:	H	W	D	Rear view:	H	W	D

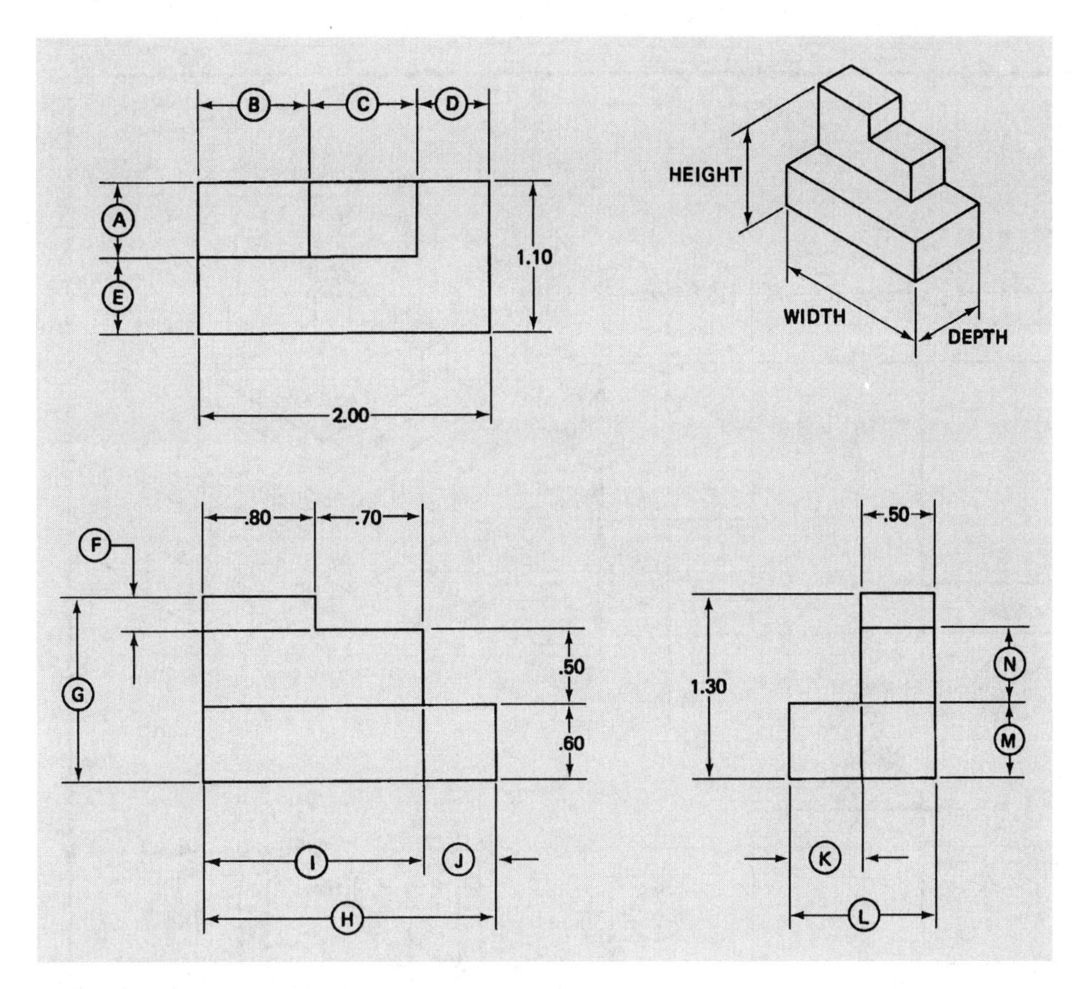

Dimension Calculations 1

INSTRUCTIONS: Enter the corresponding dimensions in the spaces provided.

1. Height dimension

   ____

2. Width dimension

   ____

3. Depth dimension

   ____

Ⓐ ______________________

Ⓑ ______________________

Ⓒ ______________________

Ⓓ ______________________

Ⓔ ______________________

Ⓕ ______________________

Ⓖ ______________________

Ⓗ ______________________

Ⓘ ______________________

Ⓙ ______________________

Ⓚ ______________________

Ⓛ ______________________

Ⓜ ______________________

Ⓝ ______________________

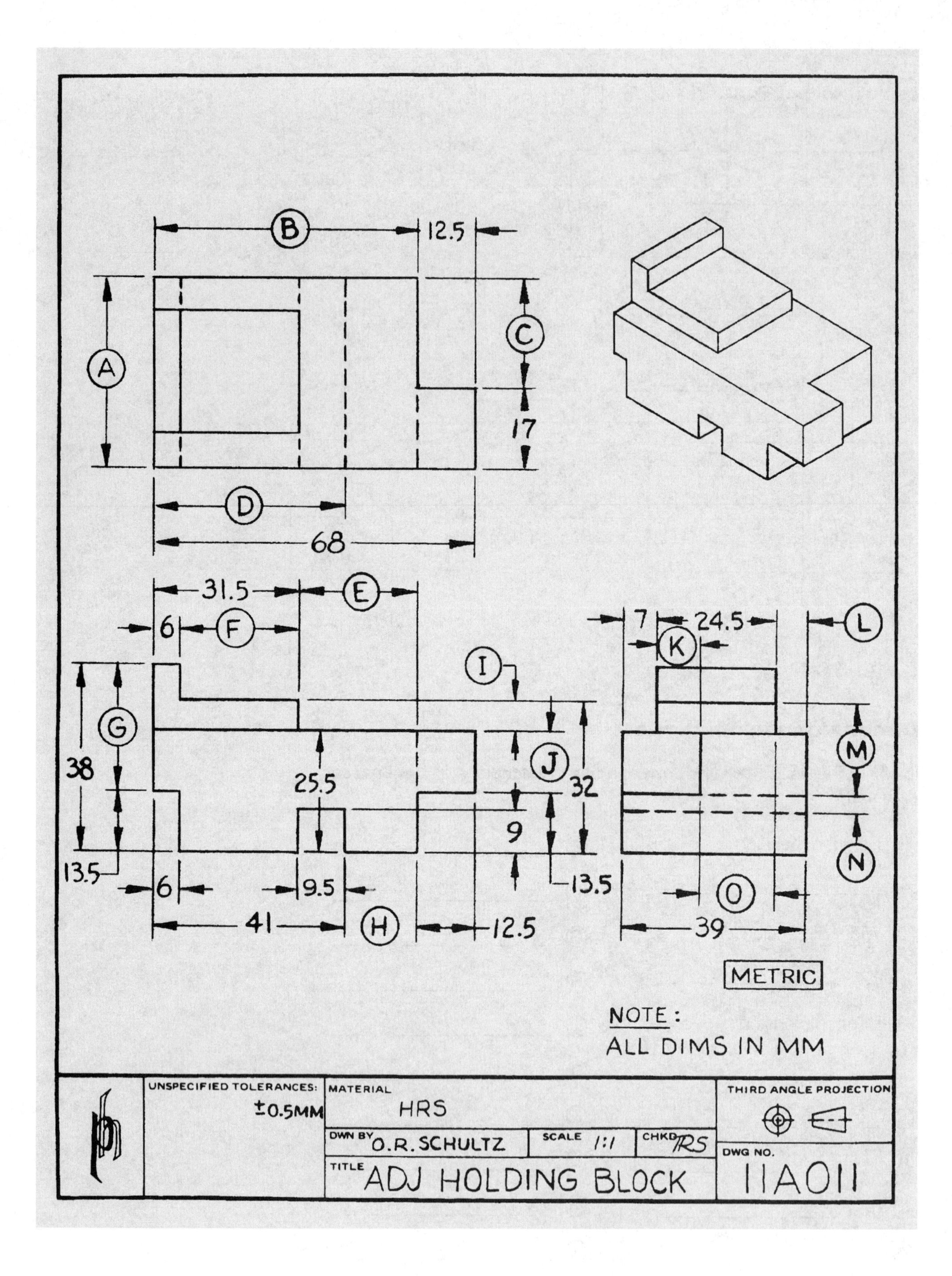
B
12.5
A
C
17
D
68
31.5
E
6
F
I
G
38
J
25.5
32
9
13.5
13.5
6
9.5
41
H
12.5
7
24.5
L
K
M
N
O
39
METRIC
NOTE :
ALL DIMS IN MM
UNSPECIFIED TOLERANCES:
±0.5MM
MATERIAL
HRS
DWN BY O.R.SCHULTZ
SCALE 1:1
CHKD RS
TITLE ADJ HOLDING BLOCK
THIRD ANGLE PROJECTION
DWG NO.
11A011

PRECEDENCE OF LINES

When visible lines, hidden lines, or centerlines coincide in the same view, a drafter will follow the precedence of lines listed below to illustrate the preferred line.

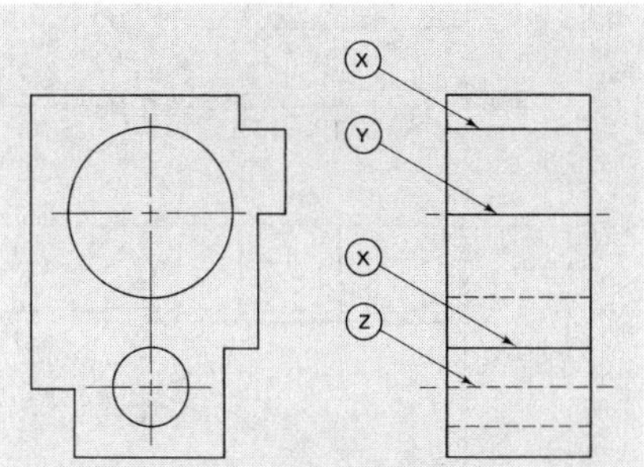

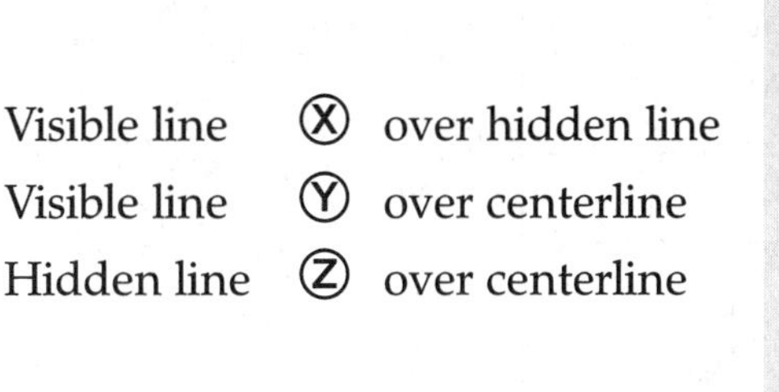

Visible line Ⓧ over hidden line
Visible line Ⓨ over centerline
Hidden line Ⓩ over centerline

INSTRUCTIONS: Refer to drawing 11A011 to answer the following questions.

1. What is the name of the object? 1. ____________
2. Of what material is the object made? (Do not abbreviate.) 2. ____________
3. To what scale (half, full, double) is the object drawn? 3. ____________
4. What system of dimensioning (fractional, decimal, metric) is used? 4. ____________
5. What dimensional increments (inches, centimeters, millimeters) are used? 5. ____________
6. What angle of projection (first, second, third) is used? 6. ____________
7. What tolerance is assigned to each dimension? 7. ____________
8. What is the height dimension? 8. ____________
9. What is the width dimension? 9. ____________
10. What is the depth dimension at maximum material condition? 10. ____________

INSTRUCTIONS: Enter the dimensions from drawing 11A011 for the following letters.

Ⓐ ____________ Ⓕ ____________ Ⓚ ____________

Ⓑ ____________ Ⓖ ____________ Ⓛ ____________

Ⓒ ____________ Ⓗ ____________ Ⓜ ____________

Ⓓ ____________ Ⓘ ____________ Ⓝ ____________

Ⓔ ____________ Ⓙ ____________ Ⓞ ____________

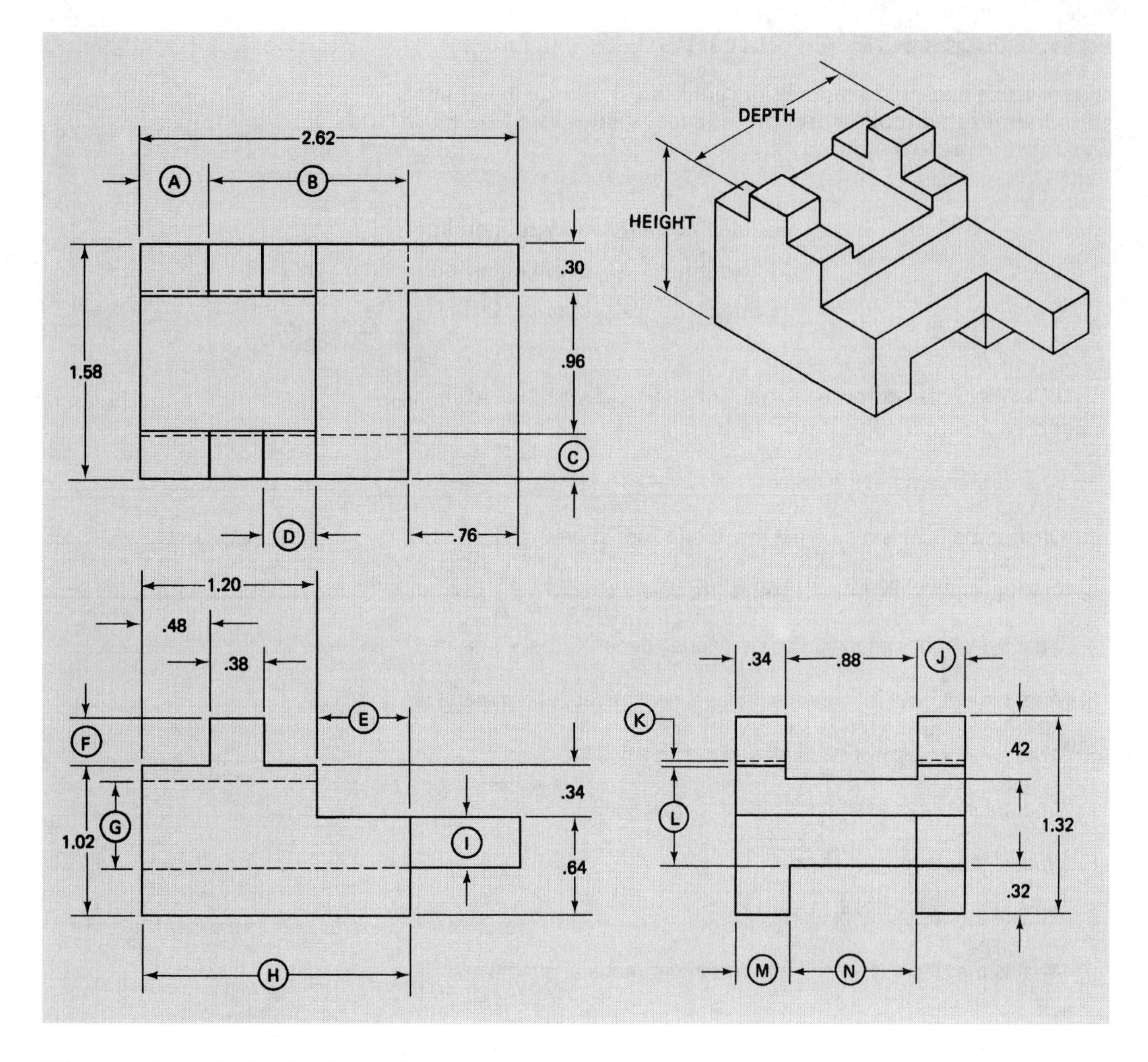

Dimension Calculations 2

INSTRUCTIONS: Enter the corresponding dimensions in the spaces provided.

1. Height dimension

____

2. Depth dimension

____

Ⓐ _______________

Ⓑ _______________

Ⓒ _______________

Ⓓ _______________

Ⓔ _______________

Ⓕ _______________

Ⓖ _______________

Ⓗ _______________

Ⓘ _______________

Ⓙ _______________

Ⓚ _______________

Ⓛ _______________

Ⓜ _______________

Ⓝ _______________

ACCUMULATED TOLERANCES

A drawing is usually dimensioned in a manner that will control the tolerances of important features. To avoid double dimensioning, the drafter will leave out certain dimensions that the machinist might prefer to have. These dimensions can be calculated mathematically by manipulating the given dimensions, such as you have been doing with the drawings in this book. You must remember, however, that the calculated dimension will accumulate the tolerance of every dimension used to arrive at it, even when subtracting.

DIMENSIONING METHODS

Illustrated are three different methods of dimensioning the same step shaft. As you can see, the length of each step is 25 mm on all three examples, but the tolerance for each step *length* may vary depending on the dimensioning method used. (Length and diameter are terms commonly used to describe cylindrical shapes.)

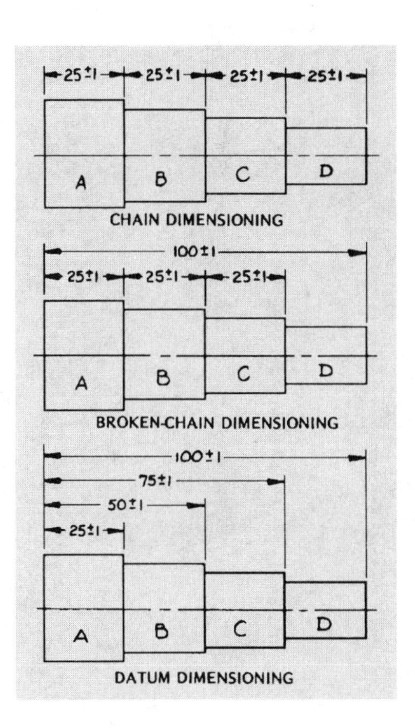

Chain Dimensioning

This method includes the length of each step, but does not include the overall length. When each step is dimensioned, the tolerance of each individual length is controlled; however, the tolerance of the overall length is accumulated. In other words, each step length has a tolerance of ±1mm, but the overall length is allowed a tolerance of ±4mm (accumulation of four tolerances).

Broken-Chain Dimensioning

This method includes the overall length, plus all except one of the step lengths. Tolerance on the overall length and all but the least important step length are thereby controlled. The undimensioned step length accumulates the tolerance of all the dimensions used to calculate it (if three steps are added together and then subtracted from the overall length, you have used four dimensions, so the missing dimension would carry a tolerance of ±4mm).

Datum Dimensioning

This method has each of the step lengths measured from a common surface, or baseline. Accumulation is limited to not more than two tolerances, since any individual step length may be calculated by subtracting the dimensions affecting each end of the step. This method of dimensioning has increased in usage in recent years.

Method	Step A	Step B	Step C	Step D	Overall
Chain	±1	±1	±1	±1	±4
Broken-Chain	±1	±1	±1	±4	±1
Datum	±1	±2	±2	±2	±1

This table compares the difference in the tolerance accumulation caused by the three dimensioning methods used above.

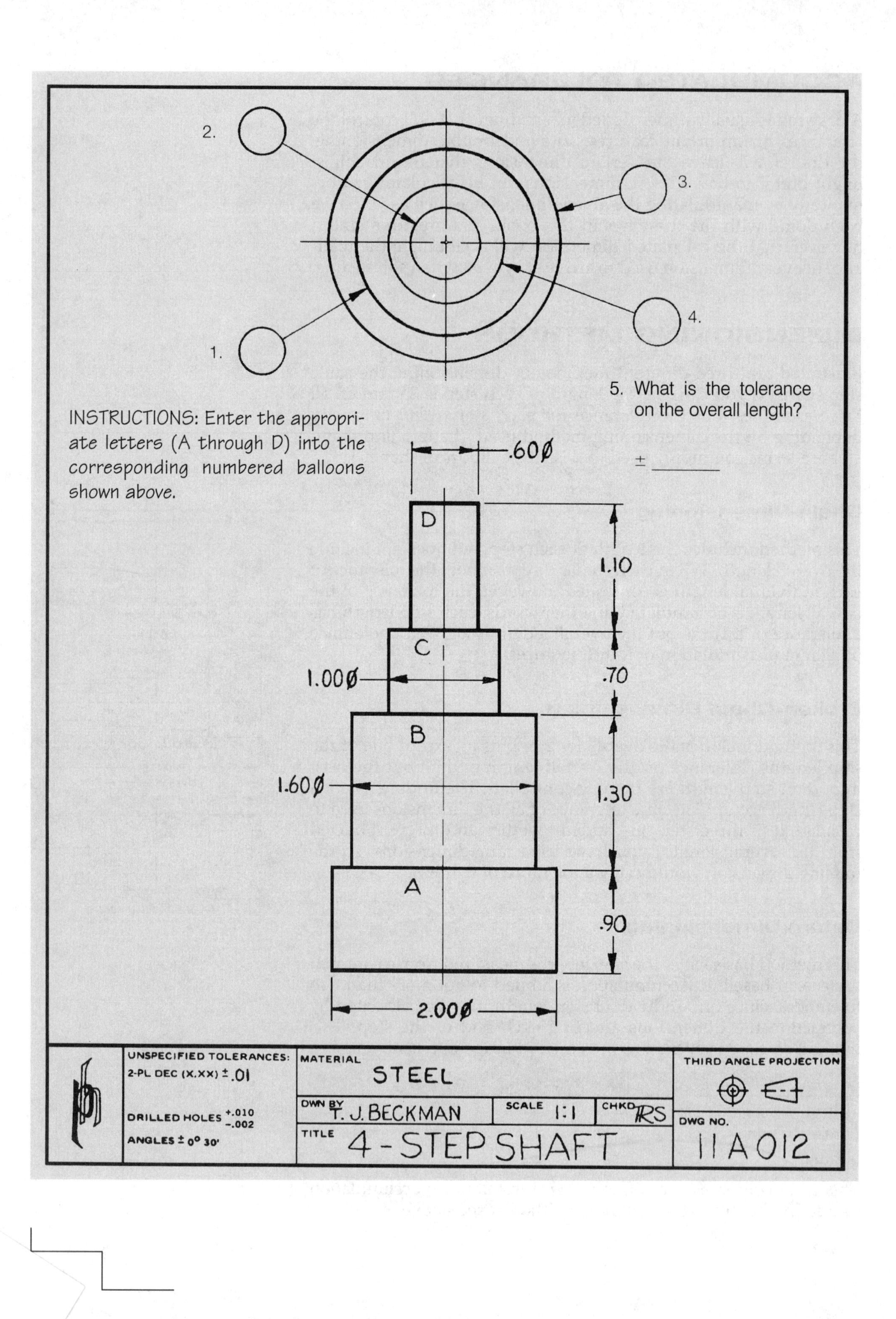
2.
3.
4.
1.
5. What is the tolerance on the overall length?
±
INSTRUCTIONS: Enter the appropriate letters (A through D) into the corresponding numbered balloons shown above.
.60Ø
D
1.10
C
1.00Ø
.70
B
1.60Ø
1.30
A
.90
2.00Ø
UNSPECIFIED TOLERANCES:
2-PL DEC (X.XX) ± .01
DRILLED HOLES +.010 −.002
ANGLES ± 0° 30'
MATERIAL
STEEL
DWN BY T. J. BECKMAN
SCALE 1:1
CHKD RS
TITLE 4-STEP SHAFT
THIRD ANGLE PROJECTION
DWG NO.
11 A 012

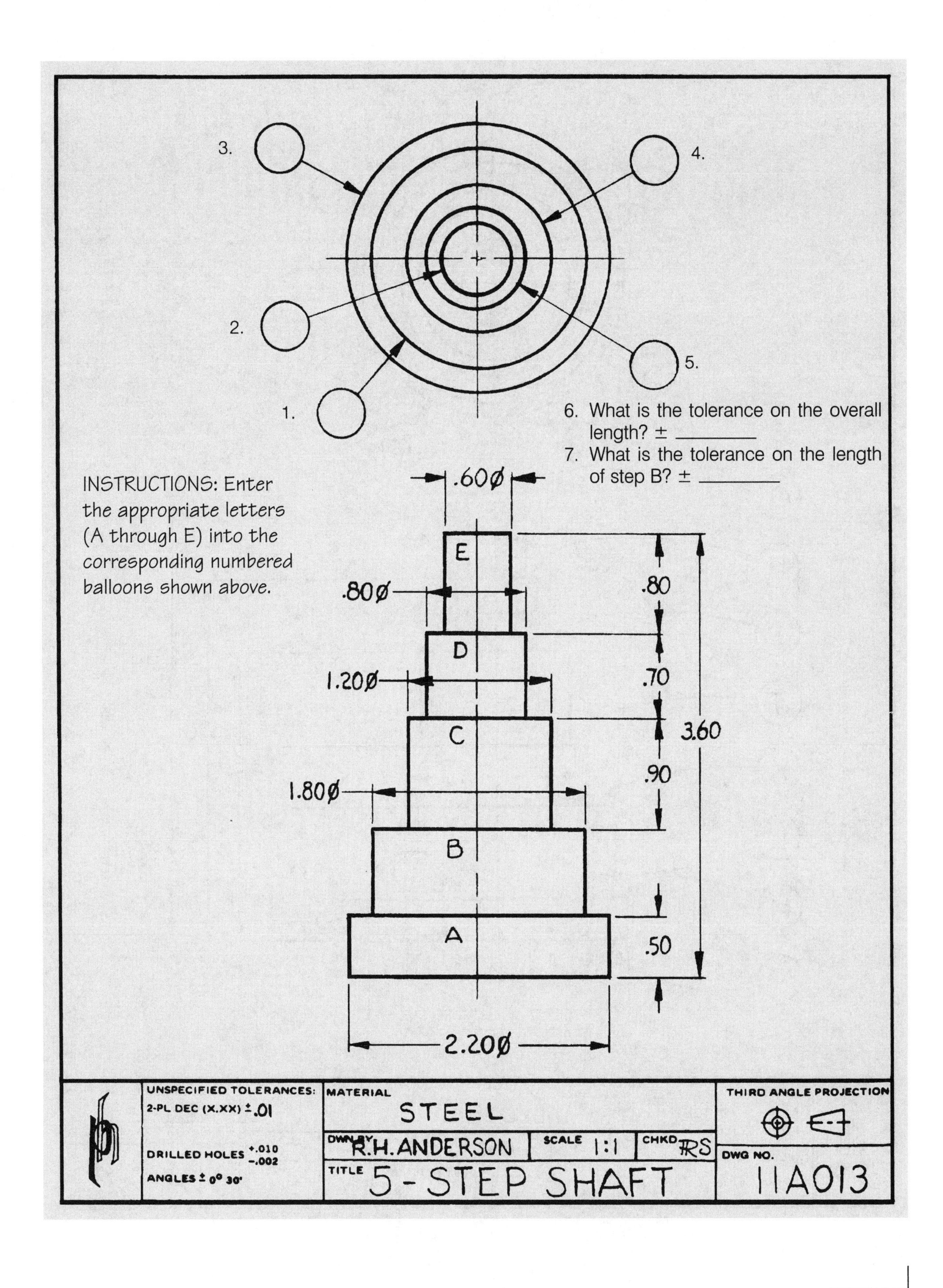

3.
4.
2.
5.
1.
6. What is the tolerance on the overall length? ± ________
7. What is the tolerance on the length of step B? ± ________
INSTRUCTIONS: Enter the appropriate letters (A through E) into the corresponding numbered balloons shown above.
.60Ø
E
.80Ø
.80
D
1.20Ø
.70
C
3.60
.90
1.80Ø
B
A
.50
2.20Ø
UNSPECIFIED TOLERANCES:
2-PL DEC (X.XX) ± .01
DRILLED HOLES +.010 −.002
ANGLES ± 0° 30'
MATERIAL
STEEL
DWN BY R.H. ANDERSON
SCALE 1:1
CHKD RS
TITLE 5 - STEP SHAFT
THIRD ANGLE PROJECTION
DWG NO.
11A013

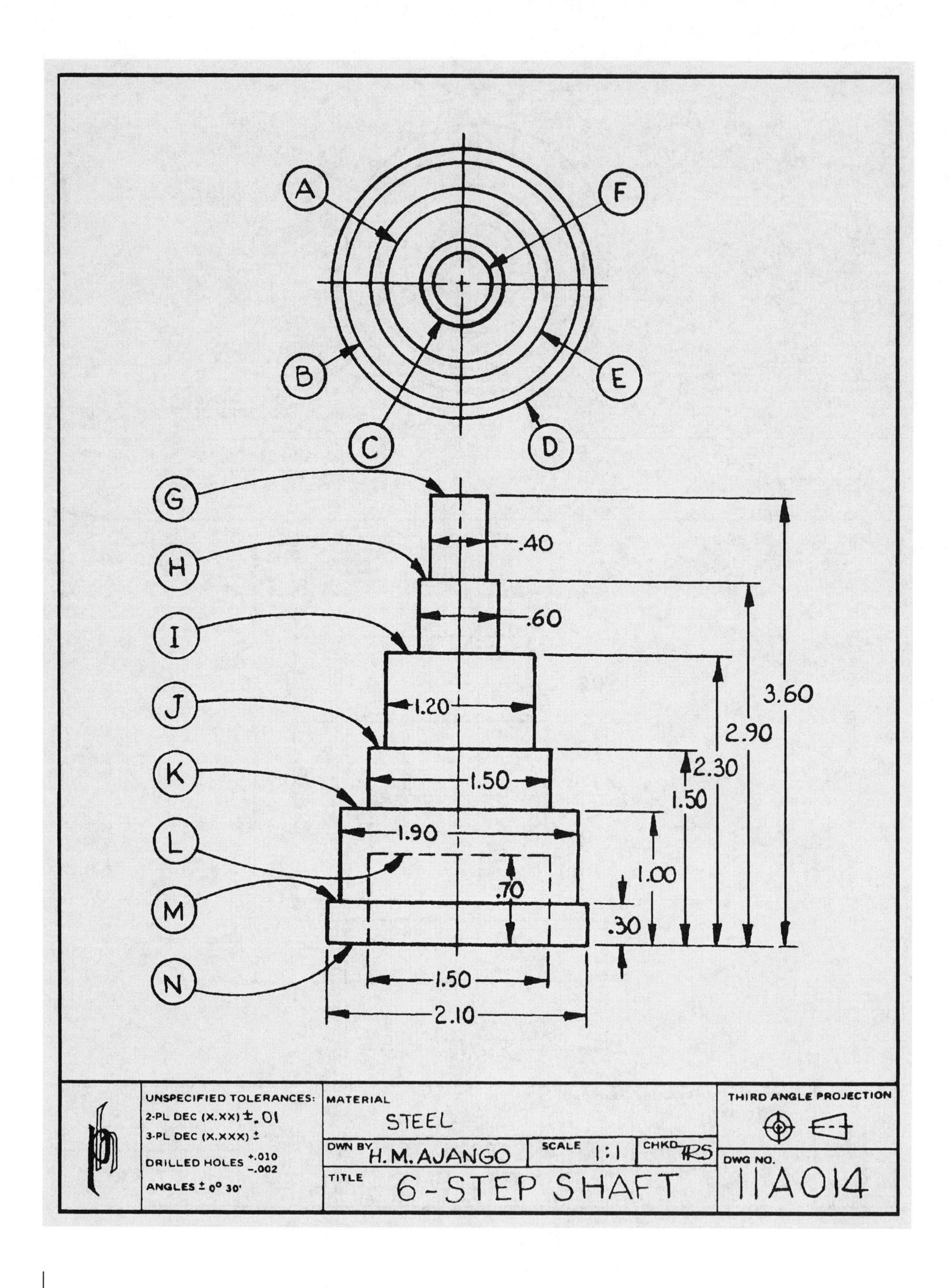
A
B
C
D
E
F
G
H
I
J
K
L
M
N
.40
.60
1.20
1.50
1.90
.70
1.50
2.10
3.60
2.90
2.30
1.50
1.00
.30
UNSPECIFIED TOLERANCES:
2-PL DEC (X.XX) ±.01
3-PL DEC (X.XXX) ±
DRILLED HOLES +.010 −.002
ANGLES ± 0° 30'
MATERIAL
STEEL
DWN BY H. M. AJANGO
SCALE 1:1
CHKD RS
TITLE 6-STEP SHAFT
THIRD ANGLE PROJECTION
DWG NO. 11A014

INSTRUCTIONS: Refer to drawing 11A014 to answer the following questions.

1. Is (a) first-angle or (b) third-angle projection used? 1. ____________
2. Is the circular view (a) a top or (b) a bottom view? 2. ____________
3. Is the part (a) concentric or (b) eccentric? 3. ____________
4. Is the part (a) symmetrical or (b) nonsymmetrical? 4. ____________
5. What is the print tolerance for each dimension shown? 5. ____________
6. What is the overall length of the shaft? 6. ____________
7. What is the MMC of shoulder Ⓔ OD? 7. ____________
8. What is the MMC of the bore ID? 8. ____________
9. What is the tolerance on the length of shoulder Ⓐ? 9. ____________
10. What is the tolerance on the length of shoulder Ⓓ? 10. ____________
11. Why are the answers to questions 9 and 10 different? Explain below. 11. ____________
12. Why is there no hidden line representing the 1.50 bore diameter in the circular view? Explain below. 12. ____________

INSTRUCTIONS: Refer to drawing 11A014. Enter the appropriate letters (G through N) into the corresponding balloons below.

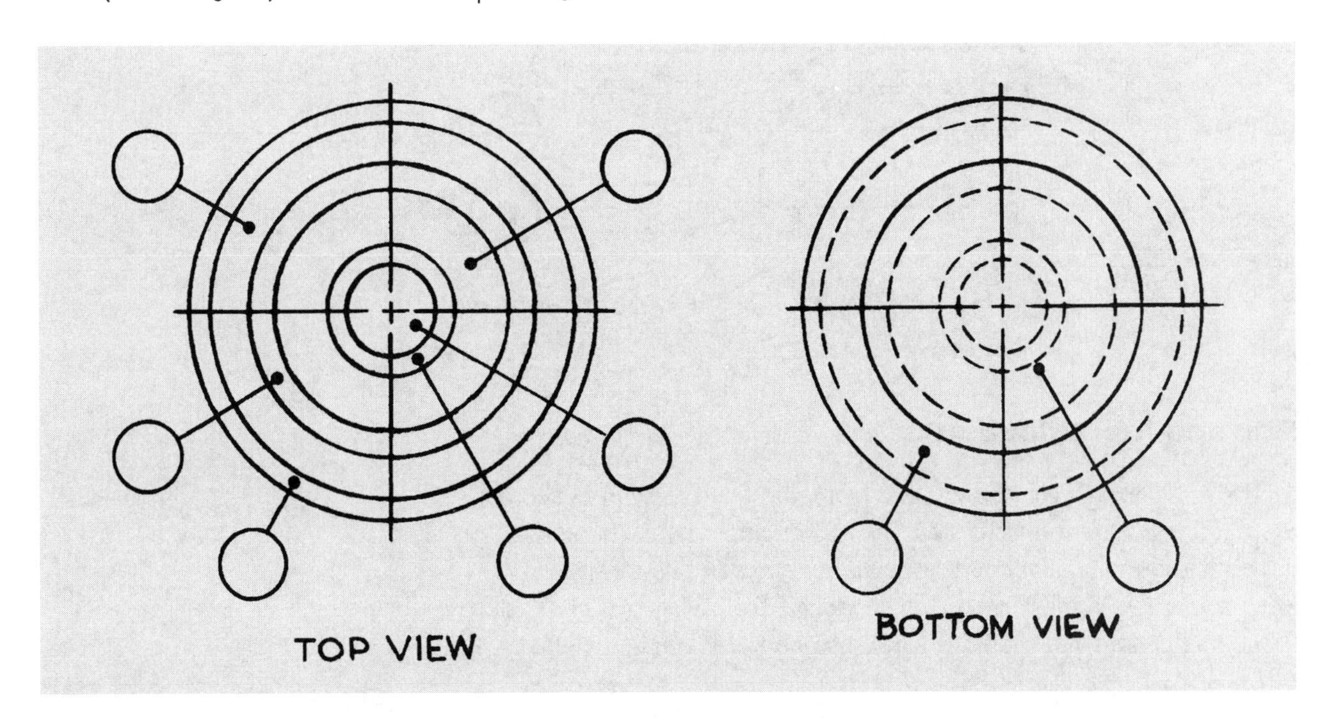

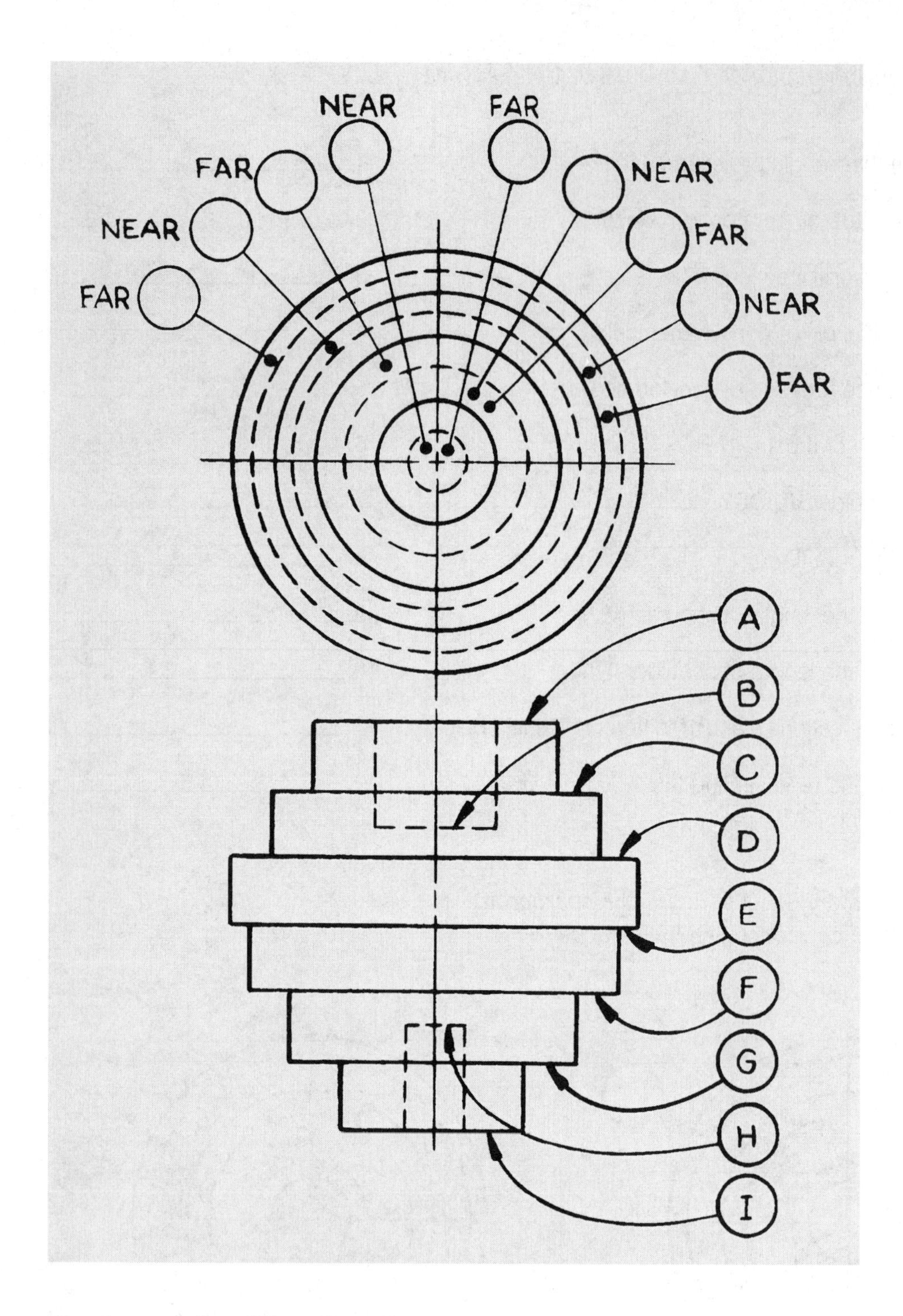

Surface Identification 3

INSTRUCTIONS: Enter the letters from the front view into the corresponding balloons in the top view. Remember that "near" designates the surface closest to the viewer's eye in the view where the word appears. "Far" surfaces are blocked from view by "near" surfaces. A bottom view would reverse the near-far designations.

INSTRUCTIONS: Illustrated below are two methods of dimensioning the same object. The broken-chain method allows tolerance to accumulate, while the datum method controls tolerance accumulation. Compare the two by calculating the tolerance between the following surfaces. Remember to add a tolerance for every dimension you use to calculate your answer. *Note:* All dimensions are assigned a tolerance of ±.01 each.

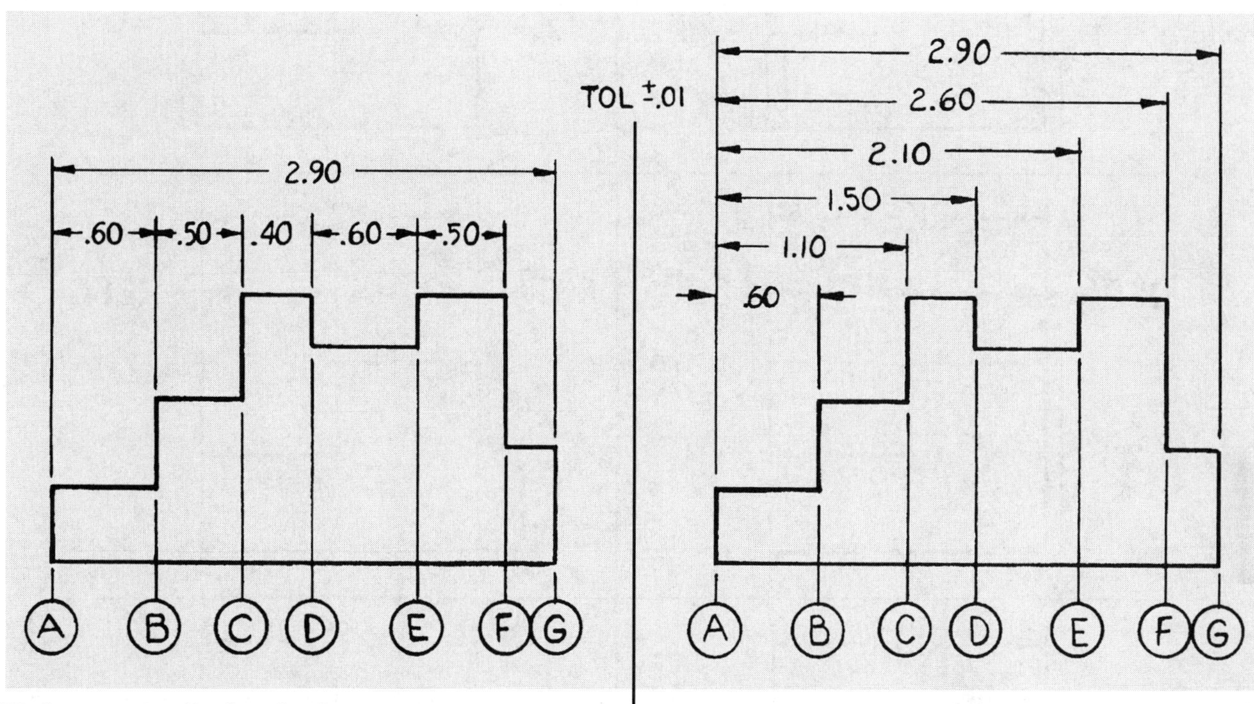

Tolerance Calculations 1

Tolerance between:

A and B ± ____________________

A and C ± ____________________

A and E ± ____________________

B and F ± ____________________

C and F ± ____________________

B and G ± ____________________

C and G ± ____________________

E and G ± ____________________

F and G ± ____________________

What is the maximum amount of tolerance that can accumulate between any two surfaces in the example above?

± ____________________

Tolerance between:

A and B ± ____________________

A and C ± ____________________

A and E ± ____________________

B and F ± ____________________

C and F ± ____________________

B and G ± ____________________

C and G ± ____________________

E and G ± ____________________

F and G ± ____________________

What is the maximum amount of tolerance that can accumulate between any two surfaces in the example above?

± ____________________

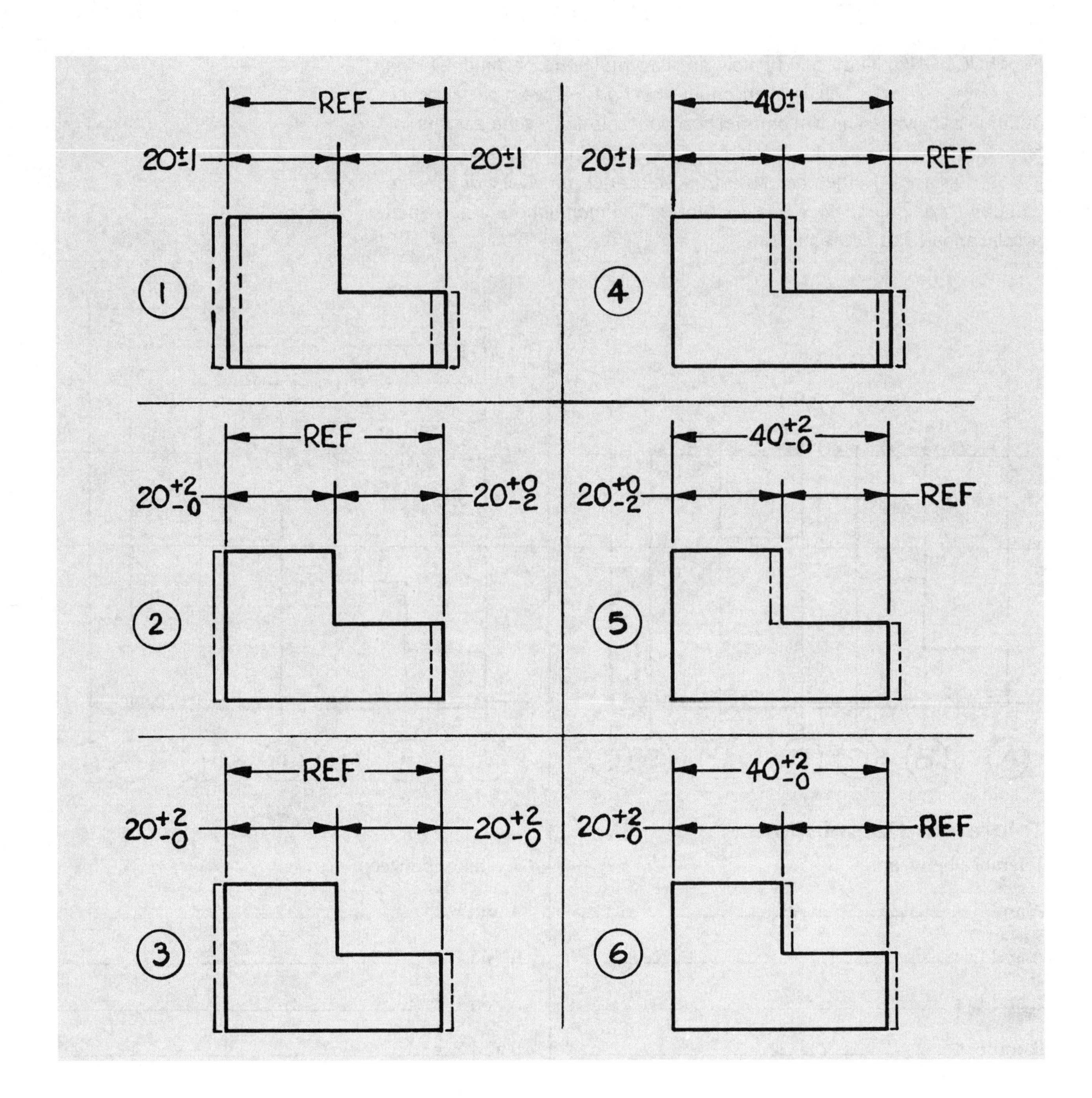

Tolerance Calculations 2

INSTRUCTIONS: Calculate the reference dimensions and their accumulated tolerances for the six illustrations above. *Tip:* Tolerances always add, even when dimensions are subtracted. Enter your answers in the spaces below. (Phantom lines represent the limits.)

① __________ ± __________ ④ __________ ± __________

② __________ ± __________ ⑤ __________ $^{+}_{-}$ __________

③ __________ $^{+}_{-}$ __________ ⑥ __________ ± __________

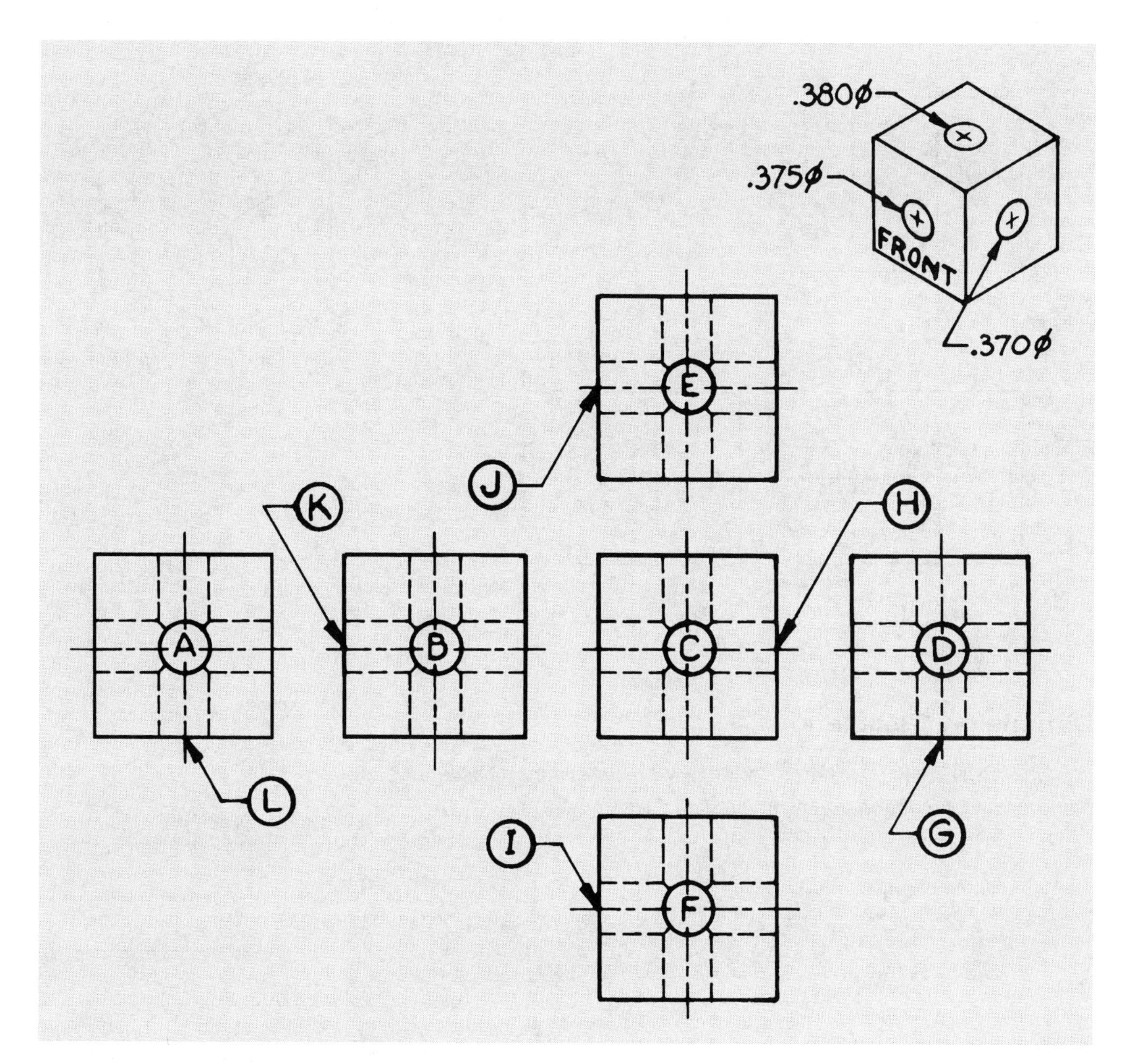

Hole Identification 1

INSTRUCTIONS: Study the pictorial view to determine the hole diameter that corresponds with each encircled letter shown on the orthographic views.

Ⓐ ____________________ Ⓖ ____________________

Ⓑ ____________________ Ⓗ ____________________

Ⓒ ____________________ Ⓘ ____________________

Ⓓ ____________________ Ⓙ ____________________

Ⓔ ____________________ Ⓚ ____________________

Ⓕ ____________________ Ⓛ ____________________

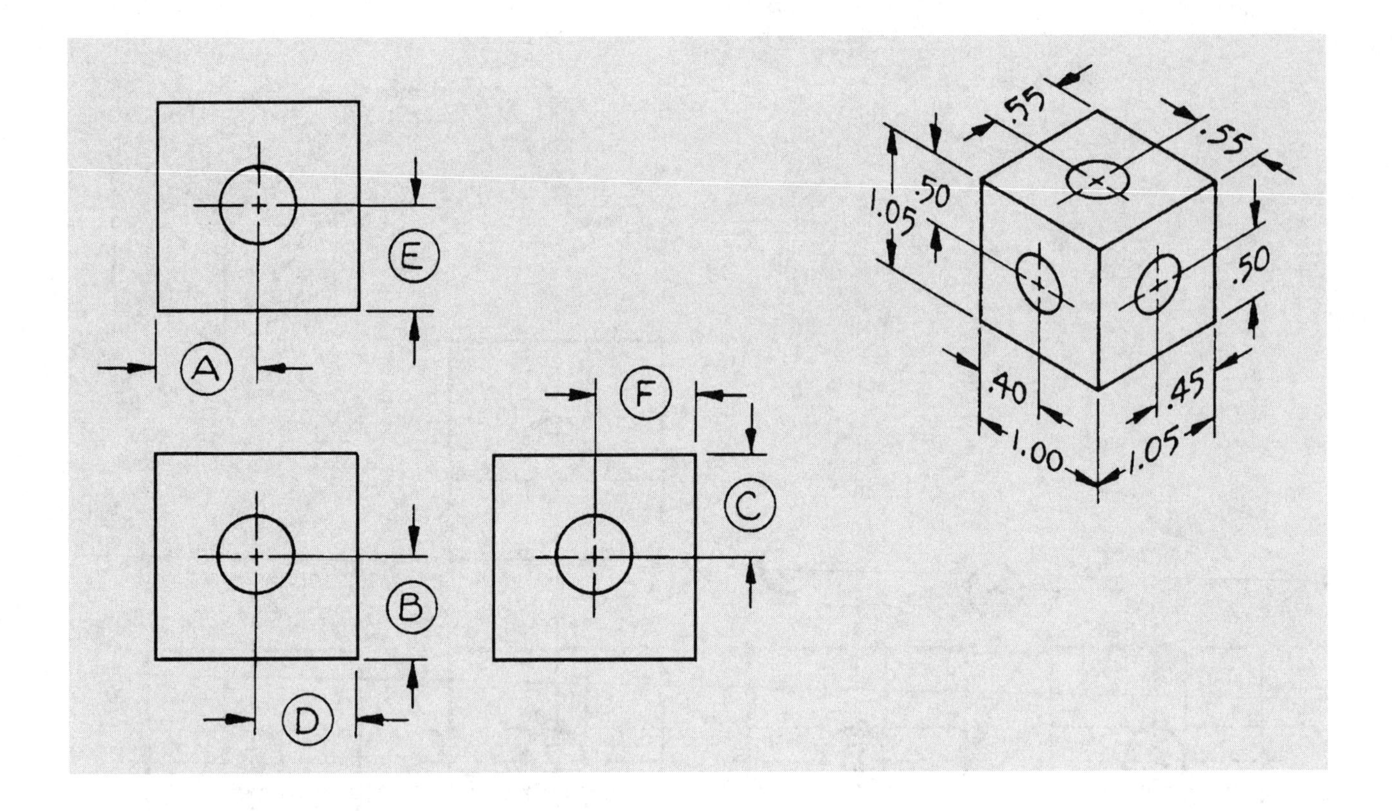

Dimension Calculations 3

INSTRUCTIONS: Study the pictorial view to determine the dimensions that correspond to the encircled letters shown below.

Ⓐ ______________

Ⓑ ______________

Ⓒ ______________

Ⓓ ______________

Ⓔ ______________

Ⓕ ______________

HEIGHT______________

WIDTH______________

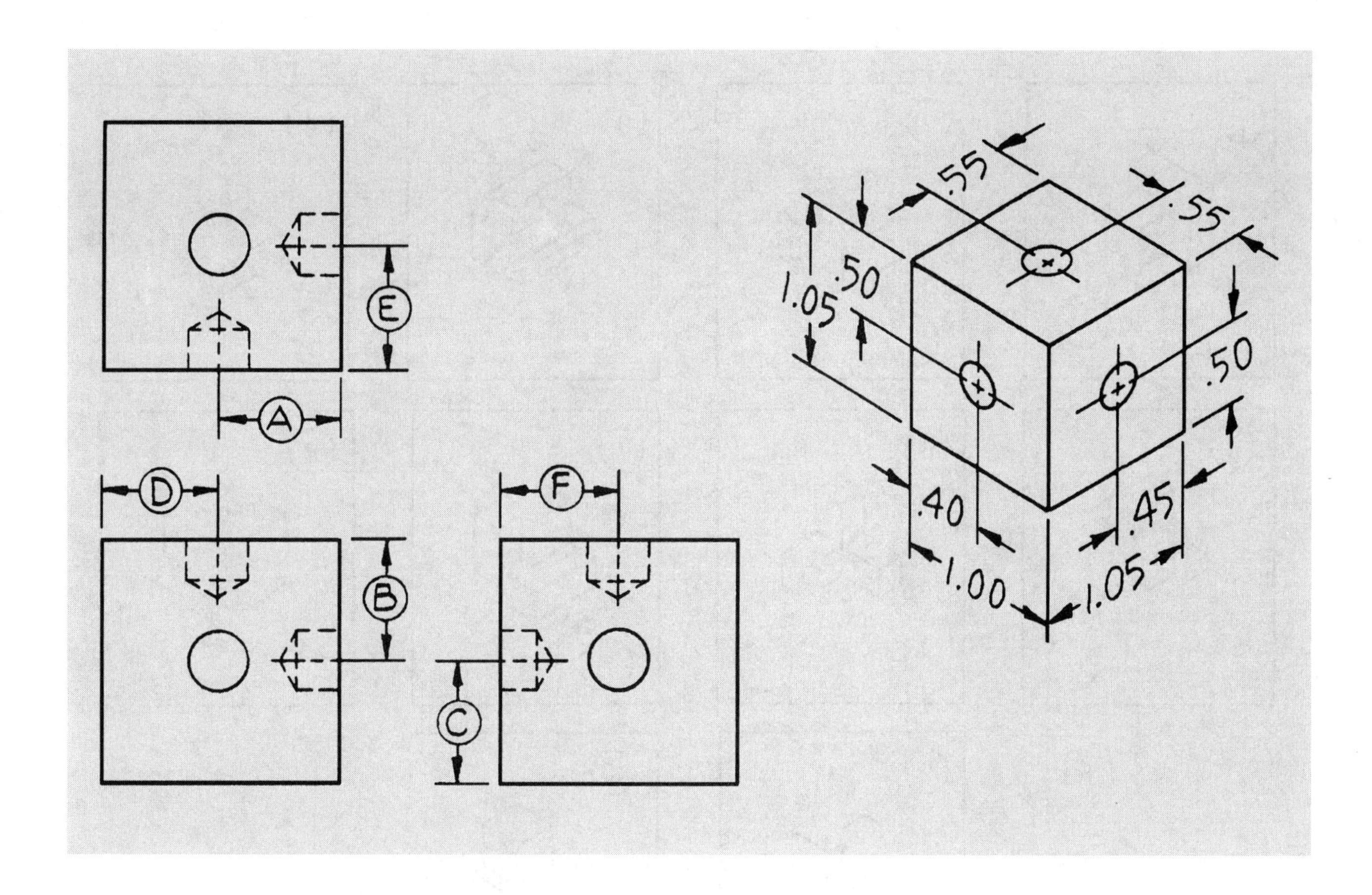

Dimension Calculations 4

INSTRUCTIONS: Study the pictorial view to determine the dimensions that correspond to the encircled letters shown below.

Ⓐ ____________________

Ⓑ ____________________

Ⓒ ____________________

Ⓓ ____________________

Ⓔ ____________________

Ⓕ ____________________

WIDTH____________________

DEPTH____________________

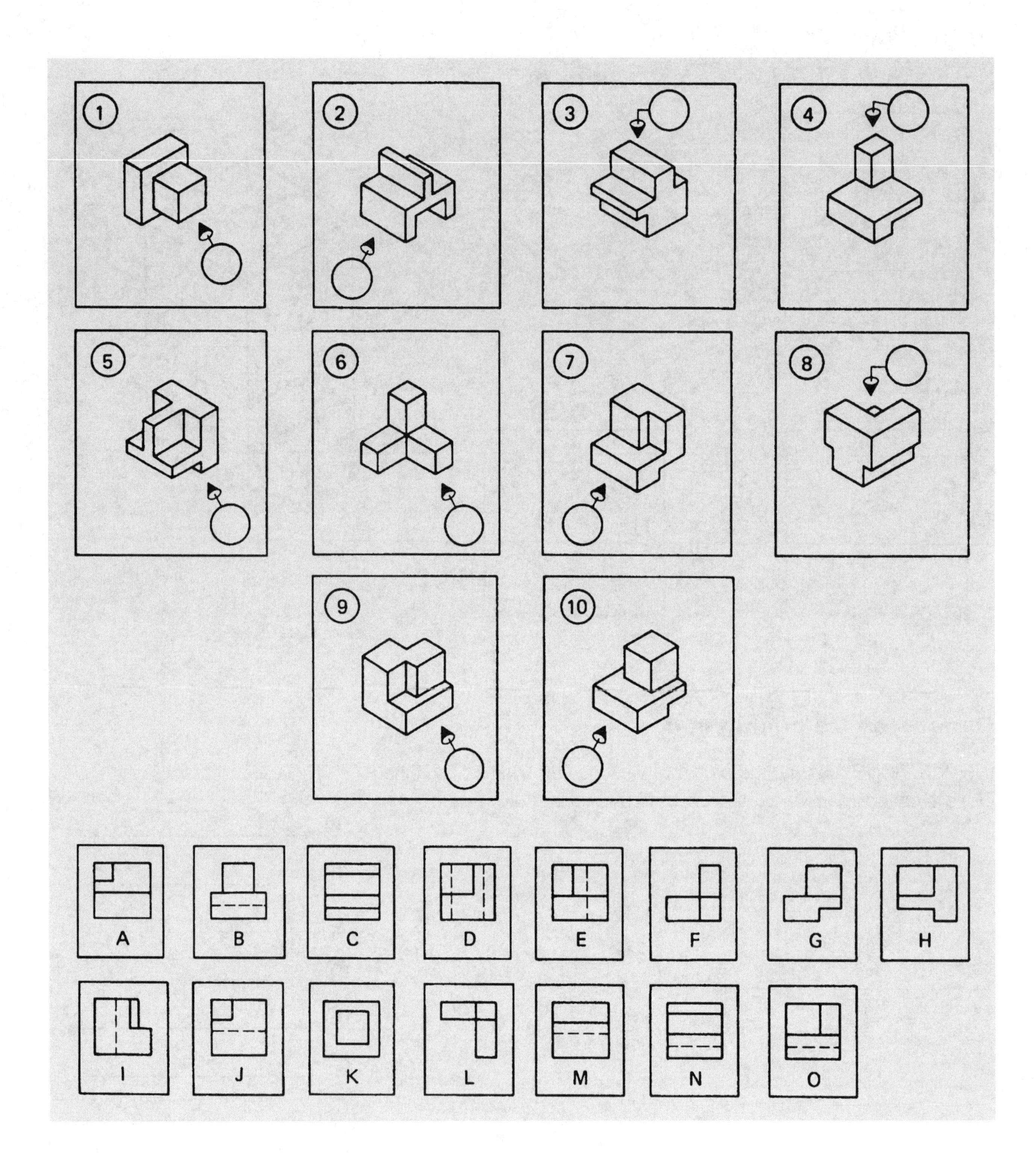

View Selection 3

INSTRUCTIONS: Select the correct orthographic view to correspond with the arrowhead for each pictorial view shown above. Enter the correct letters in the balloons provided.

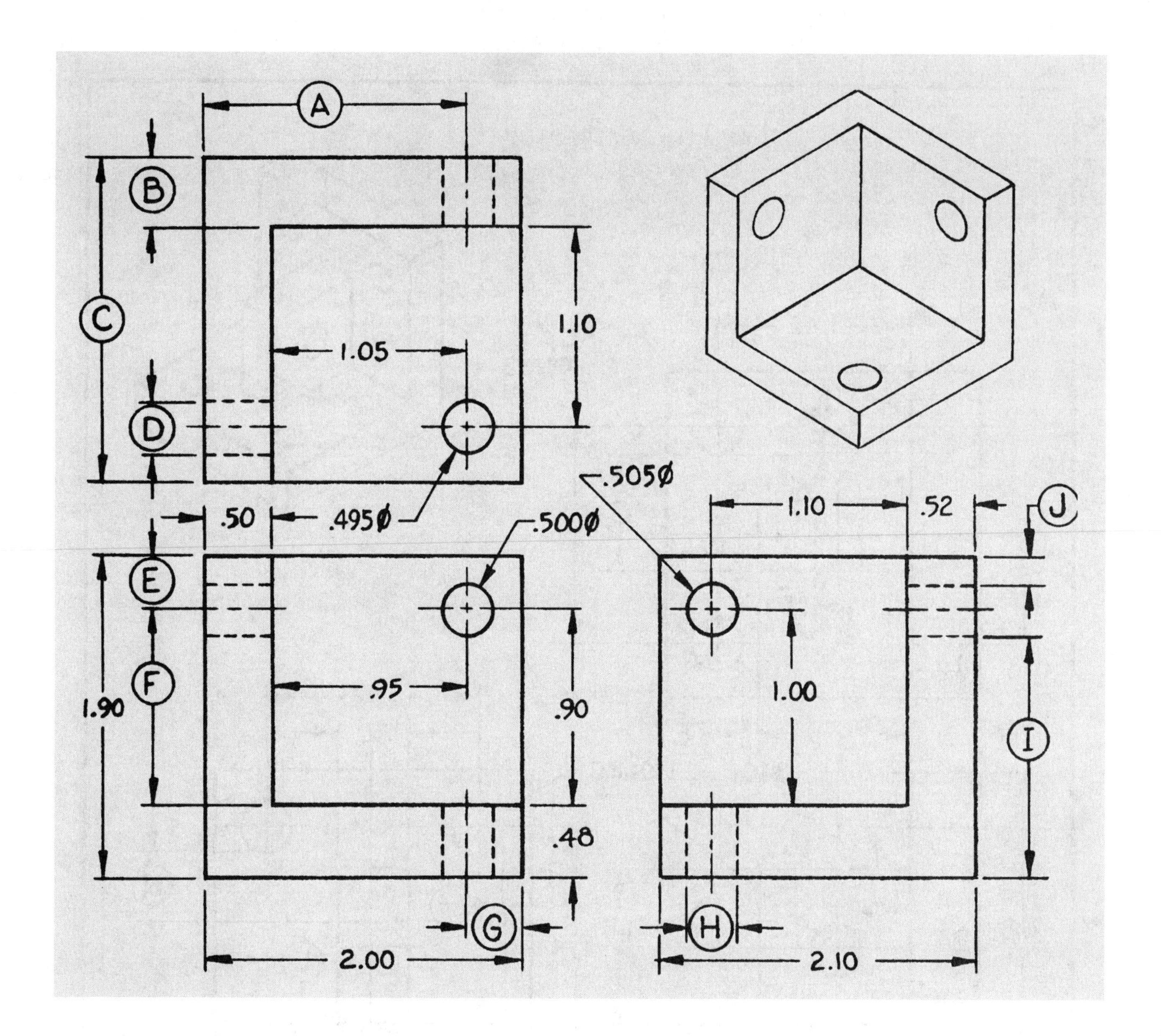

Dimension Calculations 5

INSTRUCTIONS: Enter the corresponding dimensions in the spaces provided.

1. Height dimension

____

2. Width dimension

____

Ⓐ ______________________

Ⓑ ______________________

Ⓒ ______________________

Ⓓ ______________________

Ⓔ ______________________

Ⓕ ______________________

Ⓖ ______________________

Ⓗ ______________________

Ⓘ ______________________

Ⓙ ______________________

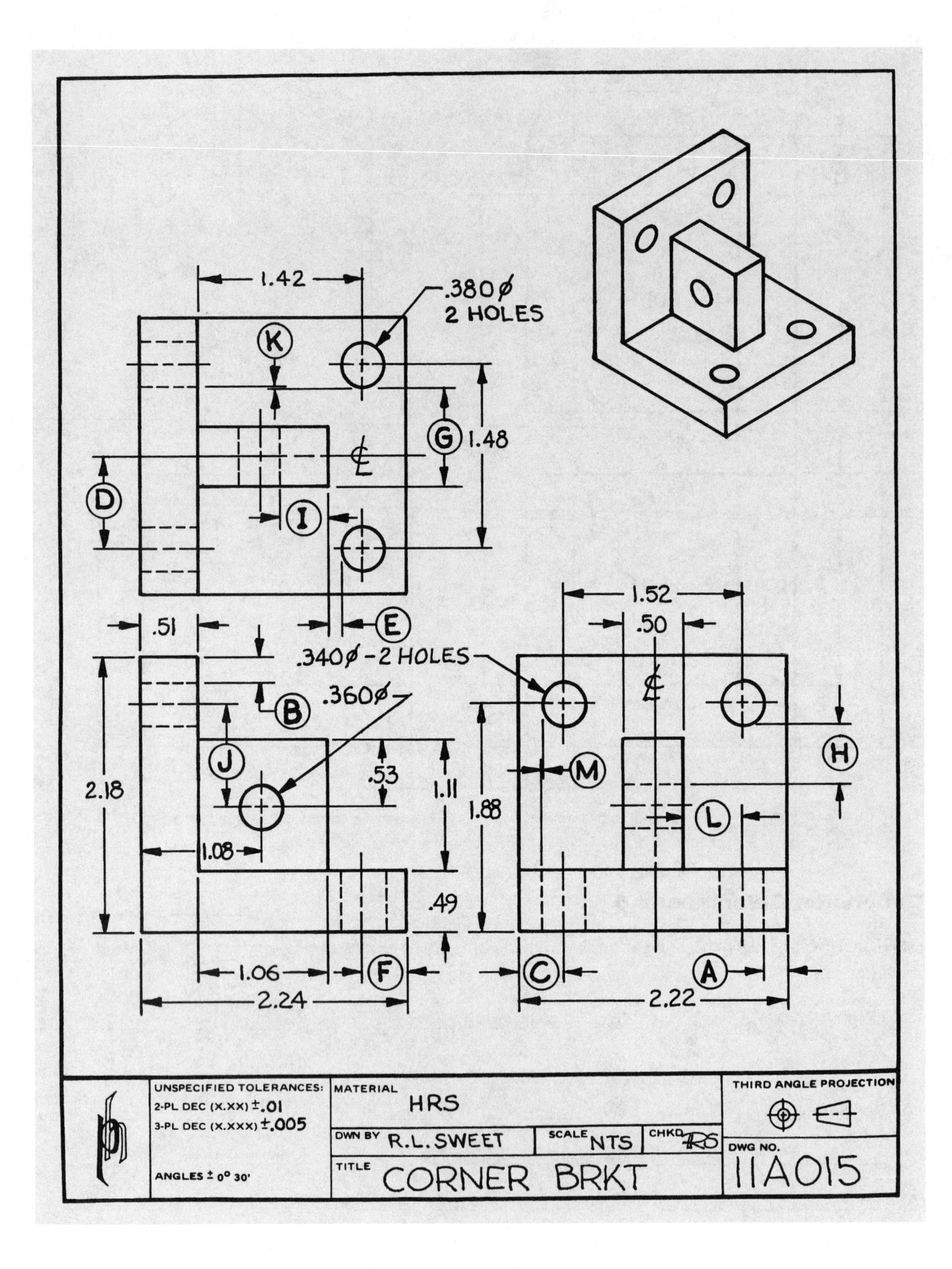

1.42
.380⌀
2 HOLES
K
G
1.48
D
I
E
.51
.340⌀ - 2 HOLES
B
.360⌀
J
2.18
.53
1.11
1.08
.49
1.06
F
2.24
1.52
.50
H
M
L
1.88
C
A
2.22
UNSPECIFIED TOLERANCES:
2-PL DEC (X.XX) ±.01
3-PL DEC (X.XXX) ±.005
ANGLES ± 0° 30'
MATERIAL
HRS
DWN BY R.L. SWEET
SCALE NTS
CHKD
TITLE CORNER BRKT
THIRD ANGLE PROJECTION
DWG NO.
11A015

INSTRUCTIONS: Refer to drawing 11A015 to answer the following questions.

1. What does BRKT in the title block abbreviate? 1. ______
2. What does HRS in the material block abbreviate? 2. ______
3. What does NTS in the scale block abbreviate? 3. ______
4. What is the overall depth dimension? 4. ______
5. What is the MMC of the overall height? 5. ______
6. What is the MMC of the .360 hole? 6. ______
7. How much tolerance can accumulate between the .380 hole centers? 7. ______

INSTRUCTIONS: Calculate the dimensions for the following letters found on drawing 11A015.

Ⓐ ______

Ⓑ ______

Ⓒ ______

Ⓓ ______

Ⓔ ______

Ⓕ ______

Ⓖ ______

Ⓗ ______

Ⓘ ______

Ⓙ ______

Ⓚ ______

Ⓛ ______

Ⓜ ______

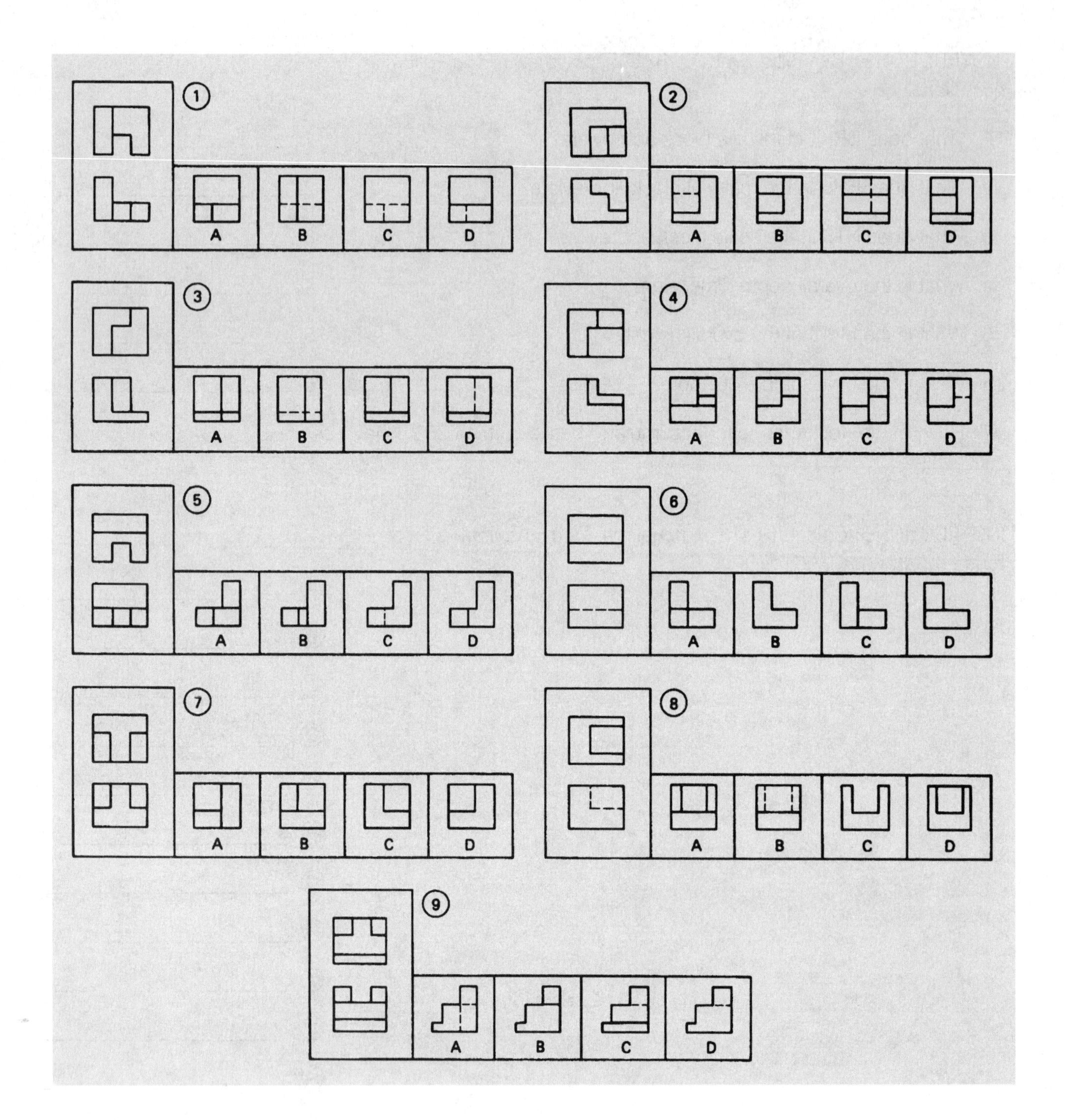

View Selection 4

INSTRUCTIONS: Circle the letter that represents the correct right-side view for each of the shapes drawn above.

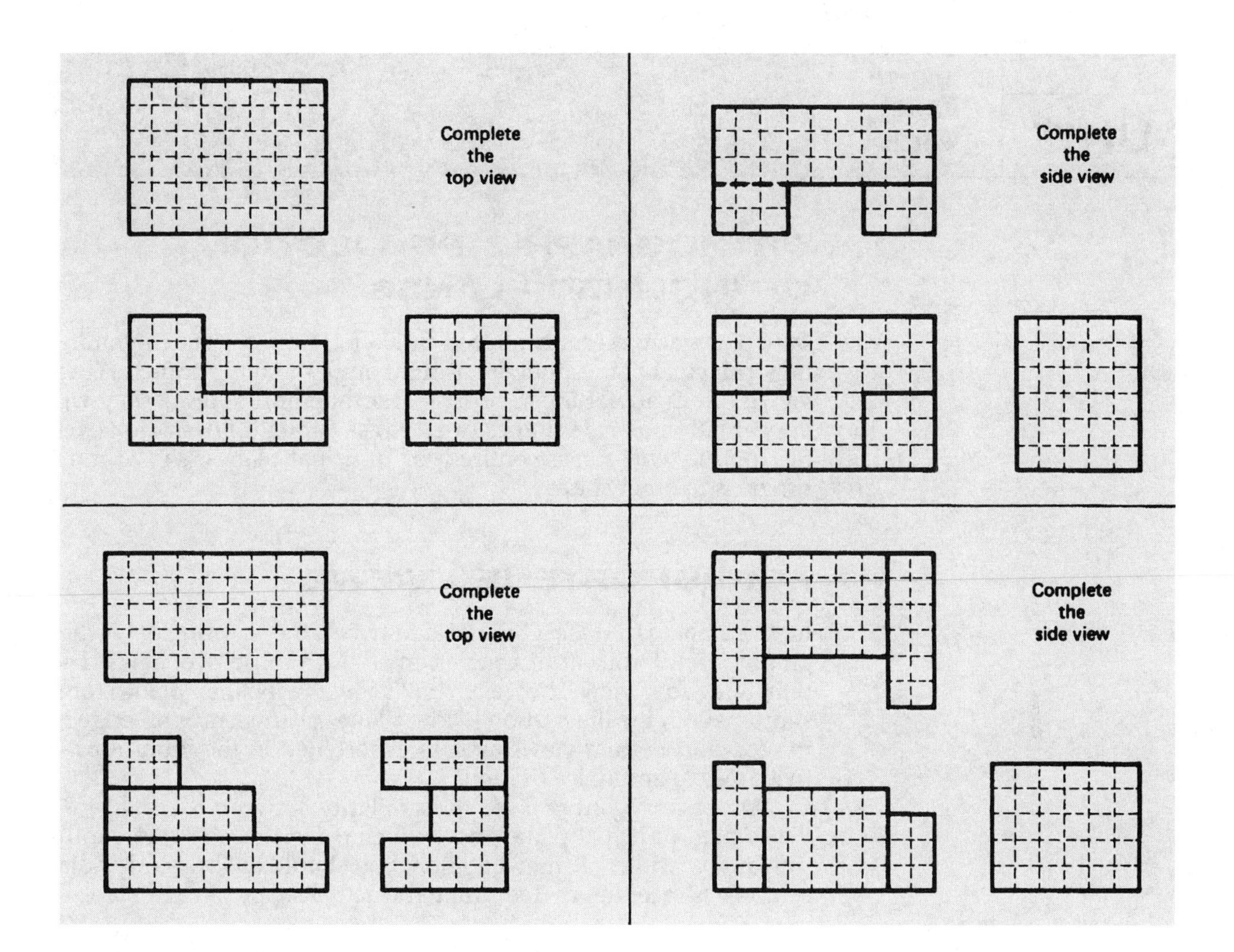

Missing Views 2

INSTRUCTIONS: Complete the designated view for each of the drawings shown above.

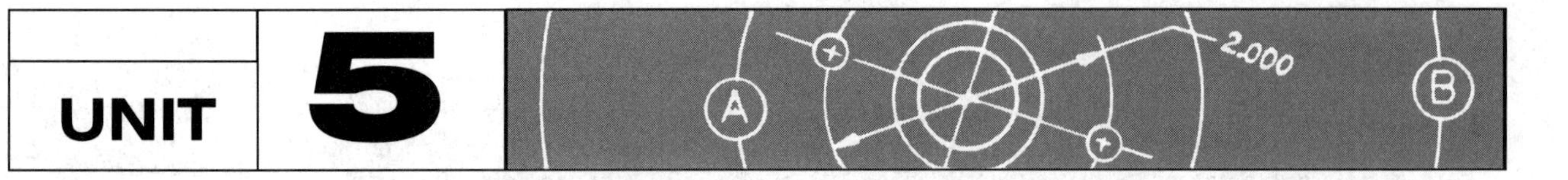

ORTHOGRAPHIC PROJECTION OF INCLINED PLANES

Thus far we have concentrated our efforts on objects containing only principal planes that are at right angles to one another. When viewed from any of the six principal orthographic views, they will appear either as true shapes or as edges (straight lines). However, few objects will consist entirely of principal planes, so we must progress to other shapes.

FORESHORTENED VIEWS

Inclined planes are at right angles to two of the six principal orthographic views, thus appearing as an edge on those two views. The remaining four views, however, will show the inclined planes foreshortened. In the illustration below, plane Ⓐ appears as an edge in the top and bottom views, and foreshortened in the other four—twice visible and twice hidden.

On page 97 you will see the similarity between the right-side views of a principal plane and an inclined plane. To avoid confusion and possible misinterpretation, you should always study adjacent views carefully to determine the actual shape of each plane.

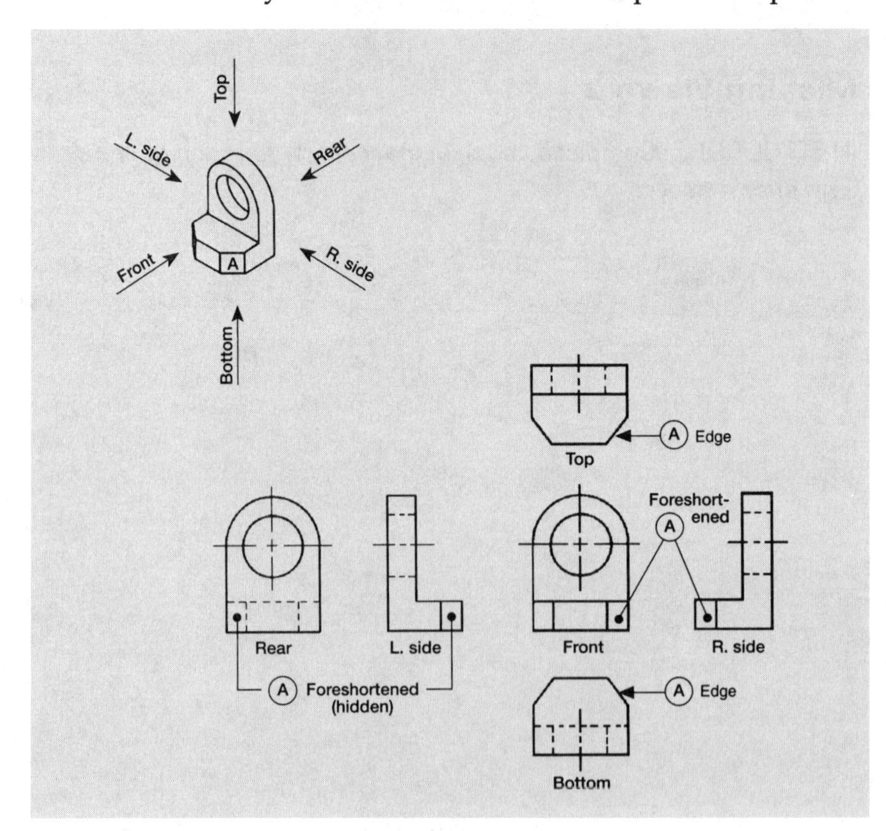

Observe the two views below that are drawn alike. Can you recognize that the middle line is representing the edge of principal plane Ⓥ on the first illustration, and the intersection of planes Ⓦ and Ⓧ on the second? By comparison, observe that there is no line of intersection created by curved surface Ⓨ running tangent with flat plane Ⓩ. This can also be seen in the illustration on page 96.

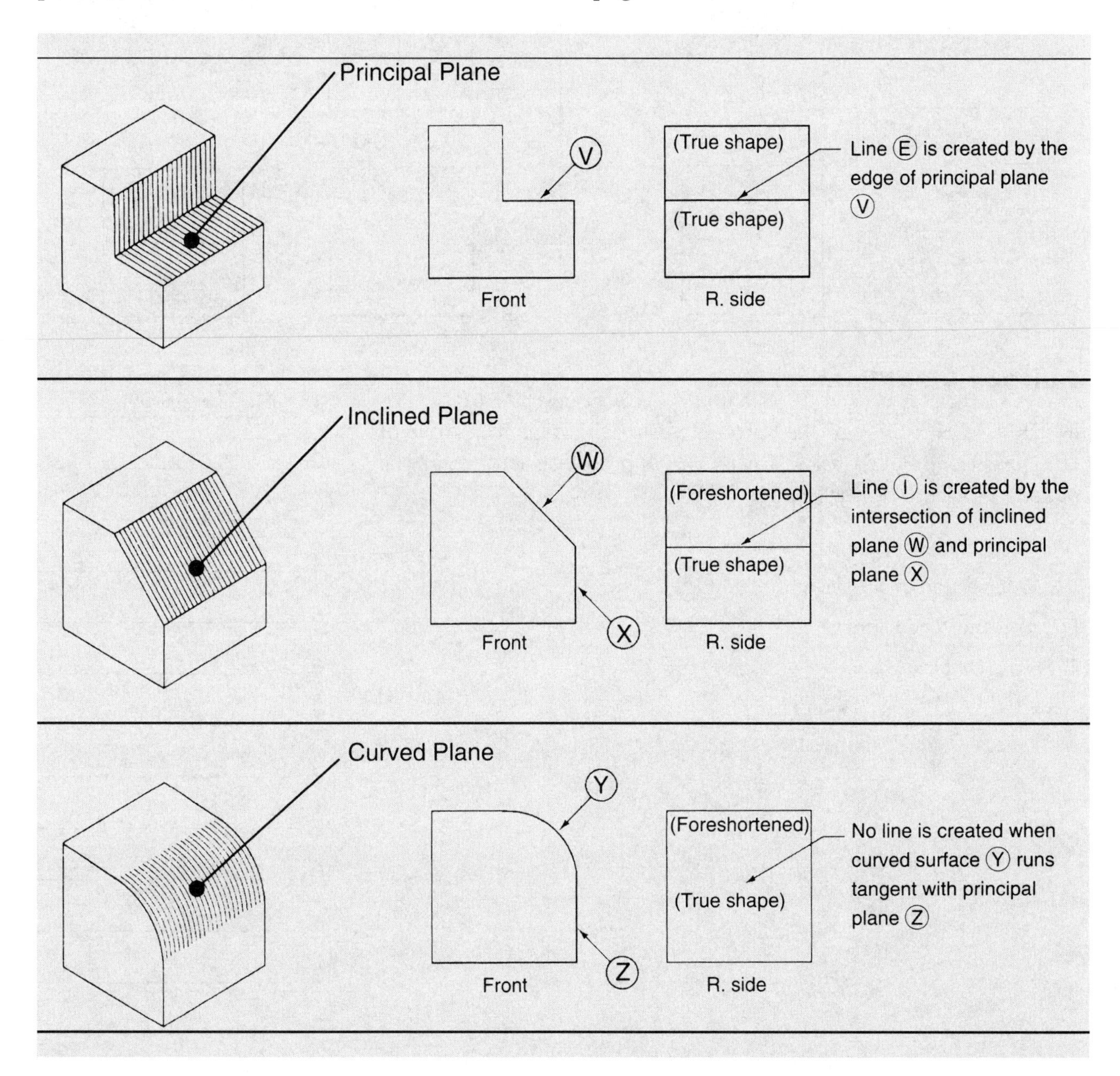

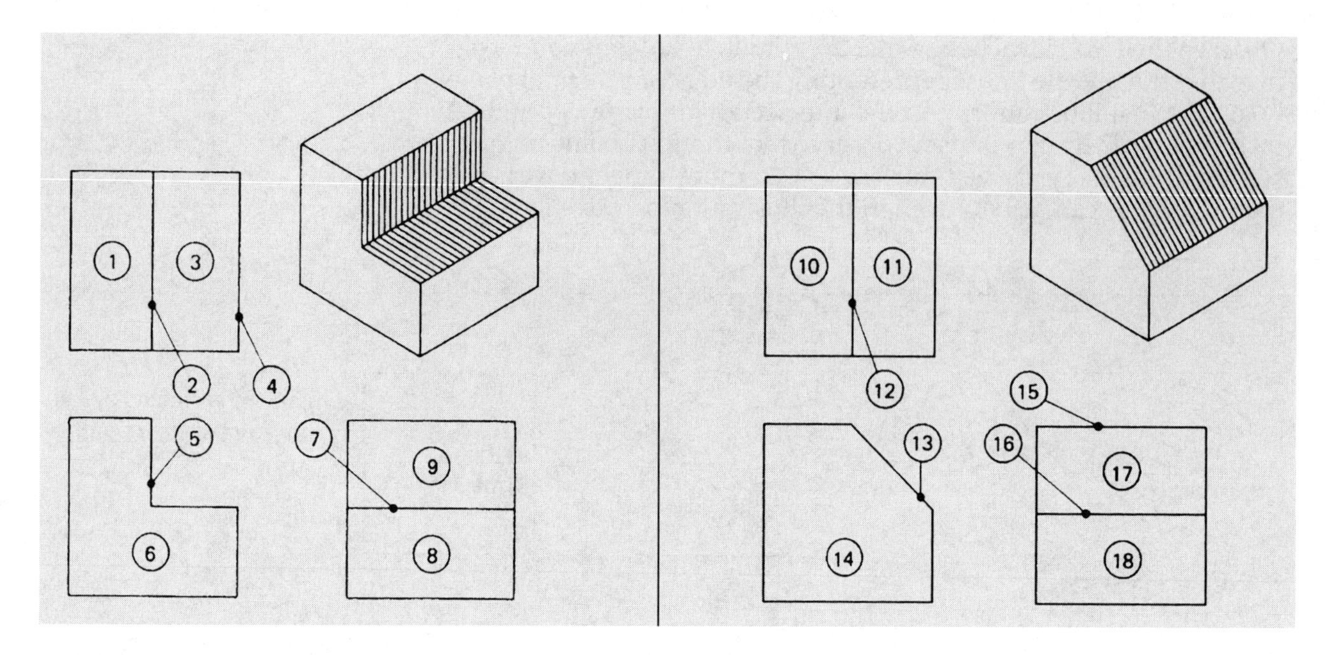

Surface Identification 4

INSTRUCTIONS: Each of the balloons shown above identifies one of the conditions listed below. Enter the appropriate letters in the spaces provided. (Refer to page 97.)

E: Edge view of surface

I: Intersection of surfaces

F: Foreshortened surface

T: True shape surface

1. T *(example)*
2. ______
3. ______
4. ______
5. ______
6. ______
7. ______
8. ______
9. ______
10. ______
11. ______
12. ______
13. ______
14. ______
15. ______
16. ______
17. ______
18. ______

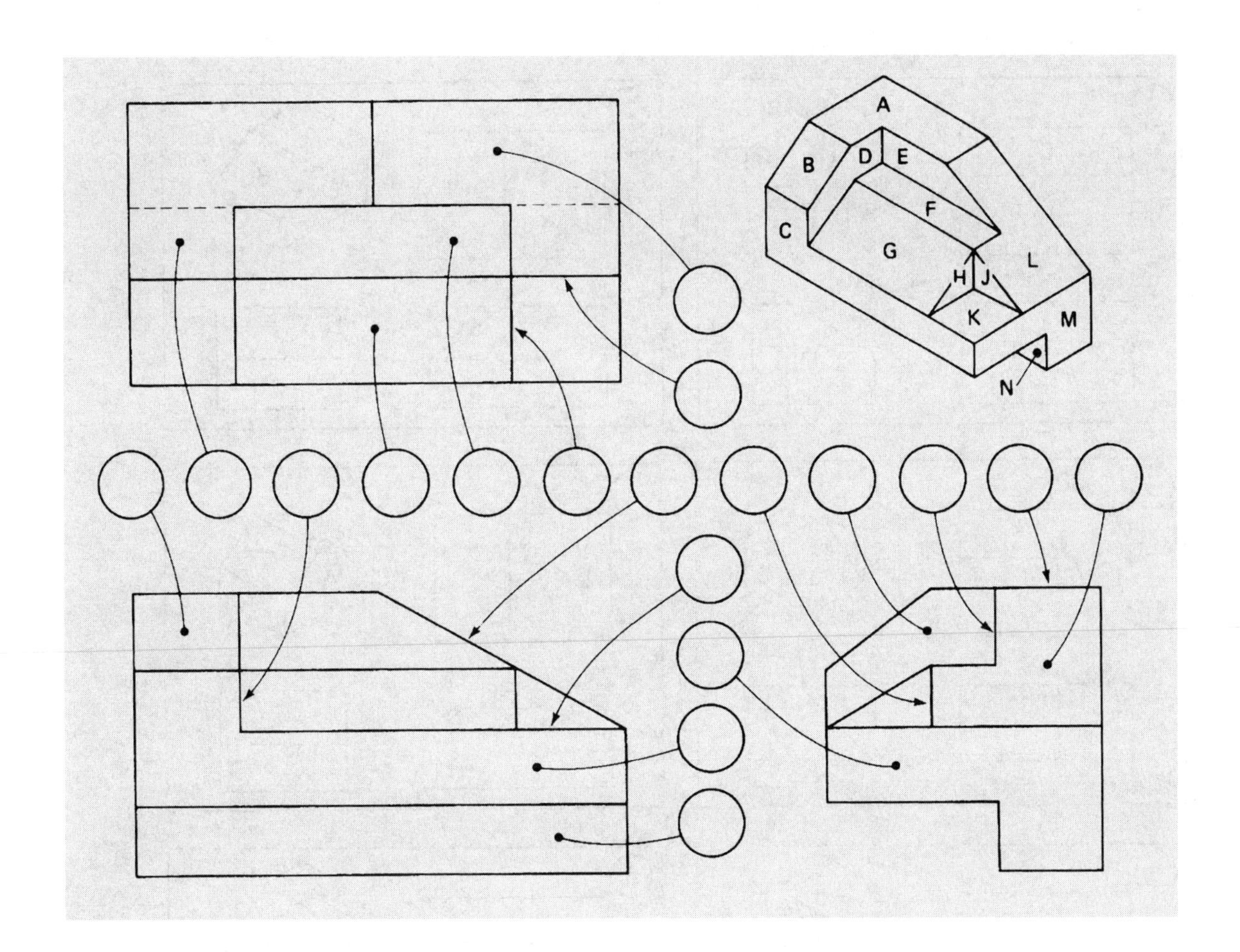

Surface Identification 5

INSTRUCTIONS: Enter the letters from the pictorial view into the corresponding balloons on the orthographic views.

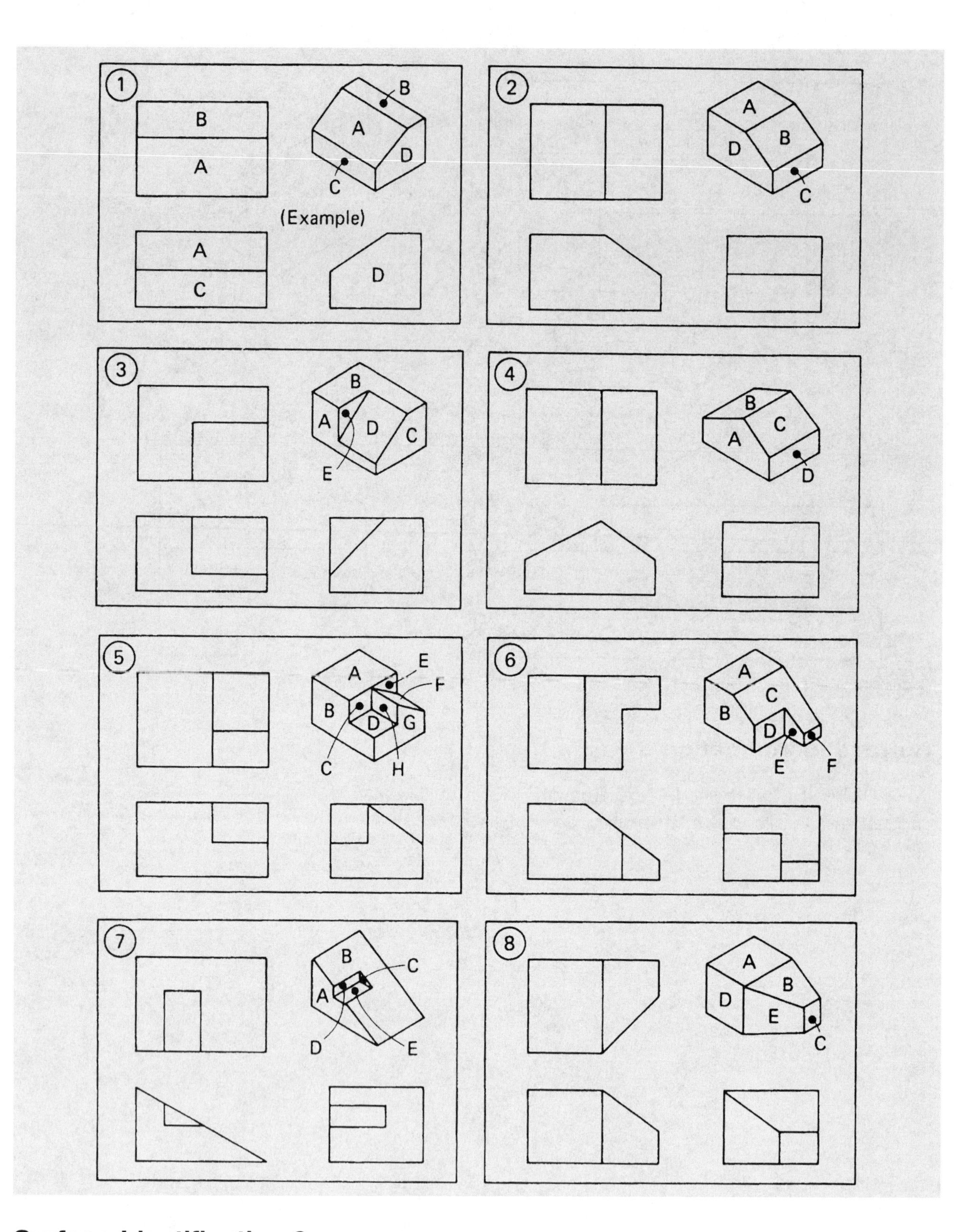

Surface Identification 6

INSTRUCTIONS: Place the letters from the pictorial views onto their respective surfaces in the orthographic views.

ANGULAR DIMENSIONS

Inclined planes may be dimensioned by the coordinate method or by the angular method. The coordinate method uses two linear dimensions to the sides of a right triangle, as in illustration (a) below. The angular method uses one linear dimension and one angular dimension, as in illustration (b). The coordinate method controls accuracy more than the angular method, because angular tolerance allows the variation to increase with the distance from the vertex. Observe the example in illustration (c).

Most angular dimensions will be held from a vertical or horizontal surface or centerline. This, then, makes it easy to calculate the other angle, by subtracting the known angle from 90°. (All triangles contain 180°, and if the right angle utilizes 90°, the sum of the other two angles must equal 90°.) For example, if an angle is dimensioned as 30° from horizontal, it must be 60° from vertical (30° + 60° = 90°). An angular dimension of 45° would obviously be the same for both vertical and horizontal (45° + 45° = 90°). Since both angles are the same, both sides of a right triangle are also the same, and this is the key to the solution of (apparently) missing dimensions on a drawing. (They actually would be double-dimensions if they were included.)

Each degree may be subdivided into 60 minutes, and each minute may be subdivided into 60 seconds. The symbols used on drawings are the (°) degree, the (′) minute, and the (″) second. An angle of less than 1° will still include the degree symbol if the minute symbol is used (for example, 0°30′). The example used may appear optionally as 1/2° on a fractionally dimensioned drawing, or as .5° or 0.5° on decimal or metric drawings.

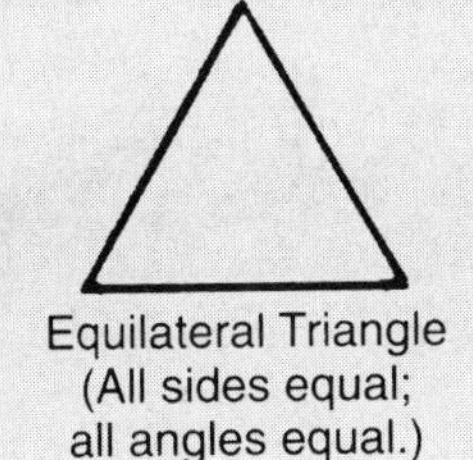

Equilateral Triangle
(All sides equal;
all angles equal.)

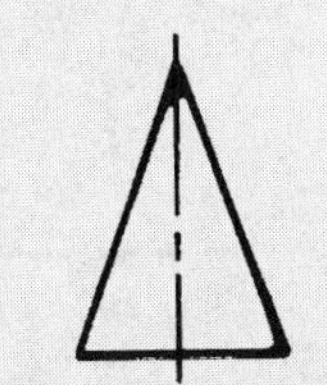

Isosceles Triangle
(2 sides equal;
2 angles equal.)

Right Triangle
(One 90° angle)

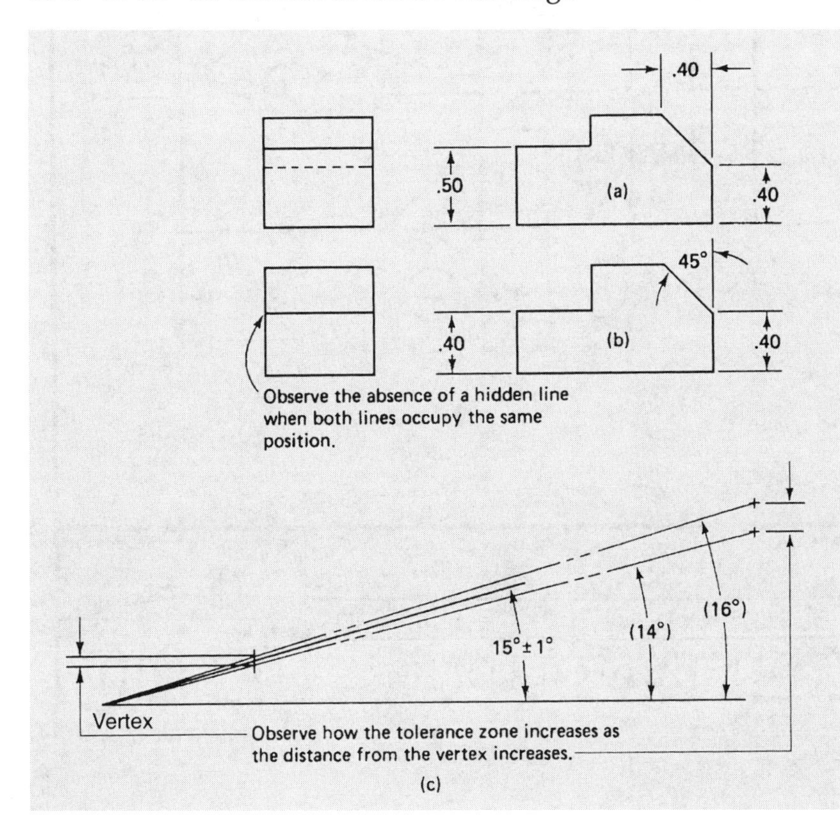

(c)

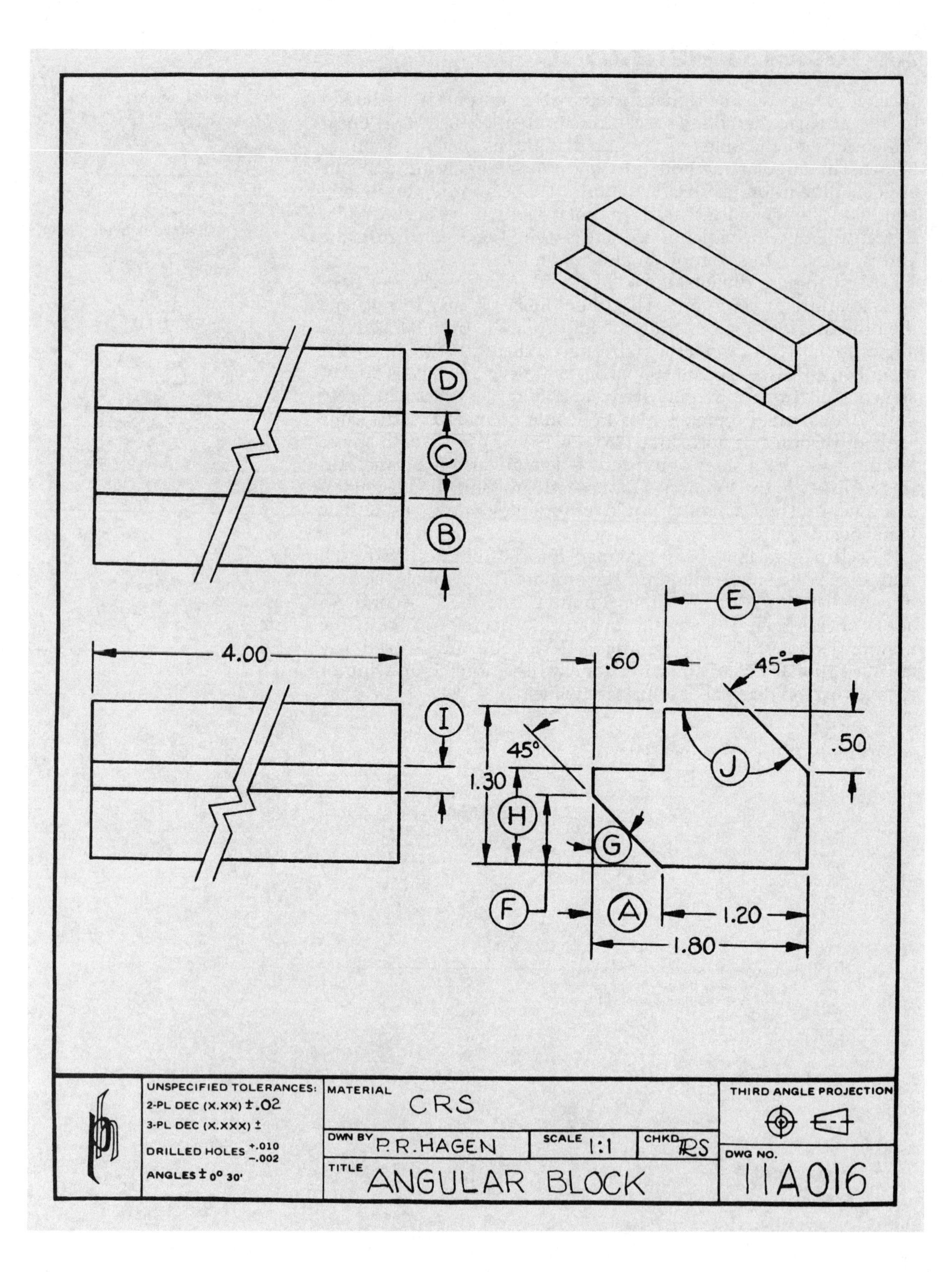

D
C
B
4.00
I
.60
E
45°
.50
J
45°
1.30
H
G
F
A
1.20
1.80
UNSPECIFIED TOLERANCES:
2-PL DEC (X.XX) ±.02
3-PL DEC (X.XXX) ±
DRILLED HOLES +.010 −.002
ANGLES ± 0° 30'
MATERIAL CRS
DWN BY P.R.HAGEN
SCALE 1:1
CHKD RS
TITLE ANGULAR BLOCK
THIRD ANGLE PROJECTION
DWG NO. 11A016

INSTRUCTIONS: Refer to drawing 11A016 to answer the following questions.

1. Are the inclined planes dimensioned by the coordinate method or the angular method? 1. ______
2. What is the tolerance assigned to angular dimensions? 2. ______
3. What is the tolerance on the linear dimensions? 3. ______
4. What type of line was drawn on the front and top views to permit them to be shortened? 4. ______
5. How many hidden lines would appear on a bottom view? 5. ______
6. How many hidden lines would appear on a left-side view? 6. ______
7. How many inclined planes does the object contain? 7. ______
8. How many principal planes does the object contain? 8. ______
9. How much tolerance accumulates with dimension Ⓐ? 9. ______
10. What is the MMC of the overall height? 10. ______

INSTRUCTIONS: Enter the dimensions for the following letters.

Ⓐ ______

Ⓑ ______

Ⓒ ______

Ⓓ ______

Ⓔ ______

Ⓕ ______

Ⓖ ______

Ⓗ ______

Ⓘ ______

Ⓙ ______

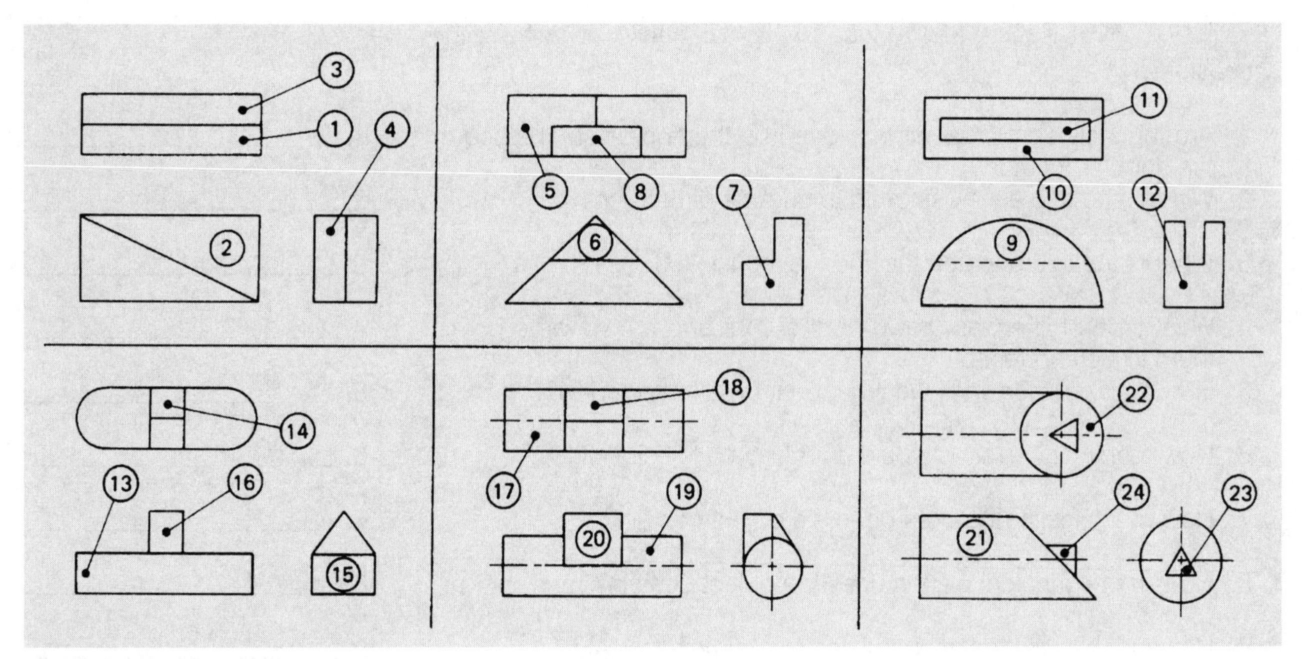

Surface Identification 7

INSTRUCTIONS: Each of the balloons shown above identifies one of the surfaces listed below. Enter the appropriate letters in the space provided.

PS: Principal surface

IS: Inclined surface

CS: Curved surface

1. IS *(example)*
2. __________
3. __________
4. __________
5. __________
6. __________
7. __________
8. __________
9. __________
10. __________
11. __________
12. __________
13. __________
14. __________
15. __________
16. __________
17. __________
18. __________
19. __________
20. __________
21. __________
22. __________
23. __________
24. __________

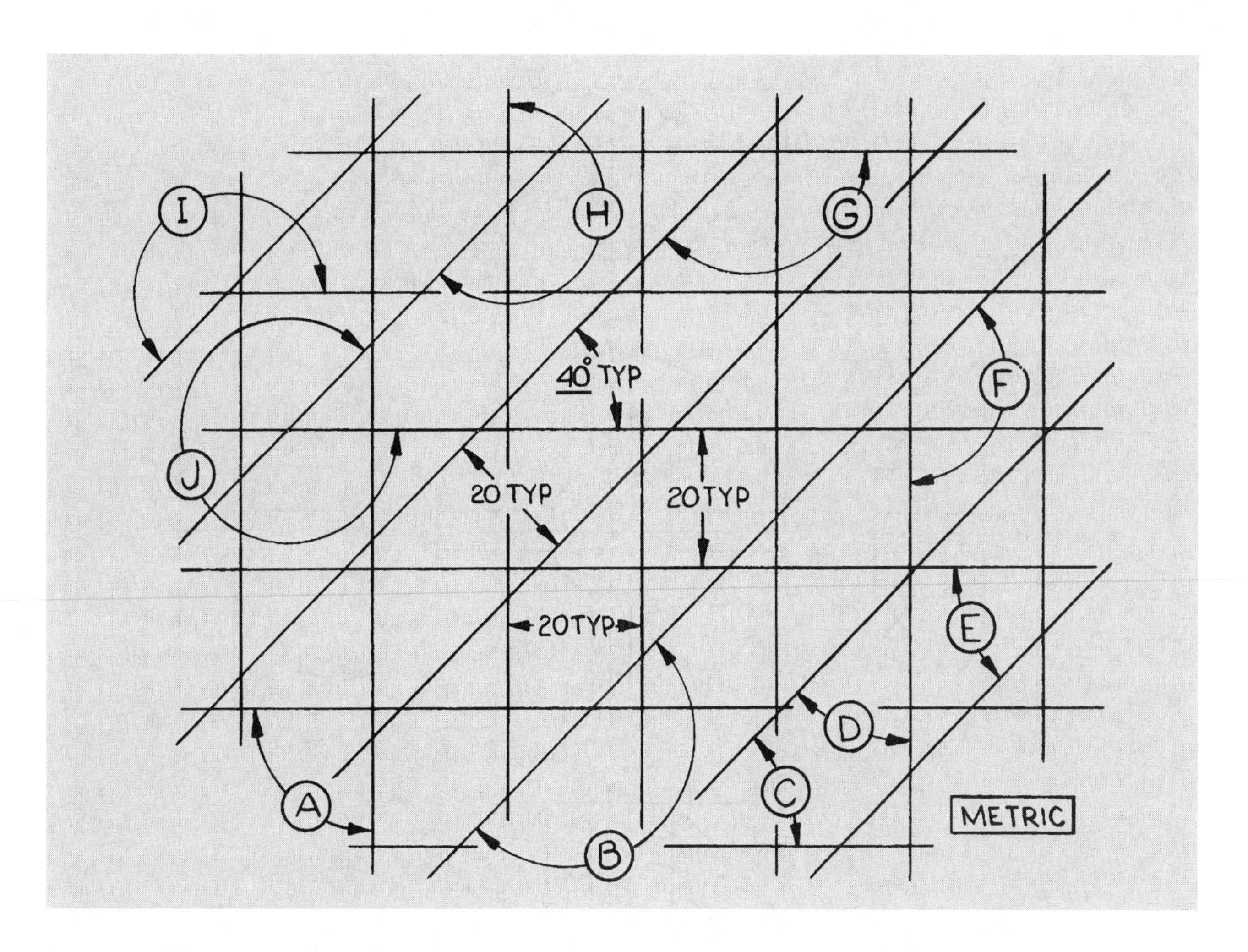

Angular Calculations 1

INSTRUCTIONS: Drawn above are three sets of parallel lines (verified by the TYP dimensions). Calculate the correct angle for each of the following letters:

Ⓐ ______________________

Ⓑ ______________________

Ⓒ ______________________

Ⓓ ______________________

Ⓔ ______________________

Ⓕ ______________________

Ⓖ ______________________

Ⓗ ______________________

Ⓘ ______________________

Ⓙ ______________________

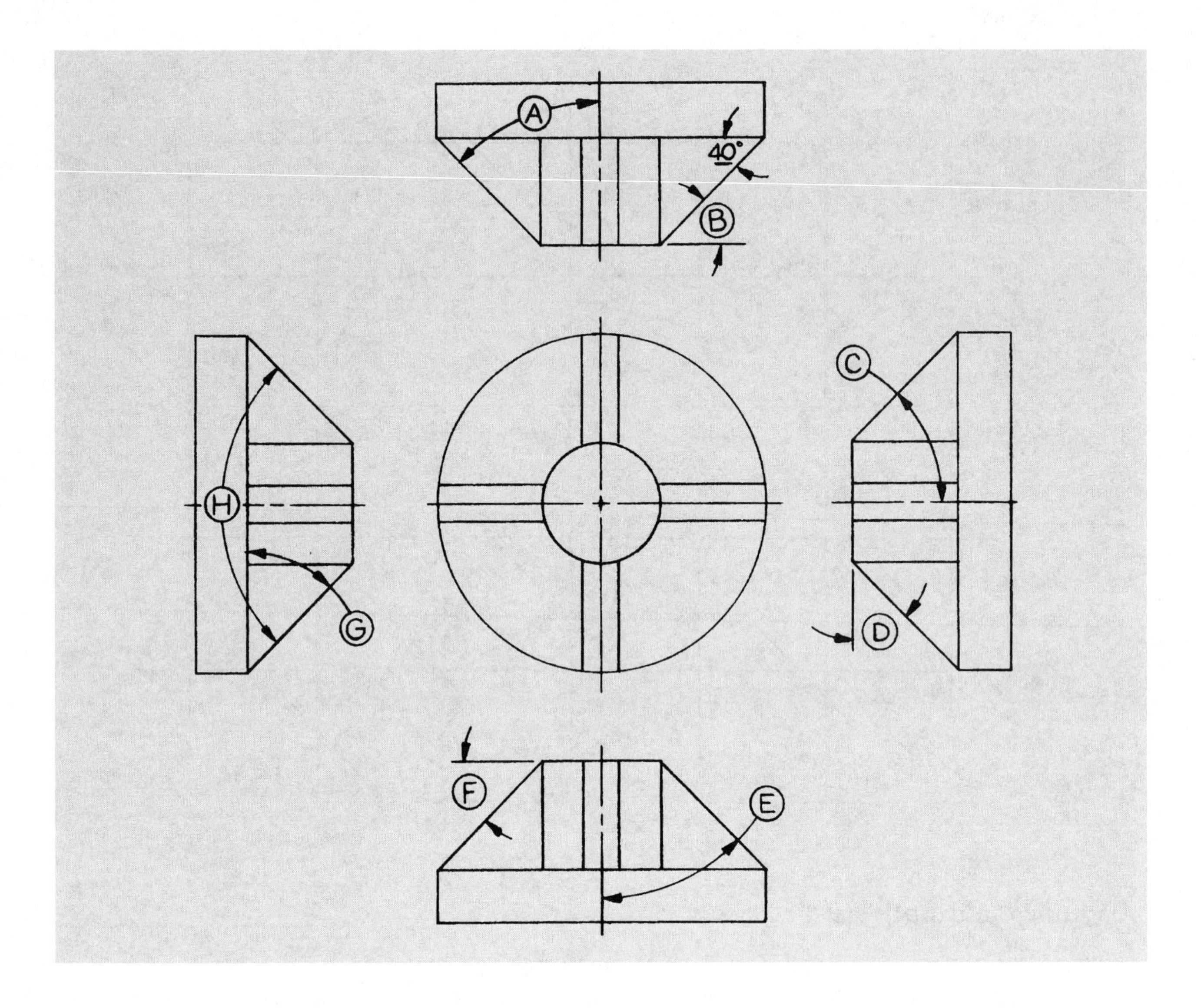

Angular Calculations 2

INSTRUCTIONS: Based on the 40° angular dimension (NTS) found on the top view, calculate the angular dimensions for the following letters.

Ⓐ ______________________

Ⓑ ______________________

Ⓒ ______________________

Ⓓ ______________________

Ⓔ ______________________

Ⓕ ______________________

Ⓖ ______________________

Ⓗ ______________________

INCLUDED ANGLE

The included angle is an angular measurement taken from one surface to another, or across a conical shape, but always less than 180°. Drill points and countersinks are commonly dimensioned by their included angles. See the examples illustrated below.

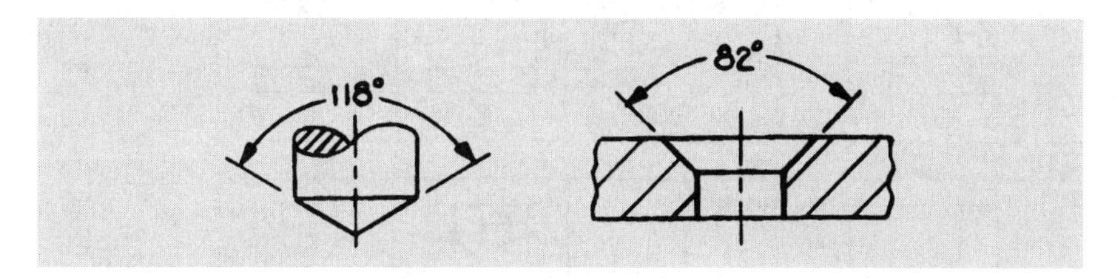

Angular Calculations 3

INSTRUCTIONS: Drawn below are several objects containing angular dimensions. Calculate the following angles:

1. Included angle between Ⓐ and Ⓑ — 1. ________
2. Included angle between Ⓐ and Ⓒ — 2. ________
3. Included angle between Ⓐ and Ⓓ — 3. ________
4. Angle Ⓔ — 4. ________
5. Angle Ⓕ — 5. ________
6. Angle Ⓖ — 6. ________
7. Angle Ⓗ — 7. ________
8. Angle Ⓘ — 8. ________
9. Angle Ⓙ — 9. ________
10. Angle Ⓚ — 10. ________

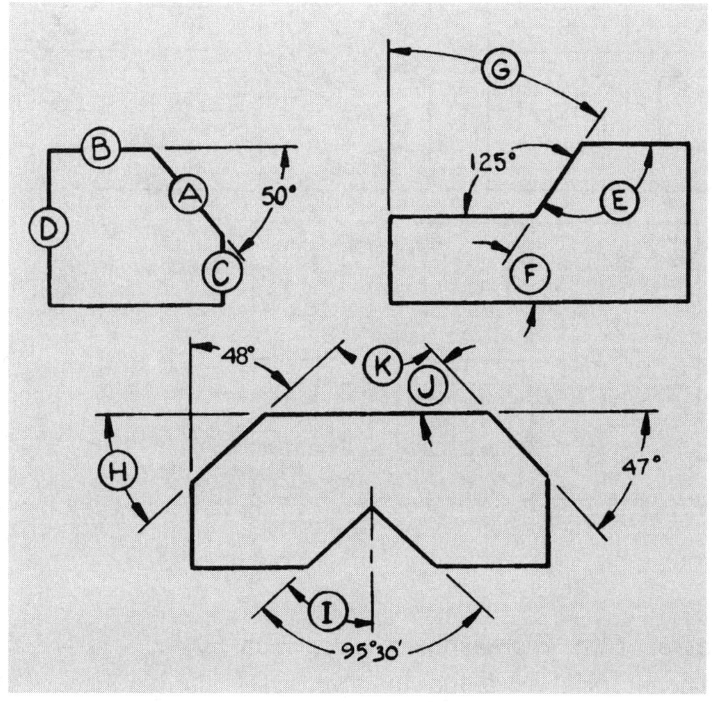

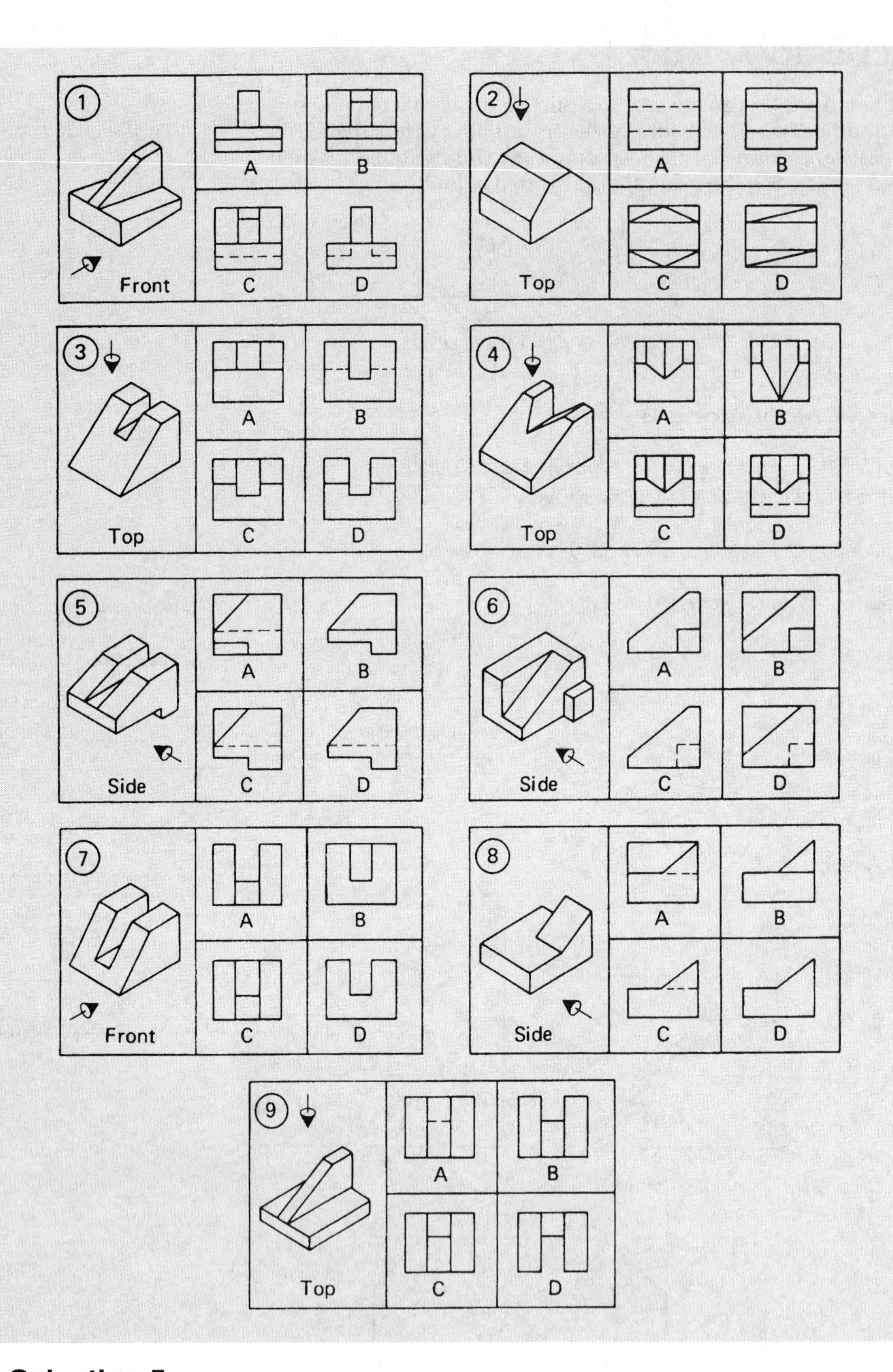

View Selection 5

INSTRUCTIONS: Circle the letter that represents the correct orthographic view for each of the pictorial views shown above.

Missing Lines 1

INSTRUCTIONS: Add the missing lines to the top and/or the right-side views only. Include the hidden lines.

SLOTS AND GROOVES

Dovetail slots and tee slots (T-slots) are both commonly used in the machine trades to provide reciprocating movement, such as the cross feed on a lathe. A slot or groove may be cut into the top, bottom, or side of an object, but its depth is always measured from the outside surface from which it is cut (see the examples below). The profile view of a slot or groove is the most descriptive view, showing both depth and width. The length of a slot or groove would be the distance over which it extends (any view adjacent to the profile view of a feature will show that feature's length). The included angle of a dovetail slot or vee groove (V-groove) would be the angle between the two sides.

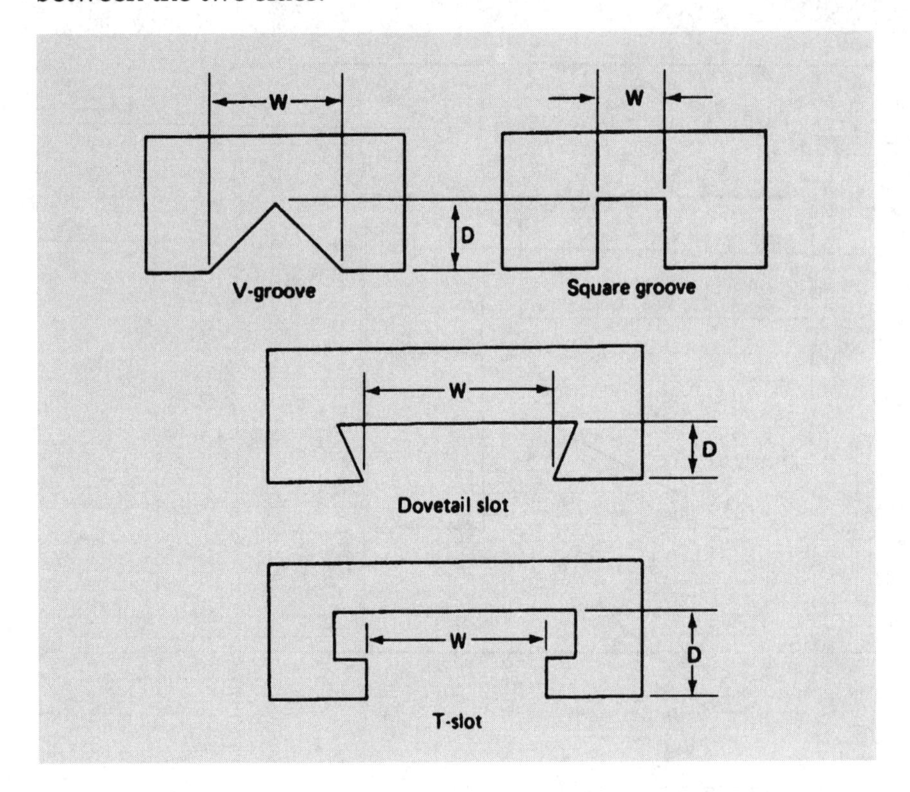

Observe that width (W) and depth (D) are both measured on the profile views of slots and grooves.

NOTE: Formulas for checking dovetails and V-grooves with pins can be found in the Mathematics Appendix beginning on p. 373.

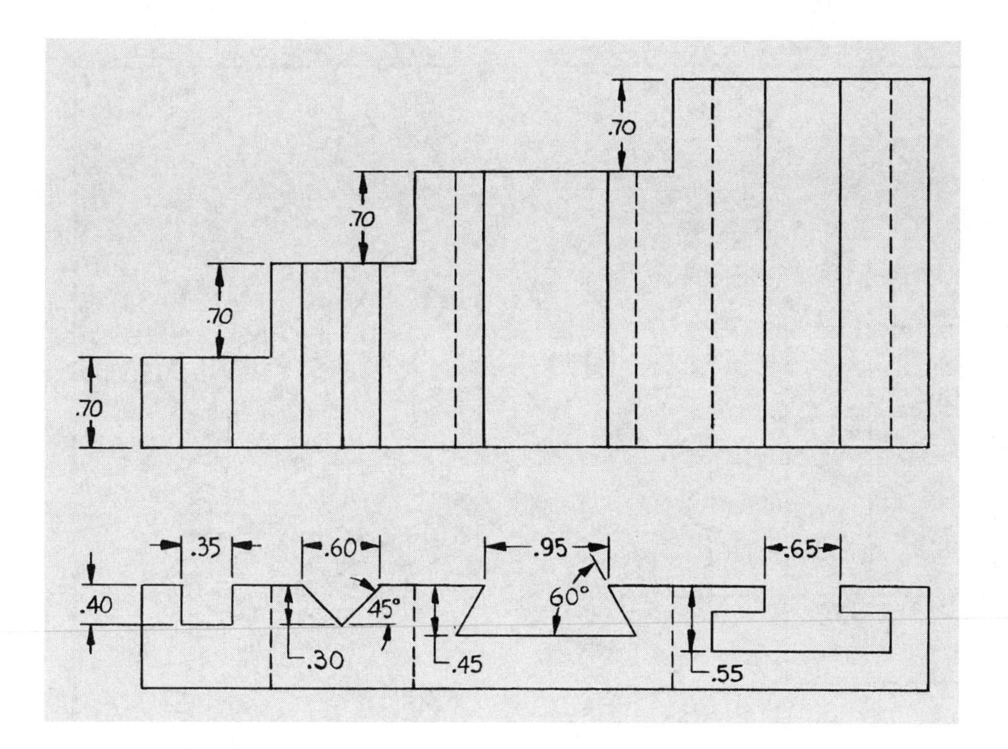

Groove and Slot Exercise

INSTRUCTIONS: Refer to the two views above, then enter the appropriate dimensions in the table below.

Groove/Slot	Width	Depth	Length
SQ Groove			
V-Groove			
Dovetail Slot			
T-Slot			

Optional Math Exercises:

1a. Using the dimensions from the dovetail slot drawn above, solve for the Measurement Between Pins (MBP) using .300∅ pins.

2a. Using the dimensions for the V-groove drawn above, determine the distance (H) from the top of a 1.00∅ pin to the top surface of this block.

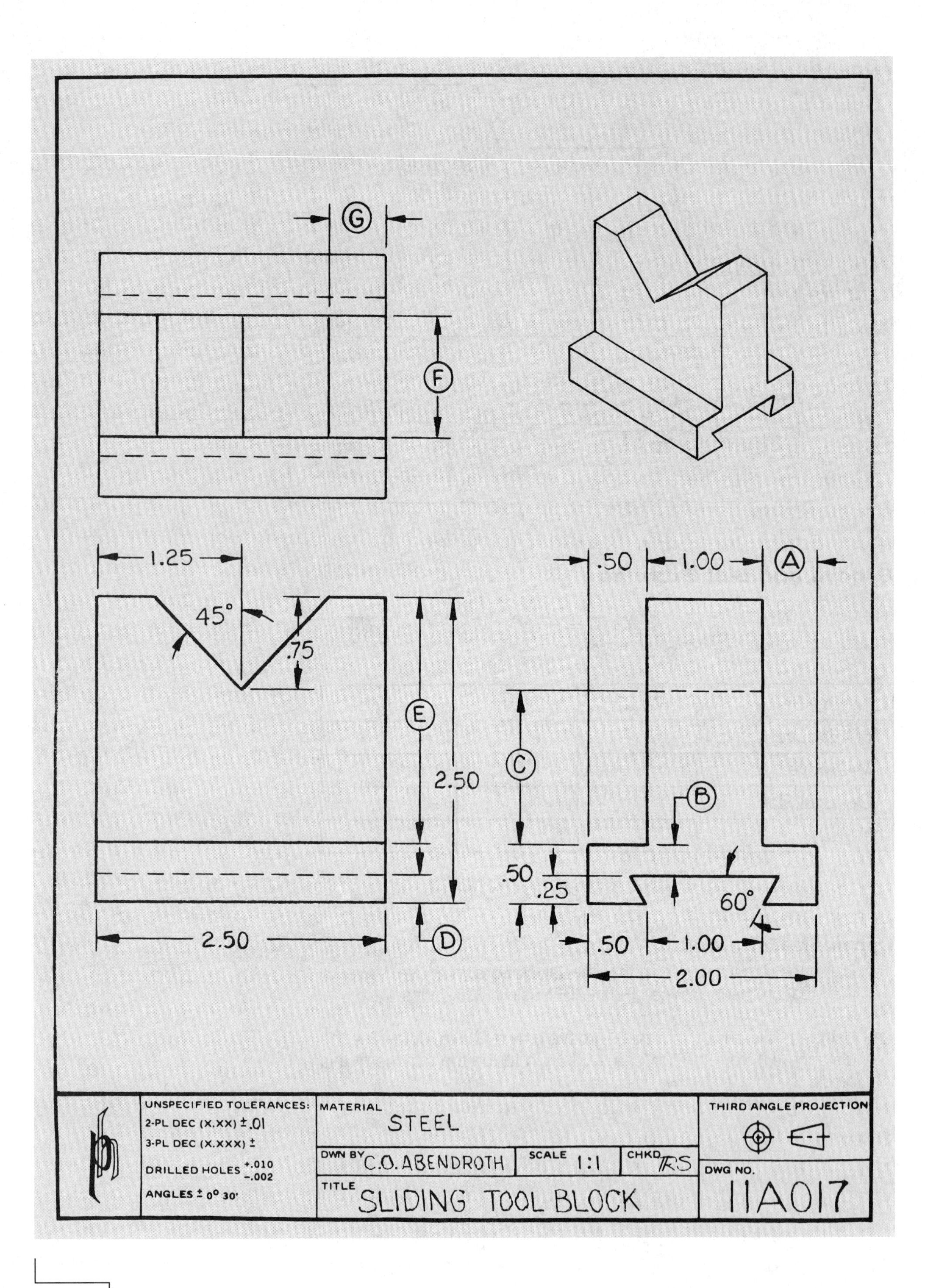
G
F
1.25
45°
.75
E
2.50
C
B
.50
.25
60°
D
2.50
.50
1.00
A
.50
1.00
2.00
UNSPECIFIED TOLERANCES:
2-PL DEC (X.XX) ± .01
3-PL DEC (X.XXX) ±
DRILLED HOLES +.010 −.002
ANGLES ± 0° 30'
MATERIAL
STEEL
DWN BY C.O. ABENDROTH
SCALE 1:1
CHKD RS
TITLE SLIDING TOOL BLOCK
THIRD ANGLE PROJECTION
DWG NO.
11A017

INSTRUCTIONS: Refer to drawing 11A017 to answer the following questions.

1. What is the overall height of the block? 1. ______
2. What is the depth of the block? 2. ______
3. What is the depth of the vee groove? 3. ______
4. What is the depth of the dovetail slot? 4. ______
5. What is the width of the vee groove? 5. ______
6. What is the length of the dovetail slot? 6. ______
7. What is the length of the vee groove? 7. ______
8. What is the included angle of the vee groove? 8. ______
9. What is the included angle of the dovetail slot? 9. ______
10. What view shows the profile of the vee groove? 10. ______
11. What view shows the profile of the dovetail slot? 11. ______
12. How many inclined surfaces does the block contain? 12. ______
13. How much tolerance accumulates with dimension Ⓐ? 13. ______

INSTRUCTIONS: Enter the dimensions for the following letters.

Ⓐ ______

Ⓑ ______

Ⓒ ______

Ⓓ ______

Ⓔ ______

Ⓕ ______

Ⓖ ______

Optional Math Exercises:

1a. Using a 1.00∅ pin in the V-groove, calculate the distance from the top of this pin to the top of the Sliding Tool Block.

2a. Using .250∅ pins, calculate the measurement between pins (MBP) after the dovetail slot is properly machined.

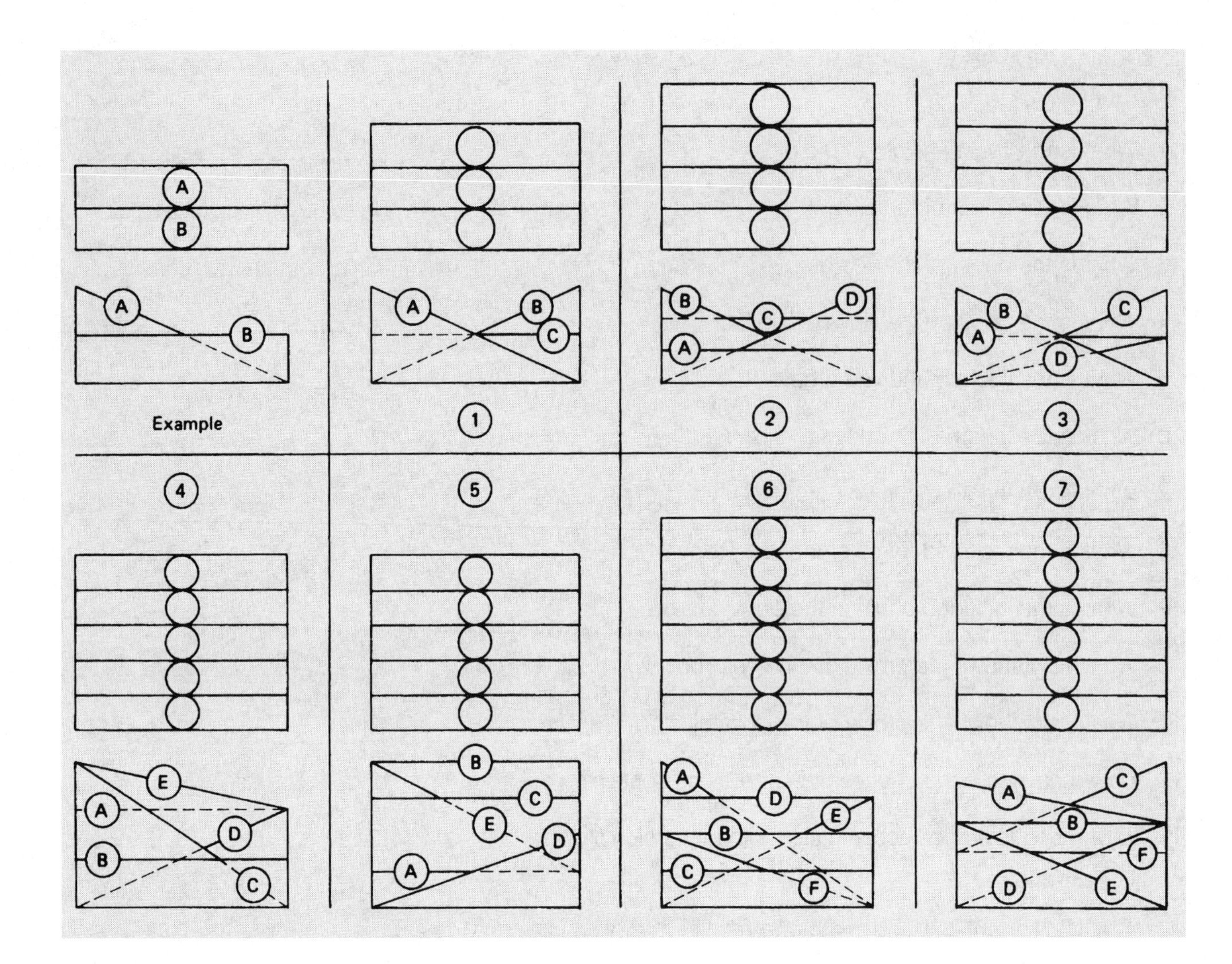

Surface Identification 8

INSTRUCTIONS: Place the letters from the front views into the balloons representing their positions in the top views. (See the example shown above.)

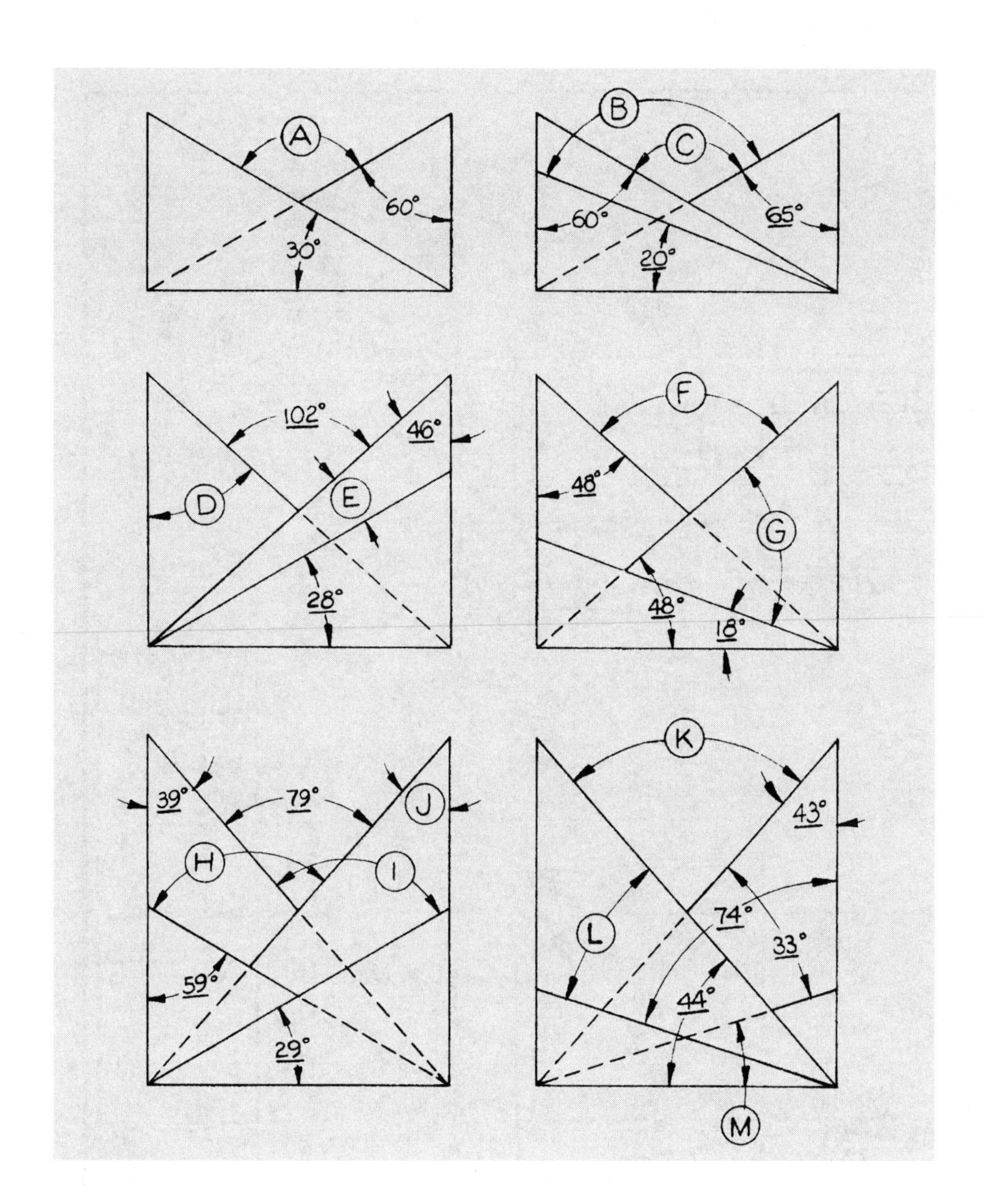

Angular Calculations 4

INSTRUCTIONS: Refer to the drawings above to calculate the angular dimensions for the following letters.

Ⓐ ____________________ Ⓗ ____________________

Ⓑ ____________________ Ⓘ ____________________

Ⓒ ____________________ Ⓙ ____________________

Ⓓ ____________________ Ⓚ ____________________

Ⓔ ____________________ Ⓛ ____________________

Ⓕ ____________________ Ⓜ ____________________

Ⓖ ____________________

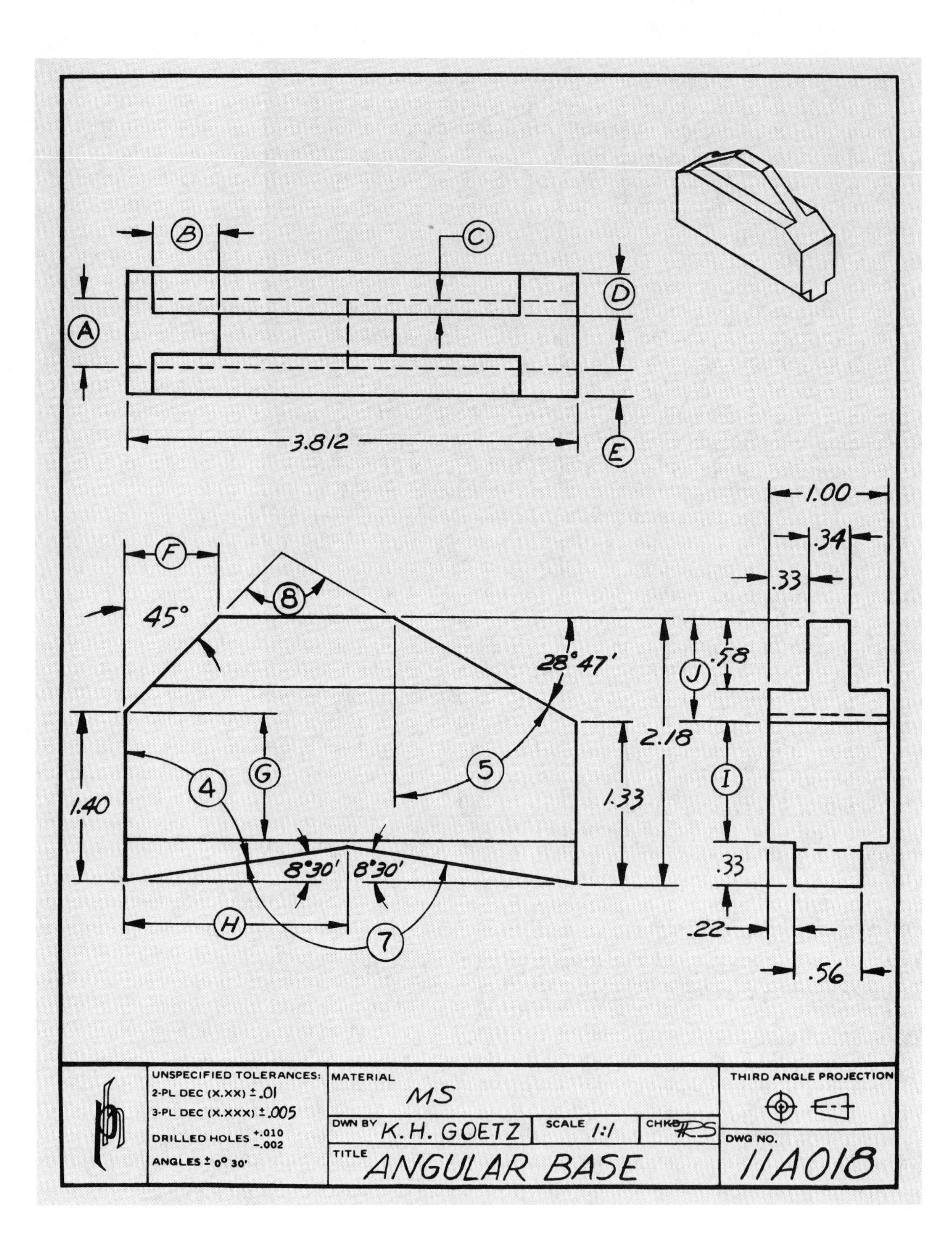

B
C
A
D
3.812
E
1.00
.34
.33
F
8
45°
28°47'
.58
J
2.18
G
4
5
1.40
1.33
I
8°30'
8°30'
.33
H
7
.22
.56
UNSPECIFIED TOLERANCES:
2-PL DEC (X.XX) ±.01
3-PL DEC (X.XXX) ±.005
DRILLED HOLES +.010 −.002
ANGLES ± 0° 30'
MATERIAL
MS
DWN BY K.H. GOETZ
SCALE 1:1
CHKD
TITLE ANGULAR BASE
THIRD ANGLE PROJECTION
DWG NO.
11A018

INSTRUCTIONS: Refer to drawing 11A018 to answer the following questions.

1. What is the tolerance on the height dimension? 1. ____________
2. What is the MMC of the 3.812 dimension? 2. ____________
3. How many inclined planes does the object contain? 3. ____________
4. What angle is the bottom of the base from vertical? (See balloon 4.) 4. ____________
5. What angle is the 28°47′ surface from vertical? (See balloon 5.) 5. ____________
6. What is the maximum angular dimension for the 28°47′ surface? (Include the tolerance in your answer.) 6. ____________
7. What is the included angle between the two bottom inclined surfaces? (See balloon 7.) 7. ____________
8. What is the included angle between the two top inclined surfaces? (See balloon 8.) 8. ____________
9. How many of the lines in the side view represent intersections of surfaces only, not edges? 9. ____________
10. How many of the lines in the top view represent intersections of surfaces only, not edges? 10. ____________

INSTRUCTIONS: Enter the dimensions for the following letters.

Ⓐ ____________

Ⓑ ____________

Ⓒ ____________

Ⓓ ____________

Ⓔ ____________

Ⓕ ____________

Ⓖ ____________

Ⓗ ____________

Ⓘ ____________

Ⓙ ____________

Optional Math Exercises:

1a. Determine the distance from the top of a 1.00∅ pin placed in the vee to the opposite side of the Angular Base.

2a. Determine the depth of the vee.

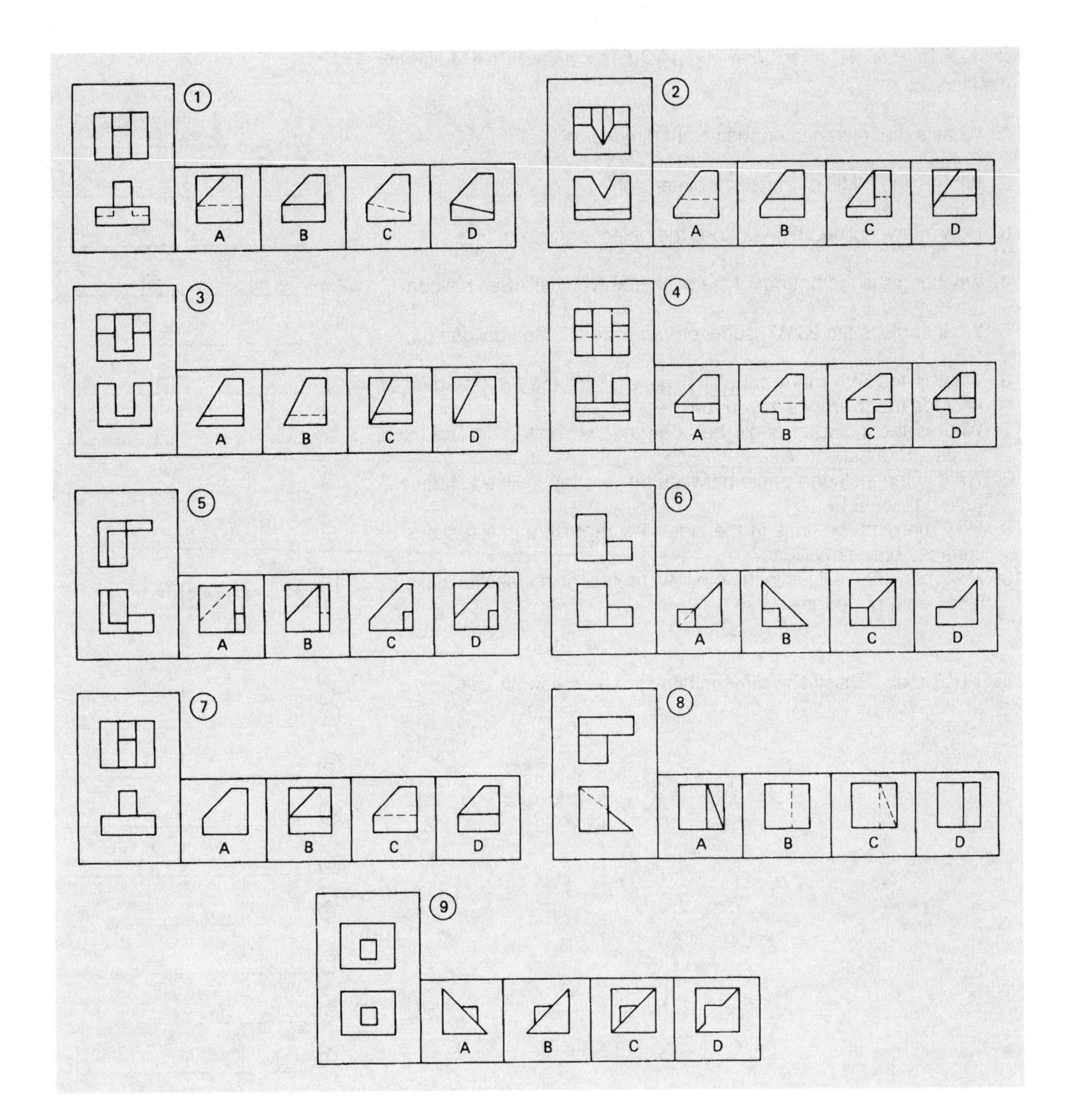

View Selection 6

INSTRUCTIONS: Circle the letter that represents the correct right-side view for each of the shapes shown above.

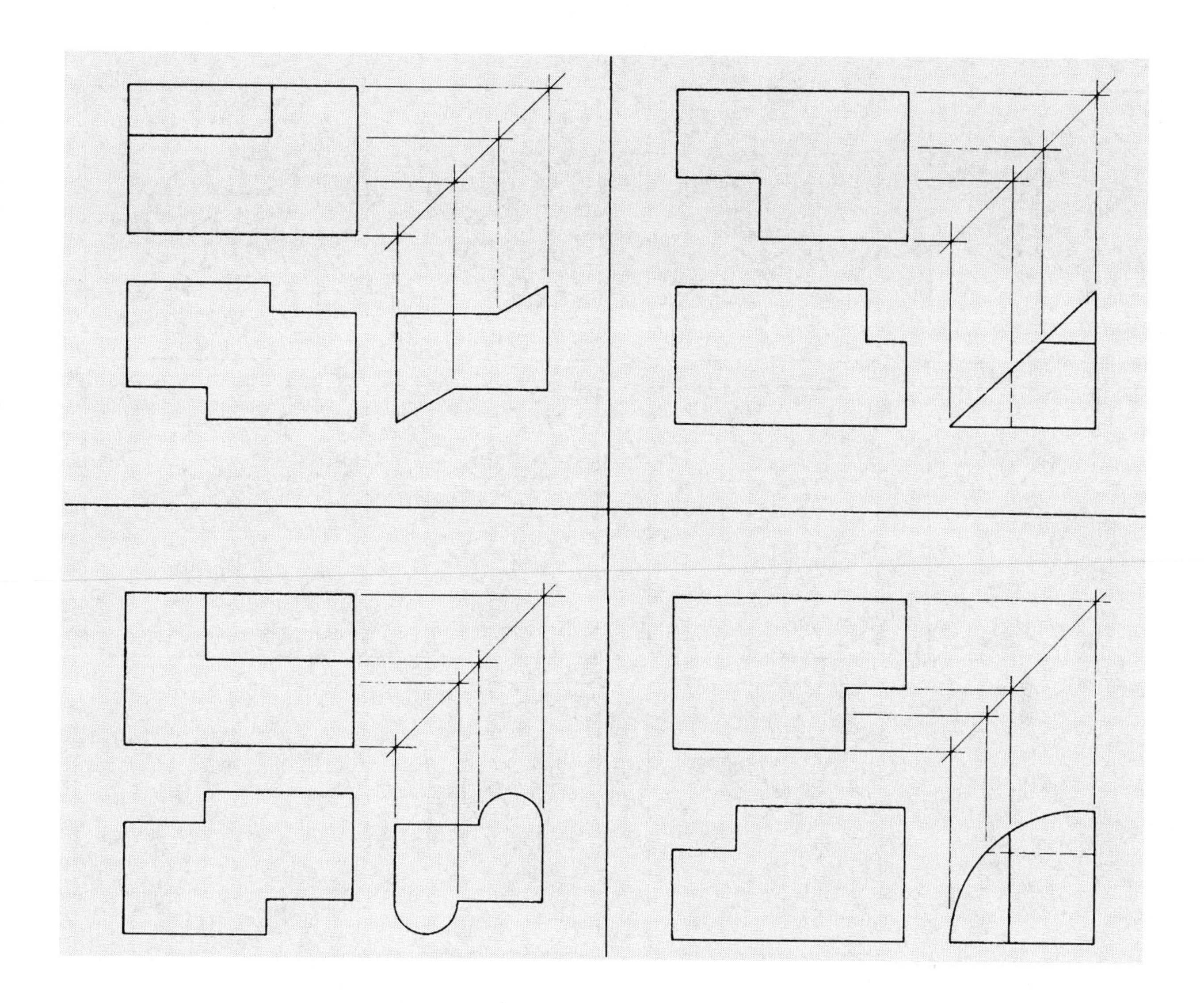

Missing Lines 2

INSTRUCTIONS: Add the missing lines to complete each view. Include all hidden lines.

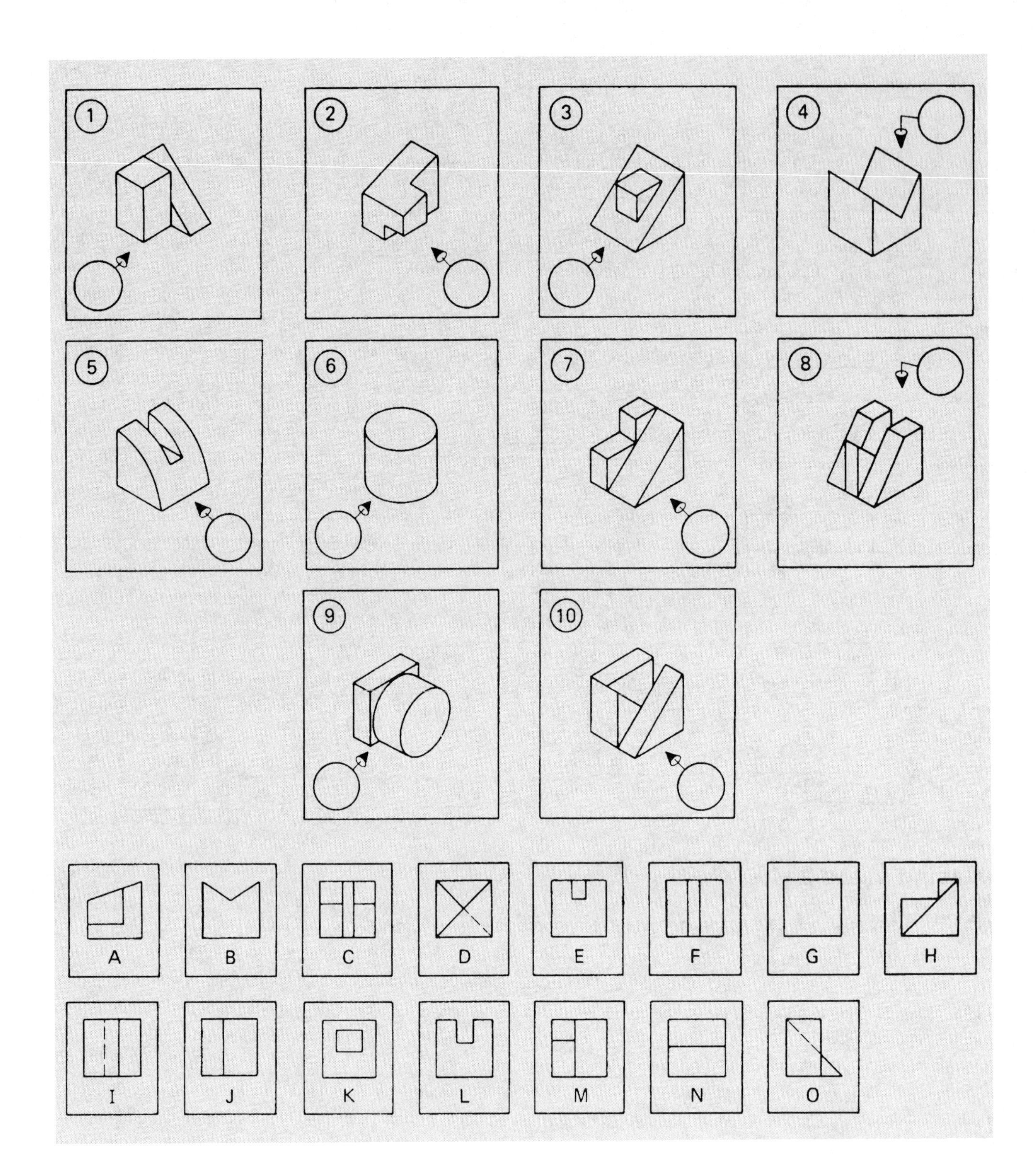

View Selection 7

INSTRUCTIONS: Select the correct orthographic view to correspond with the arrowhead for each pictorial view shown above. Enter the correct letters in the balloons provided.

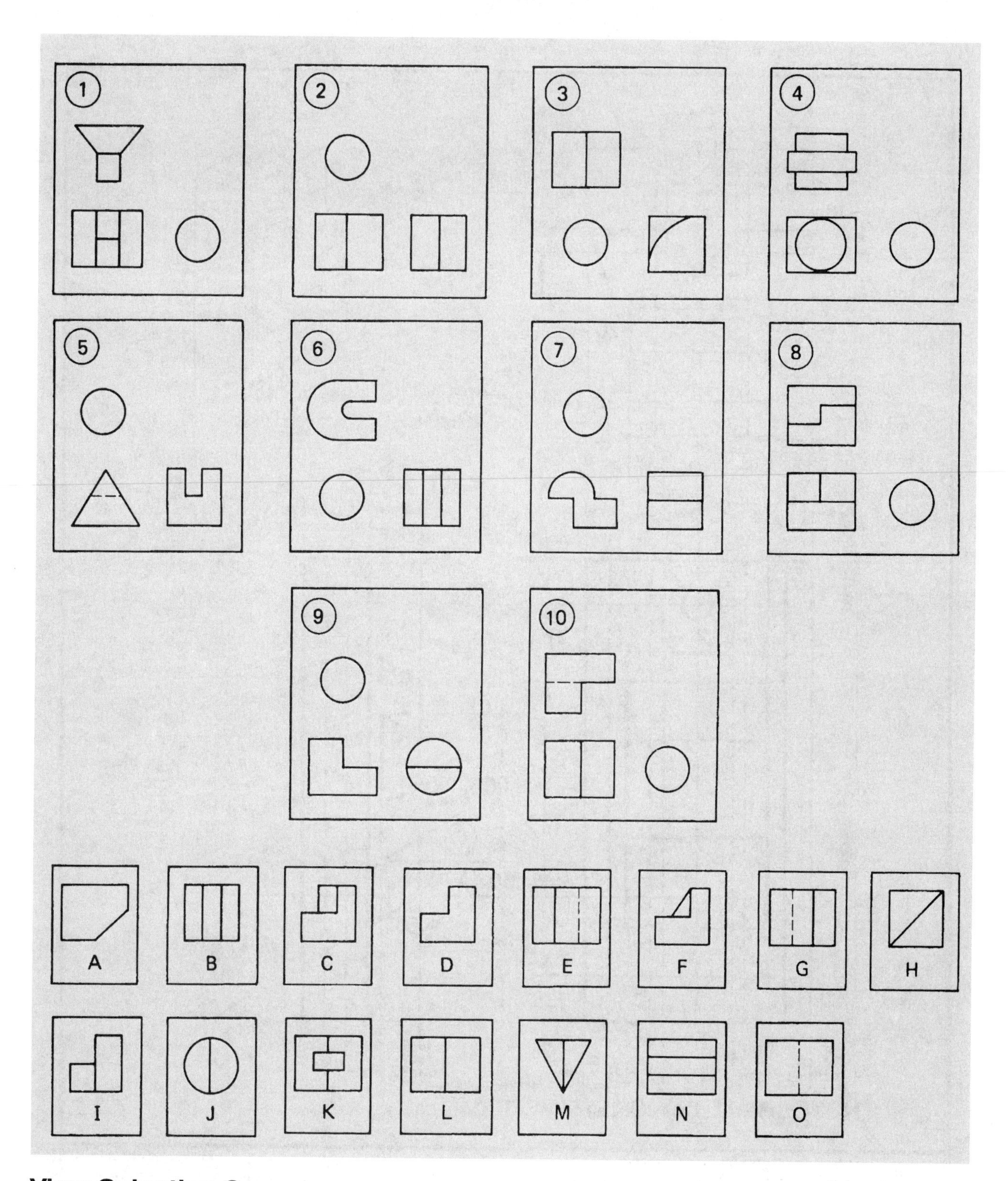

View Selection 8

INSTRUCTIONS: Each balloon represents the position of a missing view. Select the correct view and enter the corresponding letter in each balloon. The same letter may be correct for more than one view.

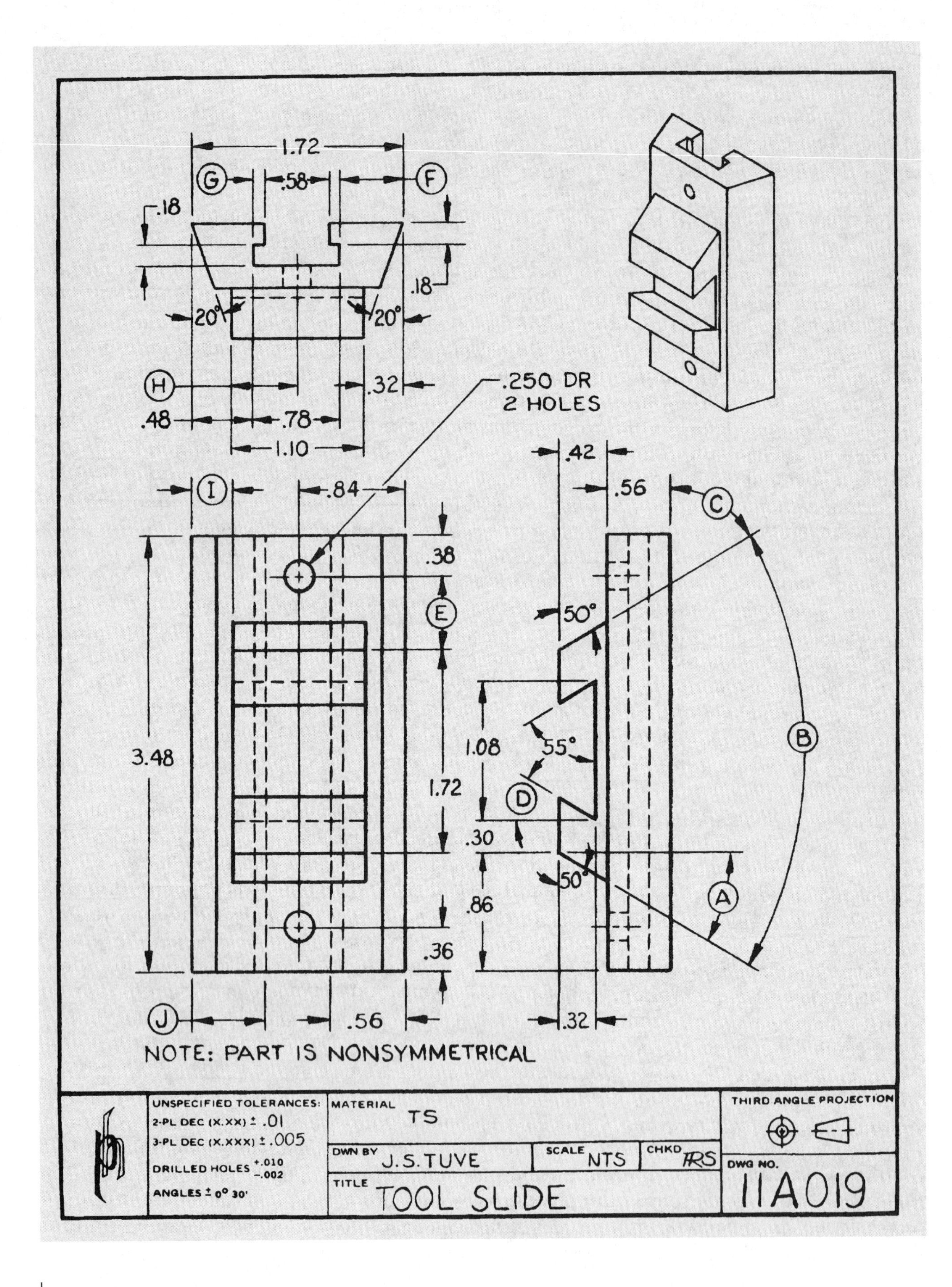
1.72
.58
.18
.18
20°
20°
.32
.250 DR
2 HOLES
.48
.78
1.10
.42
.84
.56
.38
3.48
50°
1.08
55°
1.72
.30
50°
.86
.36
.56
.32
NOTE: PART IS NONSYMMETRICAL
UNSPECIFIED TOLERANCES:
2-PL DEC (X.XX) ± .01
3-PL DEC (X.XXX) ± .005
DRILLED HOLES +.010 −.002
ANGLES ± 0° 30'
MATERIAL TS
DWN BY J.S.TUVE
SCALE NTS
CHKD
TITLE TOOL SLIDE
THIRD ANGLE PROJECTION
DWG NO. 11A019

INSTRUCTIONS: Refer to drawing 11A019 to answer the following questions.

1. What is the length of the tee slot? 1. ______________
2. What is the depth of the dovetail slot? 2. ______________
3. What is the length of the dovetail slot? 3. ______________
4. How much tolerance can accumulate in the full depth of the tee slot? 4. ______________
5. How many inclined surfaces does the tool slide contain? 5. ______________
6. What angle is created between the two sides of the dovetail slot? 6. ______________
7. What angle is created between the two outside edges in the top view? 7. ______________
8. What is the MMC of the overall depth of the tool slide? (Top and right-side views) 8. ______________
9. What is the MMC of the drilled holes? (Observe separate tolerance for drilled holes.) 9. ______________
10. How many lines (include hidden) in the front view represent surface intersections only, not edges? 10. ______________

INSTRUCTIONS: Enter the dimensions for the following letters.

Ⓐ ______________

Ⓑ ______________

Ⓒ ______________

Ⓓ ______________

Ⓔ ______________

Ⓕ ______________

Ⓖ ______________

Ⓗ ______________

Ⓘ ______________

Ⓙ ______________

Optional Math Exercises:

1a. Using pins that are .1875∅, determine the measurement between pins (MBP) after the dovetail has been properly machined.

NOTE: To properly machine the tool slide, other mathematical calculations may need to be performed.

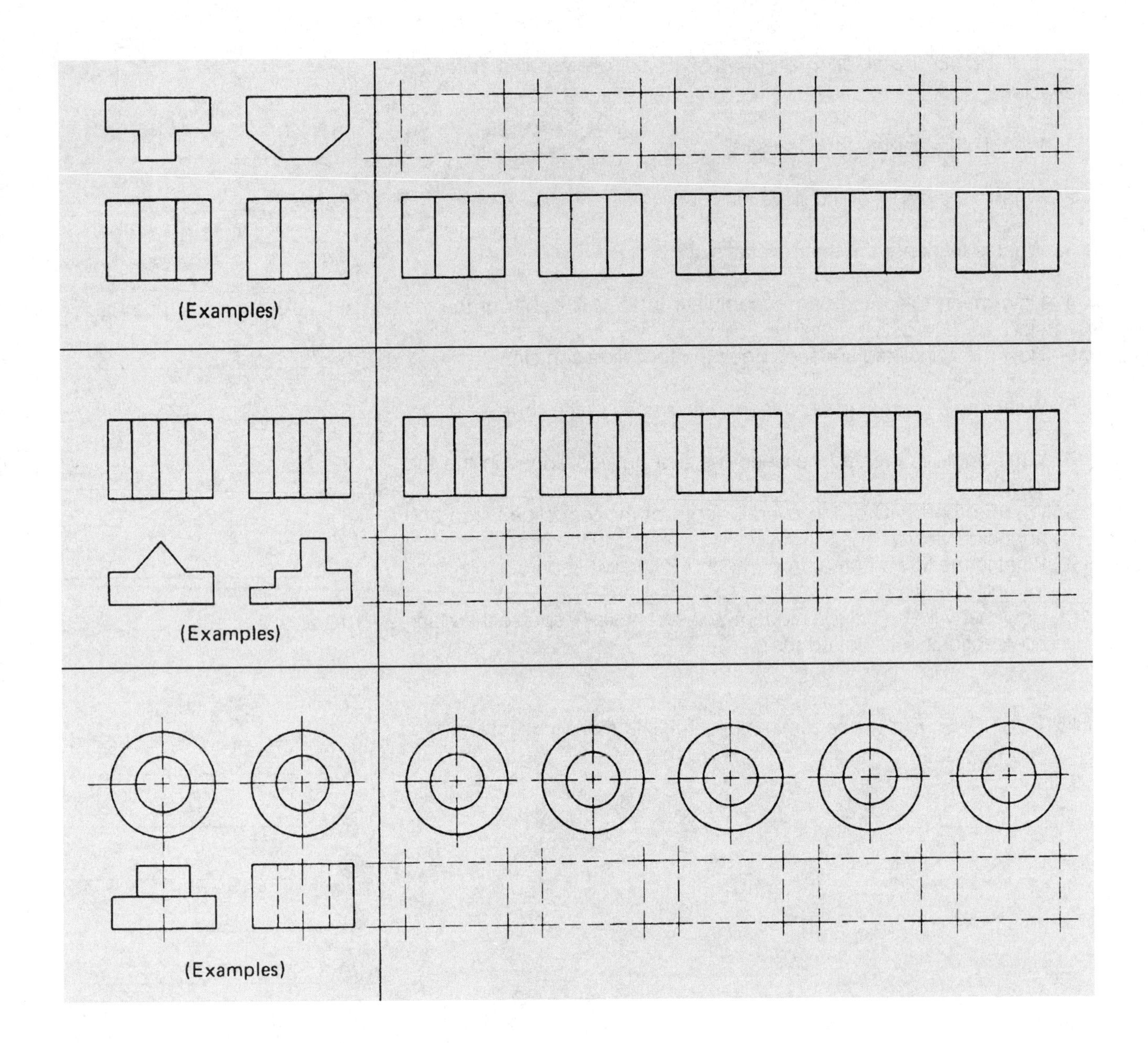

Similar View Sketches

INSTRUCTIONS: Sketch additional shapes that would correspond with the orthographic views shown above. Create shapes other than those examples shown at the left.

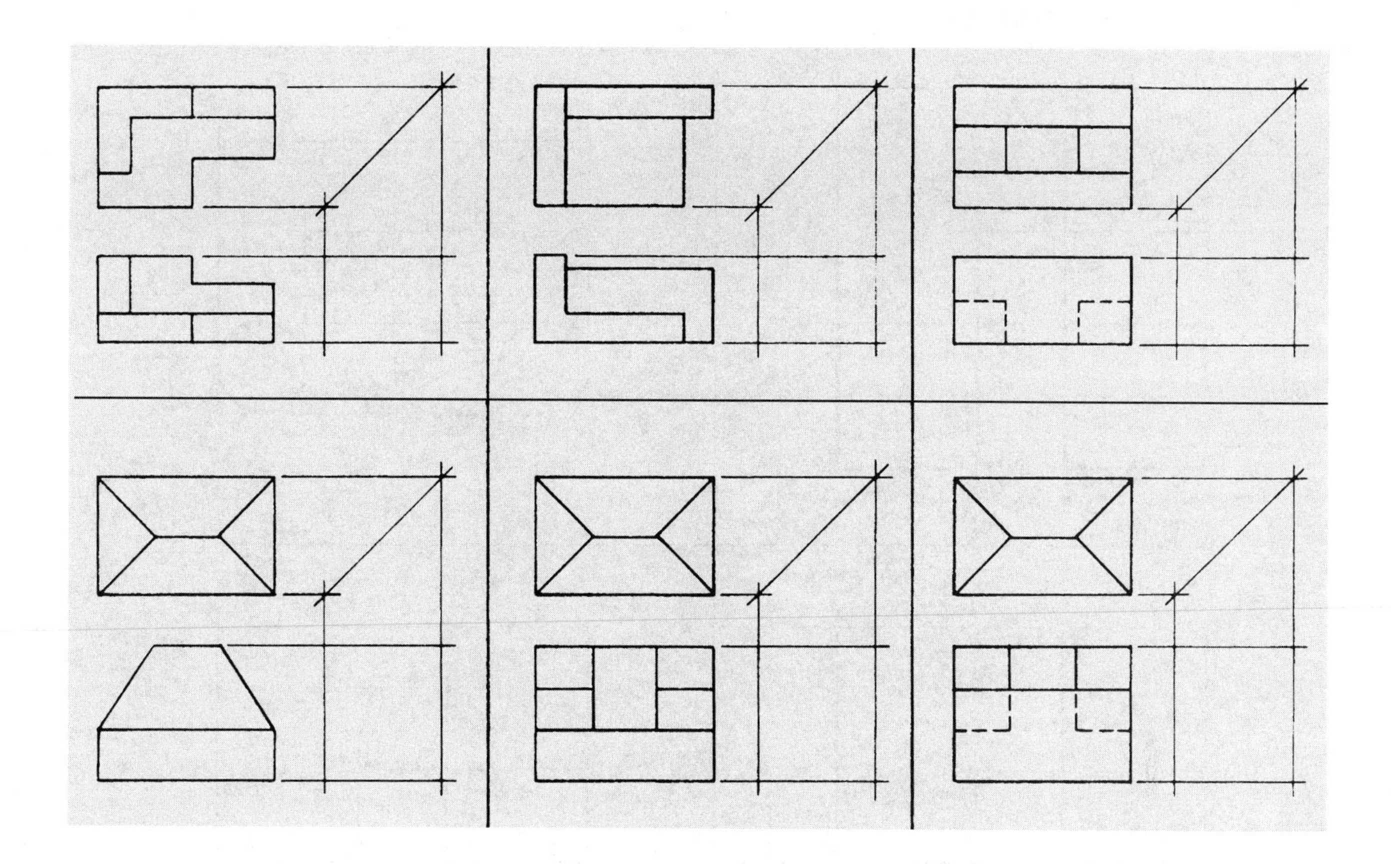

Missing Views 3

INSTRUCTIONS: Add the missing right-side view to each of the objects shown above.

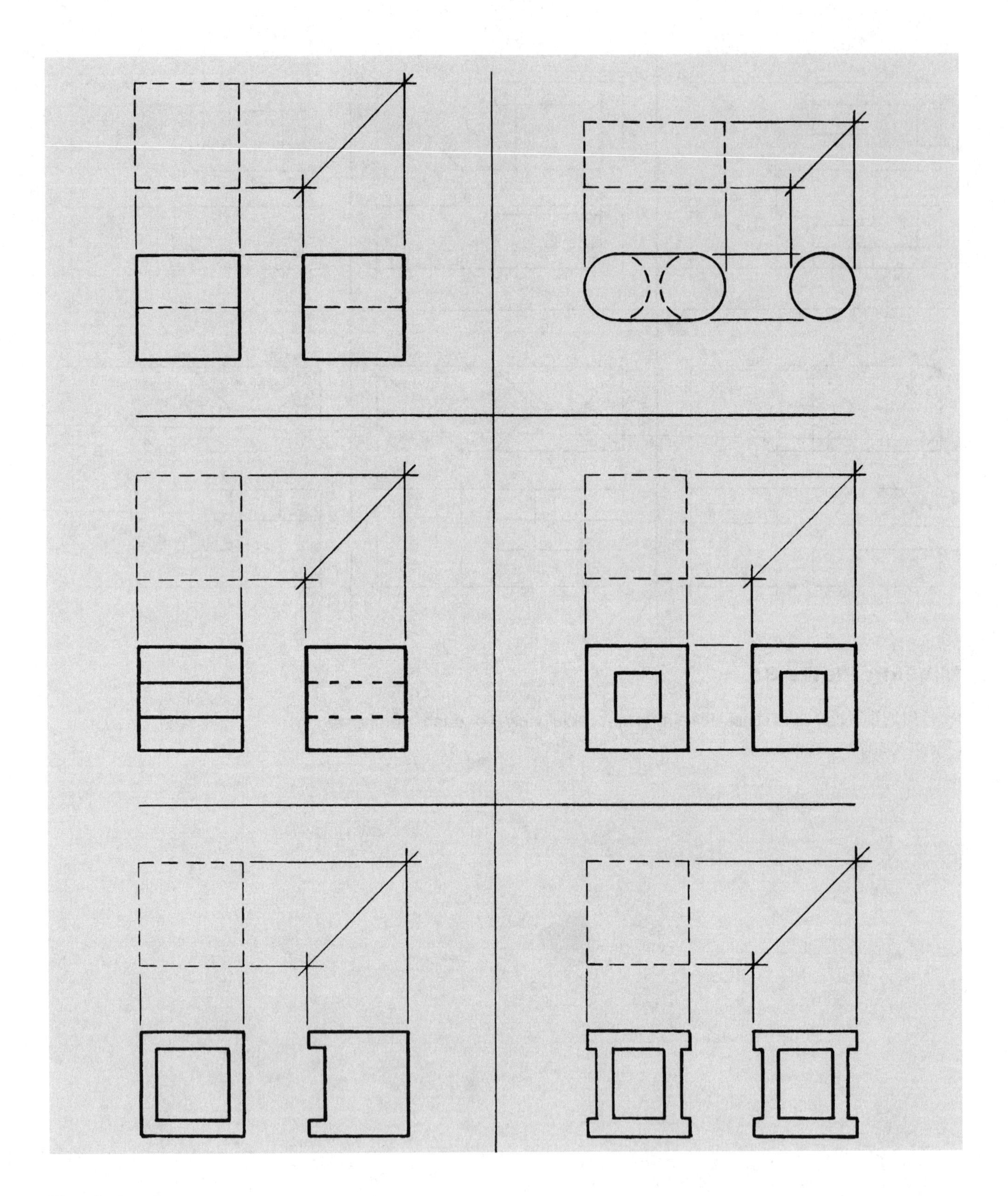

Brainteasers

INSTRUCTIONS: Shown above are six different objects. Their front views and right-side views are complete, with no missing lines. Can you solve any of their top views?

UNIT 6

REFERENCE DIMENSIONS

A dimension followed by the abbreviation REF (for reference) indicates that it is only for the convenience of the reader, and is not intended to carry a normal tolerance. In other words, it is a double dimension, since it can be calculated mathematically through normal dimensions. The tolerance of a reference dimension would be equal to the sum of all tolerances assigned to the normal dimensions used to calculate it. For example, an unbroken chain of three dimensions would cause the overall reference dimension to accumulate three tolerances. The new internationally accepted symbol for a reference dimension is to enclose the dimension figures in parentheses.

1.00 REF   (1.00)

BLIND HOLES

A blind hole is a hole that does not pass entirely through, thus its depth must be specified. When a leader is used, the diameter is listed first followed by the depth dimension. If a depth dimension is not listed, it is considered to be a through hole. A drilled hole's depth does not include the drill point. (See the illustration below.)

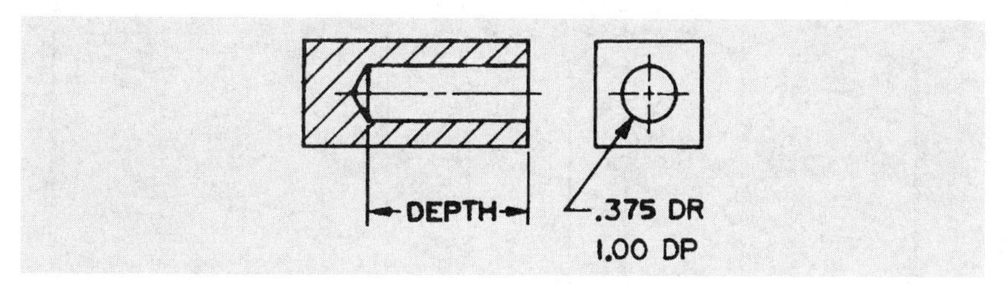

CHAMFERS

Chamfers are created on the ends of shafts and holes to eliminate sharp corners and to facilitate assembly. They are dimensioned by their depth (from the end) and their side angle. When the angle is 45° it may be dimensioned by a leader, but any other angle would require separate dimensions to avoid misinterpretation. (See examples shown below.) Chamfered holes should not be confused with countersunk holes (see page 134), which are dimensioned differently to accept flat head fasteners.

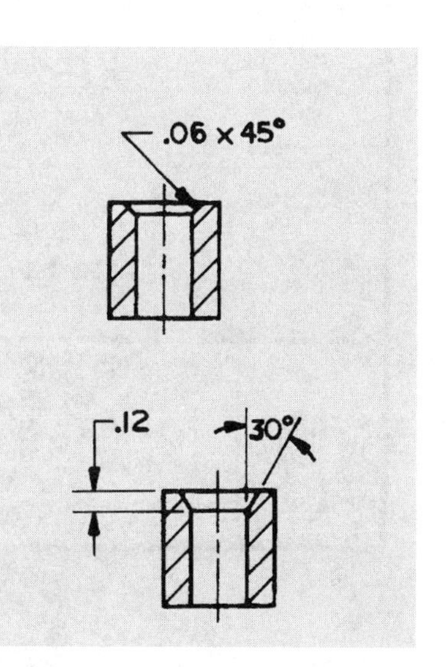

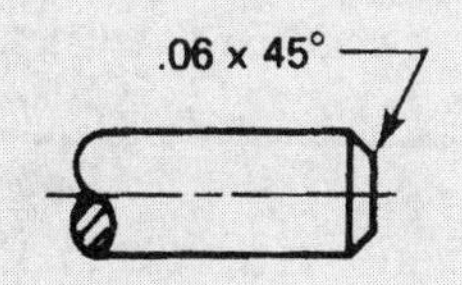

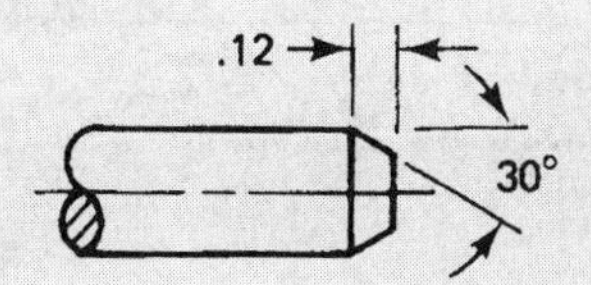

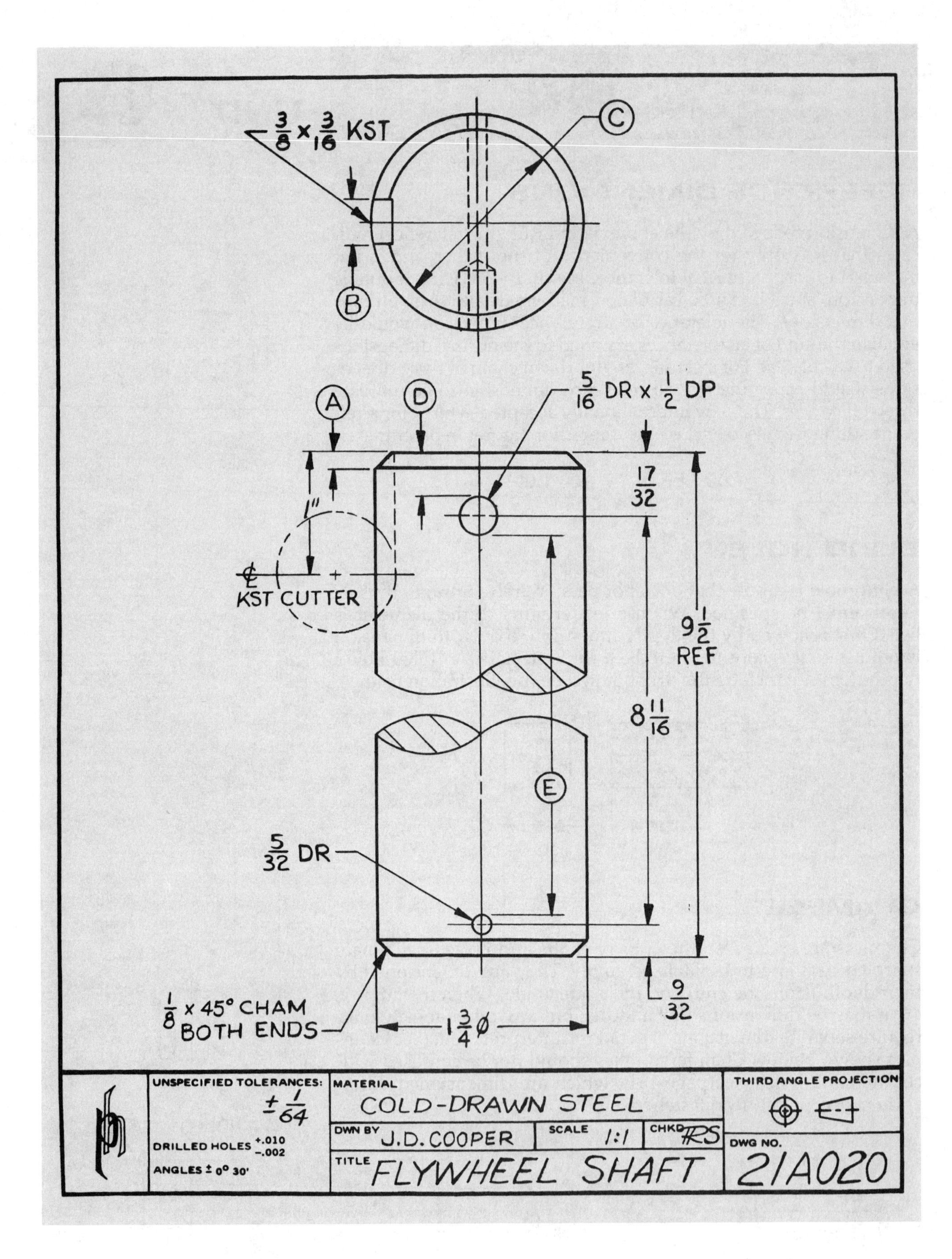
3/8 x 3/16 KST
C
B
A
D
5/16 DR x 1/2 DP
17/32
1"
℄ KST CUTTER
9 1/2 REF
8 11/16
E
5/32 DR
1/8 x 45° CHAM BOTH ENDS
1 3/4 ⌀
9/32
UNSPECIFIED TOLERANCES: ± 1/64
DRILLED HOLES +.010 −.002
ANGLES ± 0° 30'
MATERIAL COLD-DRAWN STEEL
DWN BY J.D. COOPER
SCALE 1:1
CHKD RS
TITLE FLYWHEEL SHAFT
THIRD ANGLE PROJECTION
DWG NO. 21A020

KEYSEATS AND KEYWAYS

When keyseats and keyways are dimensioned with leaders, the width dimension will appear first and the depth last. Keyseats in shafts also require a length dimension to assure full depth for a minimum distance, allowing for specific key lengths. Woodruff keyseats are dimensioned by their number and their location on a shaft. For precision fits, keyway depths in hubs are dimensioned from the opposite side of the hole, and keyseat depths in shafts are dimensioned from the opposite side of the shaft. (See examples shown below.)

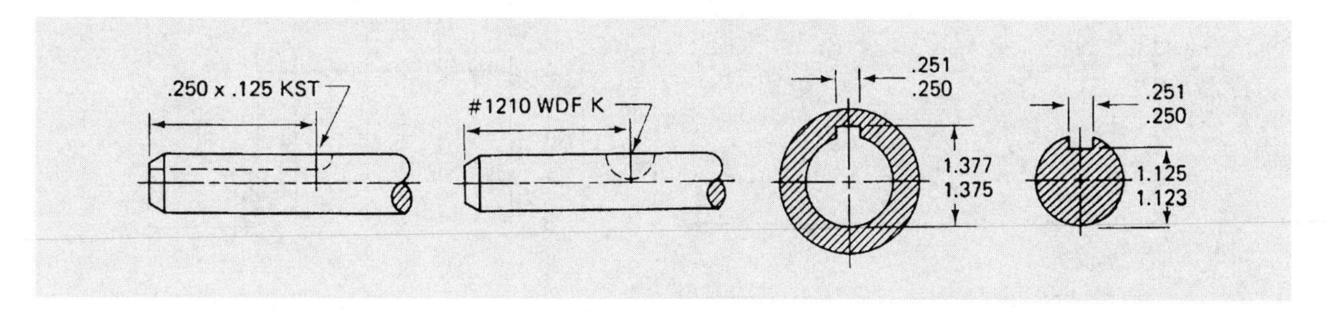

INSTRUCTIONS: Refer to drawing 21A020 to answer the following questions.

1. What system of dimensioning is used? (See page 21.) 1. ______
2. What is the tolerance on the fractional dimensions? 2. ______
3. What is the accumulated tolerance on the reference dimension? 3. ______
4. What is the tolerance on the drilled holes? 4. ______
5. What is the depth of the chamfer? 5. ______
6. What is the depth of the keyseat? 6. ______
7. What is the depth of the blind hole? 7. ______
8. What is the diameter of the through hole? 8. ______
9. What is the distance from the bottom of the keyseat to the opposite side of the shaft? 9. ______

Ⓐ ______

Ⓑ ______

Ⓒ ______

Ⓓ ______

Ⓔ ______

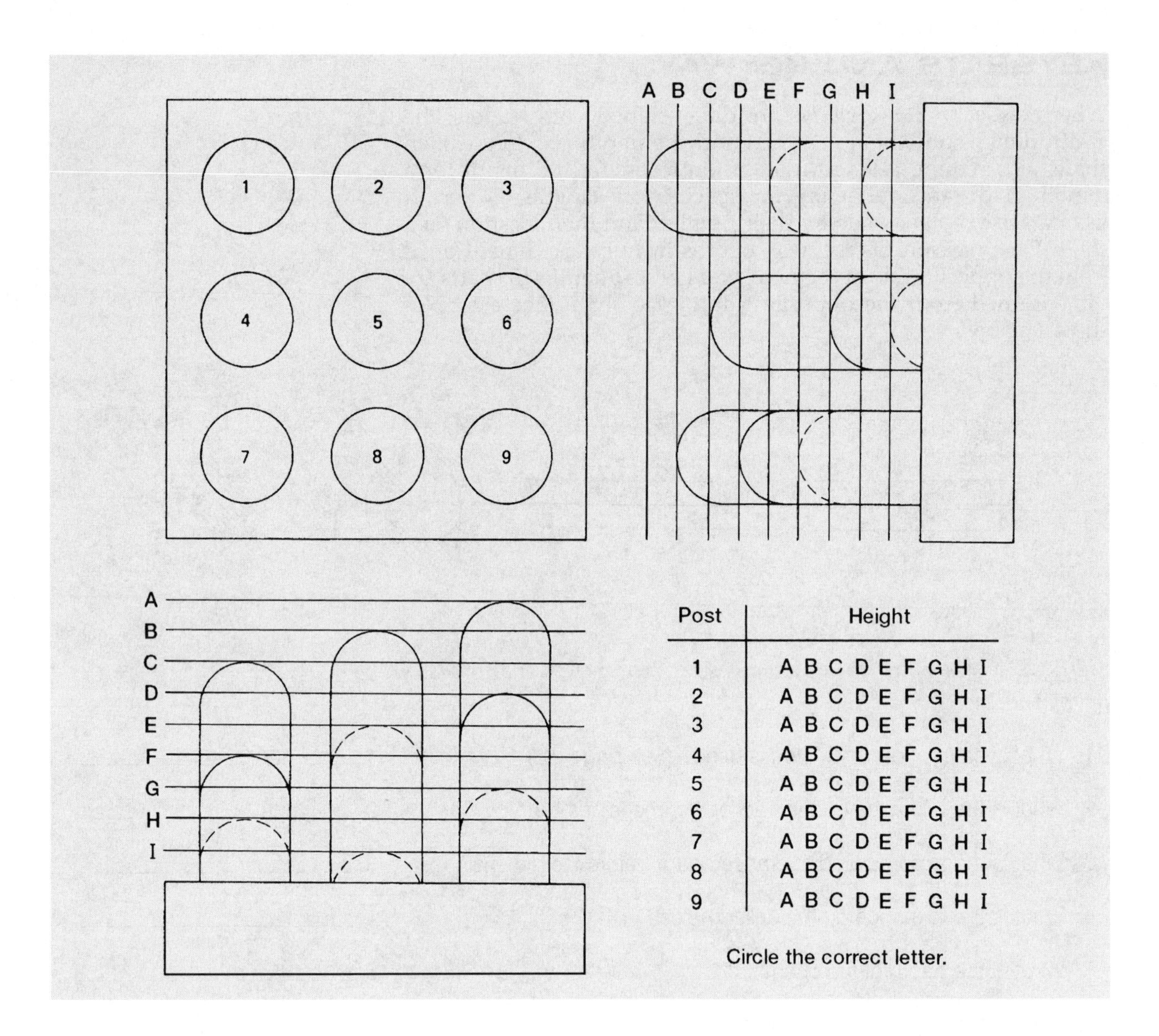

Post	Height
1	A B C D E F G H I
2	A B C D E F G H I
3	A B C D E F G H I
4	A B C D E F G H I
5	A B C D E F G H I
6	A B C D E F G H I
7	A B C D E F G H I
8	A B C D E F G H I
9	A B C D E F G H I

Circle the correct letter.

Post Identification 1

INSTRUCTIONS: Study the three views of the object shown above to determine the individual post heights. Circle the correct letter in the table.

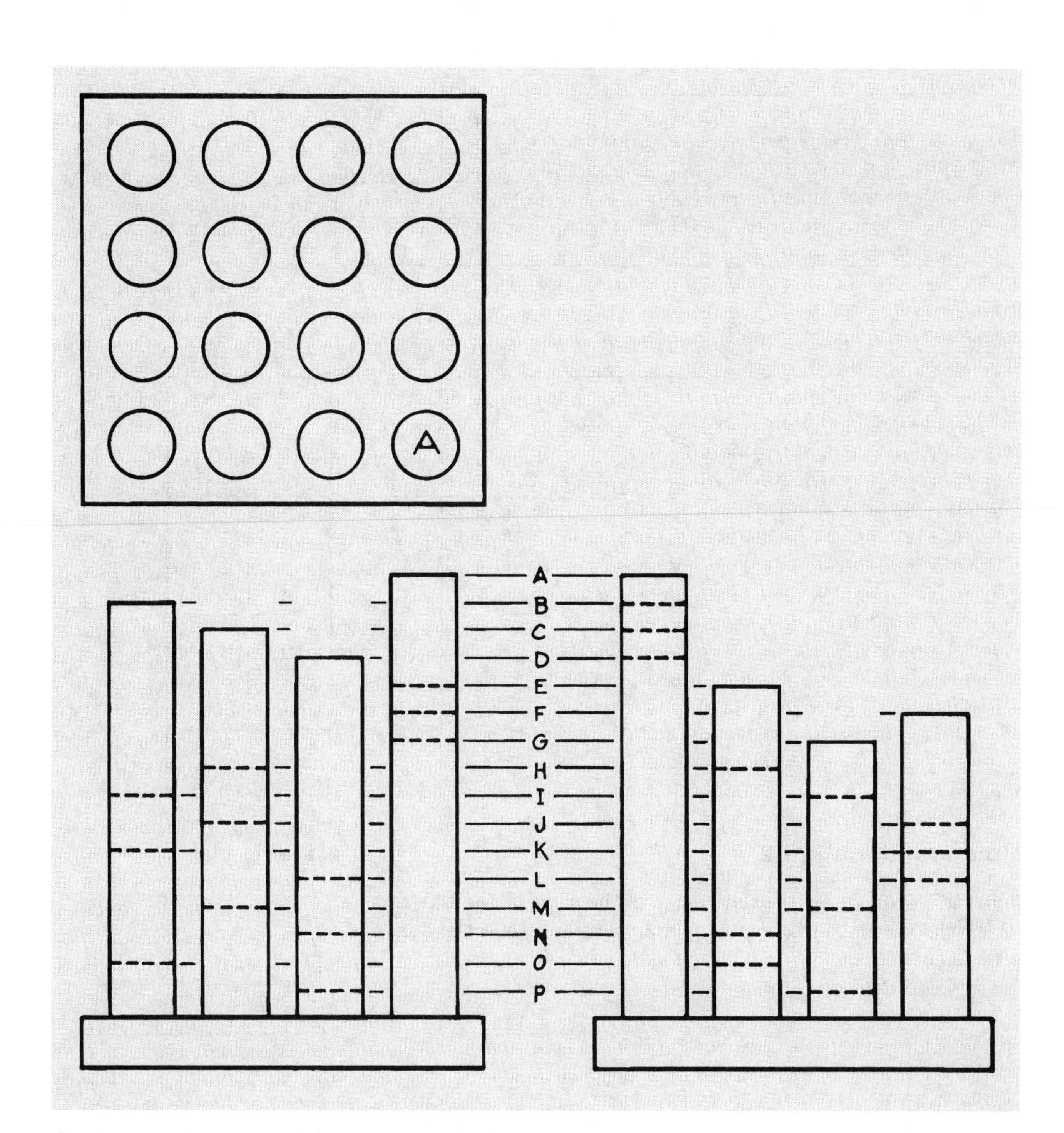

Post Identification 2

INSTRUCTIONS: Determine the position of each post in the top view. Enter the letters (B through P) into the appropriate circles.

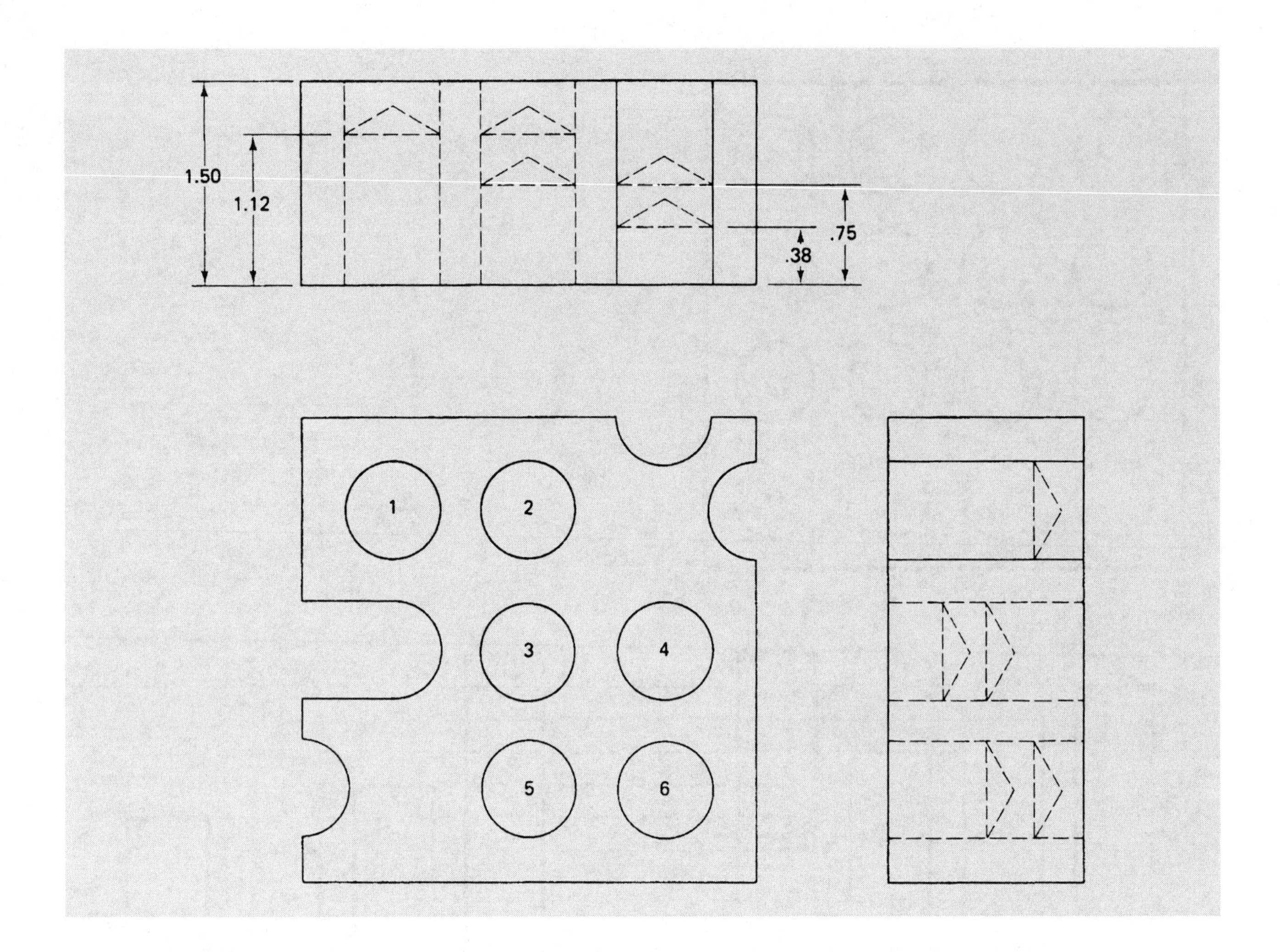

Hole Identification 2

INSTRUCTIONS: Study the three views of the plate shown above to determine individual hole depths. Write your answers in the space provided.

1. ____________________

2. ____________________

3. ____________________

4. ____________________

5. ____________________

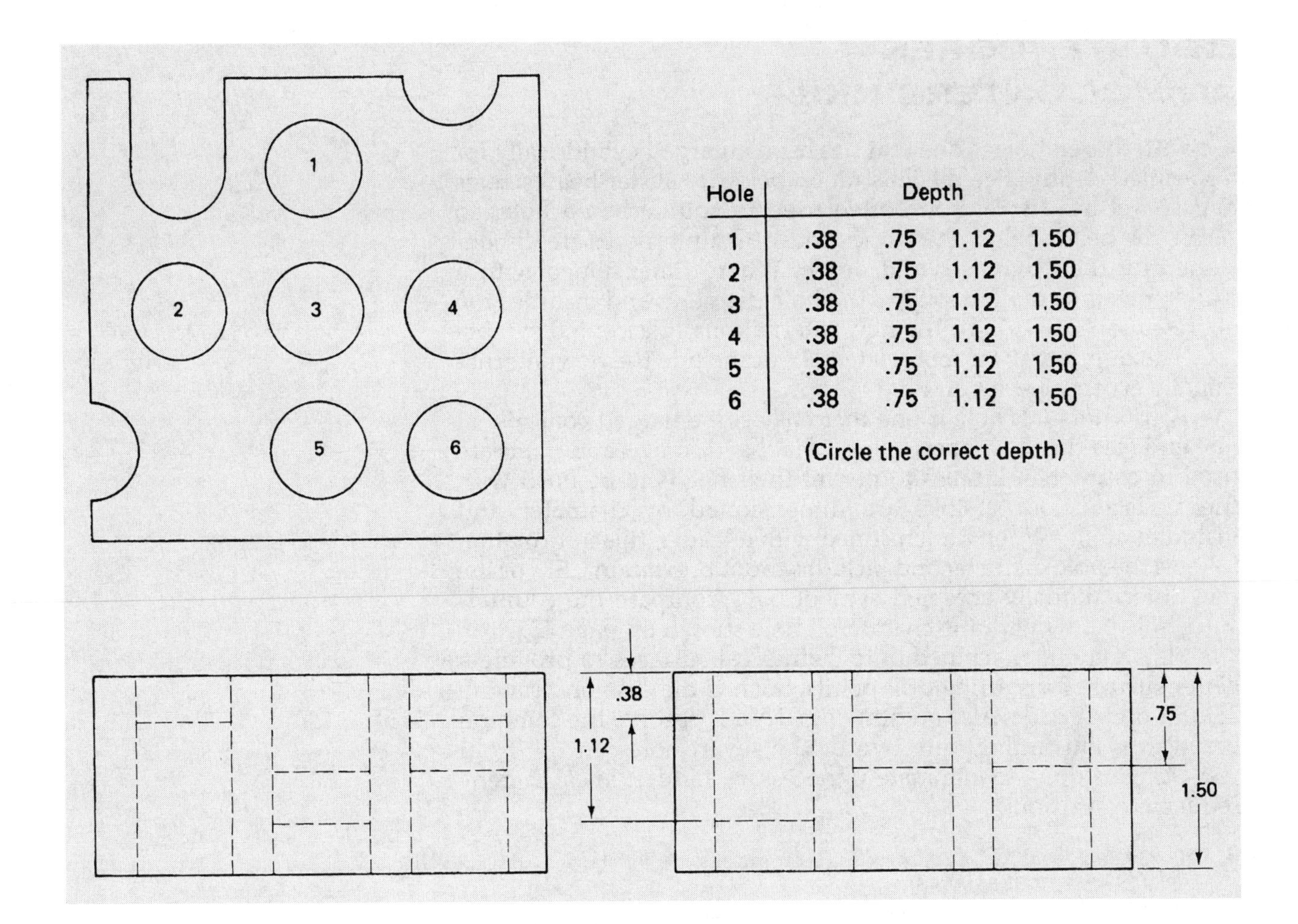

Hole	Depth			
1	.38	.75	1.12	1.50
2	.38	.75	1.12	1.50
3	.38	.75	1.12	1.50
4	.38	.75	1.12	1.50
5	.38	.75	1.12	1.50
6	.38	.75	1.12	1.50

(Circle the correct depth)

Hole Identification 3

INSTRUCTIONS: Study the three views of the plate shown above to determine individual hole depths. Circle the correct depth in the table.

COUNTERBORES AND COUNTERSINKS

A counterbored hole is one that has been enlarged cylindrically for a specified depth. (See the illustration below.) Fillister-head screws and socket-head screws are often used in counterbored holes to recess the heads below the work surface. Counterbores are dimensioned by their diameter and depth. When dimensioned with a leader, this information follows the hole diameter and includes the abbreviation CBORE, or the new internationally accepted symbol ⌴. The depth may be abbreviated DP, or include the new internationally accepted symbol ↧.

A countersunk hole is one that has been enlarged conically at one end (see the illustration below). Flat head screws and rivets are used in countersunk holes to permit their heads to be flush with the surface. Countersinks are dimensioned by diameter and included angle. When dimensioned with a leader, this information follows the hole diameter and includes the abbreviation CSK, or the new internationally accepted symbol ⌵. Compare the countersunk hole below with the chamfered hole shown on page 127.

Flats may be machined onto cylindrical surfaces to provide a better surface for starting drill points. Such is the case on drawing 21A021 on page 136. Also on drawing 21A021, observe the common practice of illustrating only two of the seven holes on the front view. This is done to eliminate unnecessary hidden lines, thereby reducing confusion.

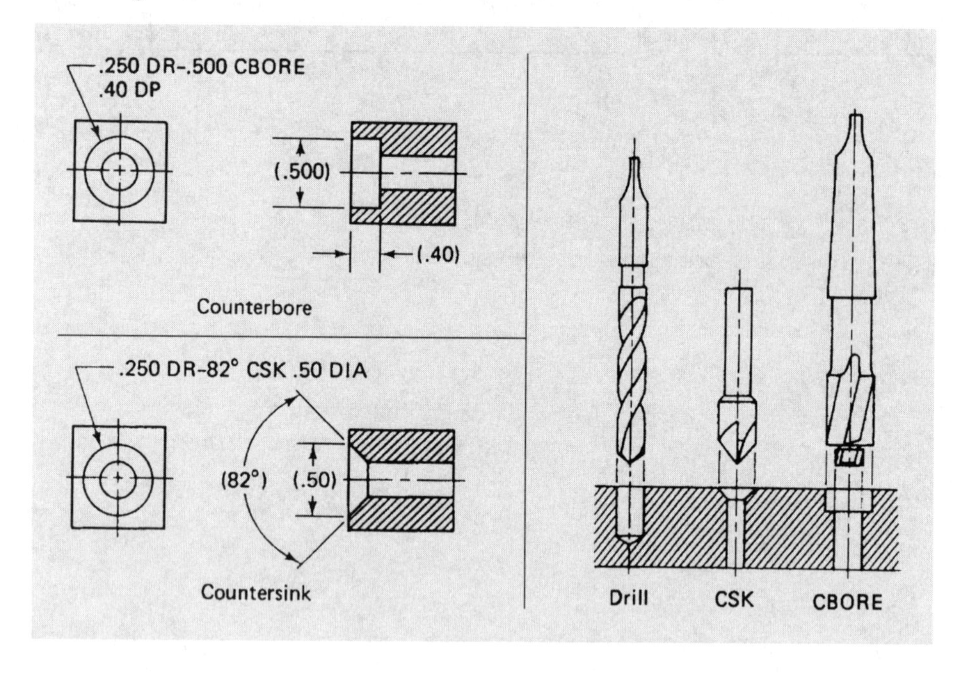

ANGULAR HOLE LOCATIONS

When holes are located angularly on bolt circles, it may be necessary to calculate some dimensions that do not appear. For example, if a drawing states that five equally spaced holes are required, you would divide 360° by 5 to find the angular spacing is actually 72°. What if the number of equally spaced holes is seven, or eleven, or thirteen? To find the answer in degrees, minutes, and seconds, multiply the remaining units by 60 and continue to divide them. (See the example at the right.)

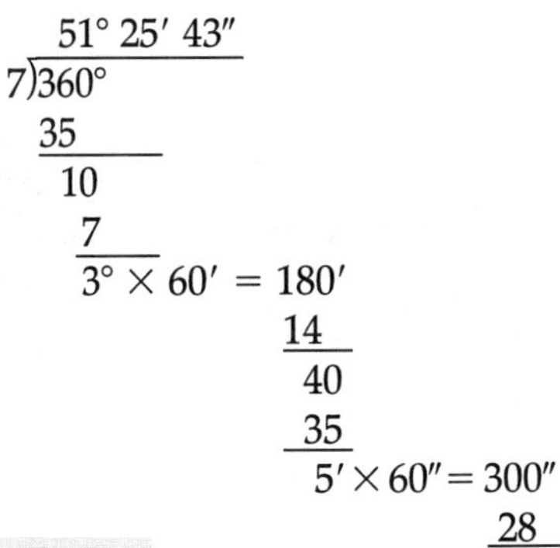

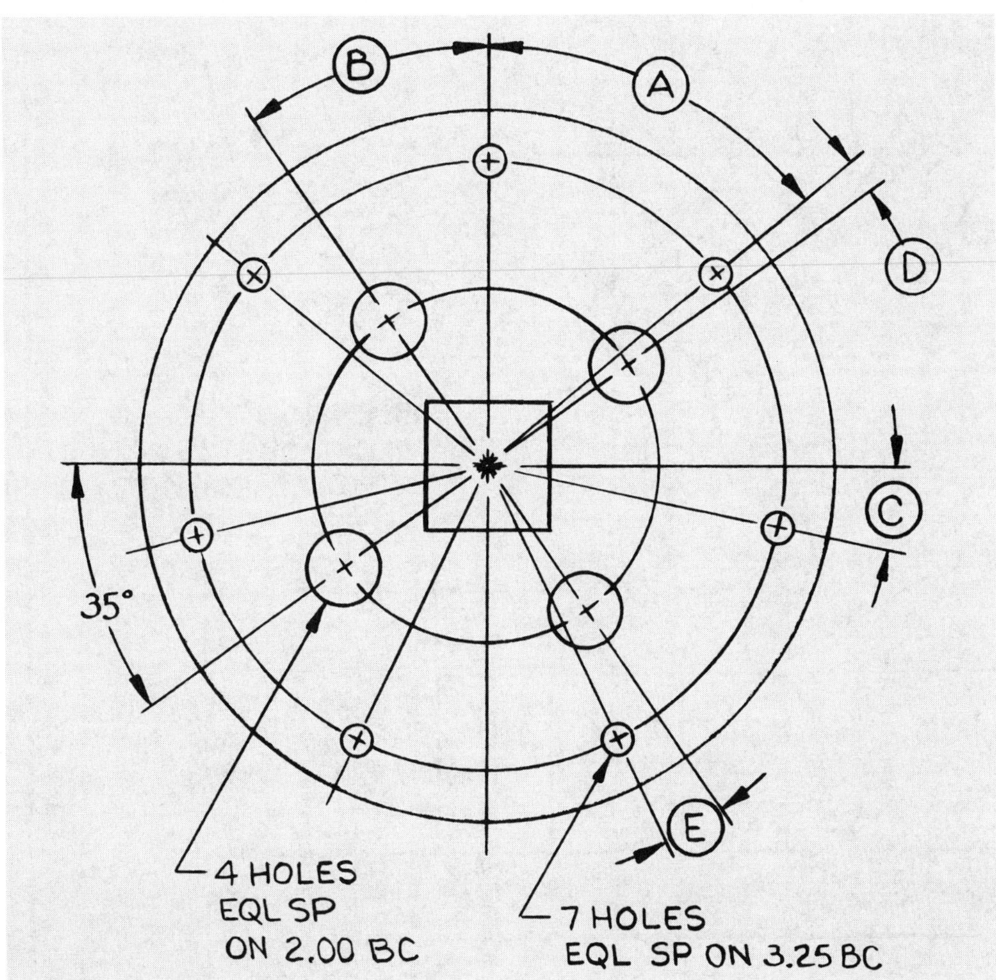

Angular Calculations 4

INSTRUCTIONS: *Calculate the angular dimension for each of the letters shown in the illustration above.*

Ⓐ ____________________

Ⓑ ____________________

Ⓒ ____________________

Ⓓ ____________________

Ⓔ ____________________

Optional Math Exercises:

1a. Calculate the Cartesian coordinates for the 4 holes located on the 2.00∅ bolt circle.

2a. Calculate the Cartesian coordinates for the 7 holes located on the 3.25∅ bolt circle.

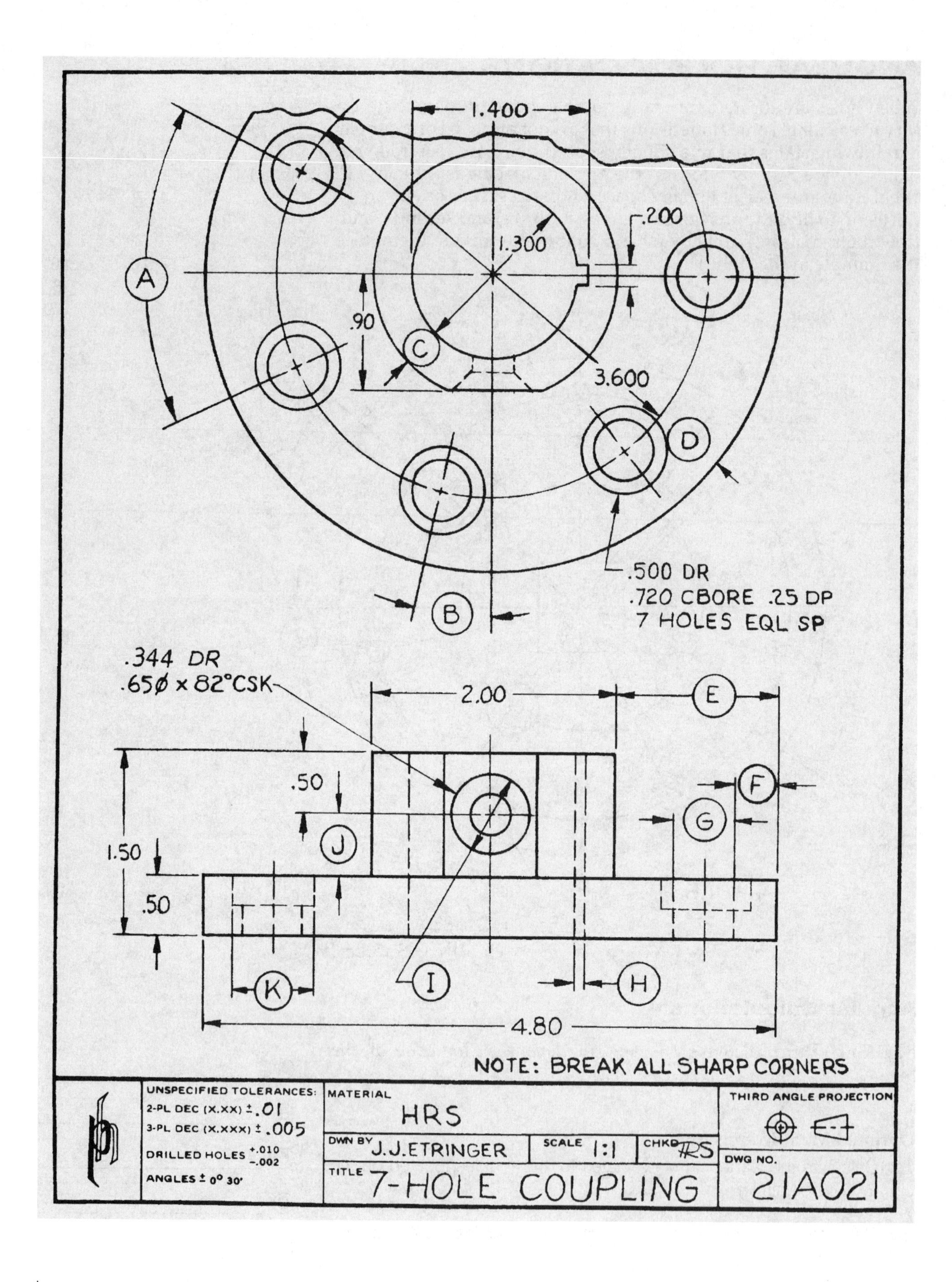
1.400
.200
1.300
A
.90
C
3.600
D
B
.500 DR
.720 CBORE .25 DP
7 HOLES EQL SP
.344 DR
.65Ø x 82°CSK
2.00
E
F
.50
G
J
1.50
.50
K
I
H
4.80
NOTE: BREAK ALL SHARP CORNERS
UNSPECIFIED TOLERANCES:
2-PL DEC (X.XX) ± .01
3-PL DEC (X.XXX) ± .005
DRILLED HOLES +.010 -.002
ANGLES ± 0° 30'
MATERIAL
HRS
DWN BY J.J.ETRINGER
SCALE 1:1
CHKD RS
TITLE 7-HOLE COUPLING
THIRD ANGLE PROJECTION
DWG NO.
21A021

INSTRUCTIONS: Refer to drawing 21A021 when calculating the following.

1. Maximum flange diameter. 1. ______
2. Minimum flange thickness. 2. ______
3. Maximum hub diameter. 3. ______
4. Minimum keyway width. 4. ______
5. Maximum counterbore depth. 5. ______
6. Included angle of countersink. 6. ______
7. Minimum bolt circle. 7. ______
8. Maximum bore diameter. 8. ______
9. Minimum diameter of drilled holes. 9. ______

INSTRUCTIONS: Enter the dimensions for the following letters.

Ⓐ ______° ______' ______"

Ⓑ ______° ______' ______"

Ⓒ ______

Ⓓ ______

Ⓔ ______

Ⓕ ______

Ⓖ ______

Ⓗ ______

Ⓘ ______

Ⓙ ______

Ⓚ ______

Optional Math Exercises:

1a. Calculate the Cartesian coordinates for the .500∅ holes.

2a. Calculate the maximum Measurement Over Pins (MOP) of two adjacent drilled holes.

3a. Calculate the minimum MOP of two adjacent drilled holes.

CASTINGS

Whereas the objects illustrated in the next two drawings (pages 140 and 142) could be machined from solid bar stock, savings in both material and labor may be obtained if they were cast. The casting process involves pouring molten metal into a hollow cavity in a sand mold. A pattern of wood or metal is used to create the cavity in the sand. After the molten metal solidifies, the casting assumes a surface roughness similar to that of the molding sand that formed it. Where this rough surface may be objectionable or where greater accuracy must be attained, a machining operation will be required. Other casting processes, such as die casting, permanent mold casting, and centrifugal casting may also be used to produce metal products.

FINISH SYMBOLS

A cast surface that is to be machined will be identified on a drawing by a finish symbol placed on the edge view of the machined surface. Each finished surface will carry its own symbol; however, when every surface of a casting is to be finished, the note "Finish All Over" or the abbreviation FAO may be substituted. See the examples below of the finish symbols in use today.

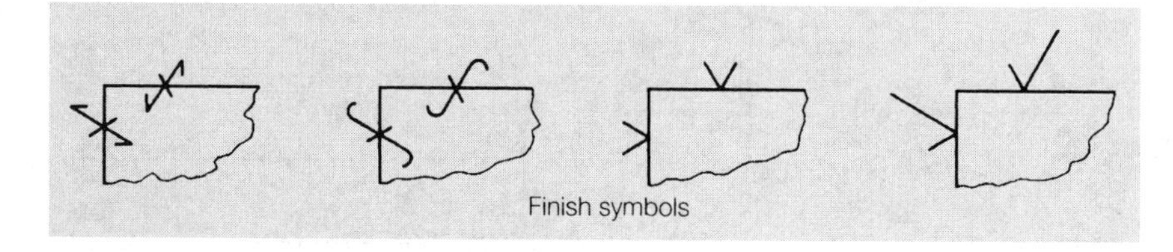
Finish symbols

FINISH ALLOWANCE

The patternmaker must allow for additional material to appear where a surface is to be machined. Called "finish allowance," this provides 1/8 in. (.12) of additional metal for the machinist to remove. Thus, a raw casting that is to be machined on opposite sides will measure approximately 1/4 in. larger than the finished product. (See the illustration shown below.)

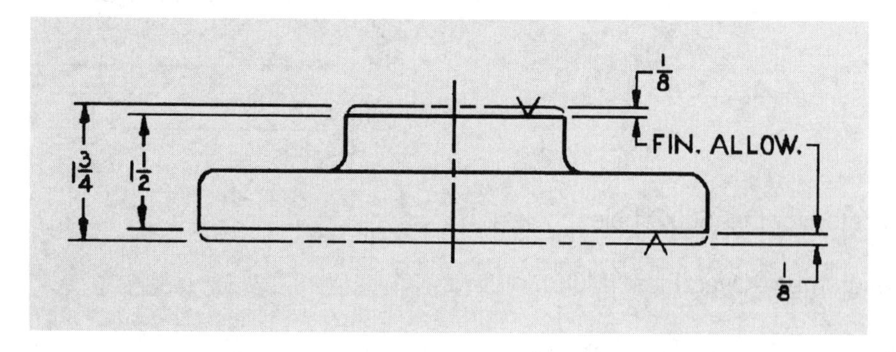

FILLETS AND ROUNDS

Since sharp corners should not be cast, the patternmaker will radius all intersecting surfaces of the pattern. These radii are called "fillets" for inside corners and "rounds" for outside corners. Normally, their sizes are specified in a general note on a casting drawing rather than as individual dimensions. (See the examples of a fillet and round shown below.)

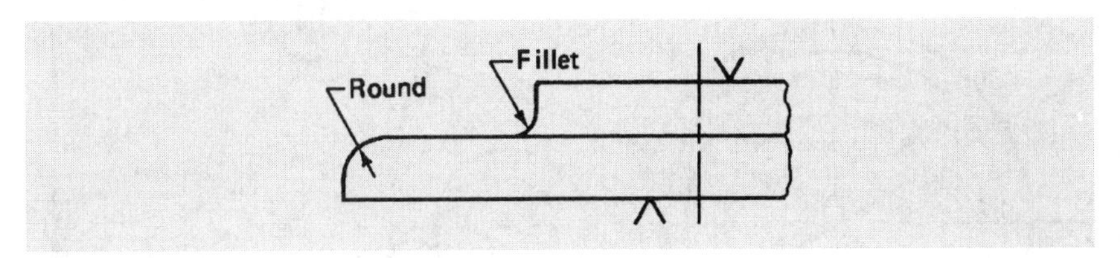

CLEARANCE HOLES

When parts are to be assembled together with fasteners such as bolts or rivets, clearance holes must be provided that are larger than the fasteners themselves. The amount of clearance will vary depending upon how critical the alignment must be. For example, dowel pins may require little or no clearance to maintain extreme accuracy. However, in normal situations for bolts and rivets, this clearance amounts to about 1/32 in. (.031). Therefore, if 1/2 in. bolts are to be used, 17/32 in. (.531) diameter holes would be specified. Conversely, if .406 diameter holes have been specified, they may be clearance holes for .375 diameter fasteners. Such is the case on drawing 21A022 on page 140.

HALF-VIEWS

When a view is symmetrical, a half-view may be drawn on one side of the centerline, as in drawing 21A023 on page 142. If additional features beyond the centerline are shown, the freehand break line is used to terminate the view. (See the example shown below.) Note the use of the new internationally accepted symbol for symmetry used on the two half-views below.

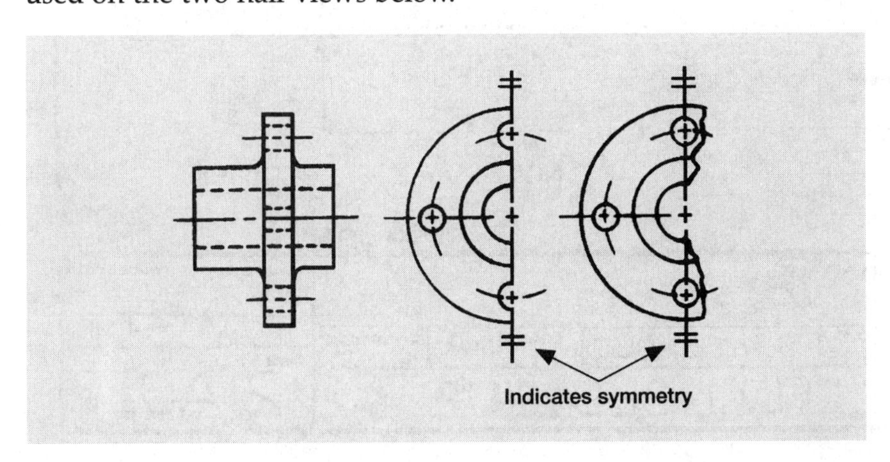

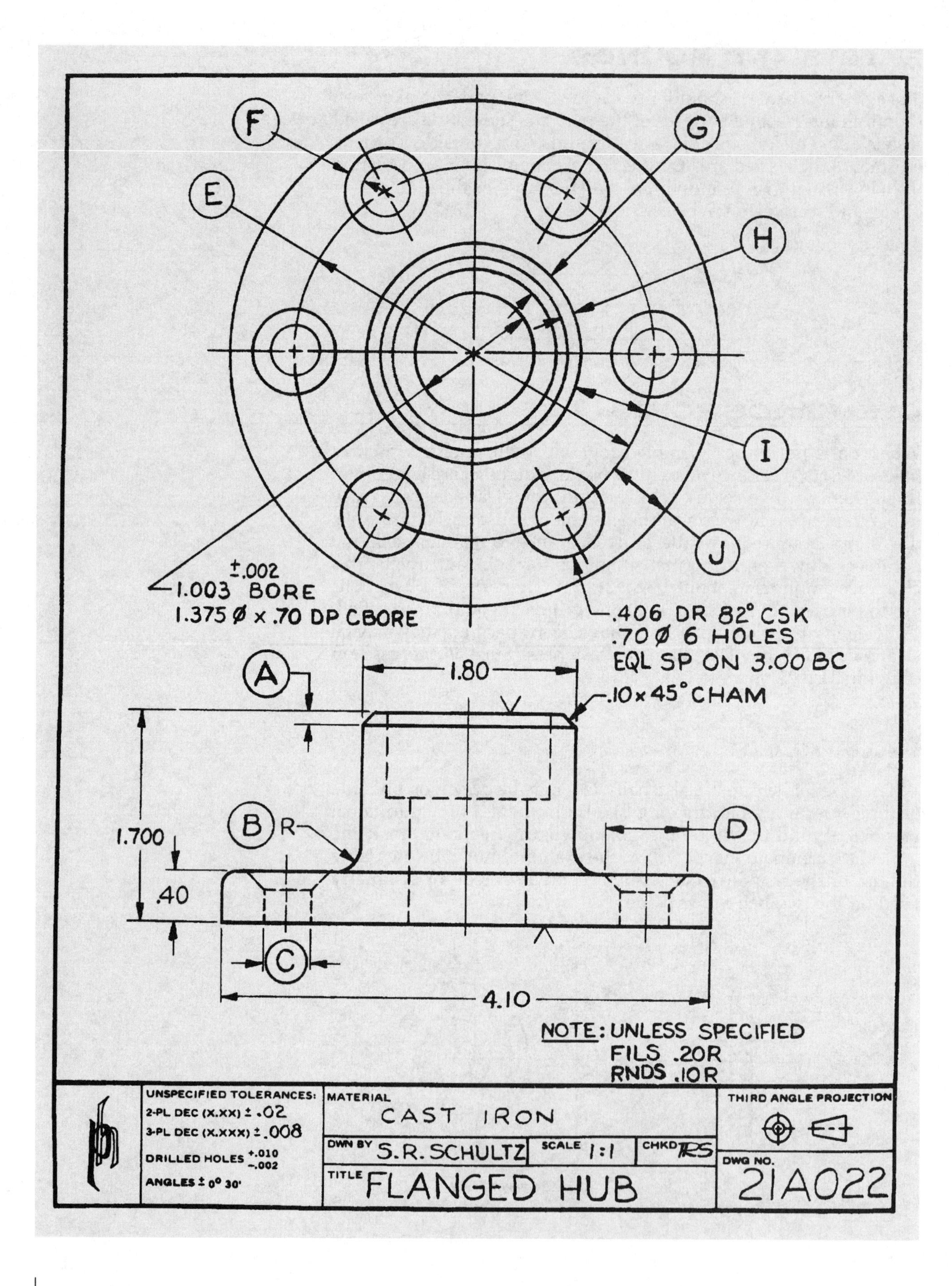
F
G
E
H
I
J
±.002
1.003 BORE
1.375 Ø x .70 DP CBORE
.406 DR 82° CSK
.70 Ø 6 HOLES
EQL SP ON 3.00 BC
A
1.80
.10 x 45° CHAM
B
R
D
1.700
.40
C
4.10
NOTE: UNLESS SPECIFIED
FILS .20R
RNDS .10R
UNSPECIFIED TOLERANCES:
2-PL DEC (X.XX) ± .02
3-PL DEC (X.XXX) ± .008
DRILLED HOLES +.010 −.002
ANGLES ± 0° 30'
MATERIAL
CAST IRON
DWN BY S.R. SCHULTZ
SCALE 1:1
CHKD TRS
TITLE FLANGED HUB
THIRD ANGLE PROJECTION
DWG NO.
21A022

INSTRUCTIONS: Refer to drawing 21A022 to answer the following questions.

1. What is the size of the fillet? 1. ____________
2. What is the size of the round? 2. ____________
3. How deep is the counterbore? 3. ____________
4. What is the bored hole tolerance? 4. ____________
5. What is the MMC of the bored hole? 5. ____________
6. What is the MMC of the drilled holes? 6. ____________
7. How many degrees apart are the drilled holes? 7. ____________
8. How many hidden-line circles would appear on a bottom view? 8. ____________
9. What would be the overall height dimension before machining? (Consider the finish allowance, page 138.) 9. ____________
10. What size fasteners are intended to fit the clearance holes in the flange? (Refer to page 139.) 10. ____________

INSTRUCTIONS: Enter the dimensions for the following letters.

Ⓐ ____________

Ⓑ ____________

Ⓒ ____________

Ⓓ ____________

Ⓔ ____________

Ⓕ ____________

Ⓖ ____________

Ⓗ ____________

Ⓘ ____________

Ⓙ ____________

Optional Math Exercises:

1a. Calculate the Cartesian coordinates for the .406∅ drilled holes.

2a. Calculate the maximum and minimum Measurement Over Pins (MOP) of the .406∅ drilled holes.

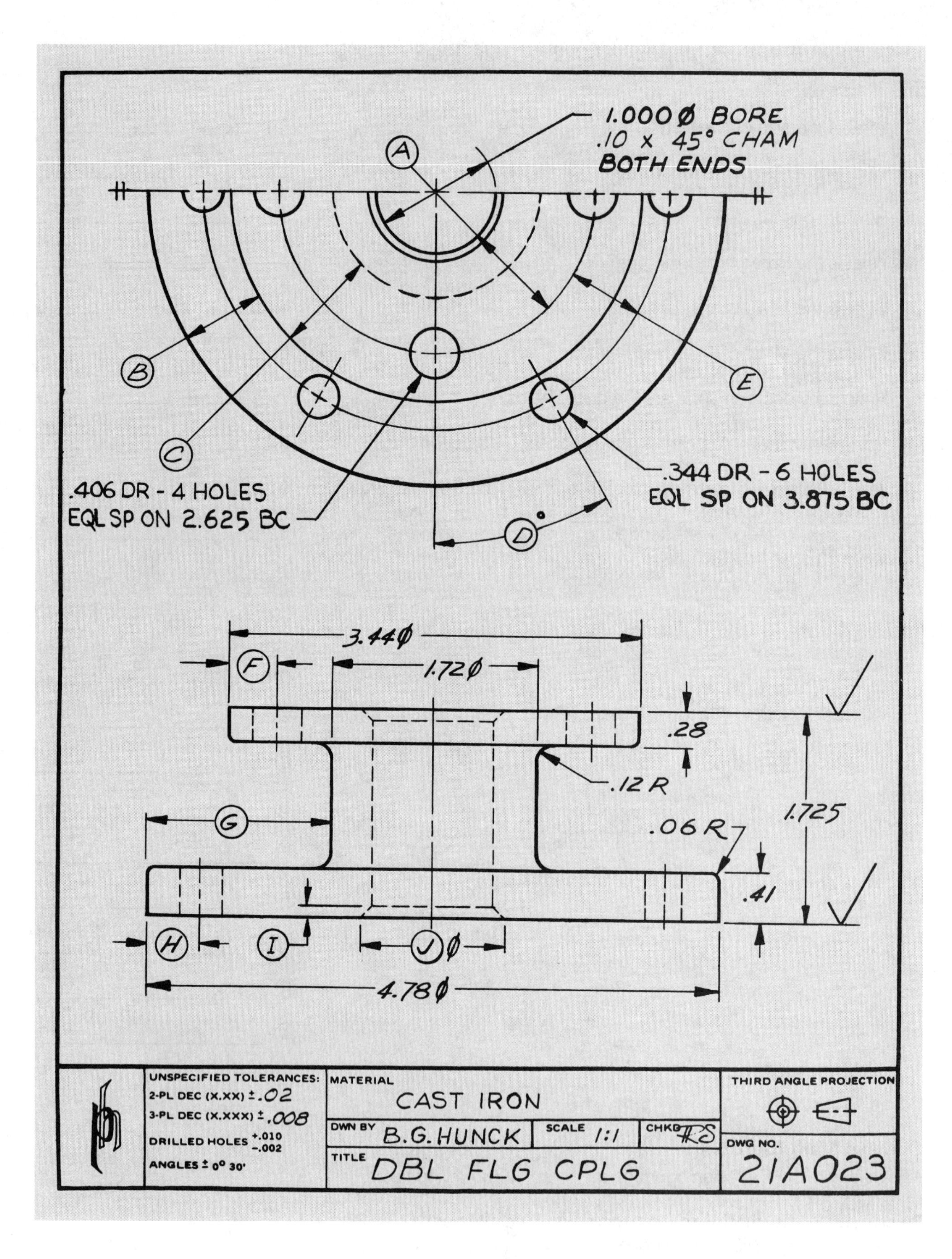
1.000 Ø BORE
.10 x 45° CHAM
BOTH ENDS
A
B
C
D°
E
.406 DR - 4 HOLES
EQL SP ON 2.625 BC
.344 DR - 6 HOLES
EQL SP ON 3.875 BC
3.44 Ø
F
1.72 Ø
.28
.12 R
1.725
G
.06 R
.41
H
I
J Ø
4.78 Ø
UNSPECIFIED TOLERANCES:
2-PL DEC (X.XX) ± .02
3-PL DEC (X.XXX) ± .008
DRILLED HOLES +.010 −.002
ANGLES ± 0° 30'
MATERIAL
CAST IRON
DWN BY B.G. HUNCK
SCALE 1:1
CHKD
TITLE DBL FLG CPLG
THIRD ANGLE PROJECTION
DWG NO.
21A023

INSTRUCTIONS: Refer to drawing 21A023 to answer the following questions.

1. What type of view is the top view? 1. ______
2. What size fillet radius is specified? 2. ______
3. What size rounds are specified? 3. ______
4. How many clearance holes are in the bottom flange? 4. ______
5. What is the tolerance on the clearance holes? 5. ______
6. What is the tolerance on the bored hole? 6. ______
7. What is the MMC of the bored hole? 7. ______
8. What size fasteners are intended for the holes in the top flange? 8. ______
9. How many surfaces are to be machined flat? 9. ______
10. What would be the overall height dimension before machining? (Consider finish allowance.) 10. ______

INSTRUCTIONS: Enter the dimensions for the following letters.

Ⓐ ______
Ⓑ ______
Ⓒ ______
Ⓓ ______
Ⓔ ______
Ⓕ ______
Ⓖ ______
Ⓗ ______
Ⓘ ______
Ⓙ ______

Optional Math Exercises:

1a. Calculate the Cartesian coordinates for the 4 .406∅ drilled holes.

2a. Calculate the maximum and minimum Measurement Over Pins (MOP) of the .406∅ drilled holes.

3a. Calculate the Cartesian coordinates for the 6 .344∅ drilled holes.

4a. Calculate the maximum and minimum MOP of the .344∅ drilled holes.

SURFACE ROUGHNESS

The presence of a finish symbol does not guarantee any particular degree of smoothness to that surface. When surface roughness height must be controlled, figures are placed above the left leg of the finish symbol as shown below. These figures represent the mathematical average deviation from the roughness centerline and are expressed in microinches (millionths of an inch) or micrometers (millionths of a meter). Thus, the figure 63 above the left leg would actually represent .000063 inches, and the metric figure 1.6 would actually represent 0.0000016 meters. These figures indicate the maximum value and any lesser value would be acceptable. When two sets of figures appear above the left leg, they represent the maximum and minimum permissible range of roughness. (See the examples shown below.)

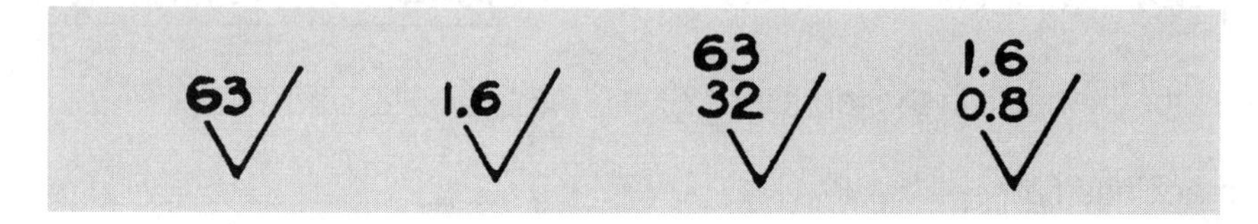

ROUGHNESS AVERAGE (R_a)

The roughness average is the arithmetic average of the absolute values of the measured profile height deviations taken within the sampling length and measured from the graphical centerline. The profile height is the maximum distance above the centerline plus the maximum distance below the centerline within the sampling length. This value is typically three or more times the roughness average, as illustrated below.

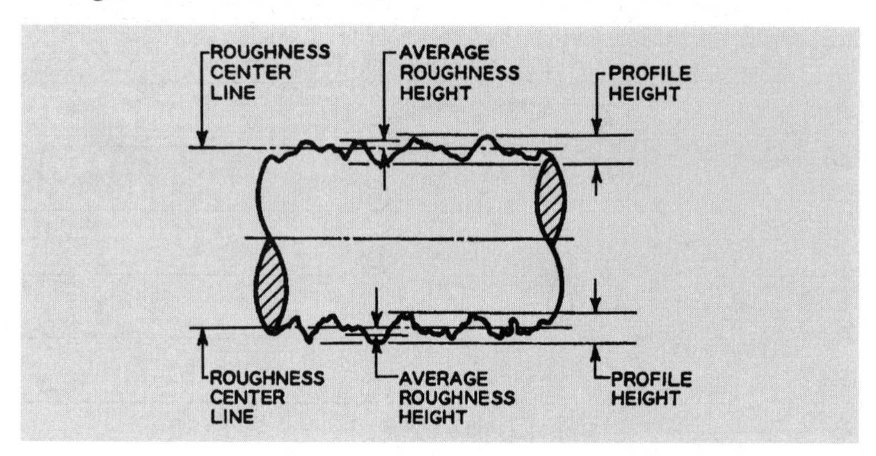

SURFACE ROUGHNESS CHART

The chart on page 145 lists the typical range of surface roughness values that can be obtained from various production processes. The solid bar indicates the average roughness produced by that process; however, higher or lower values may be obtained under special conditions. Spend a moment to compare some of the processes with which you are familiar.

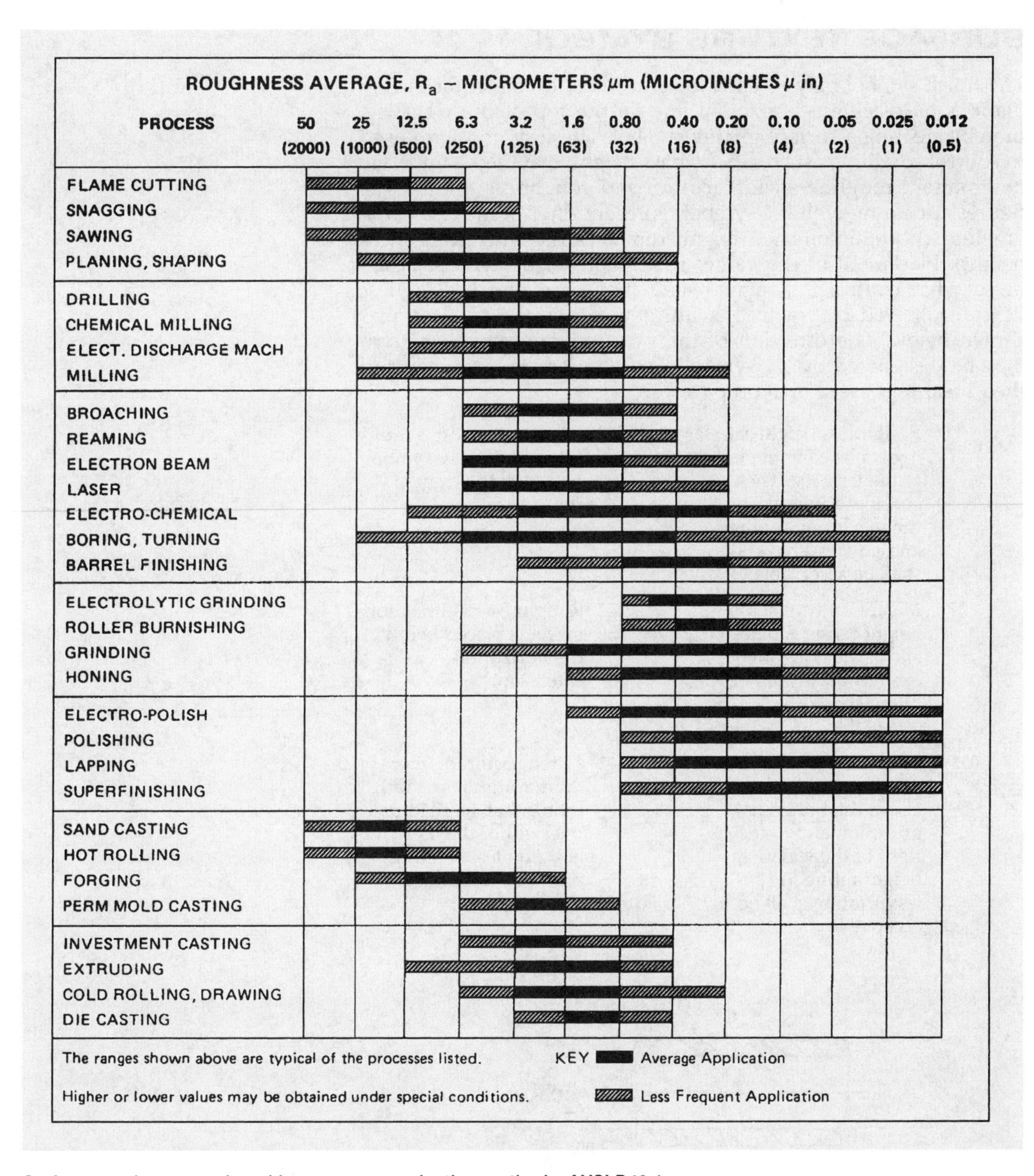

Surface roughness produced by common production methods. ANSI B46.1 (With permission of the publisher, the American Society of Mechanical Engineers)

SURFACE TEXTURE SYMBOL

The finish symbol may be used to control additional surface irregularities by placing a horizontal line on the top of the right leg, thereby creating a surface texture symbol. These symbols are used to control conditions such as waviness height, waviness width, lay designation, roughness-width cutoff, and roughness width. The figures appearing with the symbol represent units of inches or millimeters (not microinches or micrometers as with roughness height). Placement of the values in relation to the symbol designates what control is assigned; i.e., above the line is waviness height, followed by waviness width, etc. Study the six examples shown below. The direction of the predominant surface pattern may be designated by lay symbol. The standard lay symbols and their meanings are shown on page 147.

63 √ — Roughness height rating is placed at the left of the long leg. The specification indicates the maximum average and any lesser average shall be acceptable.

63 √ .002 — Maximum waviness height rating is placed above the horizontal extension. Any lesser rating shall be acceptable.

63 √ .002–2 — Maximum waviness width rating is placed above the horizontal extension and to the right of the waviness height rating. Any lesser rating shall be acceptable.

63 √ .002–2 ⊥ — Lay designation is indicated by the lay symbol placed at the right of the long leg.

63 √ .002–2 .030 ⊥ — Roughness-width cutoff rating is placed below the horizontal extension.

63 √ .002–2 .030 ⊥ .005 — When required, maximum roughness width rating shall be placed at the right of the lay symbol. Any lesser rating shall be acceptable.

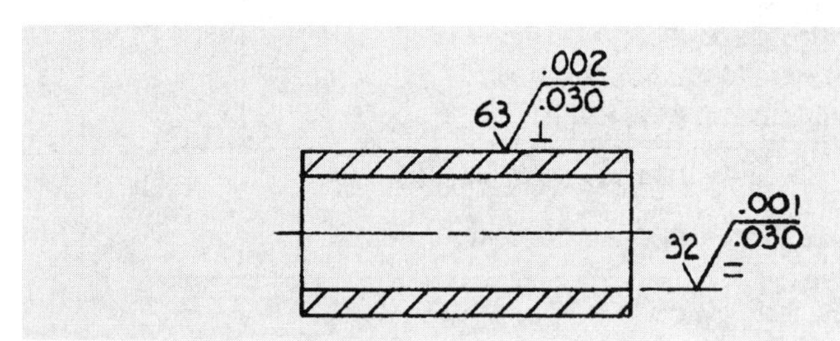

Interpretation:

Roughness height (OD)	63 μin.
Roughness height (ID)	32 μin.
Roughness-width cutoff (OD and ID)	.030 in.
Waviness height (OD)	.002 in.
Waviness height (ID)	.001 in.
Lay (OD)	Perpendicular
Lay (ID)	Parallel

Lay Symbol	Meaning	Example Showing Direction of Tool Marks
=	Lay approximately parallel to the line representing the surface to which the symbol is applied.	
⊥	Lay approximately perpendicular to the line representing the surface to which the symbol is applied.	
X	Lay angular in both directions to line representing the surface to which the symbol is applied.	
M	Lay multidirectional.	
C	Lay approximately circular relative to the center of the surface to which the symbol is applied.	
R	Lay approximately radial relative to the center of the surface to which the symbol is applied.	
P	Lay particulate, non-directional, or protuberant.	

Lay symbols. ANSI Y14.36 (With permission of the publisher, the American Society of Mechanical Engineers)

Symbol	Meaning
(a)	Basic Surface Texture Symbol. Surface may be produced by any method except when the bar or circle (Figure 1b or 1d) is specified.
(b)	Material Removal By Machining Is Required. The horizontal bar indicates that material removal by machining is required to produce the surface and that material must be provided for that purpose.
(c) 3.5	Material Removal Allowance. The number indicates the amount of stock to be removed by machining in millimeters (or inches). Tolerances may be added to the basic value shown or in a general note.
(d)	Material Removal Prohibited. The circle in the vee indicates that the surface must be produced by processes such as casting, forging, hot finishing, cold finishing, die casting, powder metallurgy or injection molding without subsequent removal of material.
(e)	Surface Texture Symbol. To be used when any surface characteristics are specified above the horizontal line or to the right of the symbol. Surface may be produced by any method except when the bar or circle (Figure b and d) is specified.

Surface texture symbols (metric). ANSI Y14.36 (With permission of the publisher, the American Society of Mechanical Engineers)

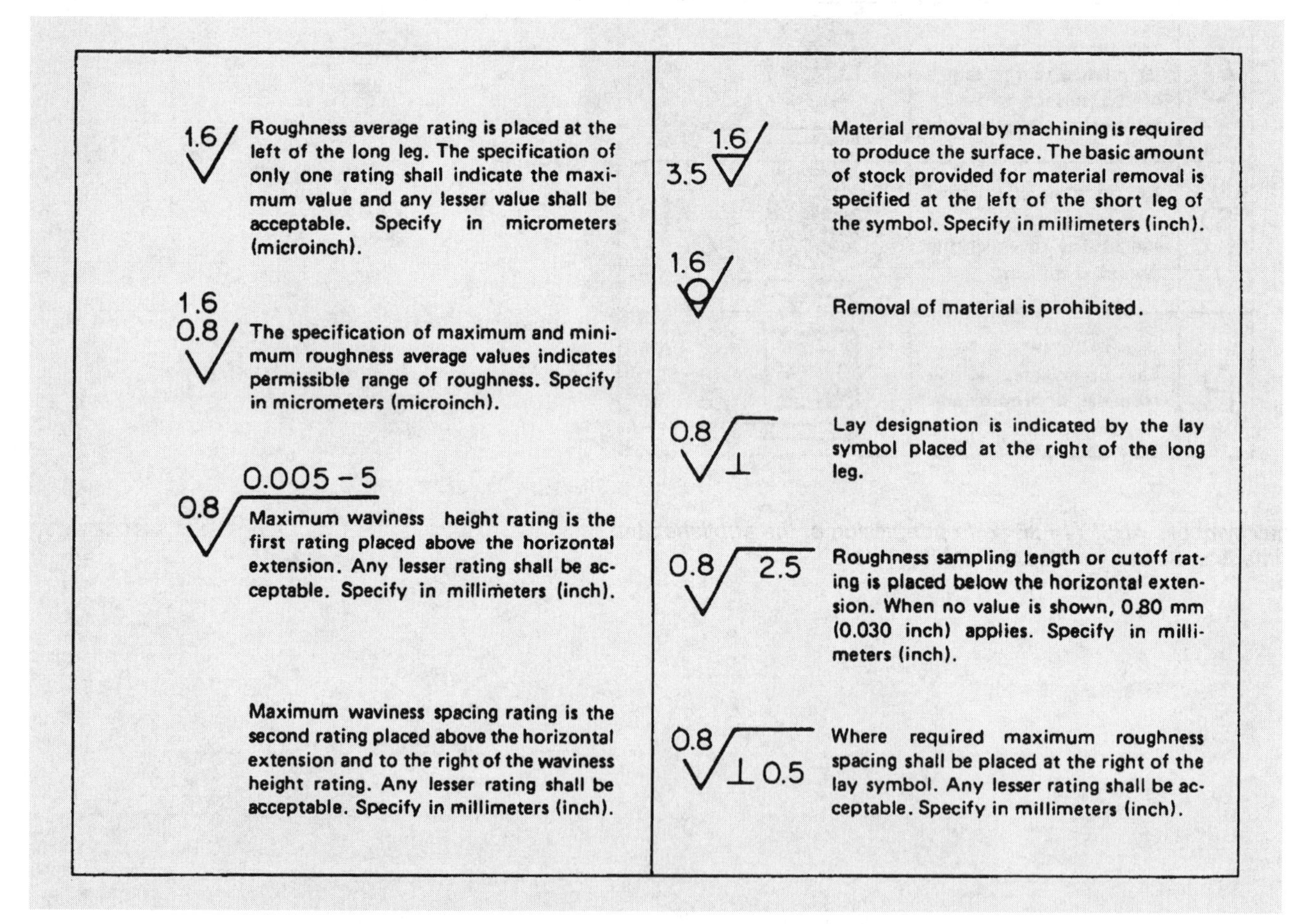

1.6 — Roughness average rating is placed at the left of the long leg. The specification of only one rating shall indicate the maximum value and any lesser value shall be acceptable. Specify in micrometers (microinch).

1.6 / 0.8 — The specification of maximum and minimum roughness average values indicates permissible range of roughness. Specify in micrometers (microinch).

0.005 – 5 / 0.8 — Maximum waviness height rating is the first rating placed above the horizontal extension. Any lesser rating shall be acceptable. Specify in millimeters (inch).

Maximum waviness spacing rating is the second rating placed above the horizontal extension and to the right of the waviness height rating. Any lesser rating shall be acceptable. Specify in millimeters (inch).

1.6 / 3.5 — Material removal by machining is required to produce the surface. The basic amount of stock provided for material removal is specified at the left of the short leg of the symbol. Specify in millimeters (inch).

1.6 — Removal of material is prohibited.

0.8 ⊥ — Lay designation is indicated by the lay symbol placed at the right of the long leg.

0.8 / 2.5 — Roughness sampling length or cutoff rating is placed below the horizontal extension. When no value is shown, 0.80 mm (0.030 inch) applies. Specify in millimeters (inch).

0.8 ⊥ 0.5 — Where required maximum roughness spacing shall be placed at the right of the lay symbol. Any lesser rating shall be acceptable. Specify in millimeters (inch).

Application of surface texture values to symbols (metric). ANSI Y14.36 (With permission of the publisher, the American Society of Mechanical Engineers)

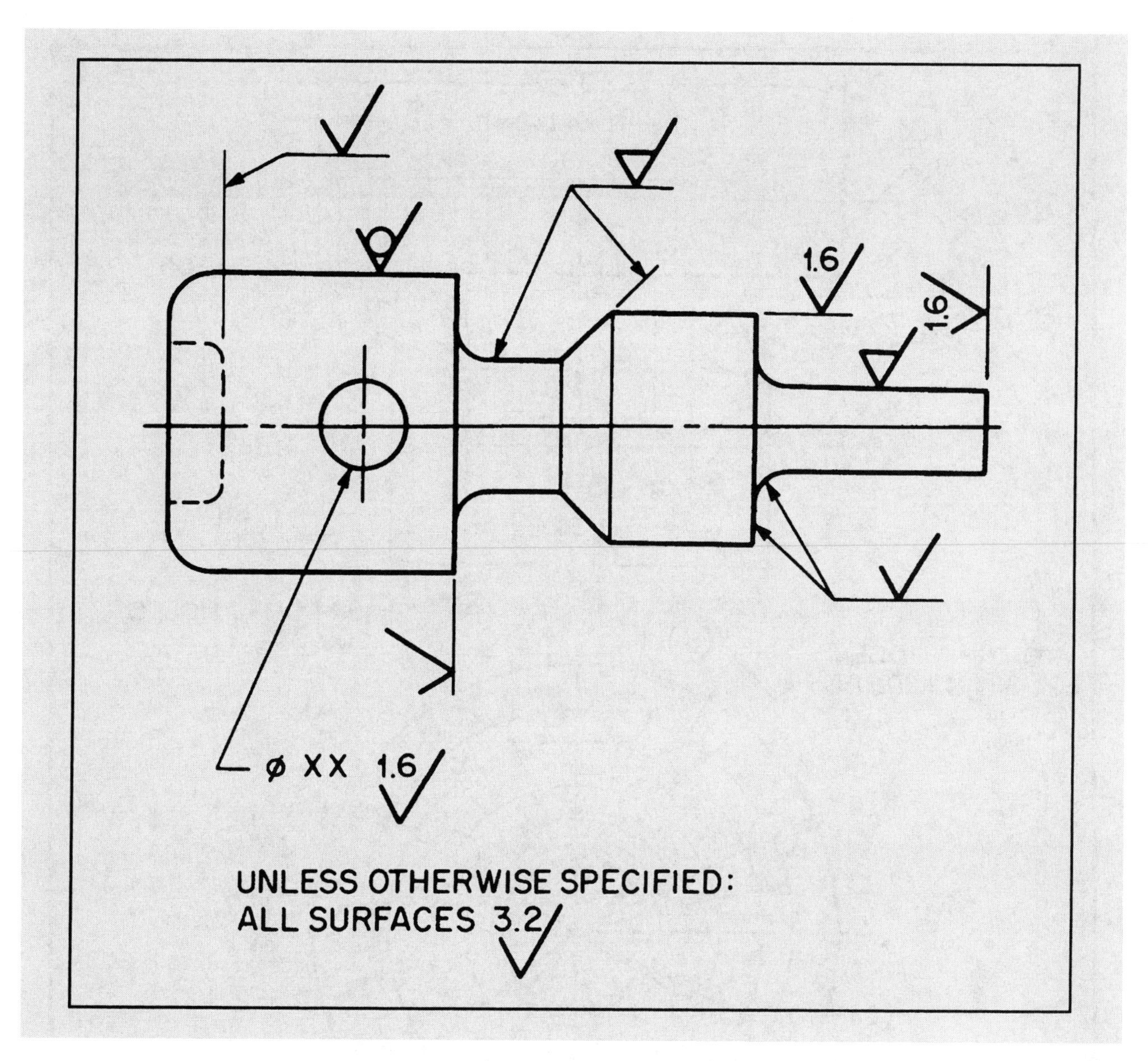

Application of surface texture symbols (metric). ANSI Y14.36 (With permission of the publisher, the American Society of Mechanical Engineers)

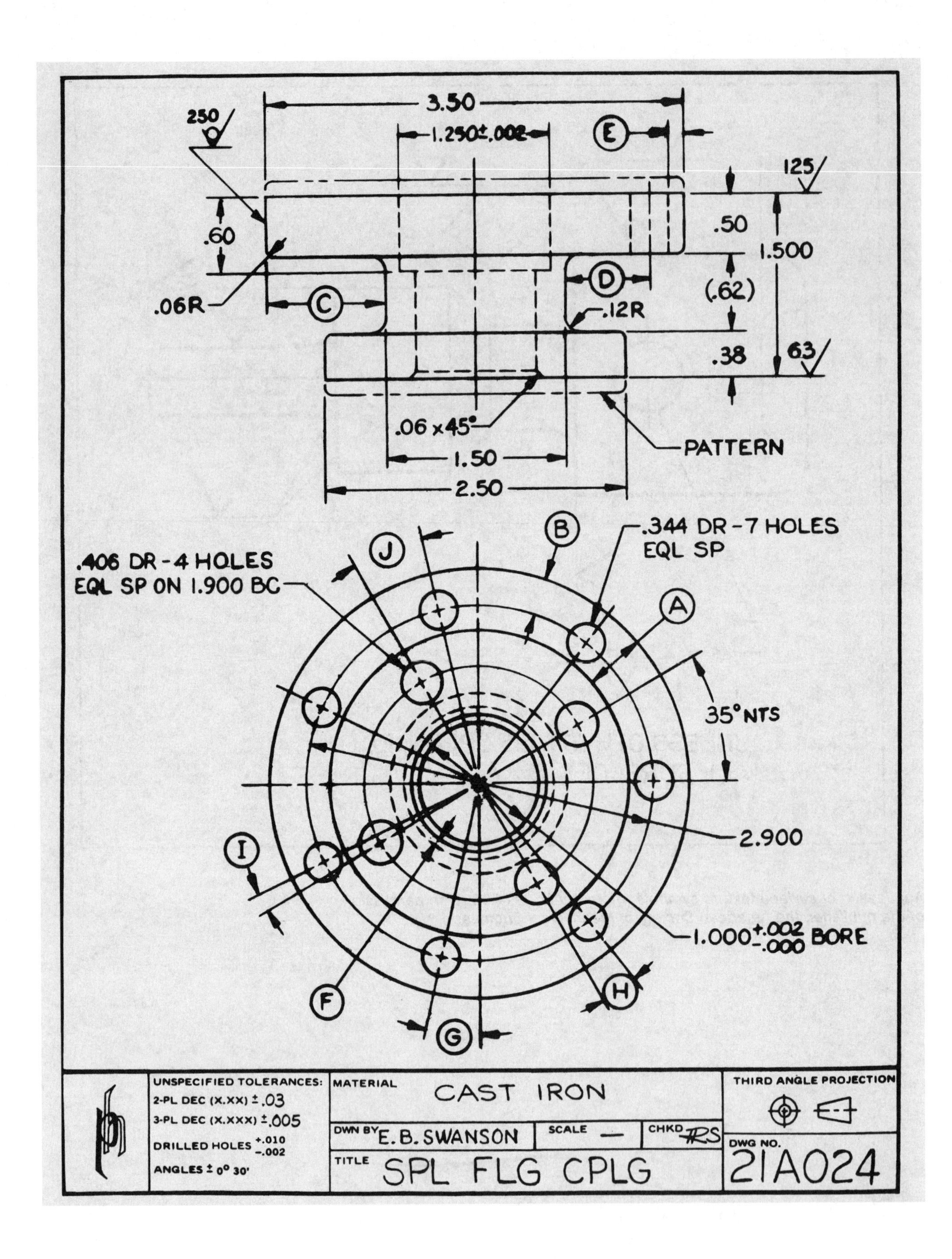
3.50
1.250±.002
E
250
125
.60
.50
1.500
(.62)
D
.06R
C
.12R
.38
63
.06 x 45°
PATTERN
1.50
2.50
B
.344 DR - 7 HOLES
EQL SP
J
.406 DR - 4 HOLES
EQL SP ON 1.900 BC
A
35° NTS
2.900
I
1.000 +.002 -.000 BORE
F
G
H
UNSPECIFIED TOLERANCES:
2-PL DEC (X.XX) ±.03
3-PL DEC (X.XXX) ±.005
DRILLED HOLES +.010 -.002
ANGLES ± 0° 30'
MATERIAL
CAST IRON
DWN BY E. B. SWANSON
SCALE —
CHKD RS
TITLE SPL FLG CPLG
THIRD ANGLE PROJECTION
DWG NO.
21A024

INSTRUCTIONS: Refer to drawing 21A024. Enter the letter that represents the correct completion of each statement.

1. The limits of the bore diameter are: (a) 1.000–1.002, (b) .998–1. 000, or (c) .998–1.002 — 1. ______
2. The MMC of the counterbore diameter is: (a) 1.252, (b) 1.250, or (c) 1.248 — 2. ______
3. Surface roughness height of the large flange face is: (a) .00125, (b) .000125, or (c) .0000125. — 3. ______
4. The bolt circles and the bore are: (a) concentric, or (b) eccentric. — 4. ______
5. The type of line drawn to represent the pattern is: (a) centerline, (b) hidden line, or (c) phantom line. — 5. ______
6. The smoothest surface designated is the: (a) small flange face, (b) large flange face, or (c) large flange diameter. — 6. ______
7. The surface texture symbol on the large diameter designates: (a) material removal by machining required, or (b) material removal prohibited. — 7. ______
8. The parentheses enclosing the .62 dimension indicate that it is: (a) a reference dimension, (b) revised, or (c) not-to-scale. — 8. ______
9. The sides of the holes are closest to the outside edge in the: (a) large flange, or (b) small flange. — 9. ______
10. The fasteners intended for use in the small flange would be: (a) .437, (b) .406, or (c) .375. — 10. ______
11. The minimum acceptable size of the seven equally spaced holes is: (a) .339, (b) .342, or (c) .404. — 11. ______
12. The bore diameter is toleranced: (a) bilaterally, (b) unilaterally, or (c) by limits. — 12. ______
13. Normal finish allowance for a finished surface is: (a) .000125, (b) .012, or (c) .12. — 13. ______
14. The ID at the small flange face is: (a) chamfered, (b) countersunk, or (c) counterbored. — 14. ______
15. The .62 dimension accumulates a tolerance of: (a) ± .03, (b) ± .06, or (c) ± .065. — 15. ______

INSTRUCTIONS: Enter the dimensions for the following letters.

Ⓐ ______ Ⓔ ______ Ⓗ ______

Ⓑ ______ Ⓕ ______ Ⓘ ______

Ⓒ ______ Ⓖ ______ Ⓙ ______

Ⓓ ______

Optional Math Exercises:

1a. Calculate the Cartesian coordinates for the 4 .406∅ drilled holes.

2a. Calculate the Measurement Over Pins (MOP) of two adjacent .406∅ holes if they are at MMC.

3a. Calculate the Cartesian coordinates for the 7 .344∅ drilled holes.

4a. Calculate the MOP of two adjacent .344∅ holes if they are at LMC.

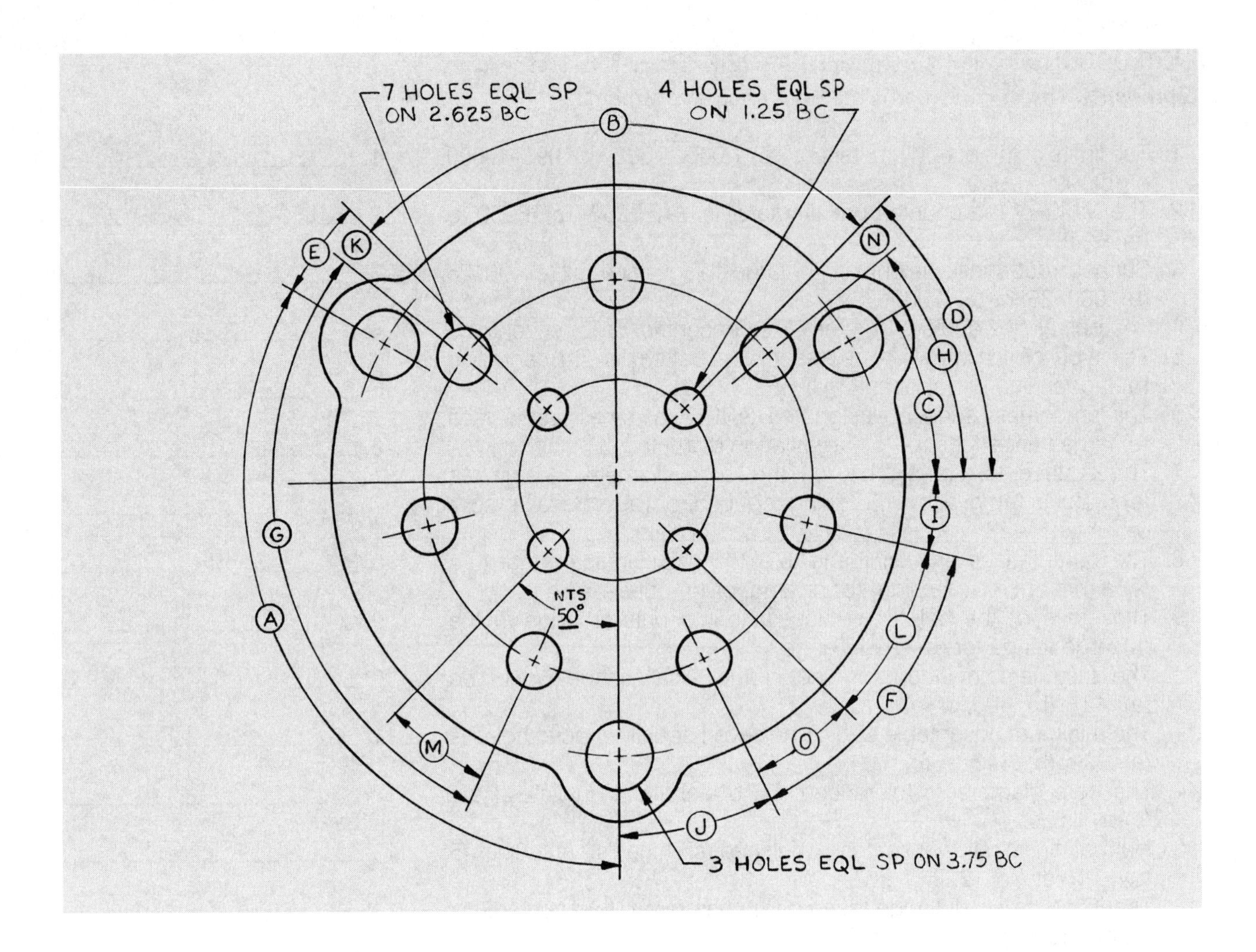

Angular Calculations 5

INSTRUCTIONS: Calculate the angular dimension for each of the letters shown in the illustration above.

Ⓐ ____________ Ⓕ ____________ Ⓚ ____________

Ⓑ ____________ Ⓖ ____________ Ⓛ ____________

Ⓒ ____________ Ⓗ ____________ Ⓜ ____________

Ⓓ ____________ Ⓘ ____________ Ⓝ ____________

Ⓔ ____________ Ⓙ ____________ Ⓞ ____________

Optional Math Exercise:

1a. Calculate the Cartesian coordinates for each of the holes in the illustration above.

BOSSES AND PADS

A common practice in casting design is to provide additional metal adjacent to a hole to allow for machining a flat surface perpendicular to the axis. When the shape of the projecting surface is round, it is called a boss. When the shape is other than round, it is referred to as a pad. A pad may have a slotted hole through it, a series of holes, or perhaps no hole at all. (See the examples shown below.)

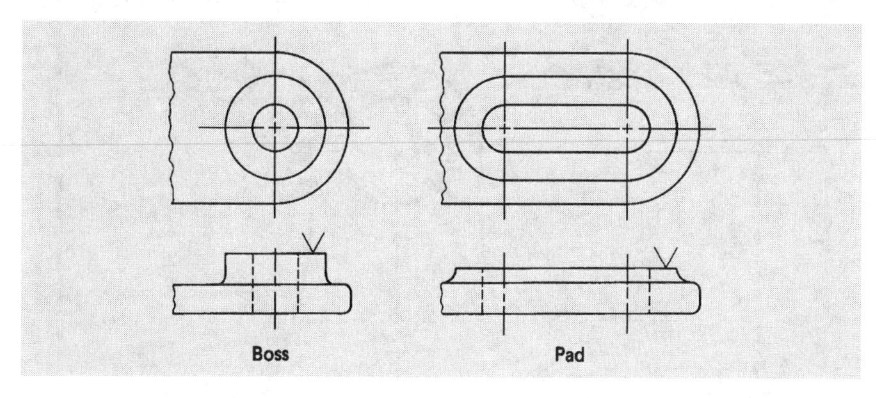

SLOTTED HOLES

Observe the differences in tolerance accumulation (shown in parentheses) between slotted holes X and Y in the illustrations below. When the width is specified (illustration X), the length is confined to two tolerances; but if the radius is specified (illustration Y), the length will accumulate three tolerances. Length tolerance can be limited to no accumulation if dimensioned by the third method illustrated below.

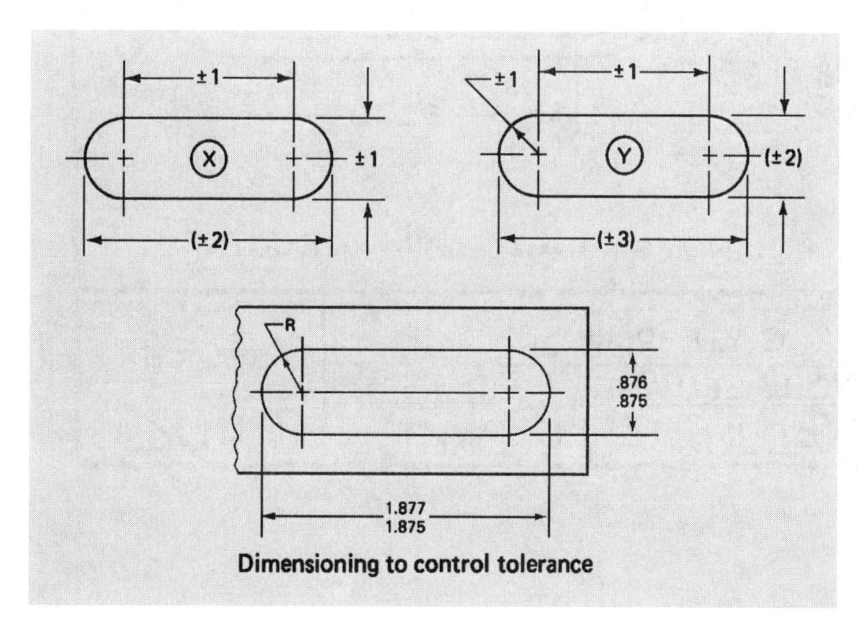

Dimensioning to control tolerance

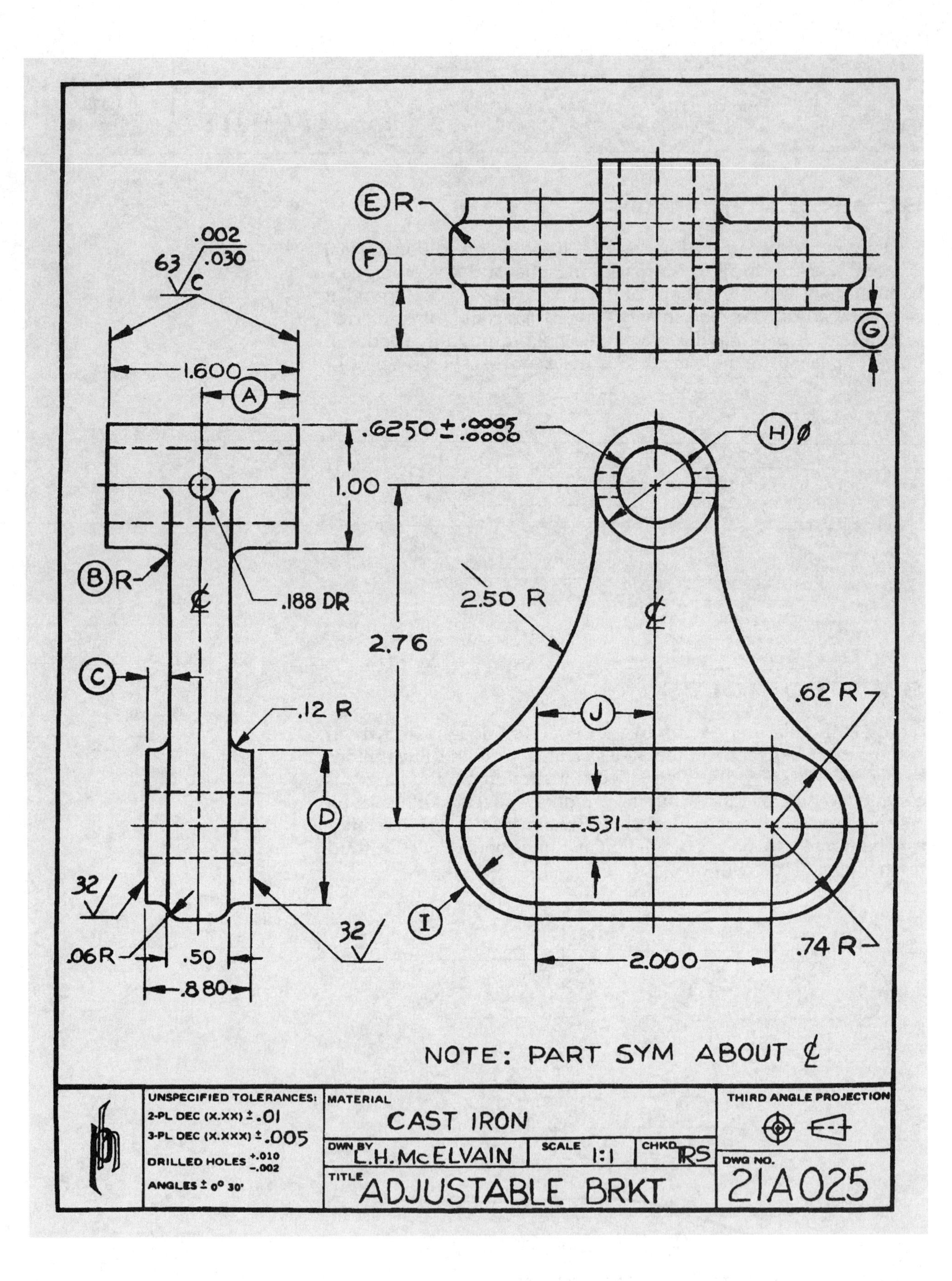
.002
63
.030
C
1.600
A
.6250 ± .0005 .0000
H Ø
1.00
B R
.188 DR
2.50 R
2.76
C
.12 R
J
.62 R
D
.531
32
I
32
.06R
.50
.74 R
2.000
.880
E R
F
G
NOTE: PART SYM ABOUT ℄
UNSPECIFIED TOLERANCES:
2-PL DEC (X.XX) ± .01
3-PL DEC (X.XXX) ± .005
DRILLED HOLES +.010 −.002
ANGLES ± 0° 30'
MATERIAL
CAST IRON
DWN BY L.H. McELVAIN
SCALE 1:1
CHKD RS
TITLE ADJUSTABLE BRKT
THIRD ANGLE PROJECTION
DWG NO. 21A025

CASTING DIMENSIONS

Dimensions found on casting drawings will usually be held to two places (decimal inch system), except for machined hole sizes or dimensions between finished surfaces, which are usually held to three or more places for closer tolerance control. Note the symbol ℄ used on the front and left-side views to indicate symmetry and thereby reduce the number of dimensions otherwise required. Also observe that the 2.50 radius does not require location dimensions because it runs tangent to two other radii.

INSTRUCTIONS: Refer to drawing 21A025 to answer the following questions.

1. What is the dimension of the fillets? 1. ____________
2. What is the overall height of the bracket? 2. ____________
3. What is the overall width of the bracket? 3. ____________
4. What would be the overall depth (1.600 dimension) before machining? 4. ____________
5. What is the tolerance on the drilled hole? 5. ____________
6. What is the maximum diameter of the large hole? 6. ____________
7. What size fasteners would be used in the slotted hole? 7. ____________
8. What is the maximum overall length of the slotted hole? (See page 153.) 8. ____________
9. What is the minimum overall length of the pad? 9. ____________
10. What is the wall thickness of the boss? 10. ____________
11. Is the surface roughness smoother on the pads or on the bosses? 11. ____________
12. Show the roughness height rating of the pads in decimal inch value (not microinch). 12. ____________
13. What does the .002 figure on the boss surface texture symbol control? (See page 146.) 13. ____________
14. What is the roughness-width cutoff rating on the boss faces? (See page 146.) 14. ____________
15. What does the letter C in the boss surface texture symbol designate? (See page 147.) 15. ____________

INSTRUCTIONS. Enter the dimensions for the following letters.

Ⓐ ____________ Ⓕ ____________

Ⓑ ____________ Ⓖ ____________

Ⓒ ____________ Ⓗ ____________

Ⓓ ____________ Ⓘ ____________

Ⓔ ____________ Ⓙ ____________

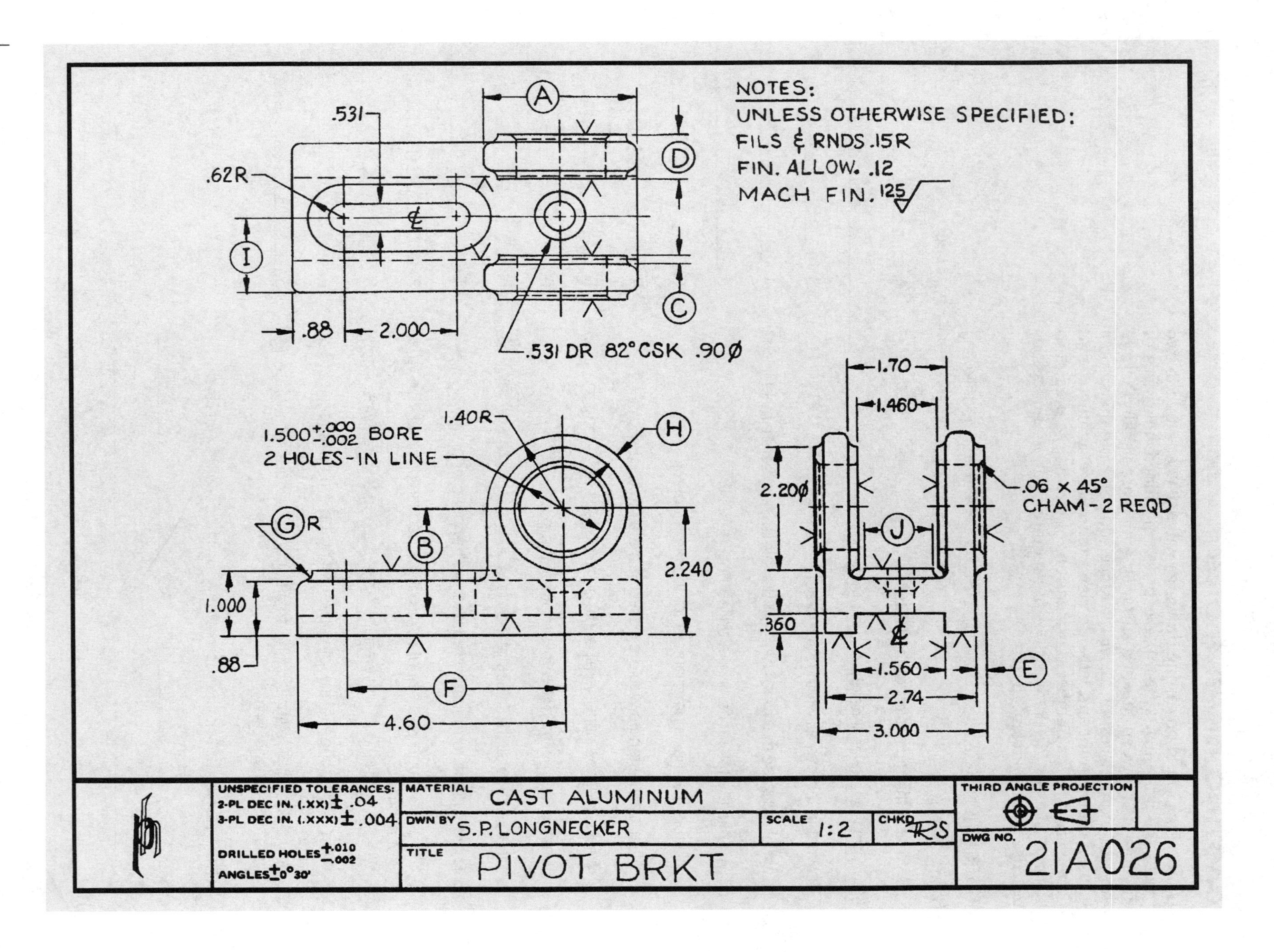

NOTES:
UNLESS OTHERWISE SPECIFIED:
FILS & RNDS .15R
FIN. ALLOW. .12
MACH FIN. 125
.531
.62R
.88
2.000
.531 DR 82° CSK .90⌀
1.500 +.000 -.002 BORE
2 HOLES-IN LINE
1.40R
2.240
1.000
.88
4.60
1.70
1.460
2.20⌀
.06 X 45°
CHAM - 2 REQD
.360
1.560
2.74
3.000
A
B
C
D
E
F
G R
H
I
J
UNSPECIFIED TOLERANCES:
2-PL DEC IN. (.XX) ± .04
3-PL DEC IN. (.XXX) ± .004
DRILLED HOLES +.010 -.002
ANGLES ± 0°30'
MATERIAL CAST ALUMINUM
DWN BY S.P. LONGNECKER
SCALE 1:2
CHKD RS
TITLE PIVOT BRKT
THIRD ANGLE PROJECTION
DWG NO. 21A026

INSTRUCTIONS: *Refer to drawing 21A026 to answer the following questions.*

1. Was the print drawn to: (a) half scale, (b) full scale, or (c) double scale? 1. ______
2. Does the surface texture symbol designate that material removal is: (a) prohibited, (b) controlled by waviness, or (c) required by machining? 2. ______
3. Is the surface roughness height designated in decimal inch value as: (a) .125, (b) .00125, or (c) .000125? 3. ______
4. Would the fastener size for the slotted hole be: (a) .500, (b) .531, or (c) .562? 4. ______
5. Is the bore dimensioned by: (a) limits, (b) unilateral tolerance, or (c) bilateral tolerance? 5. ______
6. Is the included angle of the countersink: (a) 41°, (b) 82°, or (c) 164°? 6. ______
7. Does the slotted hole length accumulate a tolerance of: (a) ±.004, (b) ±.008, or (c) ±.012? 7. ______
8. What is the MMC of the bored holes? 8. ______
9. How much material does the slotted hole pass through after machining? 9. ______
10. What would the rough casting dimension be between the inside boss faces, prior to machining the 1.460 dimension? 10. ______

INSTRUCTIONS: *Enter the dimensions for the following letters.*

Ⓐ ______

Ⓑ ______

Ⓒ ______

Ⓓ ______

Ⓔ ______

Ⓕ ______

Ⓖ ______

Ⓗ ______

Ⓘ ______

Ⓙ ______

TAPERS

Many machined shafts will include tapered portions, and there are several methods used for dimensioning them. One method includes dimensioning both the large and small diameters along with the length (distance between beginning and end of tapered portion). Another method uses a leader with a dimension and the abbreviation TPF to designate the taper per foot. When this method is used, only one of the diameters and the length are specified. If the large diameter is given, the small diameter may be calculated by the following sequence. First, the TPF dimension must be converted to inches (divide by 12), then multiply it by the tapered length and subtract the product from the large diameter. If the small diameter is given, follow the same sequence except add the product to the small diameter. See the examples shown below.

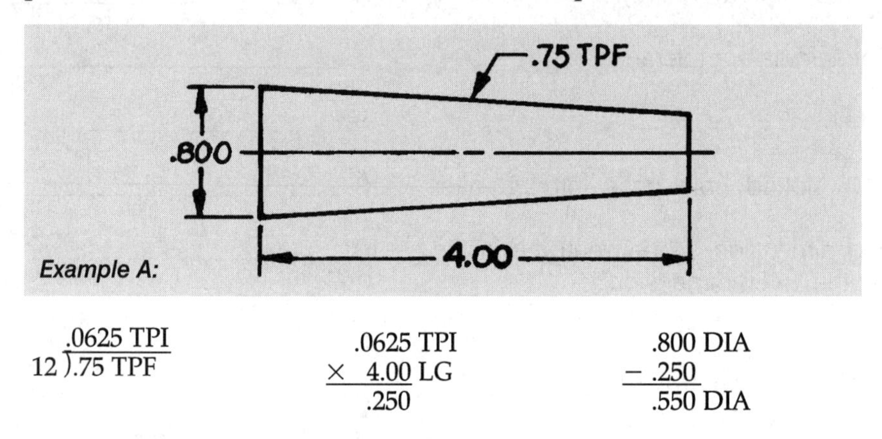

Example A:

12)‾.75 TPF = .0625 TPI	.0625 TPI × 4.00 LG = .250	.800 DIA − .250 = .550 DIA

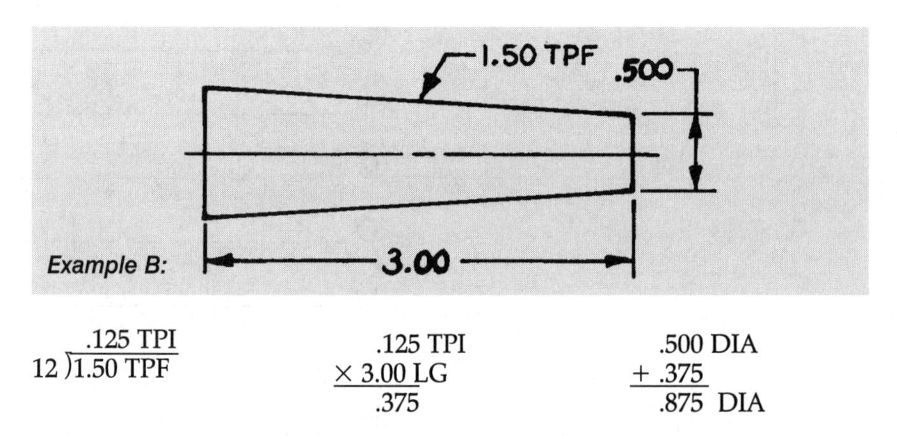

Example B:

12)‾1.50 TPF = .125 TPI	.125 TPI × 3.00 LG = .375	.500 DIA + .375 = .875 DIA

STANDARD TAPERS

Tapered shanks that fit into spindles or sockets are sometimes provided on small tools or machine parts. Standard tapers such as this may include only a leader with the note "MORSE TAPER," for example. When this occurs, refer to any of the various machinist handbooks for more information. For instance, there you will find that the Morse Taper is approximately .625 TPF, the Brown & Sharpe Taper is approximately .500 TPF, the Jarno Taper is .600 TPF, etc.

Taper Calculations

INSTRUCTIONS: Calculate the missing dimension for each of the following taper problems. Round your answers to three decimal places.

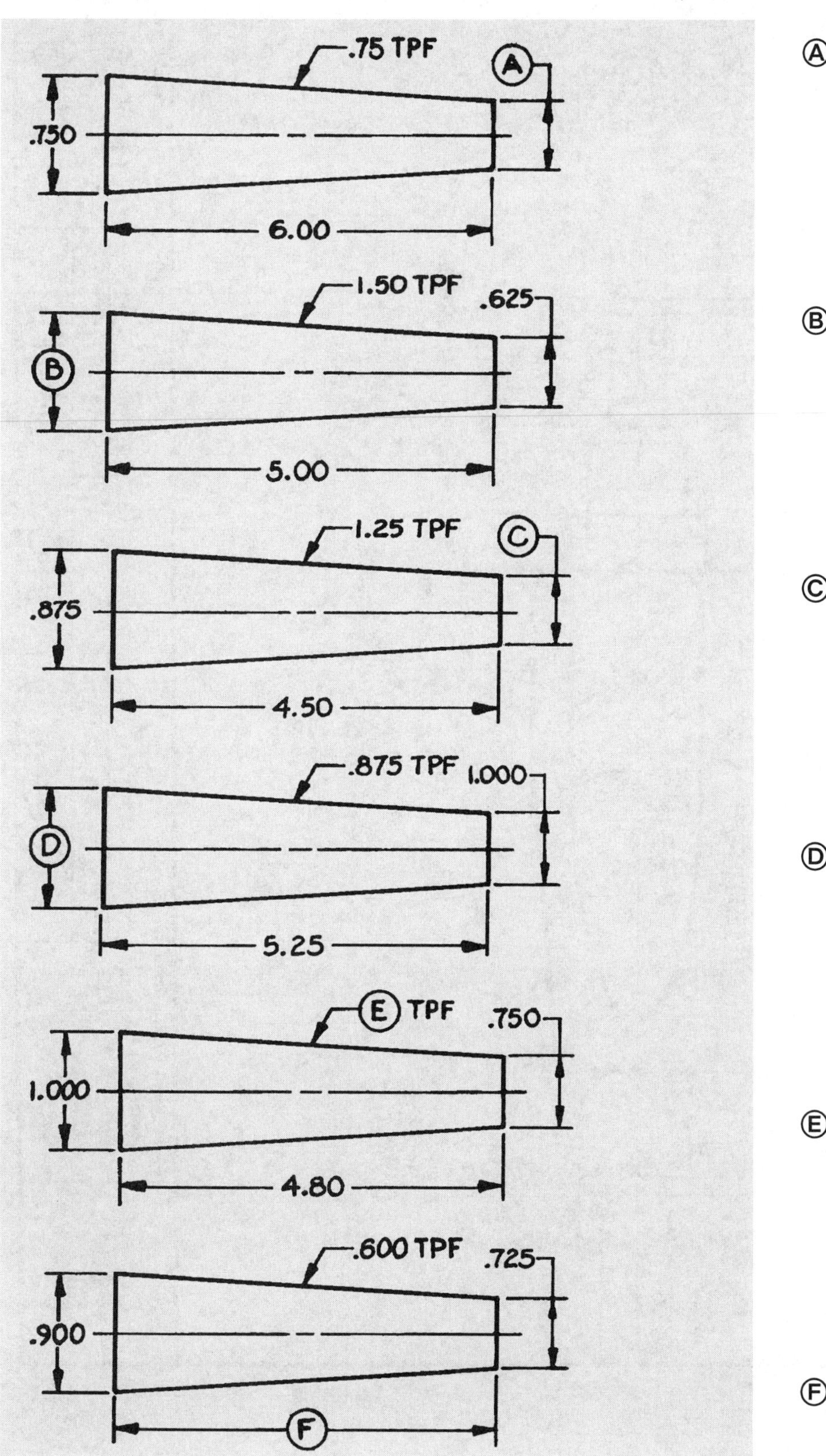

Ⓐ ______________________

Ⓑ ______________________

Ⓒ ______________________

Ⓓ ______________________

Ⓔ ______________________

Ⓕ ______________________

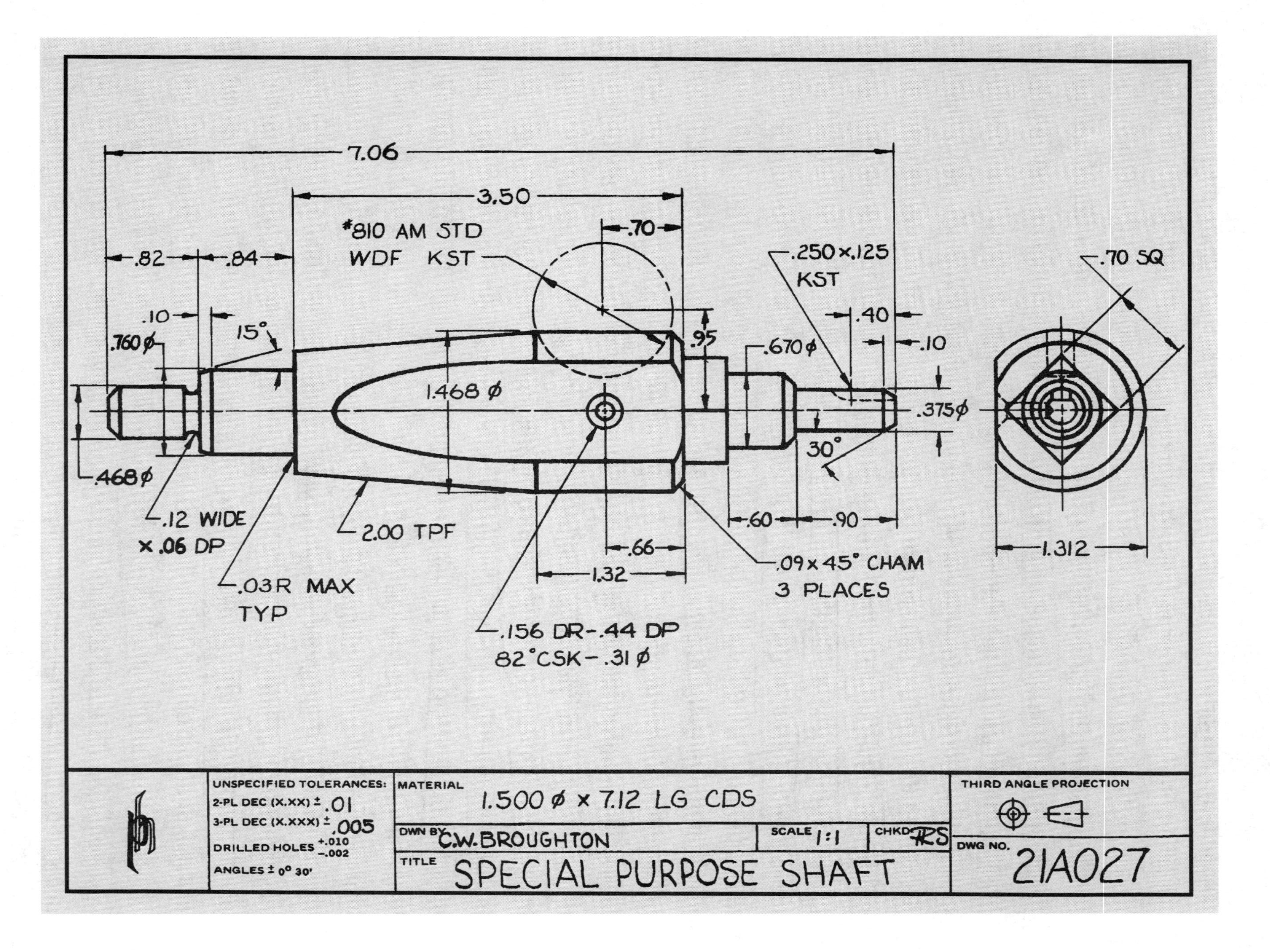

7.06
3.50
#810 AM STD
WDF KST
.70
.82
.84
.10
15°
.760ϕ
.468ϕ
.12 WIDE
x .06 DP
.03R MAX
TYP
2.00 TPF
1.468ϕ
.95
.250 x .125
KST
.40
.10
.670ϕ
.375ϕ
30°
.60
.90
.66
1.32
.09 x 45° CHAM
3 PLACES
.156 DR - .44 DP
82°CSK - .31ϕ
.70 SQ
1.312
UNSPECIFIED TOLERANCES:
2-PL DEC (X.XX) ± .01
3-PL DEC (X.XXX) ± .005
DRILLED HOLES +.010 -.002
ANGLES ± 0° 30'
MATERIAL
1.500ϕ x 7.12 LG CDS
DWN BY C.W. BROUGHTON
SCALE 1:1
CHKD RS
TITLE SPECIAL PURPOSE SHAFT
THIRD ANGLE PROJECTION
DWG NO. 21A027

NECKS

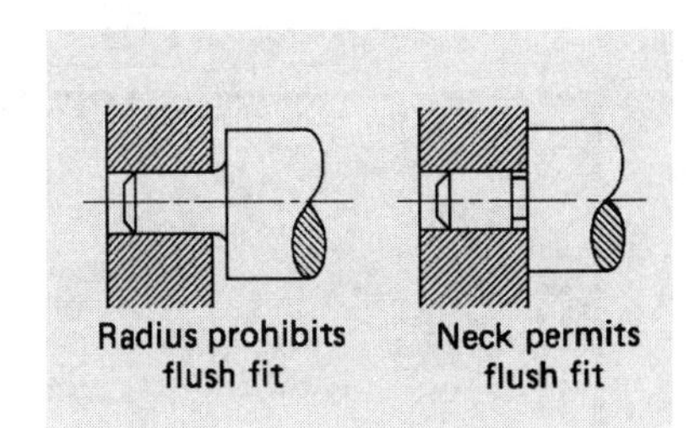

Normally, a small radius will occur between machined steps of a shaft. (See the example at the right.) Since these radii are not necessarily desirable, but must be tolerated, a maximum radial dimension may be shown. This will permit a radius but prevent it from becoming too large. When no radius can be accepted, such as with close-fitting parts, a neck may be machined to eliminate the radius. (See the example at the right.) Necks may be dimensioned by width and depth, or by width and diameter.

INSTRUCTIONS: Refer to drawing 21A027 to answer the following questions.

1. How many chamfers does the shaft contain? 1. ____________
2. What is the depth of the chamfer on the left end? 2. ____________
3. What is the included angle of the chamfer on the right end? 3. ____________
4. What is the depth of the blind hole? 4. ____________
5. What is the MMC of the blind hole diameter? 5. ____________
6. What is the countersink diameter? 6. ____________
7. What is the included angle of the countersink? 7. ____________
8. What is the MMC of the largest OD? 8. ____________
9. What Woodruff keyseat cutter number is specified? 9. ____________
10. What are the basic dimensions of the keyseat cover? (Refer to page 5.) 10. ____________
11. What is the distance from the bottom of the keyseat to the opposite side of the shaft? 11. ____________
12. What is the maximum radius permitted between steps of the shaft? 12. ____________
13. What is the diameter of the neck? 13. ____________
14. What is the dimension across flats of the square step? 14. ____________
15. What is the length of the square step? (Measured along the axis.) 15. ____________
16. How much tolerance can accumulate on the length of the square step? 16. ____________
17. What is the length of the tapered portion? 17. ____________
18. How much tolerance can accumulate on the length of the tapered portion? 18. ____________
19. What is the diameter of the small end of the tapered portion? 19. ____________
20. If the first and second steps on the left end were each machined to .83 length, would they both be within tolerance? 20. ____________

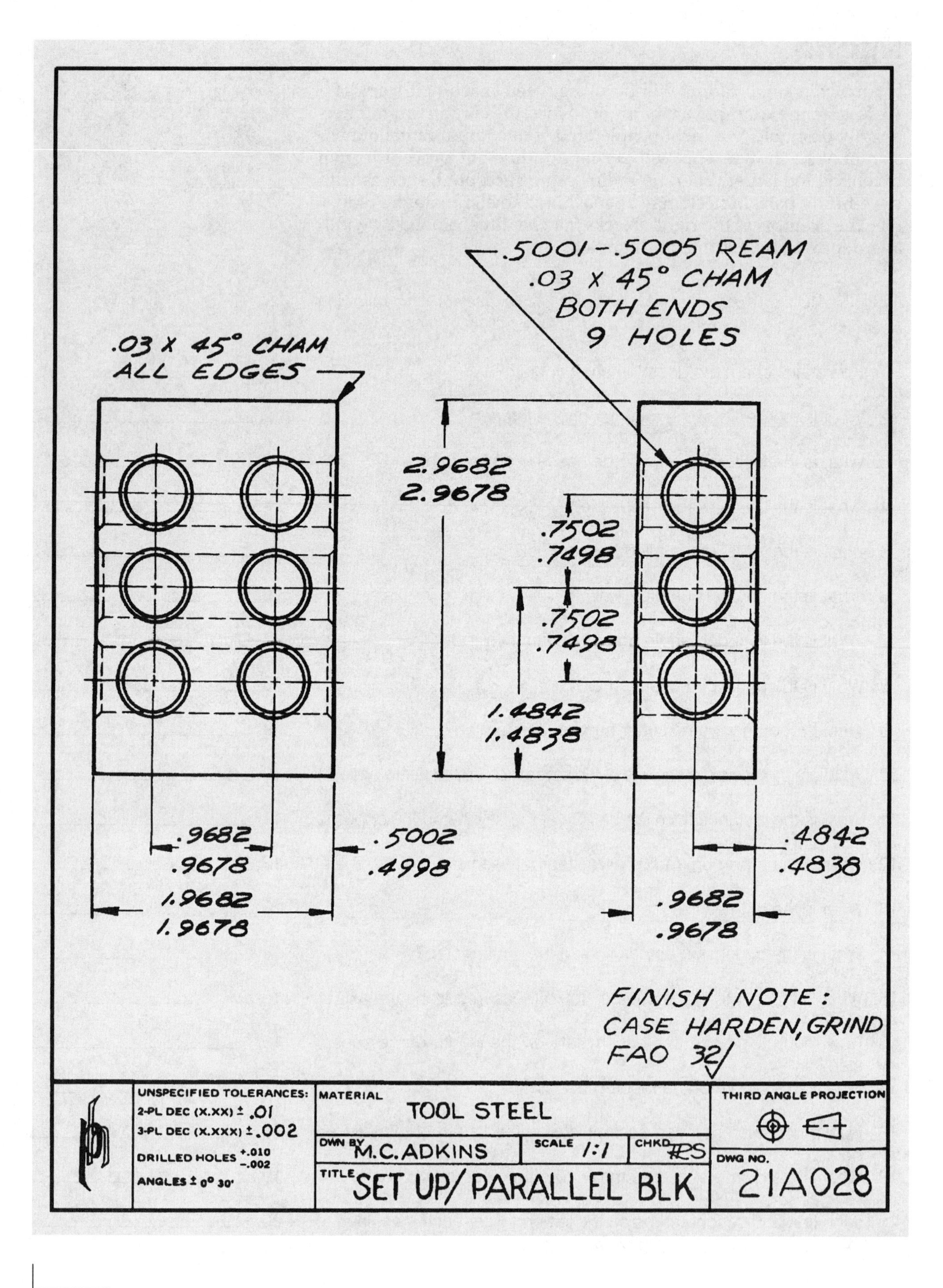
.5001-.5005 REAM
.03 X 45° CHAM
BOTH ENDS
9 HOLES
.03 X 45° CHAM
ALL EDGES
2.9682
2.9678
.7502
.7498
.7502
.7498
1.4842
1.4838
.9682
.9678
.5002
.4998
1.9682
1.9678
.4842
.4838
.9682
.9678
FINISH NOTE:
CASE HARDEN, GRIND
FAO 32
UNSPECIFIED TOLERANCES:
2-PL DEC (X.XX) ± .01
3-PL DEC (X.XXX) ± .002
DRILLED HOLES +.010 -.002
ANGLES ± 0° 30'
MATERIAL
TOOL STEEL
DWN BY M.C. ADKINS
SCALE 1:1
CHKD RS
TITLE SET UP/ PARALLEL BLK
THIRD ANGLE PROJECTION
DWG NO.
21A028

LIMITS

Limits are the maximum and minimum permissible sizes for specific dimensions. As discussed earlier, every dimension has a tolerance. These tolerances may be assigned to individual dimensions, or they may be general tolerances that apply to all untoleranced dimensions. When limits are used, the tolerance is the difference between limits. For example, the tolerance for .500–.505 is .005. (Do not attempt to assign plus or minus values.)

Any dimension may be converted into limits simply by adding and/or subtracting its assigned tolerance. For example, the limits for a .500 ±.005 dimension would be .495–.505 and the limits for .375 + .010 −.000 would be .375–.385, etc. When expressed in a single line, the low limit precedes the high limit separated by a dash. Otherwise, proper placement is for the high limit to appear above the low limit as was done on drawing 21A028.

When adding limits, the greatest accumulation of tolerance occurs by adding the maximum limits together and by adding the minimum limits together. When subtracting limits, however, the greatest tolerance accumulation occurs by subtracting the minimum limit from the maximum limit, and by subtracting the maximum limit from the minimum limit. (See the examples shown at the right.) These procedures must be followed when solving questions 7 through 10 below.

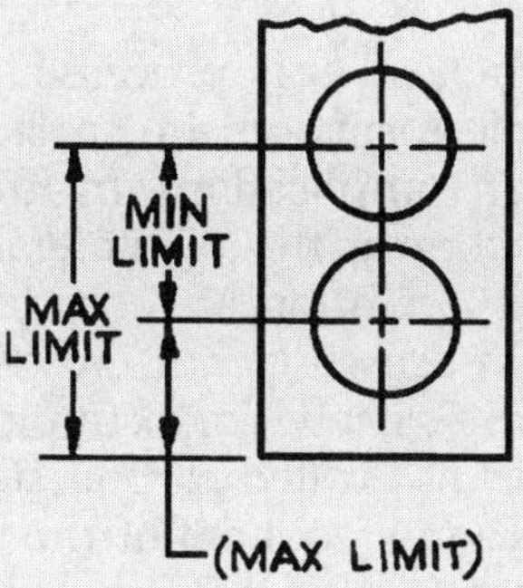

MIN LIMIT
MAX LIMIT
(MAX LIMIT)

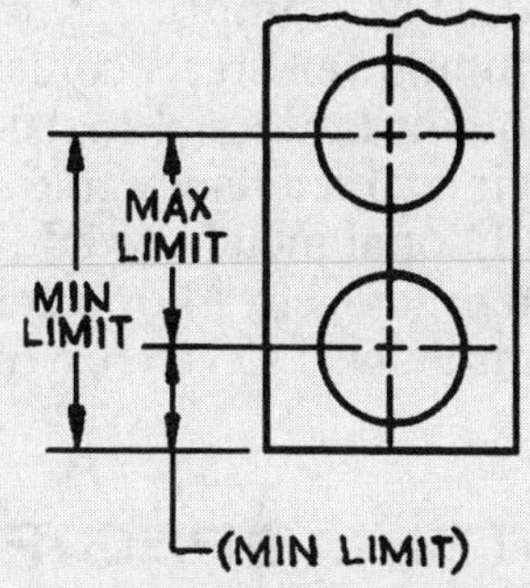

MAX LIMIT
MIN LIMIT
(MIN LIMIT)

SINGLE LIMITS

Single limit dimensions are followed by the abbreviation MAX or MIN. Other elements of design determine the other unspecified limit. Corner radii, chamfers, depth of holes, length of threads, etc. may be dimensioned by this method. Drawing 21A027 on page 160 used this method to control the radius that may occur between machined steps. By specifying .03R MAX, any radius less than that amount is acceptable.

INSTRUCTIONS: *Refer to drawing 21A028 to answer the following questions.*

1. What is the tolerance on the chamfer depth? 1. ______
2. What is the tolerance on the chamfer angle? 2. ______
3. What is the tolerance on the size of the nine reamed holes? 3. ______
4. What does the finish note FAO abbreviate? 4. ______
5. What is the surface roughness height rating? 5. ______
6. Show the block thickness dimension with a tolerance equal in both directions (bilateral). 6. ______
7. What is the maximum dimension between centers of the top row of holes and the bottom row? 7. ______
8. What is the maximum dimension between the bottom row and the bottom of the block? (See top illus. above.) 8. ______
9. What is the minimum dimension between the bottom row of holes and the bottom of the block? (See bottom illus. above.) 9. ______
10. What is the maximum dimension between the top row of holes and the top of the block? 10. ______

STEEL PROCESSING

When steel is processed hot, such as hot-rolled or hot-drawn, a scale forms on the surface, causing a roughness unacceptable for many applications. In addition, sizes may vary slightly due to contraction as the solid metal cools. However, an extra rolling or drawing operation at room temperature (referred to as cold-rolling or cold-drawing) not only eliminates surface scale, but controls size more closely. For example, manufacturing tolerances on 1-in.-diameter hot-rolled carbon steel bars are ±.009, whereas cold-finished carbon steel bars of the same diameter have a much closer tolerance, +.000–.002.

Steel is categorized into three general classifications: carbon steel, alloy steel, and tool steel. Carbon steels are the workhorses of industry, comprising more than 90% of all steel produced. Although all steels contain iron and carbon, carbon steels do not contain additional metals in amounts significant enough to alter their characteristics. In contrast, alloy steels do contain one or more additional metals, to add desirable characteristics such as strength, toughness, etc. Alloy steels are named after their major alloying element: nickel steel, chromium steel, molybdenum steel, etc.

STEEL SPECIFICATIONS

The American Iron and Steel Institute (AISI) and the Society of Automotive Engineers (SAE) have both adopted the four-digit numbering system for steel identification. The last two digits represent the percentage of carbon content in hundredths of 1%. (If a steel contains more than 1% carbon it will add a fifth digit to the system.) When the first digit is a 2 through 9, it represents the major alloying element of an alloy steel. See the code numbers on page 165. The second digit of an alloy steel specification represents the approximate percentage of the major alloying element. When the first digit is a 1, it denotes a carbon steel with no alloying element. The second digit of a carbon steel specification denotes plain carbon if it is a zero, resulfurized if it is a 1, resulfurized and rephosphorized if it is a 2, etc. (See the table on page 166.) For example, SAE 2530 designates nickel alloy steel, with approximately 5% nickel and 0.30% carbon. AISI 1117 designates resulfurized carbon steel, with 0.17% carbon.

In the SAE and AISI systems, the first number frequently, but not always, indicates the basic type of steel, as follows:

1 Carbon
2 Nickel
3 Nickel-chromium
4 Molybdenum
5 Chromium
6 Chromium-vanadium
7 Tungsten
8 Nickel-chromium-molybdenum
9 Silicon-manganese

GENERAL PROPERTIES OF ALLOY STEELS

NICKEL STEEL usually contains 3–4% nickel, rarely over 5% and from 0.20 to 0.40% carbon. The advantages of nickel steel are in its toughness and strength.

MOLYBDENUM STEEL is generally used for high speed metal cutting tools, and has air-hardening properties.

CHROMIUM STEEL is used in the construction of safes, forging dies, tools, and castings subjected to severe stresses. It has the quality of toughness and stiffness.

VANADIUM STEEL is adaptable for springs, car axles, and parts which must withstand constant vibrations and varying stresses. Vanadium is a fatigue resistor.

TUNGSTEN STEEL has properties similar to molybdenum except that a larger amount of tungsten would be required for corresponding results. It adds wear resistance to cutting tools.

MANGANESE adds strength and hardness to steel.

SILICON is an impurity and hardener. There are two types of silicon steel, one employed for electrical and the other for mechanical construction. In electrical construction it is used for transformer cores, etc., where a high degree of magnetic conductivity is desired. For mechanical purposes it is used for lead type springs.

AISI and SAE Designations of Steels (Expressed in percents)

Series	Type and Nominal Alloy Content
	Carbon steels
10XX	Plain Carbon (Mn 1.00 max)
11XX	Resulfurized
12XX	Resulfurized and Rephosphorized
15XX	Plain Carbon (max Mn range 1.00 to 1.65)
	Manganese steels
13XX	Mn 1.75
	Nickel steels
23XX	Ni 3.50
25XX	Ni 5.00
	Nickel-chromium steels
31XX	Ni 1.25; Cr 0.65 and 0.80
32XX	Ni 1.75; Cr 1.07
33XX	Ni 3.50; Cr 1.50 and 1.57
34XX	Ni 3.00; Cr 0.77
	Molybdenum steels
40XX	Mo 0.20 and 0.25
44XX	Mo 0.40 and 0.52
	Chromium-molybdenum steels
41XX	Cr 0.50, 0.80, and 0.95; Mo 0.12, 0.25, and 0.30
	Nickel-chromium-molybdenum steels
43XX	Ni 1.82; Cr 0.50 and 0.80; Mo 0.25
43BVXX	Ni 1.82; Cr 0.50; Mo 0.12 and 0.35; V 0.03 min
47XX	Ni 1.05; Cr 0.45; Mo 0.20 and 0.35
81XX	Ni 0.30; Cr 0.40; Mo 0.12
86XX	Ni 0.55; Cr 0.50; Mo 0.20
87XX	Ni 0.55; Cr 0.50; Mo 0.25
88XX	Ni 0.55; Cr 0.50; Mo 0.35
93XX	Ni 3.25; Cr 1.20; Mo 0.12
94XX	Ni 0.45; Cr 0.40; Mo 0.12
97XX	Ni 0.55; Cr 0.20; Mo 0.20
98XX	Ni 1.00; Cr 0.80; Mo 0.25
	Nickel-molybdenum steels
46XX	Ni 0.85 and 1.82; Mo 0.20 and 0.25
48XX	Ni 3.50; Mo 0.25
	Chromium steels
50XX	Cr 0.27, 0.40, 0.50, and 0.65
51XX	Cr 0.80, 0.87, 0.92, 0.95, 1.00, and 1.05
50XXX	Cr 0.50; C 1.00 min
51XXX	Cr 1.02; C 1.00 min
52XXX	Cr 1.45; C 1.00 min
	Chromium-vanadium steels
61XX	Cr 0.60, 0.80, and 0.95; V 0.10 and 0.15 min
	Tungsten-chromium steels
72XX	W 1.75; Cr 0.75
	Silicon-manganese steels
92XX	Si 1.40 and 2.00; Mn 0.65, 0.82, and 0.85; Cr 0.00 and 0.65
	High-strength low-alloy steels
9XX	Various SAE grades

Courtesy Machinery's Handbook, 24th ed. *Published by Industrial Press, New York.*

ABBREVIATIONS:
C–Carbon
Cr–Chromium
Mn–Manganese
Mo–Molybdenum
Ni–Nickel
Si–Silicon
V–Vanadium
W–Tungsten

STEEL CARBON CONTENT

Plain carbon steels containing 0.06% through 0.29% carbon are generally referred to as low carbon steel. They are usually specified when cold formability is the primary requisite. Medium carbon steels contain 0.30% through 0.59% carbon. These steels are selected for use where higher mechanical properties are needed and are frequently further hardened and strengthened by heat treatment or cold working. High carbon steels contain 0.60% carbon or more. They are used for applications where the higher carbon is needed to improve wear characteristics for cutting edges, to make springs, or for other special purposes.

Steel Designation Quiz

INSTRUCTIONS: Refer to the table on page 166 to show the composition of the following AISI-SAE steels.

1. 1045 — 1. ____________
2. 2335 — 2. ____________
3. 4024 — 3. ____________
4. 4820 — 4. ____________
5. 8615 — 5. ____________

INSTRUCTIONS: Refer to the table on page 166 to show the designated number for the following AISI-SAE steels.

6. Resulfurized carbon steel, 0.37% Carbon — 6. ____________
7. 5% Nickel, 0.50% Carbon — 7. ____________
8. 3.5% Nickel, 0.25% Molybdenum, 0.20% Carbon — 8. ____________
9. 1.75% Tungsten, 0.75% Chromium, 0.80% Carbon — 9. ____________
10. 1.45% Chromium, 1.00% min. Carbon — 10. ____________

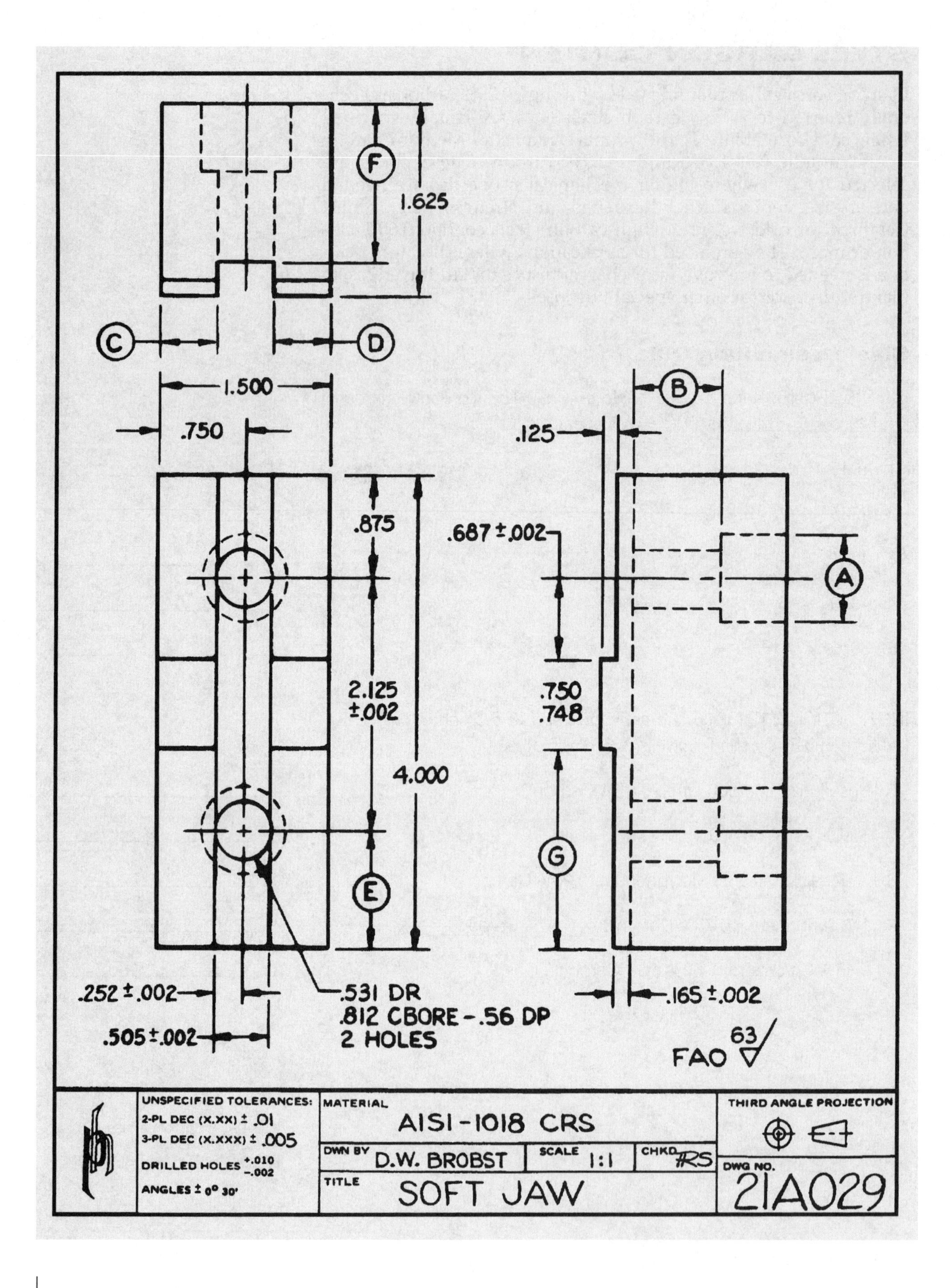
F
1.625
C
D
1.500
.750
B
.125
.875
.687 ± .002
A
2.125
±.002
.750
.748
4.000
G
E
.252 ± .002
.531 DR
.812 CBORE - .56 DP
2 HOLES
.505 ± .002
.165 ± .002
63
FAO
UNSPECIFIED TOLERANCES:
2-PL DEC (X.XX) ± .01
3-PL DEC (X.XXX) ± .005
DRILLED HOLES +.010 −.002
ANGLES ± 0° 30'
MATERIAL
AISI-1018 CRS
DWN BY
D.W. BROBST
SCALE
1:1
CHKD
RS
TITLE
SOFT JAW
THIRD ANGLE PROJECTION
DWG NO.
21A029

INSTRUCTIONS: Refer to drawing 21A029 to answer the following questions.

1. What general classification of steel (carbon or alloy) is specified? 1. ____________
2. What does the second digit, zero, in the steel specification designate? 2. ____________
3. What percentage of carbon does the steel contain? 3. ____________
4. Does the carbon content place the steel in the low-, medium-, or high-carbon range? 4. ____________
5. Convert the .748–.750 limits to a dimension with a bilateral tolerance. 5. ____________
6. Convert the overall height dimension into limits. 6. ____________
7. What is the MMC of the .531 holes? 7. ____________
8. What is the MMC of the .505 slot? 8. ____________
9. What size fasteners are intended for the clearance holes? 9. ____________
10. Show the roughness height rating as a decimal inch value. 10. ____________

INSTRUCTIONS: Enter the dimensions for the following letters.

Ⓐ ____________

Ⓑ ____________

Ⓒ ____________

Ⓓ ____________

Ⓔ MAX: ____________

Ⓔ MIN: ____________

Ⓕ MAX: ____________

Ⓕ MIN: ____________

Ⓖ MAX: ____________

Ⓖ MIN: ____________

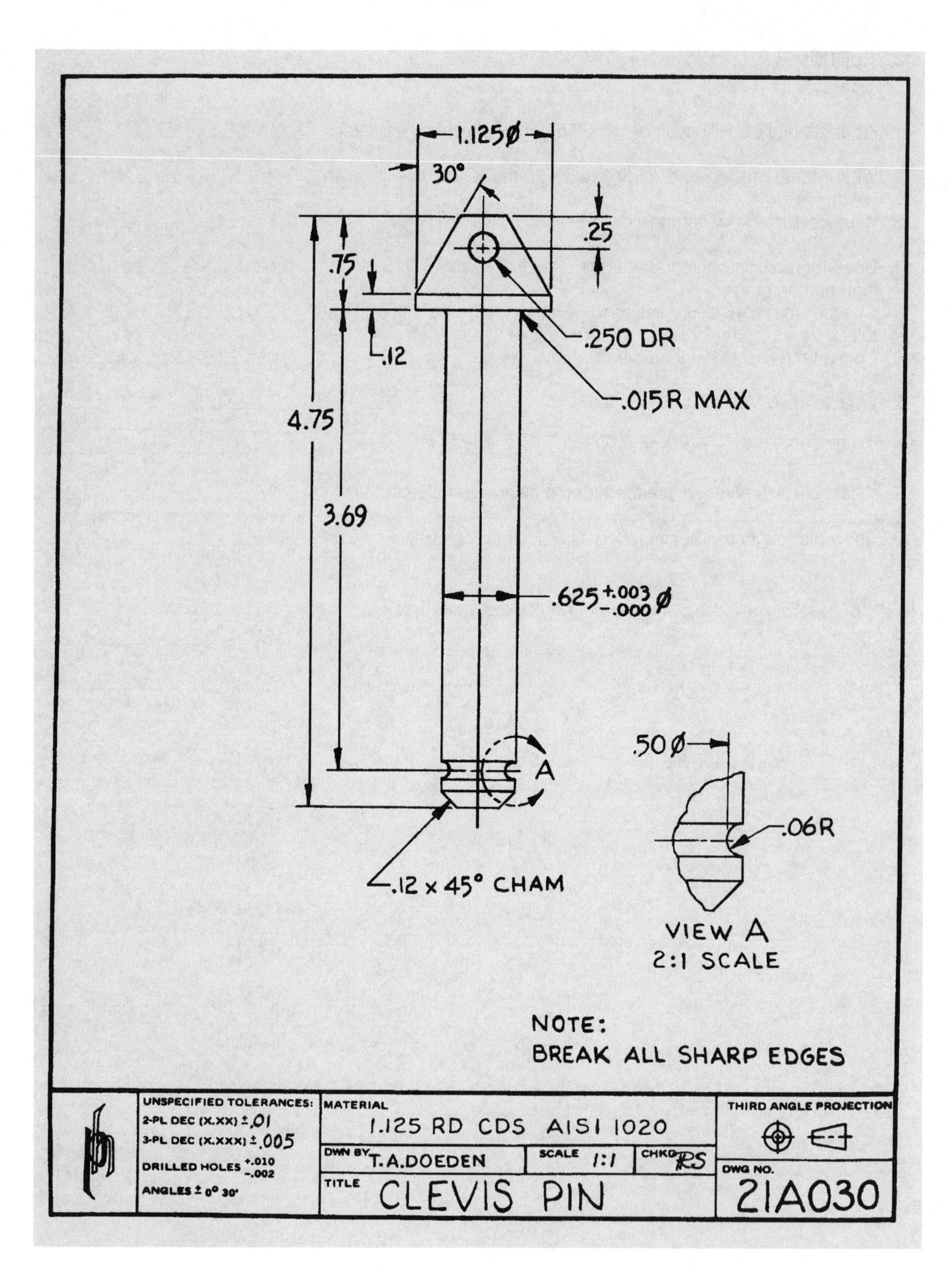
1.125Ø
30°
.25
.75
.12
.250 DR
.015R MAX
4.75
3.69
.625 +.003 -.000 Ø
A
.50Ø
.06R
.12 x 45° CHAM
VIEW A
2:1 SCALE
NOTE:
BREAK ALL SHARP EDGES
UNSPECIFIED TOLERANCES:
2-PL DEC (X.XX) ± .01
3-PL DEC (X.XXX) ± .005
DRILLED HOLES +.010 -.002
ANGLES ± 0° 30'
MATERIAL
1.125 RD CDS AISI 1020
DWN BY T.A.DOEDEN
SCALE 1:1
CHKD RS
TITLE CLEVIS PIN
THIRD ANGLE PROJECTION
DWG NO.
21A030

PARTIAL ENLARGED VIEWS

When only a portion of a view is congested or too small to dimension easily, it is not necessary to draw the entire view to a larger scale. Instead, a partial enlarged view is drawn, such as the one shown in drawing 21A030. This permits dimensions to be placed in an area where they may be read without confusion. The scale is shown underneath the enlarged view. (See the example shown below.)

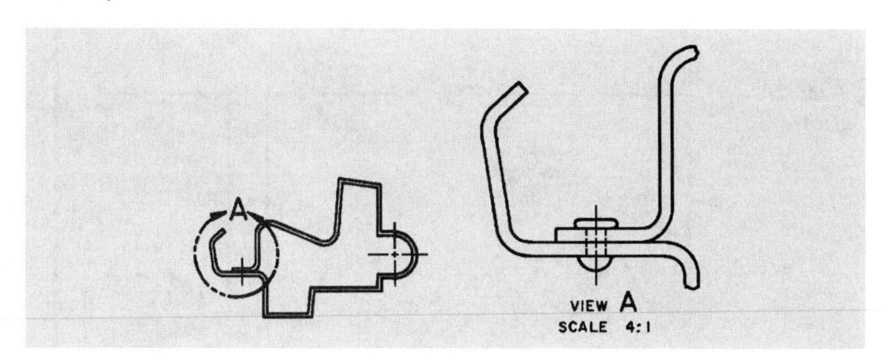

INSTRUCTIONS: Refer to drawing 21A030 to answer the following questions.

1. What general classification of steel (carbon or alloy) is specified? 1. ______
2. What is the carbon content of the steel specified? 2. ______
3. What does the steel specification CDS abbreviate? 3. ______
4. What is the carbon range (low, medium, or high) of this steel? 4. ______
5. Is a .02 radius acceptable under the clevis pin head? 5. ______
6. What is the smallest acceptable radius under the head? 6. ______
7. What is the MMC of the hole diameter? 7. ______
8. What is the MMC of the clevis pin shank? 8. ______
9. What is the width of the groove? 9. ______
10. How far is the center of the groove from the bottom end of the pin? 10. ______
11. What is the pin diameter at the center of the groove? 11. ______
12. What is the scale of the enlarged view? 12. ______
13. How much material remains between the side of the hole and the top of the pin head? 13. ______
14. Show the overall pin length in limit form. 14. ______
15. How much of the .625 shank diameter remains between the groove and the chamfer? 15. ______
16. How much tolerance can accumulate between the hole center and the groove center? 16. ______

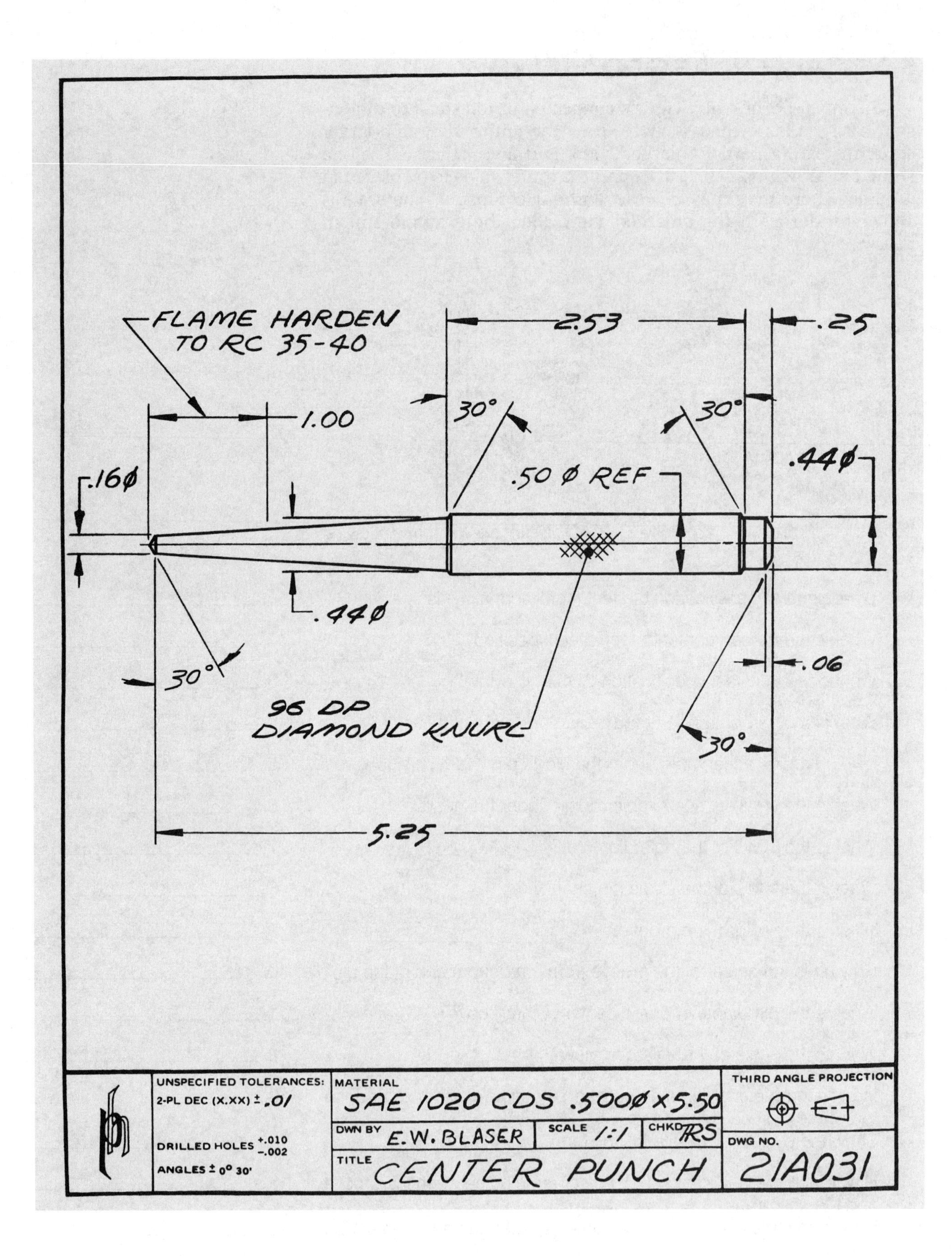
FLAME HARDEN
TO RC 35-40
2.53
.25
1.00
30°
30°
.16Ø
.50 Ø REF
.44Ø
.44Ø
.06
30°
96 DP
DIAMOND KNURL
30°
5.25
UNSPECIFIED TOLERANCES:
2-PL DEC (X.XX) ± .01
DRILLED HOLES +.010 -.002
ANGLES ± 0° 30'
MATERIAL
SAE 1020 CDS .500Ø X 5.50
DWN BY E.W. BLASER
SCALE 1:1
CHKD RS
TITLE CENTER PUNCH
THIRD ANGLE PROJECTION
DWG NO. 21A031

STOCK SIZES

When stock sizes are specified in the material block, they are often omitted on the drawing itself. Therefore, if a major diameter does not appear as a dimension on the drawing, you can assume that it is the same diameter as the stock size. When the stock size is repeated as a dimension on the drawing, it is usually followed by the abbreviation REF or enclosed in parentheses to indicate that it does not carry the standard print tolerance. Instead, it carries the manufacturer's tolerance for that particular material.

KNURLING

Knurling may be used to provide a hand grip on a cylindrical surface, or to provide for a press fit between two cylindrical parts. Basically, there are three styles of knurling: diamond, diagonal, and straightline. The knurling may be fully or partially illustrated on the view, or omitted altogether, since a leader and a note will provide the necessary specifications. This information includes the outside diameter, width, diametral pitch (DP), and style. The diameter of a knurled hand grip is usually not critical, but if the knurling is done for a press fit, the toleranced diameter before knurling and the minimum diameter after knurling must both be specified. (See the examples on the right.)

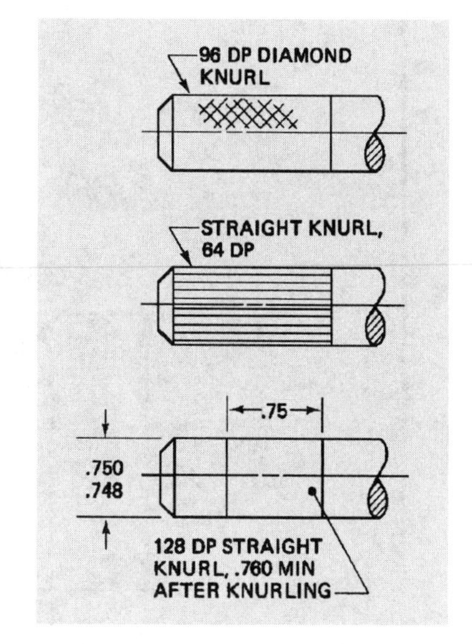

HEAT-TREAT NOTES

Usually heat-treat notes appear near the right-side border along with other general print notes. However, when the heat treatment affects only a portion of the object, it is necessary to place the note near the affected area for clarity. Observe on drawing 21A031 the flame hardening of only a portion of the center punch. The hardness rating specification is for the Rockwell C-scale.

INSTRUCTIONS: *Refer to drawing 21A031 to answer the following questions.*

1. Is the punch made from low-, medium-, or high-carbon steel? 1. ____________
2. What is the stock cutoff length? (See the material block.) 2. ____________
3. What style of knurl is specified? 3. ____________
4. What are the limits of hardness? (Include the Rockwell scale.) 4. ____________
5. What is the included angle of the punch point? 5. ____________
6. What is the length of the knurled area? (Include the chamfers.) 6. ____________
7. What is the length of the unknurled cylindrical portion after chamfering? 7. ____________
8. What is the length of the tapered portion? (Exclude the point.) 8. ____________
9. How much tolerance accumulates on the length of the tapered portion? 9. ____________
10. What is the TPF of the tapered portion? 10. ____________

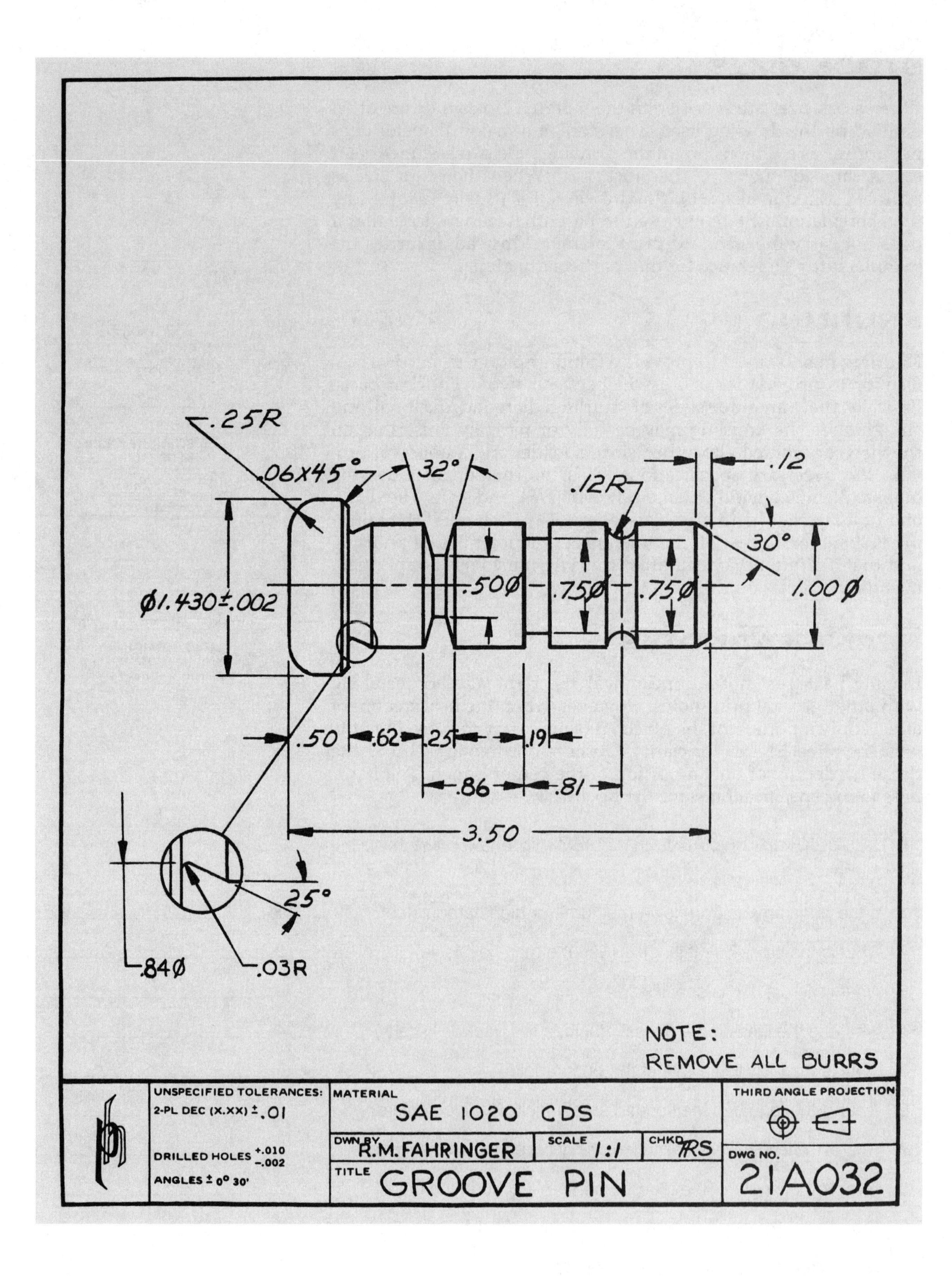
.25R
.06X45°
32°
.12R
.12
30°
Ø1.430±.002
.50Ø
.75Ø
.75Ø
1.00Ø
.50
.62
.25
.19
.86
.81
3.50
25°
.84Ø
.03R
NOTE:
REMOVE ALL BURRS
UNSPECIFIED TOLERANCES:
2-PL DEC (X.XX) ±.01
DRILLED HOLES +.010 −.002
ANGLES ± 0° 30'
MATERIAL
SAE 1020 CDS
DWN BY R.M.FAHRINGER
SCALE 1:1
CHKD RS
TITLE GROOVE PIN
THIRD ANGLE PROJECTION
DWG NO.
21A032

INSTRUCTIONS: Refer to drawing 21A032 to answer the following questions.

1. What are the limits of the pin head diameter? 1. ____________
2. What is the MMC of the pin body diameter? 2. ____________
3. What is the minimum diameter of the neck? (underneath the pin head) 3. ____________
4. What type of view was drawn to dimension the neck? 4. ____________
5. Is the material (a) hot-drawn or (b) cold-drawn? 5. ____________
6. Do the specifications indicate (a) carbon steel or (b) alloy steel? 6. ____________
7. Is the carbon content of the steel (a) 20%, (b) 2%, or (c) 0.2%? 7. ____________
8. Do the specifications indicate (a) sulfurized, (b) phosphorized, or (c) plain carbon steel? 8. ____________

INSTRUCTIONS: To emphasize the tolerance buildup that occurs when manipulating dimensions, include the tolerance with each answer for the letters shown on the sketch below. Refer to drawing 21A032.

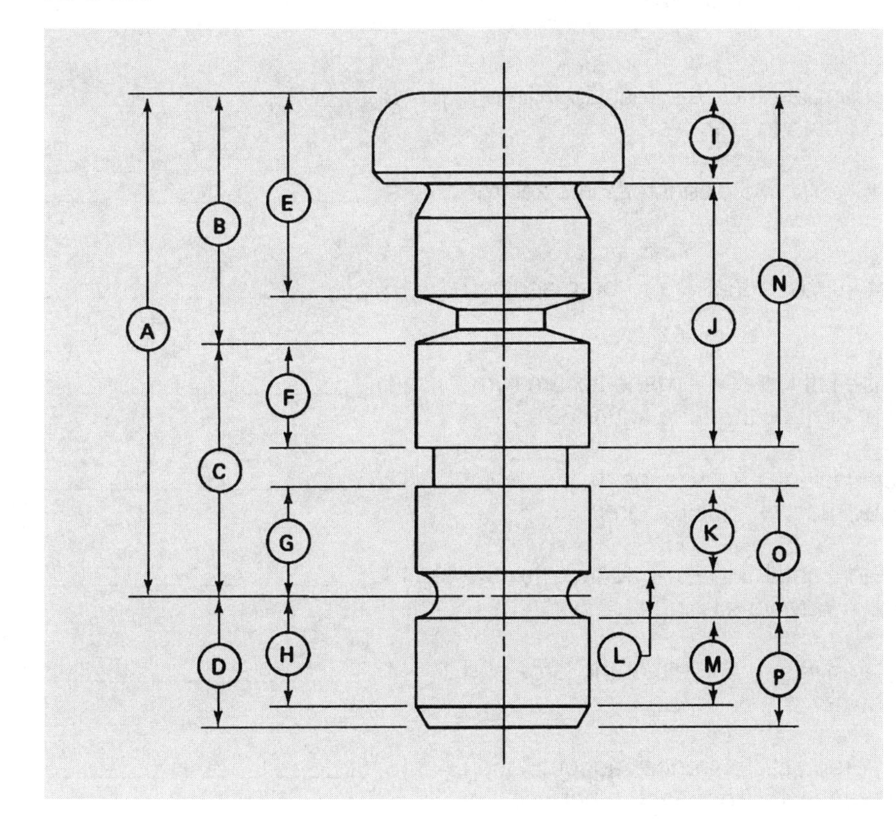

Ⓐ ________ ± ________

Ⓑ ________ ± ________

Ⓒ ________ ± ________

Ⓓ ________ ± ________

Ⓔ ________ ± ________

Ⓕ ________ ± ________

Ⓖ ________ ± ________

Ⓗ ________ ± ________

Ⓘ ________ ± ________

Ⓙ ________ ± ________

Ⓚ ________ ± ________

Ⓛ ________ ± ________

Ⓜ ________ ± ________

Ⓝ ________ ± ________

Ⓞ ________ ± ________

Ⓟ ________ ± ________

Review Quiz 1

INSTRUCTIONS: The following statements may or may not be entirely true. Read them carefully, then write either "True" or "False" in the spaces provided.

1. The depth dimension for a drilled blind hole does not include the drill point. 1. ______
2. In normal situations, a clearance hole is drilled to the same diameter as the fastener itself. 2. ______
3. An inclined plane that appears as an edge in the front view will appear foreshortened in the top and side views. 3. ______
4. The coordinate method of dimensioning inclined planes will control accuracy more closely than the angular method, particularly on large objects. 4. ______
5. The overall length of a four-step shaft dimensioned by the broken-chain method would accumulate four tolerances. 5. ______
6. The maximum amount of tolerance that can accumulate by datum dimensioning a step shaft is one tolerance per step. 6. ______
7. By orthographic projection, each view will show only two of the three overall dimensions. 7. ______
8. If one dimension is subtracted from another, the resulting dimension could accumulate the sum of both tolerances. 8. ______
9. Countersinks enlarge holes cylindrically, while counterbores enlarge holes conically. 9. ______
10. Normal finish allowance for cast surfaces that are to be machined is 1/8 inch (.12) per surface. 10. ______
11. The figure 32 appearing above the left leg of a surface texture symbol would provide for a smoother surface than the figure 63. 11. ______
12. When the surface texture symbol includes a circle inside the vee, it indicates that material removal by machining is required. 12. ______
13. A figure appearing above the horizontal line of a surface texture symbol would represent maximum waviness height. 13. ______
14. TPF can be calculated if the tapered length and either the large end or the small end diameter is known. 14. ______
15. When subtracting limits, the greatest tolerance accumulation occurs by subtracting the minimum limits from each other and the maximum limits from each other. 15. ______

UNIT 8

SECTIONAL VIEWS

When an object has internal features not clearly described by hidden lines in an exterior view, a sectional view may be used. The view is obtained by an imaginary cutting plane passing through the object perpendicular to the direction of sight. The portion of the object between the cutting plane and the sectional view is assumed to be removed and the exposed cut surfaces of the object are indicated by cross-section lines. Section lines are normally drawn at 45° so as not to be parallel with the object lines. A cutting-plane line is used to indicate where the section is taken if it is not obvious. When more than one section appears on the same drawing, identifying letters are used to match the sectional views with the cutting-plane lines. The arrowheads on a cutting-plane line will point in the direction of sight; thus normal placement of the resulting section will be directly behind those arrowheads. See the examples shown below.

FULL SECTIONS AND HALF-SECTIONS

A full section is a view that is entirely in section. If a cutting-plane line were used, it would pass entirely through a view, in effect cutting it in half. Not to be confused with that description is the half-section, which is a view showing half of the object in section and the other half showing its exterior. If a cutting-plane line were used, it would pass along the horizontal centerline and turn along the vertical centerline, in effect removing one-quarter of the object. The half-section is limited to symmetrical shapes, and is used where internal features must be shown yet external features must also be retained. See the examples of each shown below.

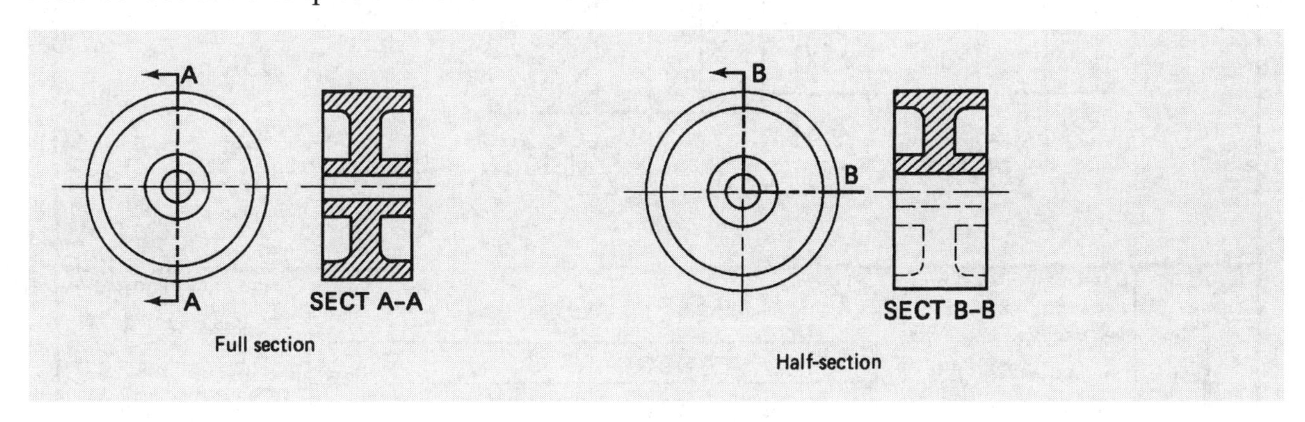

Full section

Half-section

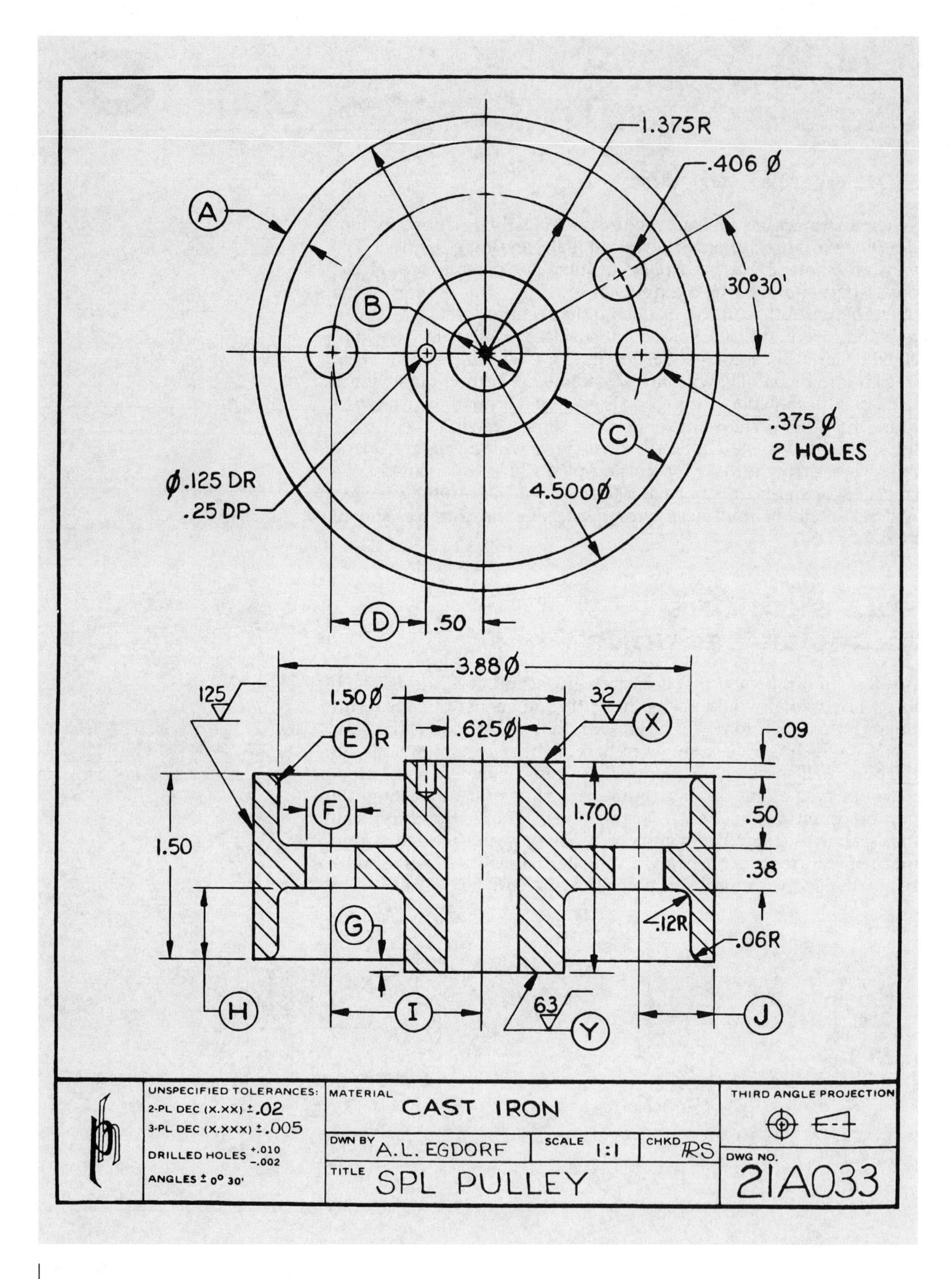
1.375R
.406 ⌀
A
30°30'
B
.375 ⌀
2 HOLES
C
⌀.125 DR
.25 DP
4.500⌀
D
.50
3.88⌀
125
1.50⌀
32
.625⌀
X
E R
.09
F
1.700
.50
1.50
.38
.12R
G
.06R
H
I
63
Y
J
UNSPECIFIED TOLERANCES:
2-PL DEC (X.XX) ±.02
3-PL DEC (X.XXX) ±.005
DRILLED HOLES +.010 −.002
ANGLES ± 0° 30'
MATERIAL
CAST IRON
DWN BY A.L. EGDORF
SCALE 1:1
CHKD RS
TITLE SPL PULLEY
THIRD ANGLE PROJECTION
DWG NO.
21A033

INSTRUCTIONS: Refer to drawing 21A033 to answer the following questions.

1. What type of sectional view is drawn? 1. ______
2. Does the cutting plane pass along the vertical CL or the horizontal CL of the top view? 2. ______
3. Does the sectional view represent the upper half or the lower half of the top view? 3. ______
4. Is surface Ⓧ smoother than surface Ⓨ? 4. ______
5. Show the surface Ⓧ roughness height rating as a decimal inch value. 5. ______
6. What size is the OD before machining? 6. ______
7. What is the diameter of the hole in the right side of the sectional view? 7. ______
8. What is the wall thickness of the hub? 8. ______
9. What is the MMC of the blind hole? 9. ______
10. At what angle to the vertical centerline is the .406 hole located? 10. ______

INSTRUCTIONS: Enter the dimensions for the following letters.

Ⓐ ______

Ⓑ ______

Ⓒ ______

Ⓓ ______

Ⓔ ______

Ⓕ ______

Ⓖ ______

Ⓗ ______

Ⓘ ______

Ⓙ ______

Optional Math Exercises:

1a. Calculate the Cartesian coordinates for the .406∅ hole.

2a. Calculate the Measurement Over Pins (MOP) for the .406∅ and .375∅ holes.

SYMBOLOGY

To bring more international standardization into graphical illustration, new symbols have been created to use in place of abbreviations. Shown below are examples using the new symbols for counterbore or spotface, countersink, depth, and square.

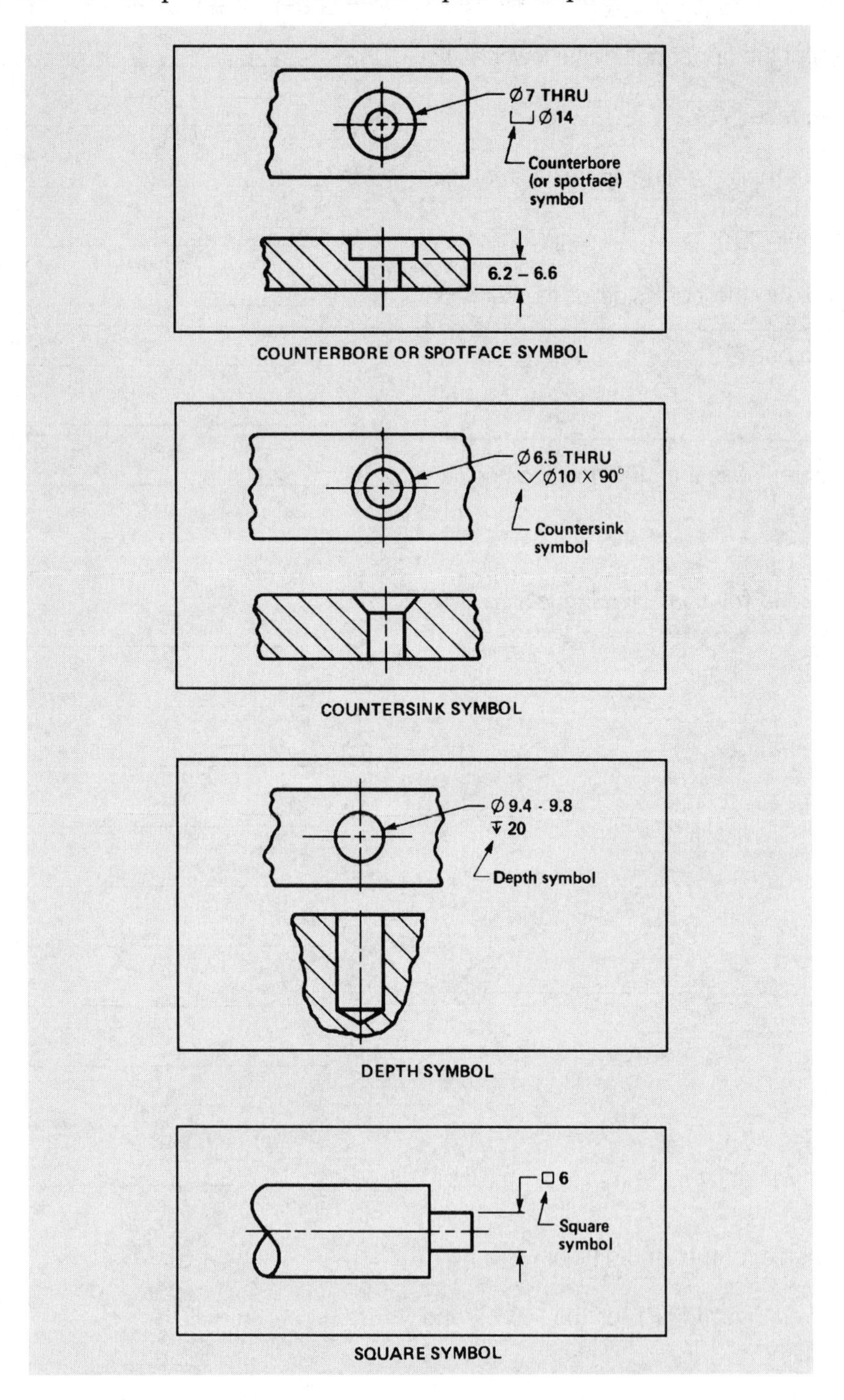

Reprinted from ASME Y14.5M–1994, by permission of The American Society of Mechanical Engineers.

ANNULAR GROOVES

Annular grooves are often specified on drawings of shafts, such as the one shown on drawing 21A034 on page 182. They may be dimensioned by diameter and width, or by depth (from the adjacent surface) and width. Inside grooves are sometimes called undercuts; outside grooves are called necks if they occur at a change in diameter.

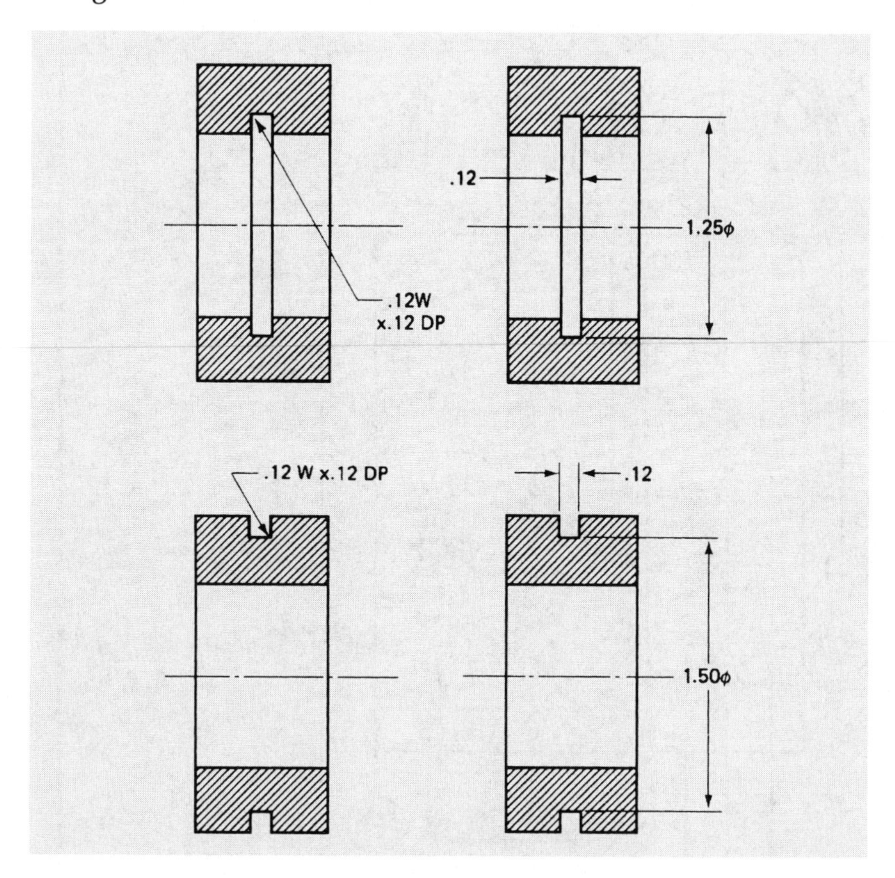

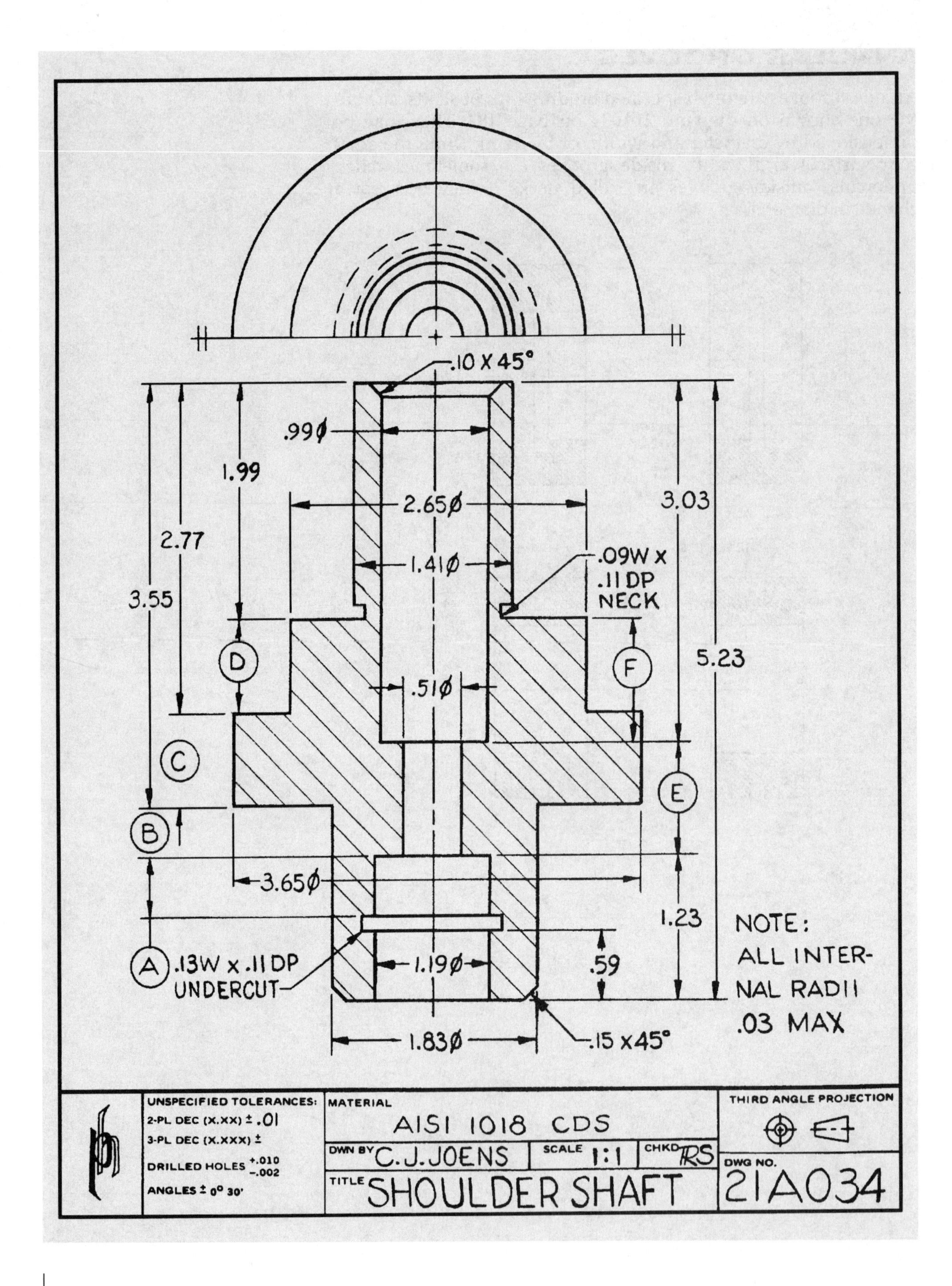
.10 X 45°
.99Ø
1.99
2.65Ø
3.03
2.77
1.41Ø
.09W x
.11 DP
NECK
3.55
D
F
5.23
.51Ø
C
E
B
3.65Ø
1.23
NOTE:
ALL INTERNAL RADII
.03 MAX
A
.13W x .11 DP
UNDERCUT
1.19Ø
.59
1.83Ø
.15 x 45°
UNSPECIFIED TOLERANCES:
2-PL DEC (X.XX) ± .01
3-PL DEC (X.XXX) ±
DRILLED HOLES +.010 -.002
ANGLES ± 0° 30'
MATERIAL
AISI 1018 CDS
DWN BY C.J.JOENS
SCALE 1:1
CHKD RS
TITLE SHOULDER SHAFT
THIRD ANGLE PROJECTION
DWG NO.
21A034

INSTRUCTIONS: Refer to drawing 21A034 to answer the following questions.

1. What type of view is the top view? 1. ______
2. What type of sectional view is drawn? 2. ______
3. What is the diameter of the undercut (inside groove)? 3. ______
4. What is the diameter of the neck (outside groove)? 4. ______
5. What is the width of the neck? 5. ______
6. What is the wall thickness at the 1.41 dia? 6. ______
7. What is the wall thickness at the 1.83 dia? 7. ______
8. What is the diameter of the inside chamfer? 8. ______
9. What is the length of the 1.41 dia after necking? 9. ______
10. What is the length of the 1.83 dia after chamfering? 10. ______
11. What is the carbon content of the steel specified? 11. ______
12. Is this steel considered a plain carbon steel? 12. ______
13. How many arcs would represent hidden surfaces on a bottom view? 13. ______
14. How many arcs would represent visible surfaces on a bottom view? 14. ______

INSTRUCTIONS: Calculate the dimensions, and their accumulated tolerances, for the following letters.

Ⓐ ______ ± ______

Ⓑ ______ ± ______

Ⓒ ______ ± ______

Ⓓ ______ ± ______

Ⓔ ______ ± ______

Ⓕ ______ ± ______

WALL THICKNESS

To obtain the wall thickness of pipe, tubing, or any other concentric diameters, the ID must be subtracted from the OD and divided by two. Maximum wall thickness may be obtained by subtracting the minimum ID (MMC) from the maximum OD (MMC) and dividing by two, To obtain minimum wall thickness, subtract the maximum ID (LMC) from the minimum OD (LMC) and divide by two. LMC is the abbreviation for Least Material Condition, which is the opposite condition of MMC. See the examples shown below.

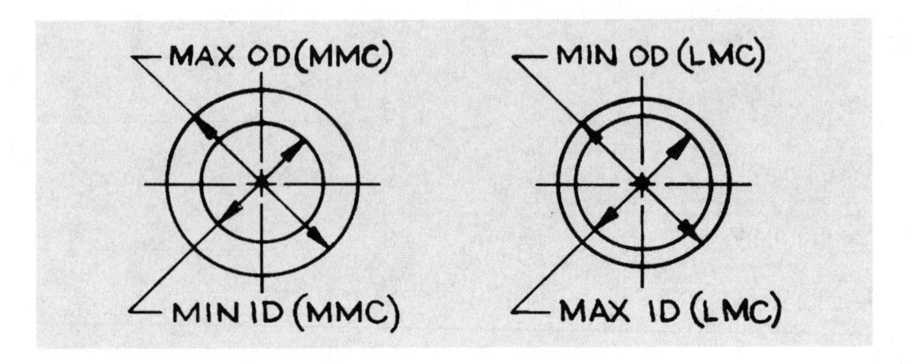

Wall Thickness Calculations

INSTRUCTIONS: Calculate the required wall thicknesses from the information provided below.

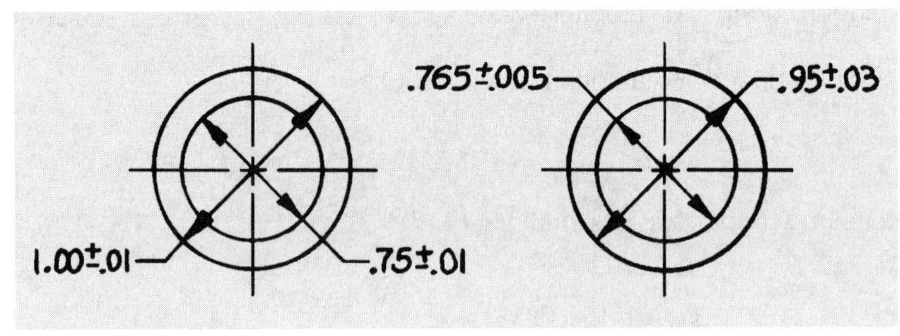

1. MAX WALL
2. MIN WALL
3. MAX WALL
4. MIN WALL

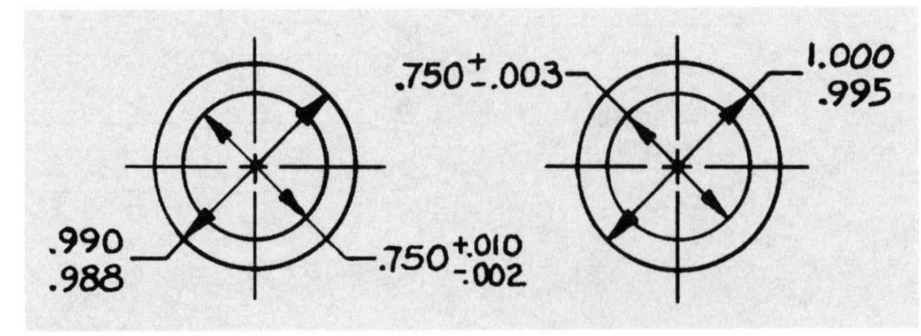

5. MAX WALL
6. MIN WALL
7. MAX WALL
8. MIN WALL

1. ____________
2. ____________
3. ____________
4. ____________
5. ____________
6. ____________
7. ____________
8. ____________

CAST IRON

Like steel, cast iron is an alloy of iron, carbon, and silicon, but it has a much higher carbon content, amounting to 2.5 to 4.0%. Basically, there are four major classifications of cast iron: gray, white, malleable, and nodular. Gray cast iron, the most widely used, receives its name from the characteristic gray fracture. It has good machinability and high compression strength. White cast iron, hard and brittle, receives its name from the characteristic silvery white fracture. Parts requiring good wear resistance, strength, and hardness are made using white cast iron. Malleable cast iron is produced from white cast iron by a process of heat treating. Malleable iron is used where even greater toughness and shock resistance are required, as in farm implements, pipe fittings, cam shafts, and brake pedals. Nodular cast iron (ductile iron) is produced by adding magnesium alloys to gray iron. This material has good machinability and can be annealed, induction-hardened, or flame-hardened.

SPOTFACING

Spotfacing is a machining operation performed on castings to provide a smooth, flat surface around a hole. The diameter and abbreviation SFACE or the symbol ⌴ follow the drill size, but the depth of a spotface is usually omitted. The drafter draws a spotface .06 in. deep, but the machinist usually spotfaces between .03 and .06 in. deep. The specified diameter should be large enough to accommodate the washer, nut, or fastener head intended to be mounted against the spotface. (See the example shown below of a spotface tool, which pilots on the drilled hole similar to a counterbore tool.)

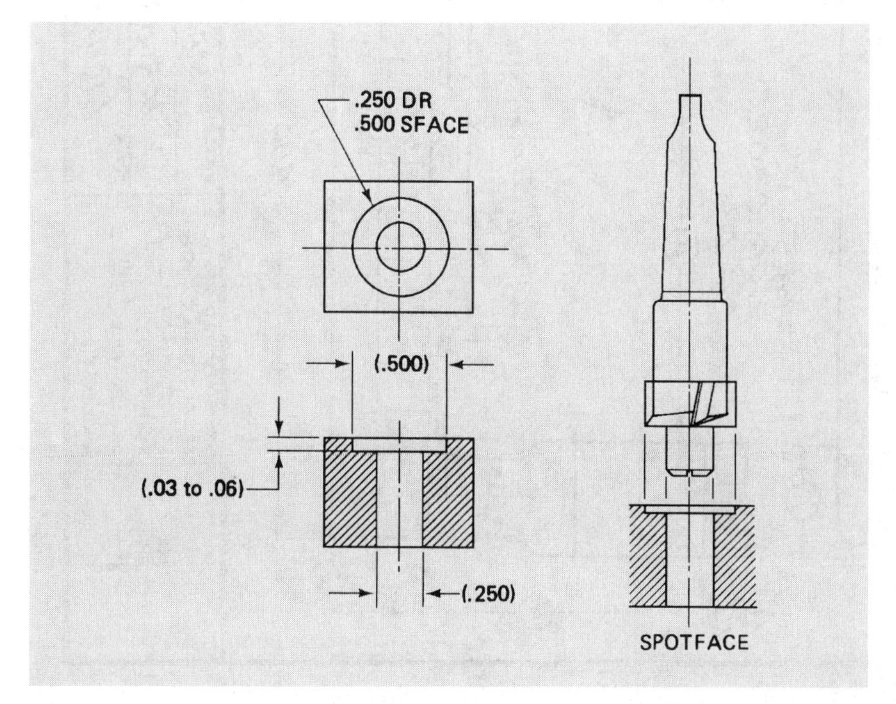

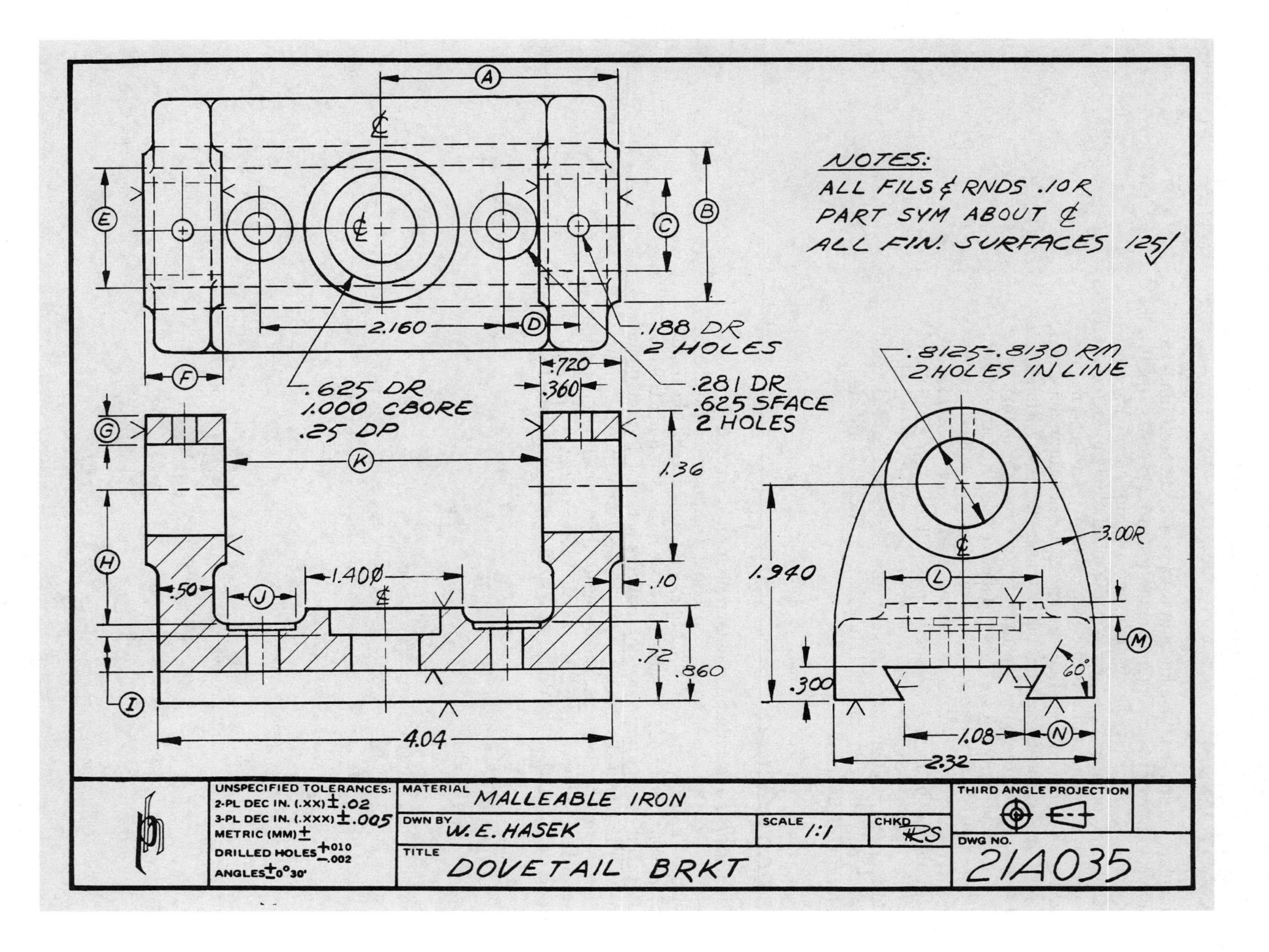
A
B
C
D
E
F
G
H
I
J
K
L
M
N
2.160
.188 DR
2 HOLES
.625 DR
1.000 CBORE
.25 DP
.281 DR
.625 SFACE
2 HOLES
.720
.360
1.36
1.400
.50
.10
.72
.860
4.04
NOTES:
ALL FILS & RNDS .10R
PART SYM ABOUT ℄
ALL FIN. SURFACES 125
.8125-.8130 RM
2 HOLES IN LINE
3.00R
1.940
.300
60°
1.08
2.32
UNSPECIFIED TOLERANCES:
2-PL DEC IN. (.XX) ±.02
3-PL DEC IN. (.XXX) ±.005
METRIC (MM) ±
DRILLED HOLES +.010 −.002
ANGLES ±0°30'
MATERIAL MALLEABLE IRON
DWN BY W. E. HASEK
SCALE 1:1
CHKD RS
TITLE DOVETAIL BRKT
THIRD ANGLE PROJECTION
DWG NO. 21A035

INSTRUCTIONS: Refer to drawing 21A035 to answer the following questions.

1. What type of sectional view is drawn? 1. ______
2. What roughness height rating is specified for finished surfaces? 2. ______
3. What is the tolerance on the reamed holes? (total) 3. ______
4. What is the MMC of the reamed holes? 4. ______
5. What is the maximum overall height after machining? 5. ______
6. What would be the overall height before machining? 6. ______
7. What is the approximate depth of the spotfaces? (See page 185.) 7. ______
8. How thick is the base where the spotfaced holes pass through? (Less dovetail.) 8. ______
9. Is the specified material produced from gray iron or white iron? (See page 185.) 9. ______

INSTRUCTIONS: Enter the dimensions for the following letters.

Ⓐ ______
Ⓑ ______
Ⓒ ______
Ⓓ ______
Ⓔ ______
Ⓕ ______
Ⓖ MIN: ______
Ⓖ MAX: ______
Ⓗ ______
Ⓘ ______
Ⓙ ______
Ⓚ ______
Ⓛ ______
Ⓜ ______
Ⓝ ______

Optional Math Exercise:

1a. Using pins that are .250∅, determine the MBP (Measurement Between Pins) after the dovetail has been properly machined.

REVOLVED SECTIONS

A revolved section is drawn directly on one of the normal views to conserve drawing space. It is the same view that would have resulted from a cutting plane passing through the object where the section is drawn. Sometimes the object lines are removed and break lines used on either side of the revolved section. (See the example shown below.) The purpose of the revolved section used on drawing 21A036 on page 190 is to show that the ribs are radiused rather than flat-faced. The actual radius is deliberately omitted to avoid a double dimension. (It can be calculated by dividing the rib thickness by 2.)

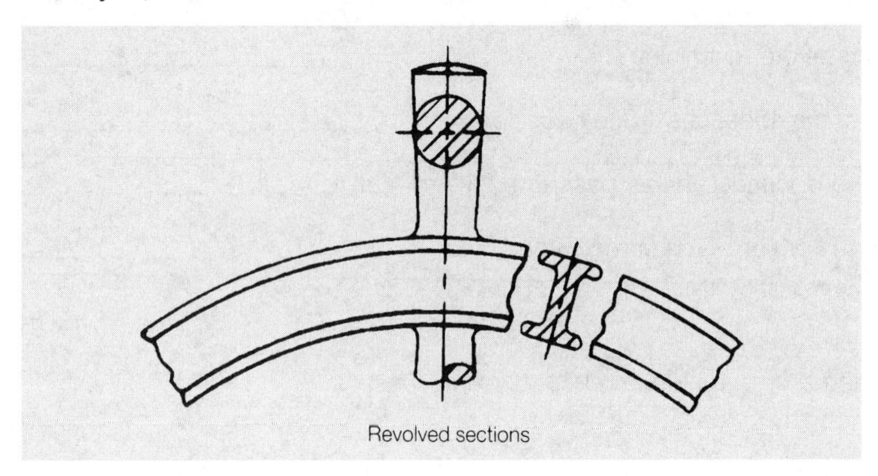

Revolved sections

BROKEN-OUT SECTIONS

Often, only a portion of an object needs to have its interior shape exposed. This can be accomplished through the use of a broken-out section. A short break line is used to separate the sectioned portion, as in drawing 21A037 on page 192, and the examples below. The cutting-plane line is usually omitted, since the sectioned portion normally corresponds with an obvious centerline.

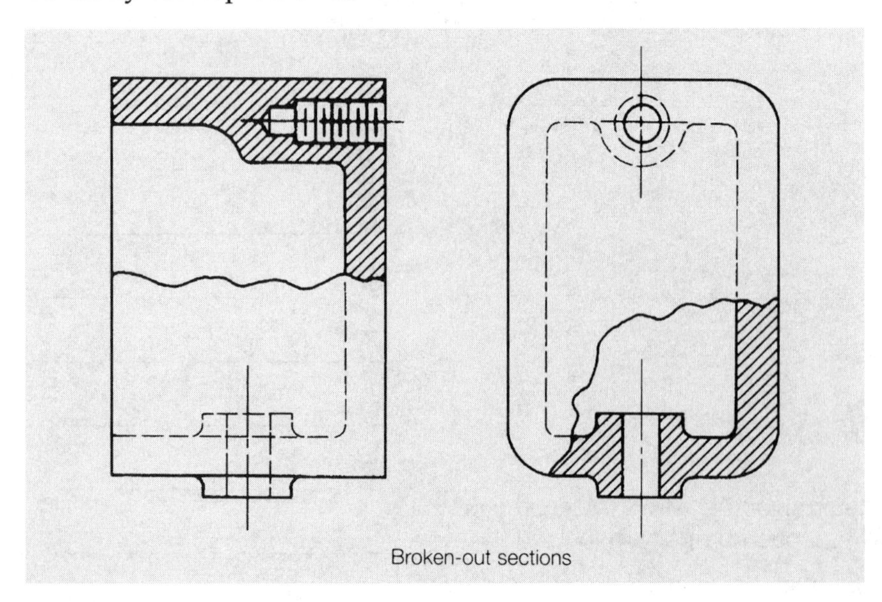

Broken-out sections

RIBS IN SECTION

When the cutting-plane passes flatwise along a rib or web, the resulting sectional view will not include section lines through the rib or web, as shown in drawing 21A036 on page 190. This is done intentionally to avoid creating the false impression of thickness and solidity. (See the example shown below.) Also observe the customary practice of showing holes in the section view even though the cutting plane does not actually pass through them. This is called "Rotation of Features" and is explained on page 195.

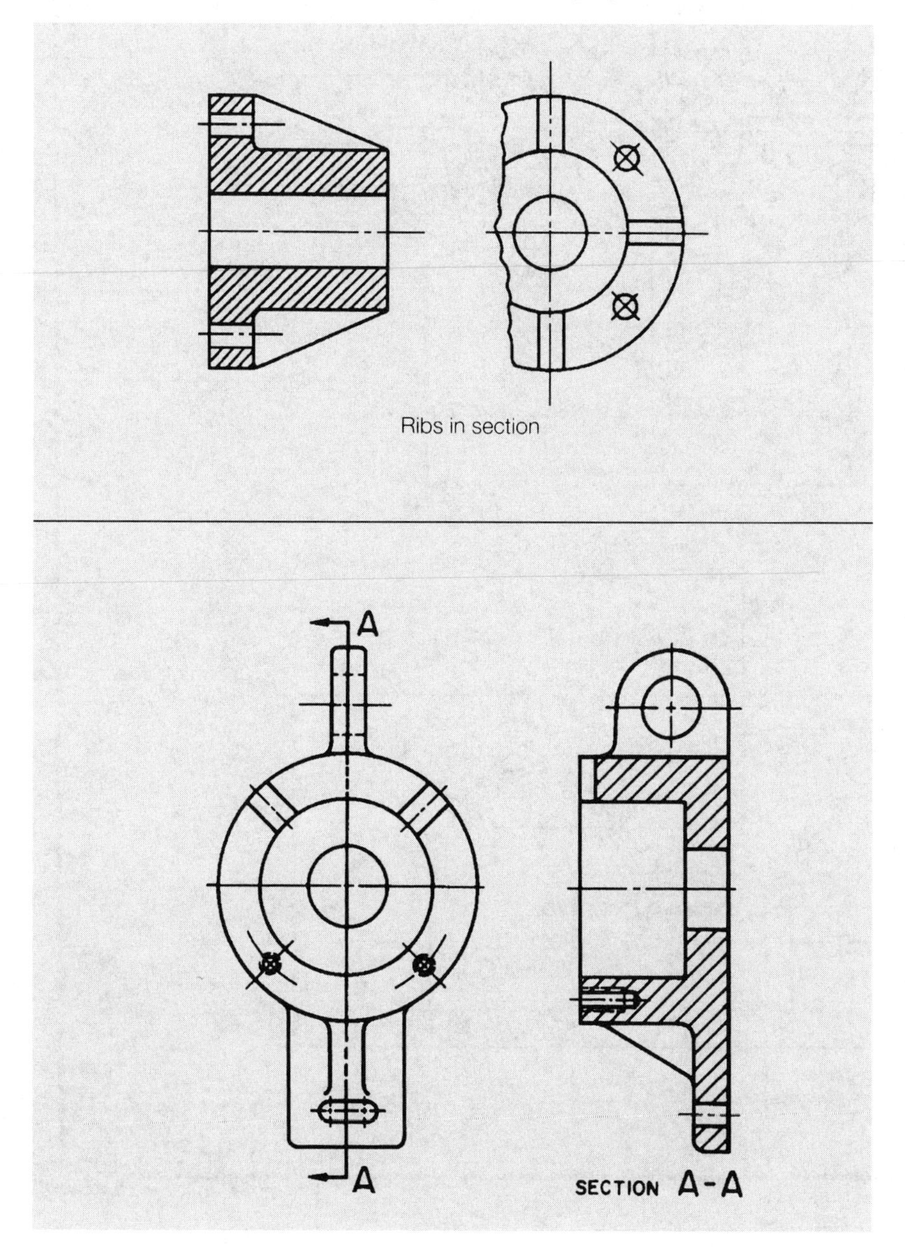

Ribs in section

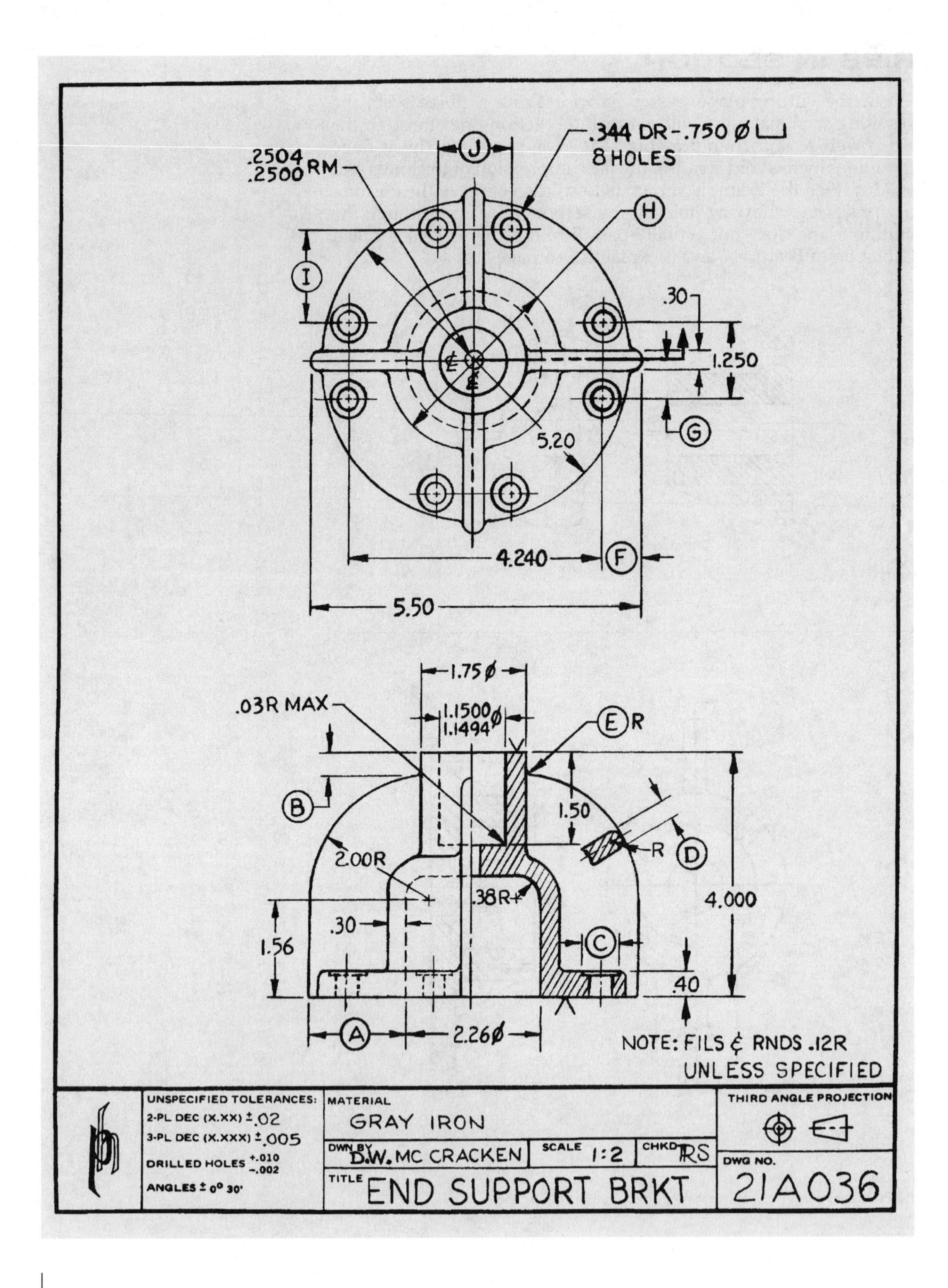

.344 DR-.750 ⌀ ⌴
8 HOLES
.2504
.2500 RM
J
H
I
.30
1.250
G
5.20
4.240
F
5.50
1.75⌀
.03R MAX
1.1500
1.1494 ⌀
E R
B
1.50
R
D
2.00R
.38R
4.000
.30
1.56
C
.40
A
2.26⌀
NOTE: FILS & RNDS .12R
UNLESS SPECIFIED
UNSPECIFIED TOLERANCES:
2-PL DEC (X.XX) ±.02
3-PL DEC (X.XXX) ±.005
DRILLED HOLES +.010 −.002
ANGLES ± 0° 30'
MATERIAL
GRAY IRON
DWN BY D.W. MC CRACKEN
SCALE 1:2
CHKD RS
TITLE END SUPPORT BRKT
THIRD ANGLE PROJECTION
DWG NO.
21A036

INSTRUCTIONS: Refer to drawing 21A036 to answer the following questions.

1. What type of section is the front view? 1. ______
2. What is the name of the partial section shown on the front view? 2. ______
3. What is the radius on the face of the ribs? (Refer to page 188.) 3. ______
4. What is the total tolerance on the bored hole? 4. ______
5. What would be the "cored" diameter of the bored hole (as cast, before machining)? 5. ______
6. What are the limits of the drilled holes? 6. ______
7. What does the symbol used with the 8 holes dimension designate? 7. ______
8. What is the approximate depth of the spotfaces? 8. ______
9. What type of dimension is the .03R MAX at the bottom of the bored hole? 9. ______
10. What is the minimum acceptable wall thickness of the boss? 10. ______

INSTRUCTIONS: Enter the dimensions for the following letters.

Ⓐ ______

Ⓑ ______

Ⓒ ______

Ⓓ ______

Ⓔ ______

Ⓕ ______

Ⓖ ______

Ⓗ ______

Ⓘ ______

Ⓙ ______

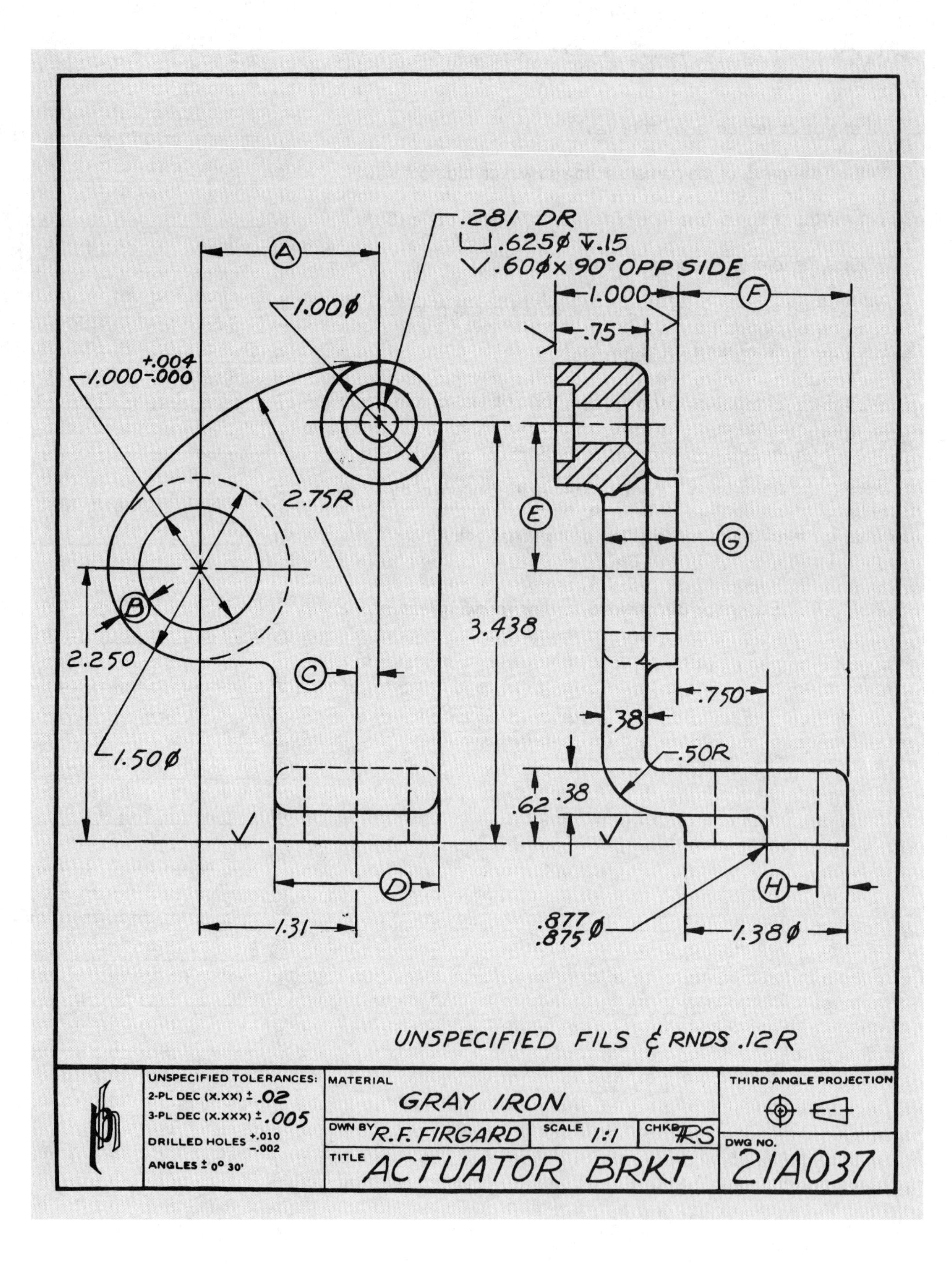
.281 DR
⌴ .625⌀ ↧.15
∨ .60⌀ × 90° OPP SIDE
A
1.00⌀
+.004
1.000 -.000
1.000
F
.75
2.75R
E
G
B
2.250
C
3.438
.750
.38
1.50⌀
.50R
.62
.38
D
H
1.31
.877
.875⌀
1.38⌀
UNSPECIFIED FILS & RNDS .12R
UNSPECIFIED TOLERANCES:
2-PL DEC (X.XX) ± .02
3-PL DEC (X.XXX) ± .005
DRILLED HOLES +.010 -.002
ANGLES ± 0° 30'
MATERIAL
GRAY IRON
DWN BY R.F. FIRGARD
SCALE 1:1
CHKD RS
TITLE ACTUATOR BRKT
THIRD ANGLE PROJECTION
DWG NO.
21A037

INSTRUCTIONS: Refer to drawing 21A037 to answer the following questions.

1. What type of section is drawn? 1. ______
2. Is the counterbore larger than the countersink? 2. ______
3. Is the countersink deeper than the counterbore? (Calculate.) 3. ______
4. What size fastener is intended for use in the drilled hole? 4. ______
5. What is the tolerance on the boss diameters? 5. ______
6. What is the tolerance on the hole in the 1.50 boss? 6. ______
7. What is the tolerance on the hole in the 1.00 boss? 7. ______
8. What is the tolerance on the hole in the 1.38 boss? 8. ______
9. What is the overall height dimension after machining? 9. ______
10. What would be the overall height before machining? 10. ______

INSTRUCTIONS: Calculate the dimensions for the following letters.

Ⓐ ______

Ⓑ MIN: ______

Ⓑ MAX: ______

Ⓒ ______

Ⓓ ______

Ⓔ ______

Ⓕ ______

Ⓖ ______

Ⓗ MIN: ______

Ⓗ MAX: ______

SECTION LINE SYMBOLS

Most detail drawings with sectional views utilize the general purpose cross-section symbol (cast iron) regardless of the actual material from which the object is produced. This is possible because the material is normally listed in the title block. However, on sectioned assembly drawings where material is not normally listed, standardized cross-section symbols are used to represent the different materials. These symbols are shown below.

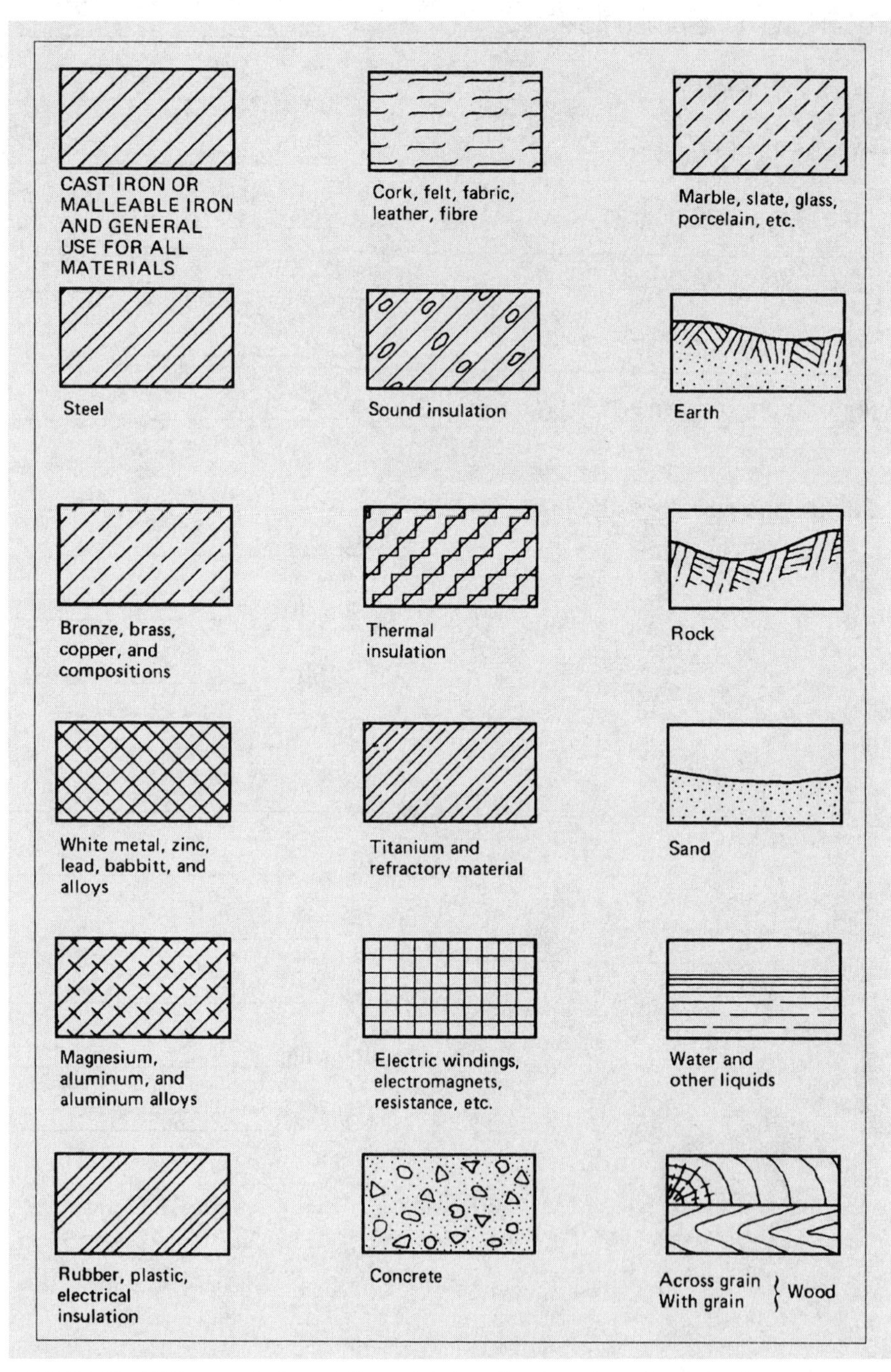

ANSI Y14.2 (With permission of the publisher, the American Society of Mechanical Engineers)

ROTATED FEATURES

When the true projection of features such as ribs, webs, or spokes would result in foreshortening, the drafter will normally rotate them to appear true shape. This practice accomplishes two purposes: reducing drafting time and (hopefully) eliminating confusion. Observe from drawing 21A038 on page 196, and the examples shown below how the features have been rotated to the centerline and then projected to the next view. This practice permits the holes to show at their true distance from center, and the ribs to show their true shape.

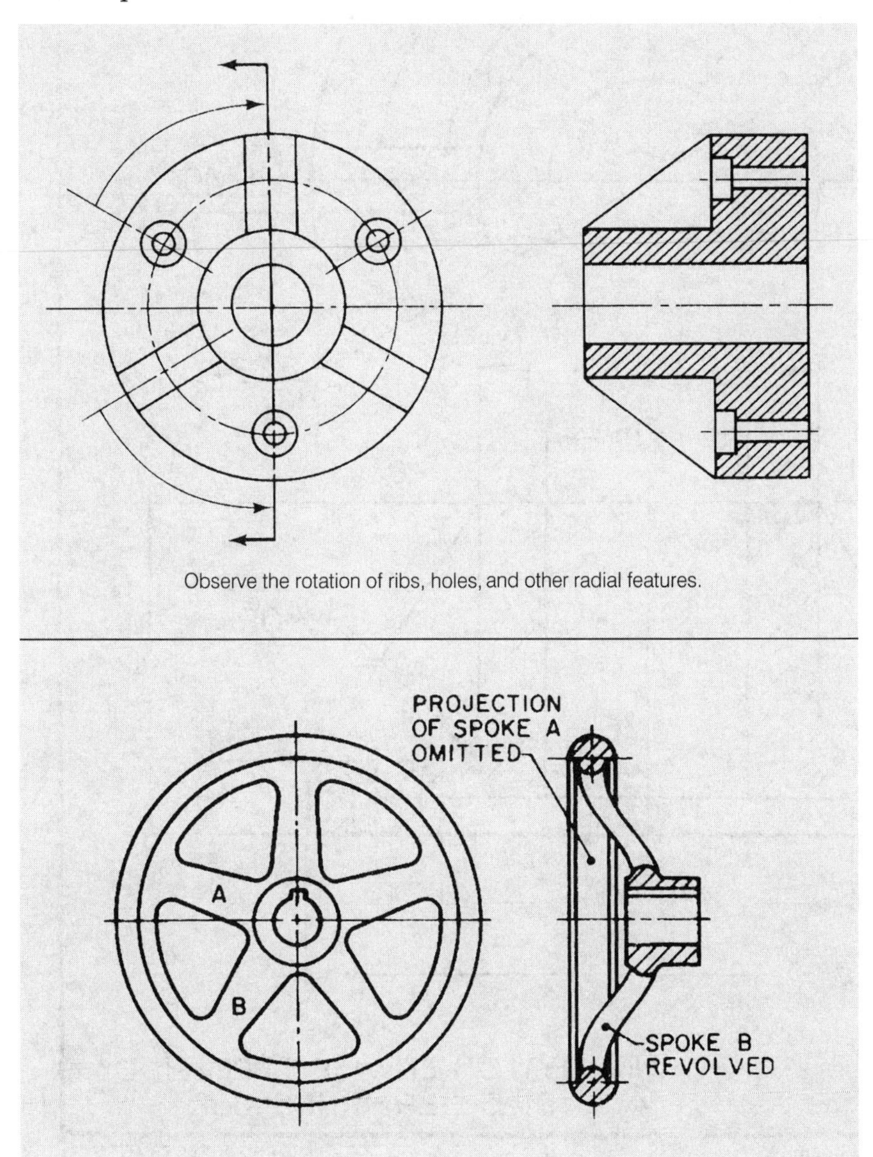

Observe the rotation of ribs, holes, and other radial features.

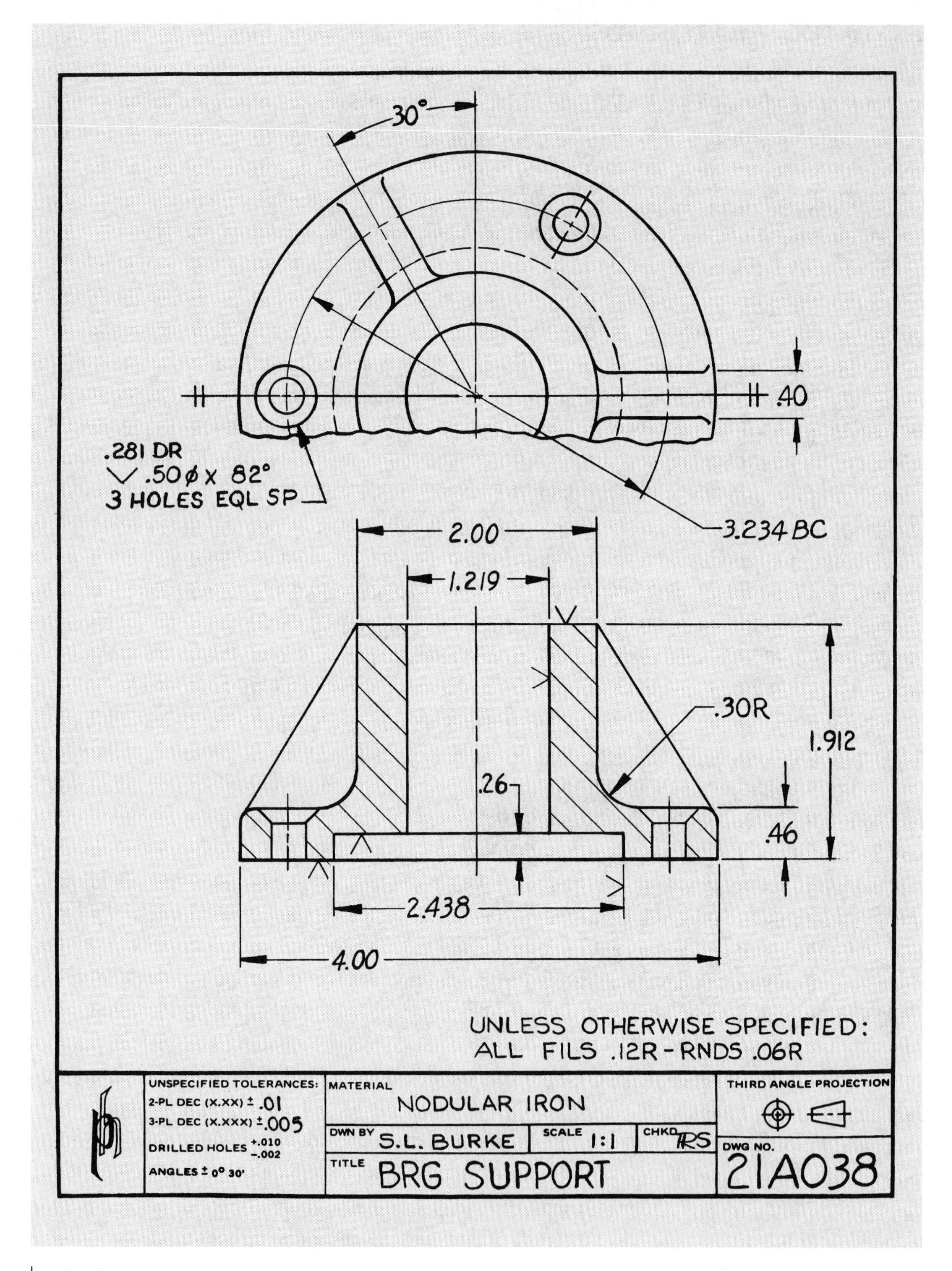
30°
.40
.281 DR
⌵ .50⌀ X 82°
3 HOLES EQL SP
3.234 BC
2.00
1.219
.30R
1.912
.26
.46
2.438
4.00
UNLESS OTHERWISE SPECIFIED:
ALL FILS .12R - RNDS .06R
UNSPECIFIED TOLERANCES:
2-PL DEC (X.XX) ± .01
3-PL DEC (X.XXX) ± .005
DRILLED HOLES +.010 −.002
ANGLES ± 0° 30'
MATERIAL
NODULAR IRON
DWN BY S.L. BURKE
SCALE 1:1
CHKD RS
TITLE BRG SUPPORT
THIRD ANGLE PROJECTION
DWG NO.
21A038

INSTRUCTIONS: Refer to drawing 21A038 to answer the following questions.

1. What type of iron is specified? 1. ______
2. What is the rib thickness? 2. ______
3. What is the minimum wall thickness of the hub? 3. ______
4. What is the maximum wall thickness of the hub? 4. ______
5. What size fasteners are intended to be used? 5. ______
6. What is the minimum thickness of the flange? 6. ______
7. What is the maximum depth of the counterbore? 7. ______
8. What is the minimum diameter of the counterbore? 8. ______
9. What size is the radius between the hub and the flange? 9. ______
10. What size is the fillet radius between the ribs and the hub? 10. ______
11. What is the maximum diameter of the flange? 11. ______
12. What is the minimum diameter of the mounting holes? 12. ______
13. What is the maximum diameter of the bolt circle? 13. ______
14. What is the minimum diameter of the countersink? 14. ______
15. How far apart (angularly) are the mounting holes? 15. ______
16. Is the section view a half section or a full section? 16. ______
17. Was rotated projection used on the section view? (See page 195.) 17. ______
18. What do the marks on the ends of the horizontal centerline signify? 18. ______
19. What would be the overall height before machining? 19. ______
20. How many circular hidden lines would appear on a full bottom view? 20. ______

Optional Math Exercises:

1a. Calculate the Cartesian coordinates for the .281∅ drilled holes.

2a. Calculate the Measurement Over Pins (MOP) of two adjacent .281∅ drilled holes at MMC.

REMOVED SECTIONS

A removed section violates the direct projection rule in that the view is not in proper alignment with related views. It may be placed anywhere on the drawing as long as it is identified with section letters corresponding with the cutting-plane letters. Also, the view must not be turned from the position established by the arrowheads of the cutting-plane line. The advantages are twofold: (1) the removed section view may be drawn to a different scale than the projected views, and (2) the size of the drawing sheet may be reduced by placing the section view into an unused open area.

Another variation of the removed section permits the view to be placed on a centerline extending from the imaginary cutting-plane. However, this practice is limited to sections that are symmetrical. See the examples shown below.

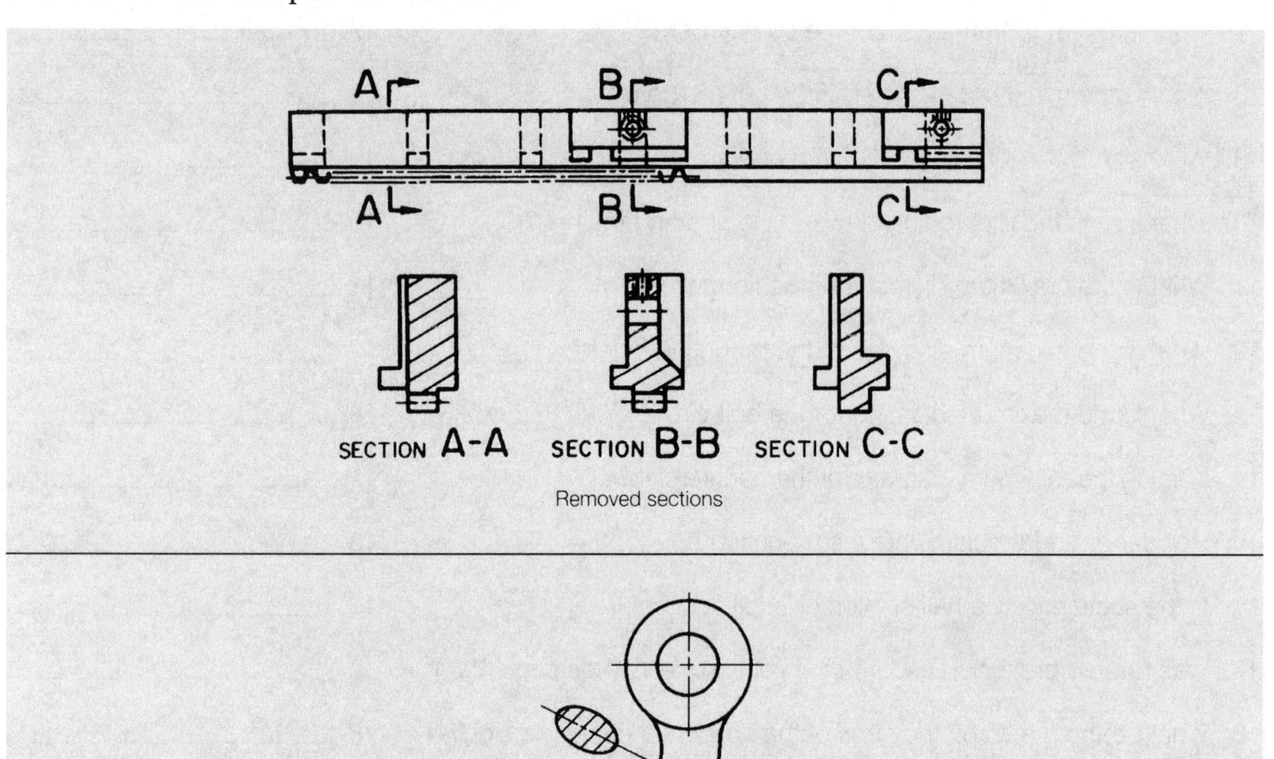

Removed sections

OFFSET SECTIONS

Often, internal features do not fall on a common plane that would permit them to be exposed in a single sectional view. To eliminate the need for additional sections, the cutting plane may be offset to expose these features all on one view. Drawing 21A039 on page 200 utilizes the offset cutting plane to show the various types of holes on a full-section view even though the holes are staggered. Notice that the changes in the cutting-plane line are not visible in the section view, since it is only an imaginary plane. Shown below are additional examples of offset sections.

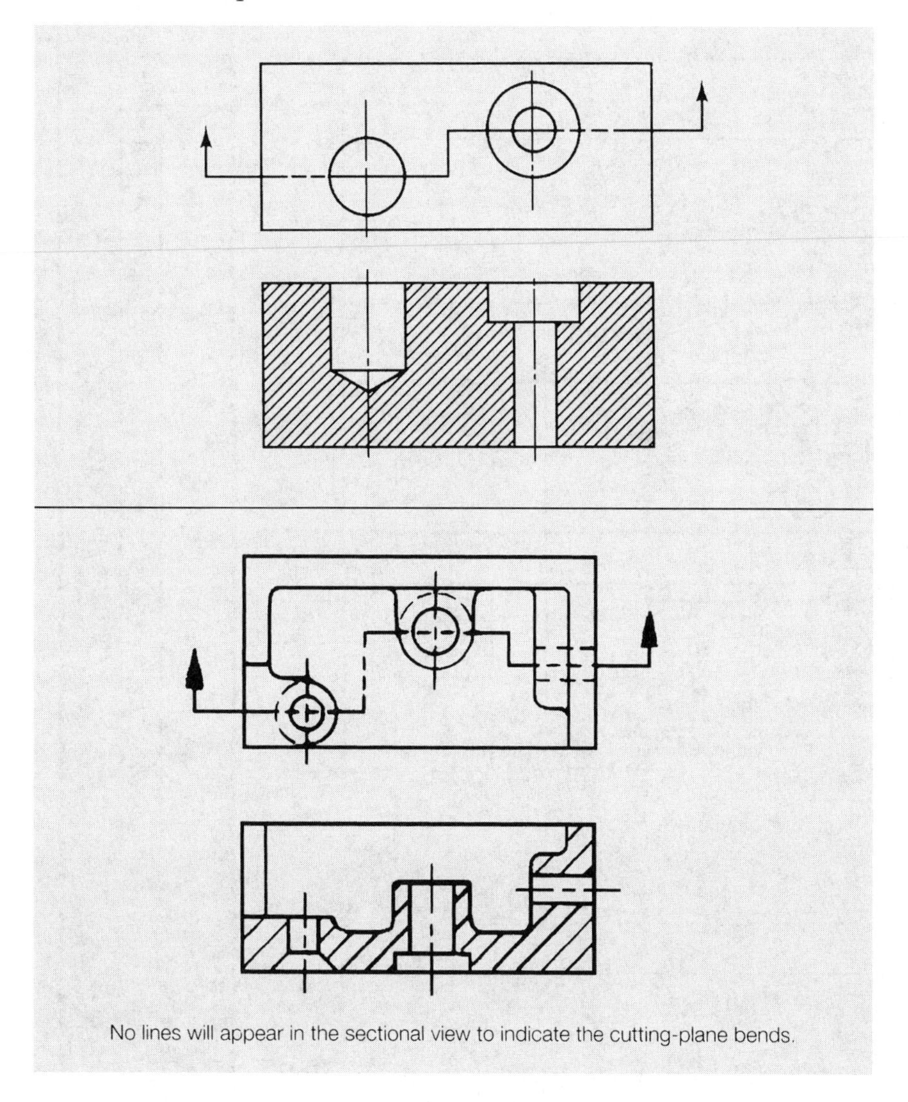

No lines will appear in the sectional view to indicate the cutting-plane bends.

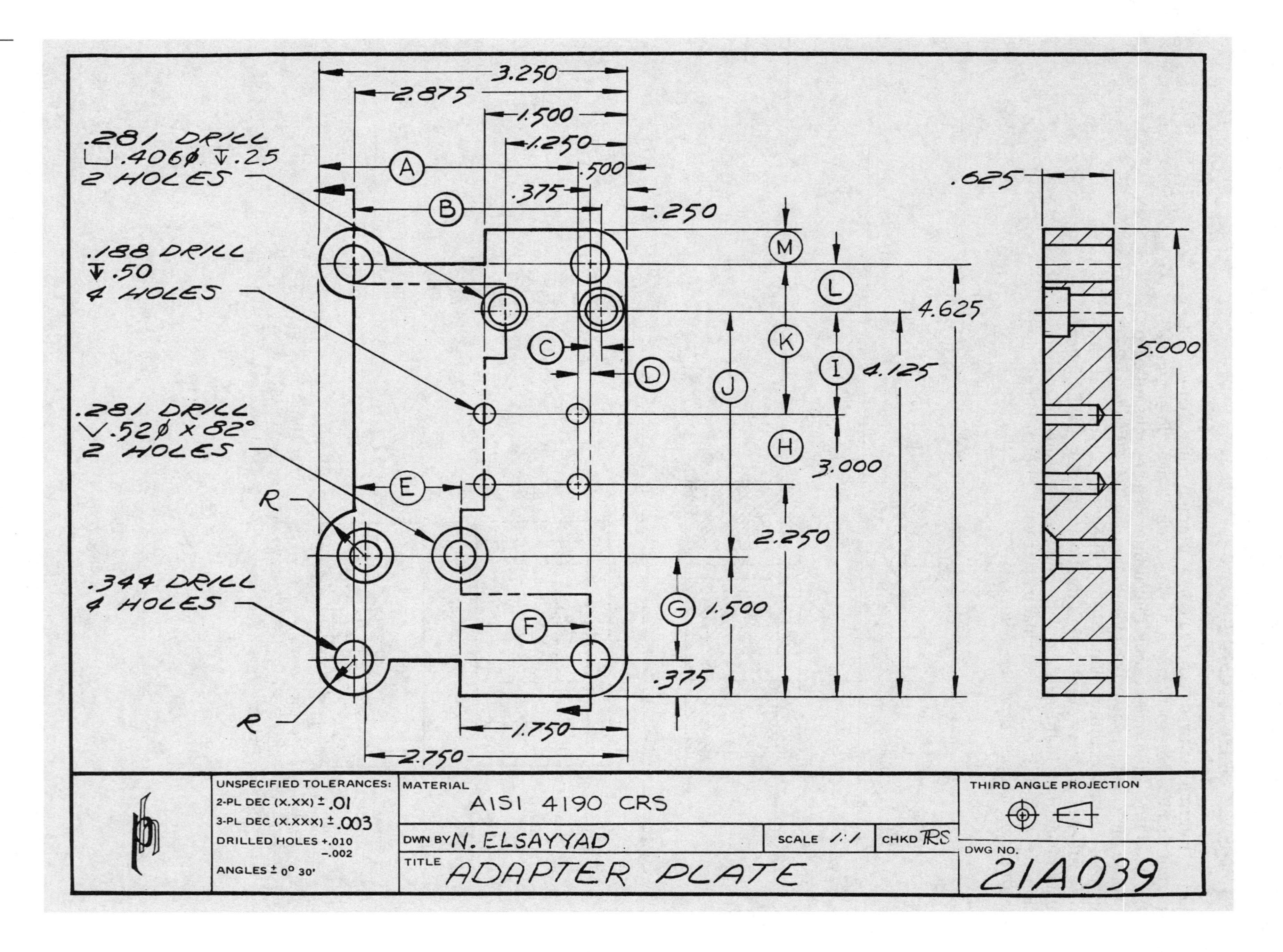
3.250
2.875
1.500
1.250
.500
.375
.250
.281 DRILL
⌴ .406⌀ ↧.25
2 HOLES
.188 DRILL
↧.50
4 HOLES
.281 DRILL
⌵ .520⌀ X 82°
2 HOLES
.344 DRILL
4 HOLES
R
R
A
B
C
D
E
F
G
H
I
J
K
L
M
4.625
4.125
3.000
2.250
1.500
.375
1.750
2.750
.625
5.000
UNSPECIFIED TOLERANCES:
2-PL DEC (X.XX) ± .01
3-PL DEC (X.XXX) ± .003
DRILLED HOLES +.010 −.002
ANGLES ± 0° 30'
MATERIAL
AISI 4190 CRS
DWN BY N. ELSAYYAD
SCALE 1:1
CHKD RS
TITLE ADAPTER PLATE
THIRD ANGLE PROJECTION
DWG NO.
21A039

INSTRUCTIONS: Refer to drawing 21A039 to answer the following questions.

1. Is the material carbon steel or alloy steel? 1. ____________
2. Is the steel cold-rolled or hot-rolled? 2. ____________
3. Is the steel carbon content 90%, 9%, or 0.9%? 3. ____________
4. How many holes are in the plate? 4. ____________
5. How many of the holes are blind holes? 5. ____________
6. What is the depth of the blind holes? 6. ____________
7. Show the limits of the plate thickness. 7. ____________
8. Show the limits of the countersink diameter. 8. ____________
9. Show the limits of the corner holes diameter. 9. ____________
10. How much material remains between a corner hole and the nearest outside edge? 10. ____________
11. What dimensioning method (chain, broken-chain, or datum) was used? 11. ____________
12. What is the maximum tolerance between any two hole locations? 12. ____________

INSTRUCTIONS: Calculate the dimensions for the following letters.

Ⓐ ____________

Ⓑ ____________

Ⓒ ____________

Ⓓ ____________

Ⓔ ____________

Ⓕ ____________

Ⓖ ____________

Ⓗ ____________

Ⓘ ____________

Ⓙ ____________

Ⓚ ____________

Ⓛ ____________

Ⓜ ____________

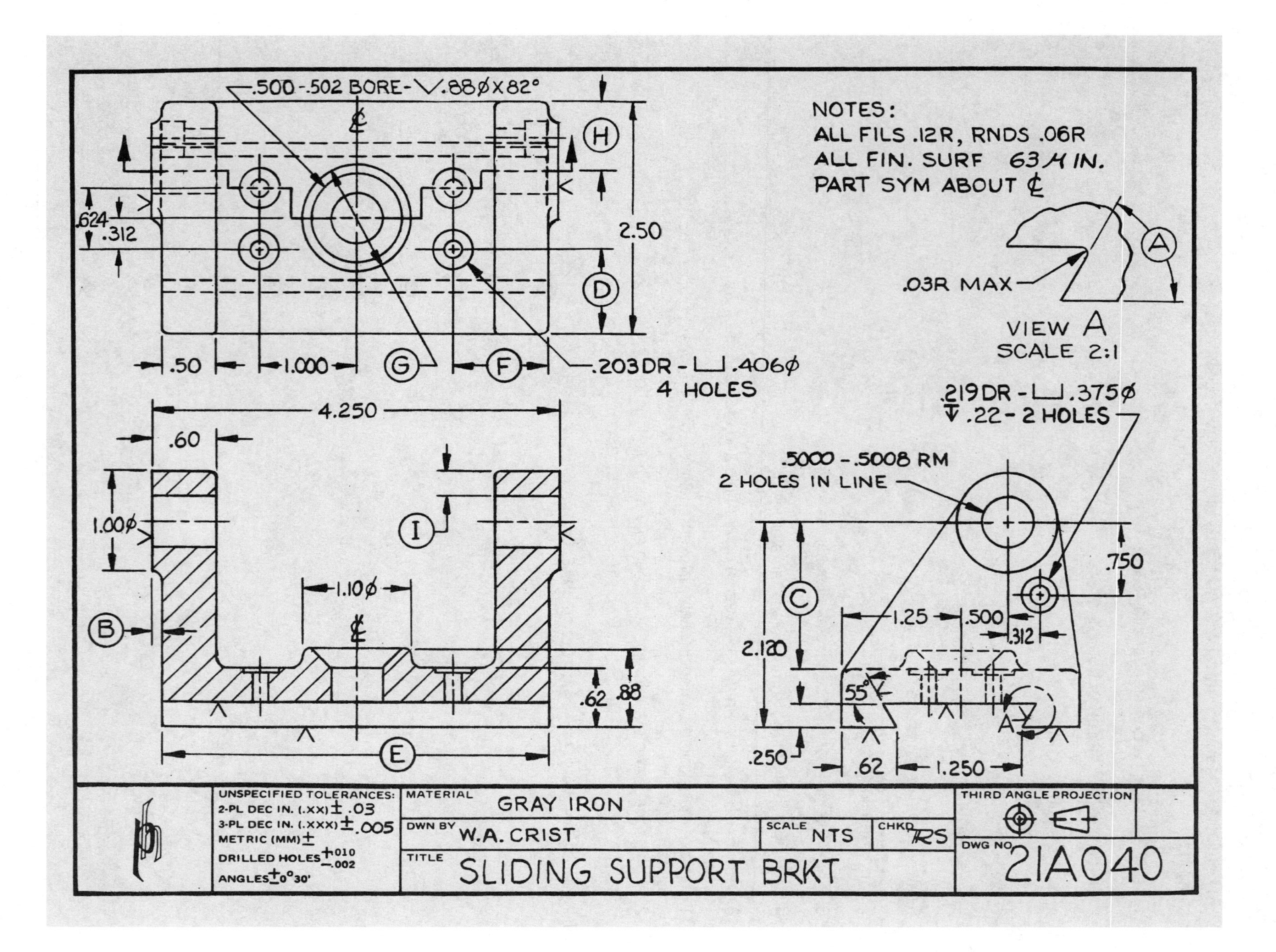

.500 - .502 BORE - ⌵ .88⌀ X 82°
℄
H
.624
.312
2.50
D
.50
1.000
G
F
.203DR - ⌴ .406⌀
4 HOLES
NOTES:
ALL FILS .12R, RNDS .06R
ALL FIN. SURF 63 µ IN.
PART SYM ABOUT ℄
A
.03R MAX
VIEW A
SCALE 2:1
.219DR - ⌴ .375⌀
▽ .22 - 2 HOLES
4.250
.60
1.00⌀
I
1.10⌀
B
.62
.88
E
.5000 - .5008 RM
2 HOLES IN LINE
.750
C
1.25
.500
.312
2.120
55°
.250
.62
1.250
UNSPECIFIED TOLERANCES:
2-PL DEC IN. (.XX) ± .03
3-PL DEC IN. (.XXX) ± .005
METRIC (MM) ±
DRILLED HOLES +.010 -.002
ANGLES ±0°30'
MATERIAL GRAY IRON
DWN BY W.A. CRIST
SCALE NTS
CHKD RS
TITLE SLIDING SUPPORT BRKT
THIRD ANGLE PROJECTION
DWG NO 21A040

INSTRUCTIONS: Refer to drawing 21A040 to answer the following questions.

1. What type of sectional view is drawn? 1. ______
2. What type of view is view A? 2. ______
3. What is the approximate spotface depth? 3. ______
4. What is the counterbore depth? 4. ______
5. What is the included angle of the countersink? 5. ______
6. What is the MMC of the bored hole? 6. ______
7. What is the overall height of the casting before machining? 7. ______
8. What is the maximum overall height after machining? 8. ______
9. What is the included angle between the two sides of the dovetail slot? 9. ______
10. What is the base thickness adjacent to the spotfaced holes after machining? 10. ______

INSTRUCTIONS: Enter the dimensions for the following letters.

Ⓐ ______

Ⓑ ______

Ⓒ ______

Ⓓ ______

Ⓔ ______

Ⓕ ______

Ⓖ ______

Ⓗ ______

Ⓘ MIN: ______

Ⓘ MAX: ______

Optional Math Exercise:

1a. Using pins that are .250∅, determine the MBP (Measurement Between Pins) after the dovetail has been properly machined.

UNIT 9

THREADED HOLE SPECIFICATIONS

A typical drawing specification for a threaded hole is as follows: .250–20 UNC-2B. The first figures represent the major thread diameter, which may be expressed decimally, fractionally, or by wire gauge number (for sizes smaller than 1/4 in.). The second set of figures represents the number of threads per inch, followed by the thread form and series. Next is the class of fit followed by either an A (external thread) or a B (internal thread). If the thread specified is left-hand, it will have the abbreviation LH after the class of fit. A right-hand thread is assumed if LH is omitted. If the threaded hole is a blind hole, the thread depth will be specified. A through hole is assumed if the depth is omitted. Normally, the recommended tap drill size is included with the thread specifications; however, when it is omitted you may refer to tables such as the one on page 206.

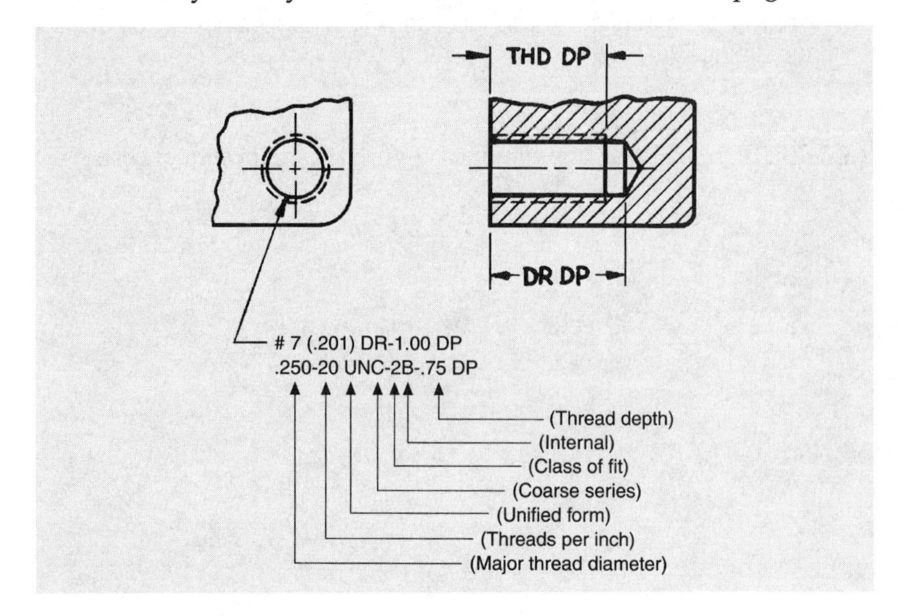

UNIFIED THREAD FORM

Whereas a number of different thread forms are used for various applications, those most commonly used for fasteners are the Unified or their predecessor, the American National. The Unified is actually a compromise between the American National and the British Whitworth threads. It is, however, mechanically interchangeable with American National threads of the same diameter and pitch. The unified thread profile is illustrated below. Additional thread forms may be viewed on page 219.

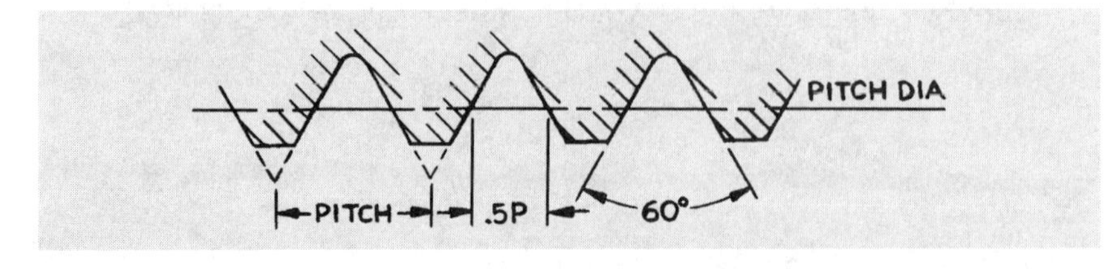

THREAD SERIES

Thread series are groups of diameter-pitch combinations distinguished from each other by the number of threads per inch applied to a specific diameter. There are three standard series available in the Unified National Threads: coarse, fine, and extra-fine.

The coarse series, abbreviated UNC, is the one most commonly used in the bulk production of bolts, screws, nuts, and other general engineering applications. It is applicable for rapid assembly or disassembly, or if corrosion or slight damage is possible.

The fine series, abbreviated UNF, is suitable for the production of bolts, screws, nuts, and other applications where the coarse series is not applicable. It is used where the length of engagement is short, where a smaller lead angle is desired, or where the wall thickness demands a fine pitch.

The extra-fine series, abbreviated UNEF, is applicable where even finer pitches are desirable, as for short lengths of engagement and for thin-walled tubes, nuts, ferrules, or couplings.

CLASSES OF FIT

Three standard classes of fit are available. Class 1 is the loosest, and is used for applications requiring the most clearance between mating threads. Class 2 is the standard for most fastener applications. Class 3 has the closest fit between mating threads and is used for more critical applications. When the class of fit has been omitted from the specs, class 2 may be assumed., The letters A or B may also be omitted when the thread is visibly external or internal.

THREADED HOLE ILLUSTRATIONS

Threaded holes are easily distinguished from unthreaded holes on drawings by the presence of a circle of dashes surrounding a solid circle. The dashed circle represents the major thread diameter; the solid circle represents the minor diameter. Sometimes the threaded hole is chamfered, which causes the dashed circle to be replaced with another solid circle. The side view of threaded holes includes two pairs of hidden lines to represent the major and minor thread diameters. However, when the side view is in section, threads may be represented either by the schematic method or the more commonly used simplified method. (See the examples of both shown below.)

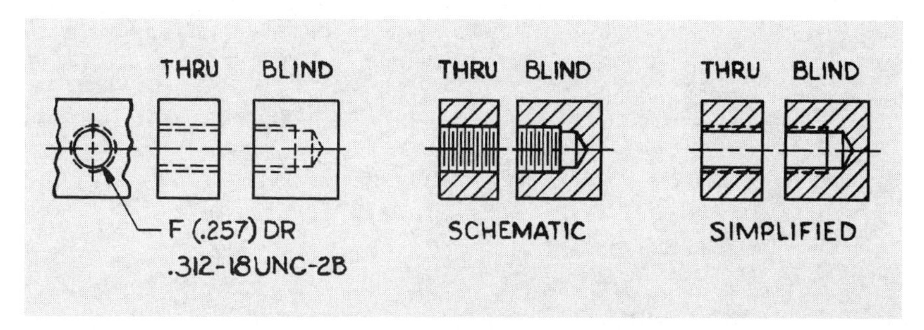

UNIFIED THREAD TABLE

Major Thread Diameter		Coarse Thread TPI	Coarse Thread Tap Drill		Fine Thread TPI	Fine Thread Tap Drill		Extra-Fine Thread TPI	Extra-Fine Thread Tap Drill	
0	(.060)				80	3/64	(.0469)			
1	(.073)	64	53	(.0595)	72	53	(.0595)			
2	(.086)	56	50	(.070)	64	50	(.070)			
3	(.099)	48	47	(.0785)	56	45	(.082)			
4	(.112)	40	43	(.089)	48	42	(.0935)			
5	(.125)	40	38	(.1015)	44	37	(.104)			
6	(.138)	32	36	(.1065)	40	33	(.113)			
8	(.164)	32	29	(.136)	36	29	(.136)			
10	(.190)	24	25	(.1495)	32	21	(.159)			
12	(.216)	24	16	(.177)	28	14	(.182)	32	13	(.185)
1/4	(.250)	20	7	(.201)	28	3	(.213)	32	7/32	(.219)
5/16	(.312)	18	F	(.257)	24	I	(.272)	32	9/32	(.281)
3/8	(.375)	16	5/16	(.312)	24	Q	(.332)	32	11/32	(.343)
7/16	(.438)	14	U	(.368)	20	25/64	(.391)	28	13/32	(.406)
1/2	(.500)	13	27/64	(.422)	20	29/64	(.453)	28	15/32	(.469)
9/16	(.562)	12	31/64	(.484)	18	33/64	(.316)	24	33/64	(.516)
5/8	(.625)	11	17/32	(.531)	18	37/64	(.578)	24	37/64	(.578)
3/4	(.750)	10	21/31	(.656)	16	11/16	(.688)	20	45/64	(.703)
7/8	(.875)	9	49/64	(.766)	14	13/16	(.812)	20	53/64	(.828)
1"	(1.000)	8	7/8	(.875)	12	59/64	(.922)	20	61/64	(.953)
1-1/8	(1.125)	7	63/64	(.984)	12	1-3/64	(1.047)	18	1-5/64	(1.078)
1-1/4	(1.250)	7	1-7/64	(1.109)	12	1-11/64	(1.172)	18	1-3/16	(1.188)
1-3/8	(1.375)	6	1-7/32	(1.219)	12	1-19/64	(1.297)	18	1-5/16	(1.312)
1-1/2	(1.500)	6	1-11/32	(1.344)	12	1-27/64	(1.422)	18	1-7/16	(1.438)
1-3/4	(1.750)	5	1-9/16	(1.562)				16	1-11/16	(1.688)
2"	(2.000)	4-1/2	1-25/32	(1.781)				16	1-15/16	(1.938)
2-1/4	(2.250)	4-1/2	2-1/32	(2.031)						
2-1/2	(2.500)	4	2-1/4	(2.250)						
2-3/4	(2.750)	4	2-1/2	(2.500)						
3"	(3.000)	4	2-3/4	(2.750)						
3-1/4	(3.250)	4	3"	(3.000)						
3-1/2	(3.500)	4	3-1/4	(3.250)						
3-3/4	(3.750)	4	3-1/2	(3.500)						
4"	(4.000)	4	3-3/4	(3.750)						

NUMBERED AND LETTERED DRILLS

No.	Dia.	No.	Dia.	No.	Dia.	Ltr.	Dia.
80	.0135	53	.0595	26	.1470	A	.2340
79	.0145	52	.0635	25	.1495	B	.2380
78	.0160	51	.0670	24	.1520	C	.2420
77	.0180	50	.0700	23	.1540	D	.2460
76	.0200	49	.0730	22	.1570	E	.2500
75	.0210	48	.0760	21	.1590	F	.2570
74	.0225	47	.0785	20	.1610	G	.2610
73	.0240	46	.0810	19	.1660	H	.2660
72	.0250	45	.0820	18	.1695	I	.2720
71	.0260	44	.0860	17	.1730	J	.2770
70	.0280	43	.0890	16	.1770	K	.2810
69	.0292	42	.0935	15	.1800	L	.2900
68	.0310	41	.0960	14	.1820	M	.2950
67	.0320	40	.0980	13	.1850	N	.3020
66	.0330	39	.0995	12	.1890	O	.3160
65	.0350	38	.1015	11	.1910	P	.3230
64	.0360	37	.1040	10	.1935	Q	.3320
63	.0370	36	.1065	9	.1960	R	.3390
62	.0380	35	.1100	8	.1990	S	.3480
61	.0390	34	.1110	7	.2010	T	.3580
60	.0400	33	.1130	6	.2040	U	.3680
59	.0410	32	.1160	5	.2055	V	.3770
58	.0420	31	.1200	4	.2090	W	.3860
57	.0430	30	.1285	3	.2130	X	.3970
56	.0465	29	.1360	2	.2210	Y	.4040
55	.0520	28	.1405	1	.2280	Z	.4130
54	.0550	27	.1440				

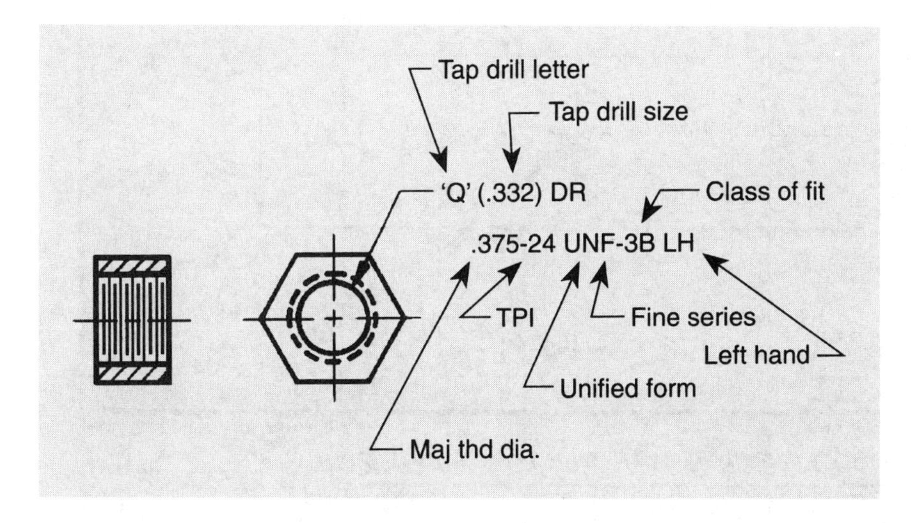

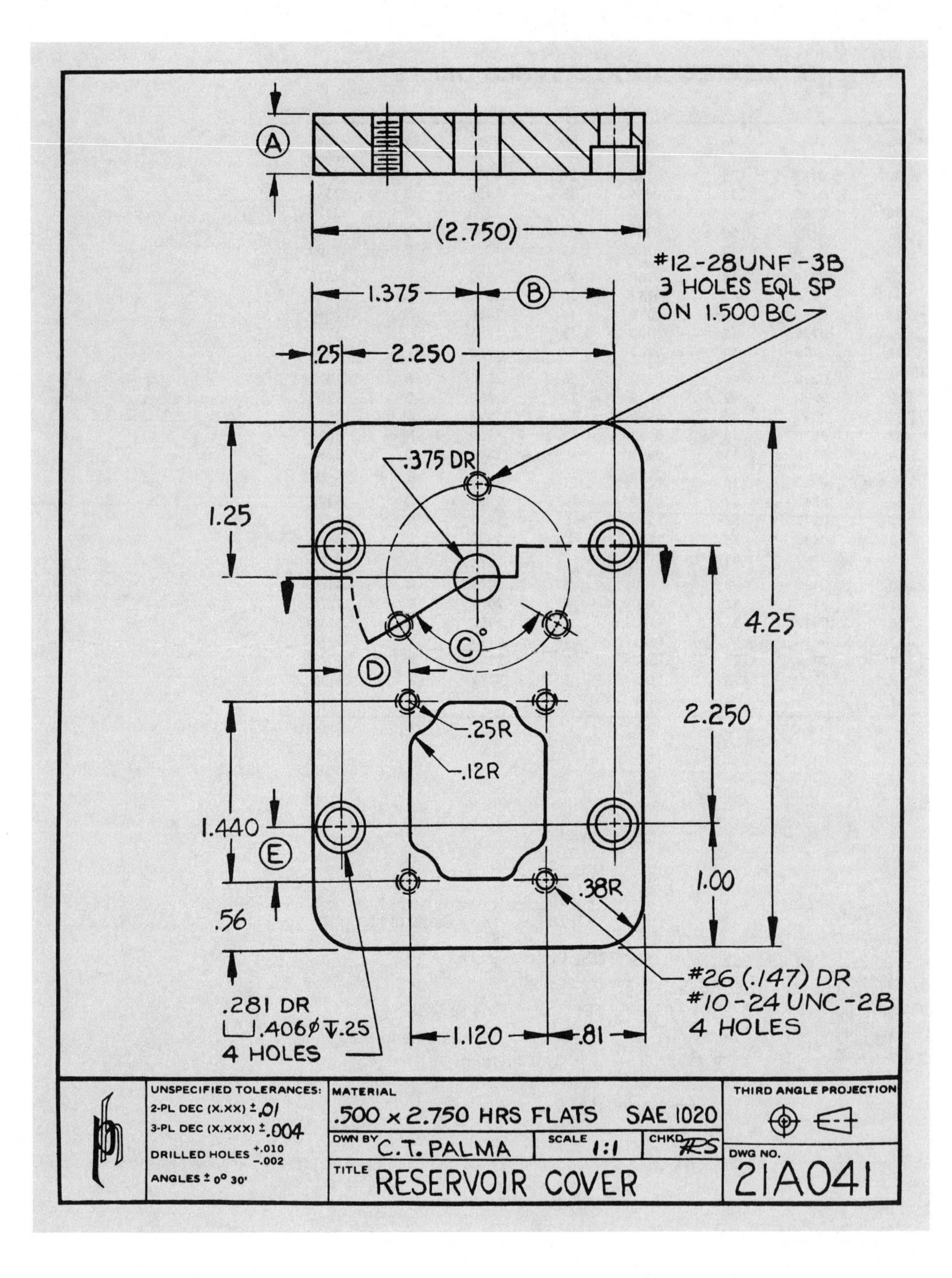
A
(2.750)
#12-28UNF-3B
3 HOLES EQL SP
ON 1.500 BC
1.375
B
.25
2.250
.375 DR
1.25
4.25
C°
D
2.250
.25R
.12R
1.440
E
1.00
.38R
.56
#26 (.147) DR
#10-24 UNC-2B
4 HOLES
.281 DR
⌴ .406⌀ ↧.25
4 HOLES
1.120
.81
UNSPECIFIED TOLERANCES:
2-PL DEC (X.XX) ±.01
3-PL DEC (X.XXX) ±.004
DRILLED HOLES +.010 −.002
ANGLES ± 0° 30'
MATERIAL
.500 x 2.750 HRS FLATS SAE 1020
DWN BY C.T. PALMA
SCALE 1:1
CHKD RS
TITLE RESERVOIR COVER
THIRD ANGLE PROJECTION
DWG NO.
21A041

THREAD CALCULATIONS

Thread pitch is the distance measured from a point on one thread to a corresponding point on the next thread. Pitch may be calculated by dividing threads per inch (TPI) into 1 in. *Example:* 20 TPI = .05 pitch.

It may sometimes be necessary to determine the number of threads a tapped hole contains. To calculate this information, you may either divide the thread depth by its pitch, or multiply its thread depth by TPI. *Examples:* .50 depth ÷ .05 pitch = 10 threads, or .50 depth × 20 TPI = 10 threads.

INSTRUCTIONS: *Refer to drawing 21A041 to answer the following questions.*

1. What type of section is shown? (Observe the cutting-plane line.) — 1. ____________
2. How many threaded holes does the part contain? — 2. ____________
3. What representation method is used to illustrate the threads in the top view? — 3. ____________
4. Are the threads right-hand or left-hand? — 4. ____________
5. What thread form is specified? — 5. ____________
6. What thread series is specified on the No. 12 holes? — 6. ____________
7. What class of fit is specified on the No. 12 holes? — 7. ____________
8. What does the "B" after the class of fit designate? — 8. ____________
9. What is the major diameter of the No. 10 threaded holes? (Consult the table on page 206). — 9. ____________
10. Is the proper tap drill specified for the No. 10 tapped holes? (Consult the table on page 206). — 10. ____________
11. How many full threads will each No. 10 hole contain? — 11. ____________
12. How many full threads will each No. 12 hole contain? — 12. ____________
13. What diameter tap drill is recommended for the No. 12 tapped holes? (Consult the table on page 206). — 13. ____________
14. How much material remains between a counterbore and the nearest outside edge? — 14. ____________
15. What do the parentheses around the width dimension represent? — 15. ____________

INSTRUCTIONS: *Calculate the dimensions for the following letters.*

Ⓐ ____________

Ⓑ ____________

Ⓒ ____________

Ⓓ ____________

Ⓔ ____________

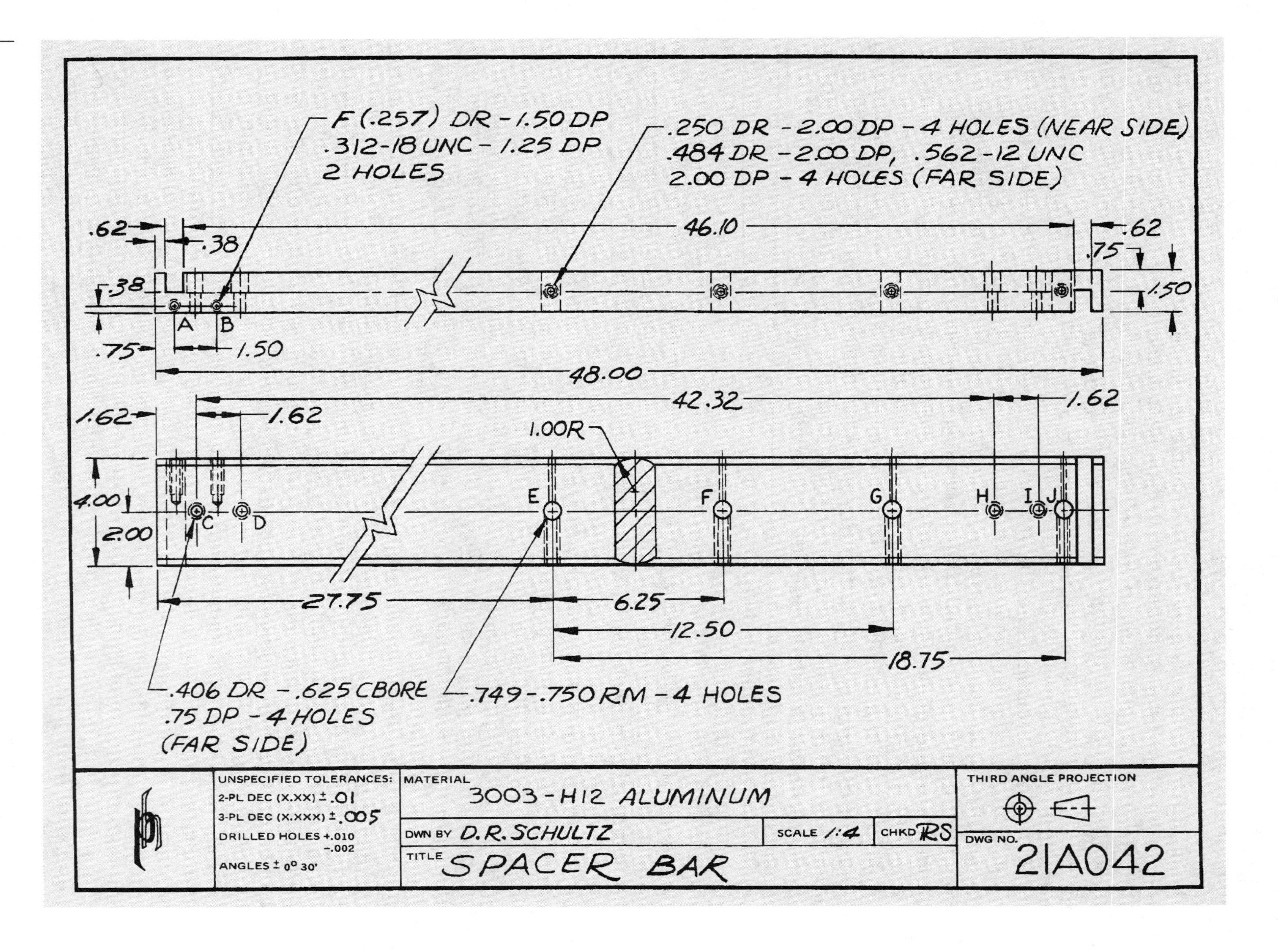

F (.257) DR - 1.50 DP
.312-18 UNC - 1.25 DP
2 HOLES
.250 DR - 2.00 DP - 4 HOLES (NEAR SIDE)
.484 DR - 2.00 DP, .562-12 UNC
2.00 DP - 4 HOLES (FAR SIDE)
.62
.38
46.10
.62
.75
.38
1.50
A
B
.75
1.50
48.00
42.32
1.62
1.62
1.62
1.00R
4.00
2.00
E
F
G
H
I
J
C
D
27.75
6.25
12.50
18.75
.406 DR - .625 CBORE
.75 DP - 4 HOLES
(FAR SIDE)
.749-.750 RM - 4 HOLES
UNSPECIFIED TOLERANCES:
2-PL DEC (X.XX) ± .01
3-PL DEC (X.XXX) ± .005
DRILLED HOLES +.010 -.002
ANGLES ± 0° 30'
MATERIAL
3003-H12 ALUMINUM
DWN BY D.R. SCHULTZ
SCALE 1:4
CHKD RS
TITLE SPACER BAR
THIRD ANGLE PROJECTION
DWG NO. 21A042

INSTRUCTIONS: Refer to drawing 21A042 to answer the following questions.

1. What type of section view was used? 1. ____________
2. What type of aluminum was specified? 2. ____________
3. How many threaded holes does the object contain? 3. ____________
4. Why do holes C and D appear to be threaded holes in their circular view? 4. ____________
5. Are the slots equal distance from the ends of the bar? (Calculate.) 5. ____________
6. How many different drill sizes are specified? 6. ____________
7. What is the tolerance on the reamed holes? 7. ____________
8. Are the reamed holes equally spaced? 8. ____________
9. How far apart (center-to-center) are holes E and H? 9. ____________
10. How much tolerance may accumulate between the centers of holes E and H? 10. ____________
11. What is the maximum center spacing between holes B and D? 11. ____________
12. What is the maximum center spacing between holes G and H? 12. ____________
13. What is the minimum wall thickness between the side of hole J and the adjacent slot? 13. ____________
14. What thread form is specified for the tapped holes? 14. ____________
15. What thread series is specified? 15. ____________
16. What class of fit may be assumed for the threaded holes? 16. ____________
17. Are the threads right-hand or left-hand? 17. ____________
18. What is the tap *drill* depth in hole A? 18. ____________
19. How may full threads will hole A contain? 19. ____________
20. What size fastener is intended for use in hole C? 20. ____________

Thread Quiz 1

INSTRUCTIONS: Refer to the tables on pages 206 and 207 to answer the following questions.

1. What tap drill diameter is recommended for a 7/16-14 UNC threaded hole? 1. ________
2. What tap drill diameter is recommended for a 1/4-28 UNF threaded hole? 2. ________
3. What tap drill diameter is recommended for a 5/16-32 UNEF threaded hole? 3. ________
4. What is the major thread diameter of a No. 4 Unified thread? 4. ________
5. What numbered thread size is closest to the fraction 3/16? 5. ________
6. What thread requires a 5/16 diameter tap drill? 6. ________
7. What is the pitch (decimally) of a 1″ UNEF thread? 7. ________
8. How thick would a threaded member have to be to provide 3 full threads using a 9/16 UNC thread? 8. ________
9. How far would a threaded member advance with 3 revolutions on a shaft threaded 1″ UNF? 9. ________
10. How much difference in diameters is there between a letter I drill and a number 1 drill? 10. ________
11. What is the closest numbered drill to 1/8 inch diameter? 11. ________
12. What is the closest numbered drill to 3/16 inch diameter? 12. ________
13. Is a No. 12 drill larger or smaller than a No. 10 drill? 13. ________
14. Is a No. 12 thread larger or smaller than a No. 10 thread? 14. ________
15. Which has the largest major diameter: a No. 6 thread or a No. 6 drill? 15. ________

CONSTANT PITCH SERIES

In addition to the coarse, fine, and extra-fine series, there is also another Unified series. Called the Constant Pitch Series, the pitch remains the same as the diameter varies. Designated UN-4, 6, 8, 12, 16, 20, 28, or 32 threads per inch, these combinations may be used where threads in the coarse, fine, and extra-fine series do not meet the particular requirements of the design.

Preference is given to the 8-, 12-, and 16-thread series. The 8-thread series (8UN) is used as a substitute for the coarse thread for diameters larger than one inch. The 12-thread series (12UN) is used as a continuation of the fine thread for diameters larger than 1 1/2 inches. The 16-thread series (16UN) is used as a continuation of the extra-fine thread for diameters larger than 1 11/16 inches.

When design considerations require pitches that are nonstandard, the abbreviation UNS (for Unified National Special) is used. *Example:* .250–40 UNS.

Thread Quiz 2

INSTRUCTIONS: Read the thread information printed below, then arrange it in the proper order as it should appear on a drawing. (Disregard tap drill specifications.) Refer to pages 204–206.

1. Right-hand thread, .625 major thread diameter, Unified thread form, internal threads, coarse series, class 2 fit. — 1. ____________________
2. Unified thread form, left-hand thread, class 3 fit, fine series, internal thread, .438 major diameter. — 2. ____________________
3. 32 TPI, Unified form, internal threads, right-hand, .216 major diameter, class 2. — 3. ____________________
4. Left-hand thread, 8 threads per inch, class 1 fit, 1.125 major diameter, internal threads. — 4. ____________________
5. Class 3 fit, internal threads, 40 TPI, right-hand, Unified form, .250 major diameter. — 5. ____________________

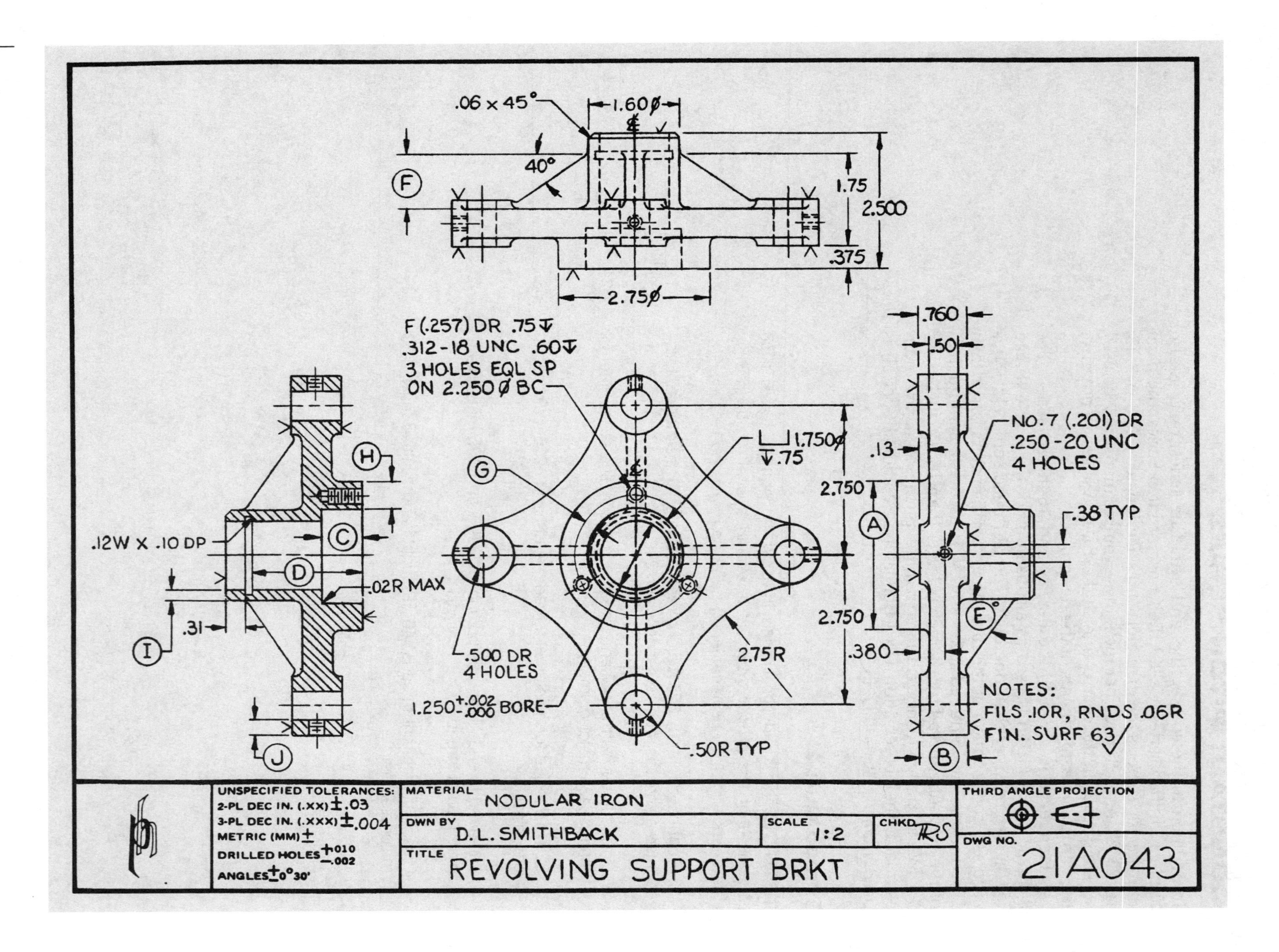
.06 x 45°
1.60∅
40°
F
1.75
2.500
.375
2.75∅
F (.257) DR .75▽
.312-18 UNC .60▽
3 HOLES EQL SP
ON 2.250∅ BC
.760
.50
1.750∅
▽.75
.13
NO. 7 (.201) DR
.250-20 UNC
4 HOLES
G
H
2.750
A
.38 TYP
.12W X .10 DP
C
D
.02R MAX
2.750
E
.31
I
.500 DR
4 HOLES
2.75R
.380
NOTES:
FILS .10R, RNDS .06R
FIN. SURF 63
1.250 +.002 -.000 BORE
.50R TYP
J
B
UNSPECIFIED TOLERANCES:
2-PL DEC IN. (.XX) ±.03
3-PL DEC IN. (.XXX) ±.004
METRIC (MM) ±
DRILLED HOLES +.010 -.002
ANGLES ±0°30'
MATERIAL NODULAR IRON
DWN BY D.L. SMITHBACK
SCALE 1:2
CHKD RS
TITLE REVOLVING SUPPORT BRKT
THIRD ANGLE PROJECTION
DWG NO. 21A043

LINE OMISSION

It is quite common for a drafter to omit hidden lines from a view if those lines are repetitious or unnecessary for proper interpretation. The hidden lines representing internal features on the right side view of drawing 21A043 have been omitted for that reason.

INSTRUCTIONS: Refer to drawing 21A043 to answer the following questions.

1. How many tapped holes does the part contain? 1. ____________
2. What thread *series* is specified for the tapped holes? 2. ____________
3. How much deeper is the tap drill than the thread depth of the blind holes? 3. ____________
4. How many full threads will each of the .250–20 UNC tapped holes contain? 4. ____________
5. What type of section was drawn for the left-side view? 5. ____________
6. What would be the overall thickness before machining? (Includes finish allowance.) 6. ____________
7. What is the diameter of the annular groove inside of the bored hole? 7. ____________
8. If the bored hole is cored into the casting, what size would it be before boring? 8. ____________
9. How much wall thickness remains after machining the annular groove? 9. ____________
10. How much tolerance can accumulate on the overall height dimension? 10. ____________

INSTRUCTIONS: Enter the dimensions for the following letters.

Ⓐ ____________

Ⓑ ____________

Ⓒ ____________

Ⓓ ____________

Ⓔ ____________

Ⓕ ____________

Ⓖ ____________

Ⓗ MIN: ____________

Ⓘ MAX: ____________

Ⓙ MIN: ____________

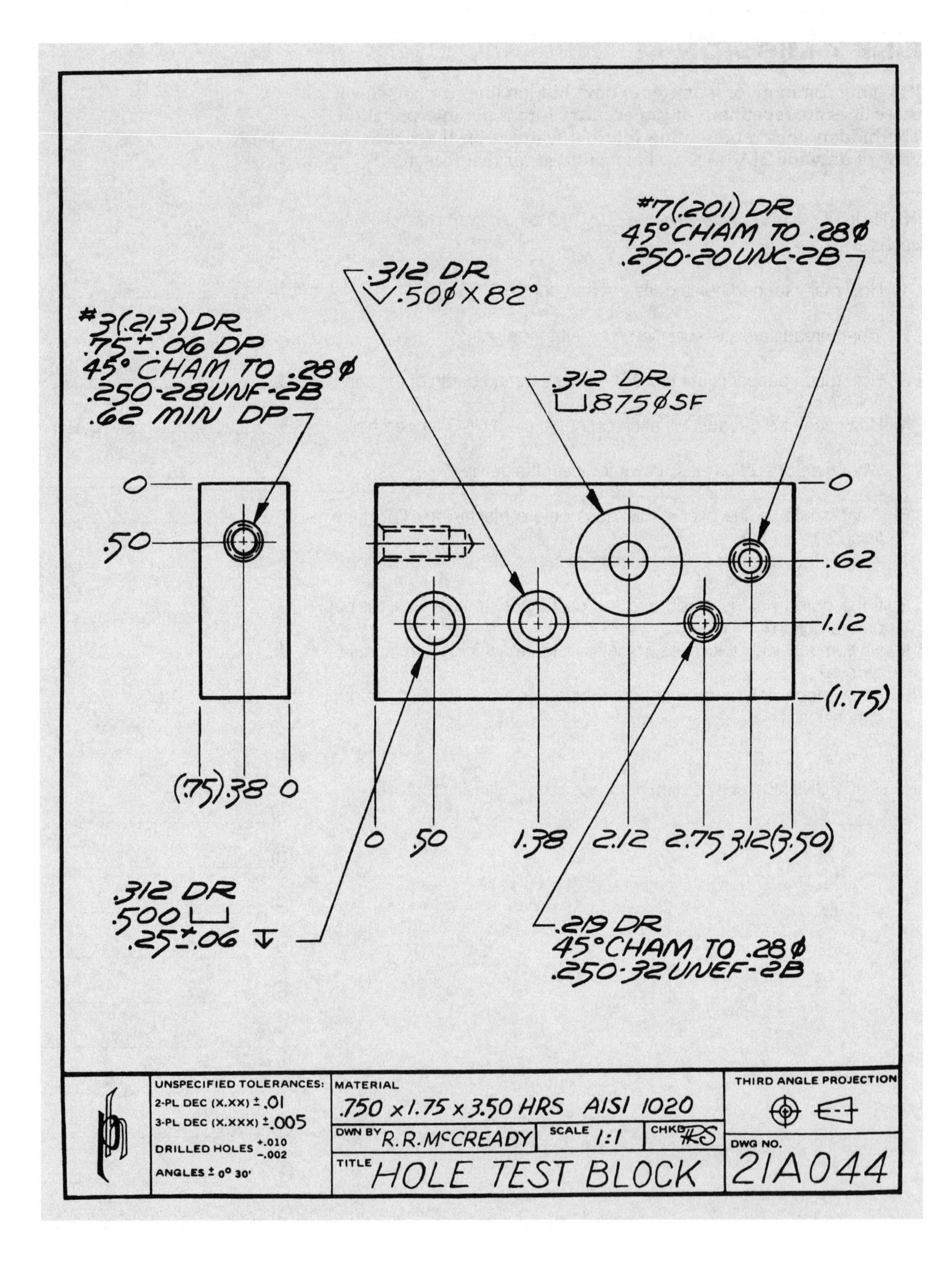
#7(.201) DR
45° CHAM TO .28⌀
.250-20UNC-2B
.312 DR
⌵.50⌀ X 82°
#3(.213) DR
.75±.06 DP
45° CHAM TO .28⌀
.250-28UNF-2B
.62 MIN DP
.312 DR
⌴.875⌀SF
0
.50
.62
1.12
(1.75)
(.75) .38 0
0 .50 1.38 2.12 2.75 3.12 (3.50)
.312 DR
.500 ⌴
.25±.06 ↧
.219 DR
45° CHAM TO .28⌀
.250-32UNEF-2B
UNSPECIFIED TOLERANCES:
2-PL DEC (X.XX) ± .01
3-PL DEC (X.XXX) ± .005
DRILLED HOLES +.010 −.002
ANGLES ± 0° 30'
MATERIAL
.750 x 1.75 x 3.50 HRS AISI 1020
DWN BY R.R.McCREADY
SCALE 1:1
CHKD RS
TITLE HOLE TEST BLOCK
THIRD ANGLE PROJECTION
DWG NO.
21A044

ARROWLESS DIMENSIONING

Drawing 21A044 lends itself to a simplified drafting technique called arrowless dimensioning. Instead of using dimension lines, the dimension figures are placed at the ends of the extension or centerlines. This method is limited to datum dimensioning, however. The datum surfaces on the drawing are identified by 0, thus each dimension figure represents its actual distance from zero.

INSTRUCTIONS: Refer to drawing 21A044 to answer the following questions.

1. What simplified method of dimensioning was used for locating the holes? 1. ______
2. Why do the three overall dimensions appear inside of parentheses? 2. ______
3. What is the MMC of the .312 diameter holes? 3. ______
4. How far apart are the counterbored hole and the countersunk hole? (C to C) 4. ______
5. What is the maximum tolerance between any two hole locations? 5. ______
6. What thread *form* is specified for the tapped holes? 6. ______
7. What class of fit is specified for the tapped holes? 7. ______
8. What is the thread depth of the blind hole? 8. ______
9. Is the tap drill smaller for the fine thread or the coarse thread? 9. ______
10. What TPI is specified for the fine thread? 10. ______
11. What is the pitch (decimally) of the coarse thread? 11. ______
12. What is the pitch (decimally) of the extra-fine thread? 12. ______
13. What is the diameter of the tapped hole chamfers? 13. ______
14. What is the maximum depth of the counterbore? 14. ______
15. What is the approximate depth of the spotface? 15. ______
16. Would print tolerance permit the use of a 7/32 tap drill for the fine thread? 16. ______
17. Would print tolerance permit the use of a 3/16 tap drill for the coarse thread? 17. ______
18. How many full threads will the coarse tapped hole contain? (Disregard chamfer.) 18. ______
19. How many full threads, at minimum depth, will the fine tapped hole contain? (Disregard chamfer.) 19. ______
20. How many full threads will the extra-fine tapped hole contain? (Disregard chamfer.) 20. ______

EXTERNAL THREADS

External threads are illustrated much the same as internal threads on drawings. The simplified method is perhaps more commonly used than either the schematic or detailed methods. Unlike internal threads, external threads are usually specified on the longitudinal view rather than the circular view. The specifications follow the same sequence, however, which is the major thread diameter, threads per inch, form and series, class of fit, and LH when applicable. The only difference is that the letter A will follow the class of fit instead of the letter B. External threads are usually chamfered to their minor diameter to facilitate assembly. (Observe the examples of external threads shown below.)

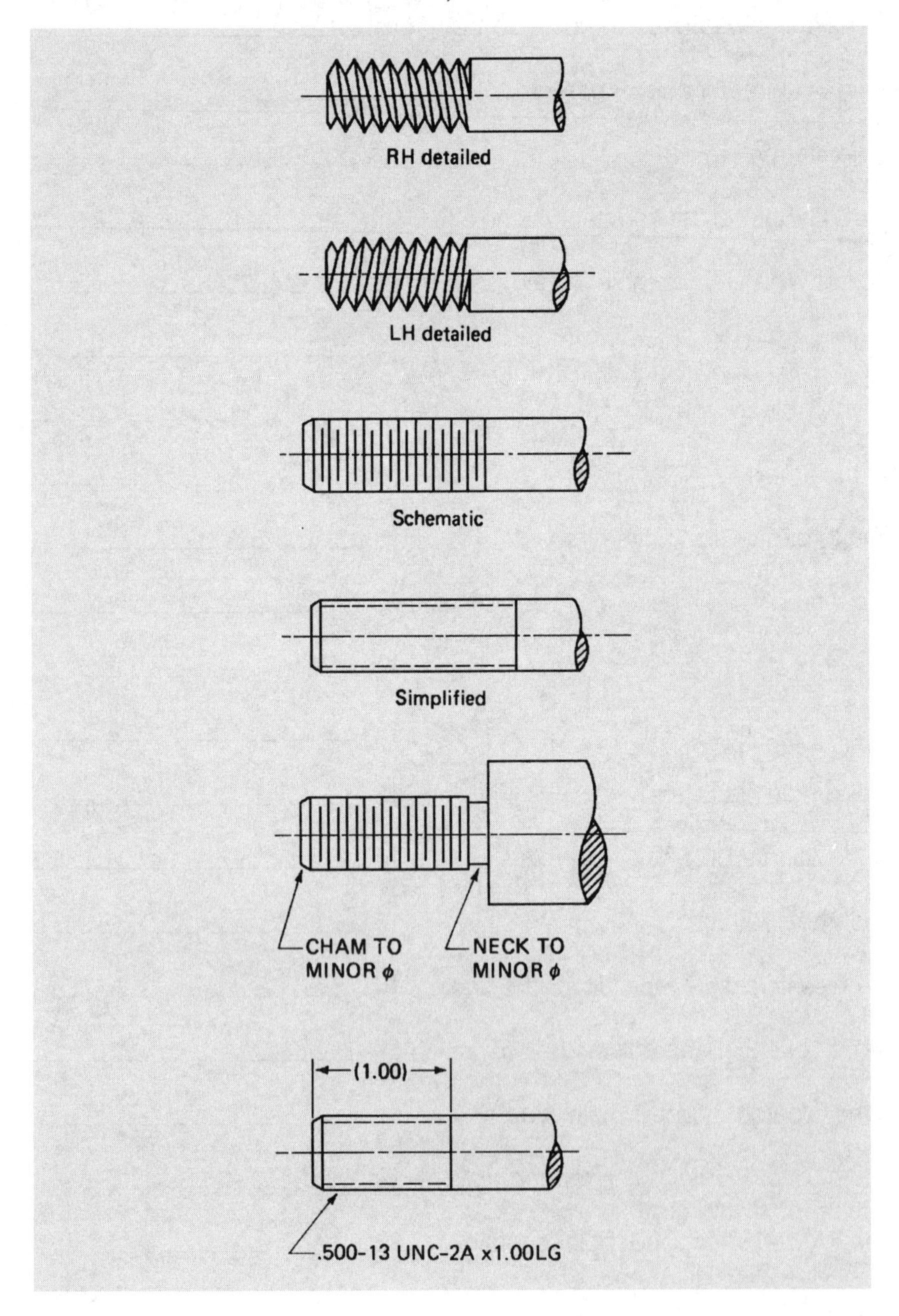

THREAD FORMS

Although the most common thread form used with fasteners is the Unified National, there are many other applications for threads. Threads may be used to transmit power, such as the square or acme thread used on a bench vise and the lead screw of a lathe. The buttress thread will transmit power in one direction only. The knuckle thread is commonly used with light bulbs. The sharp-V thread is used where increased friction from its full thread face is useful, such as on brass pipe work. (Examples of various thread forms are shown below.)

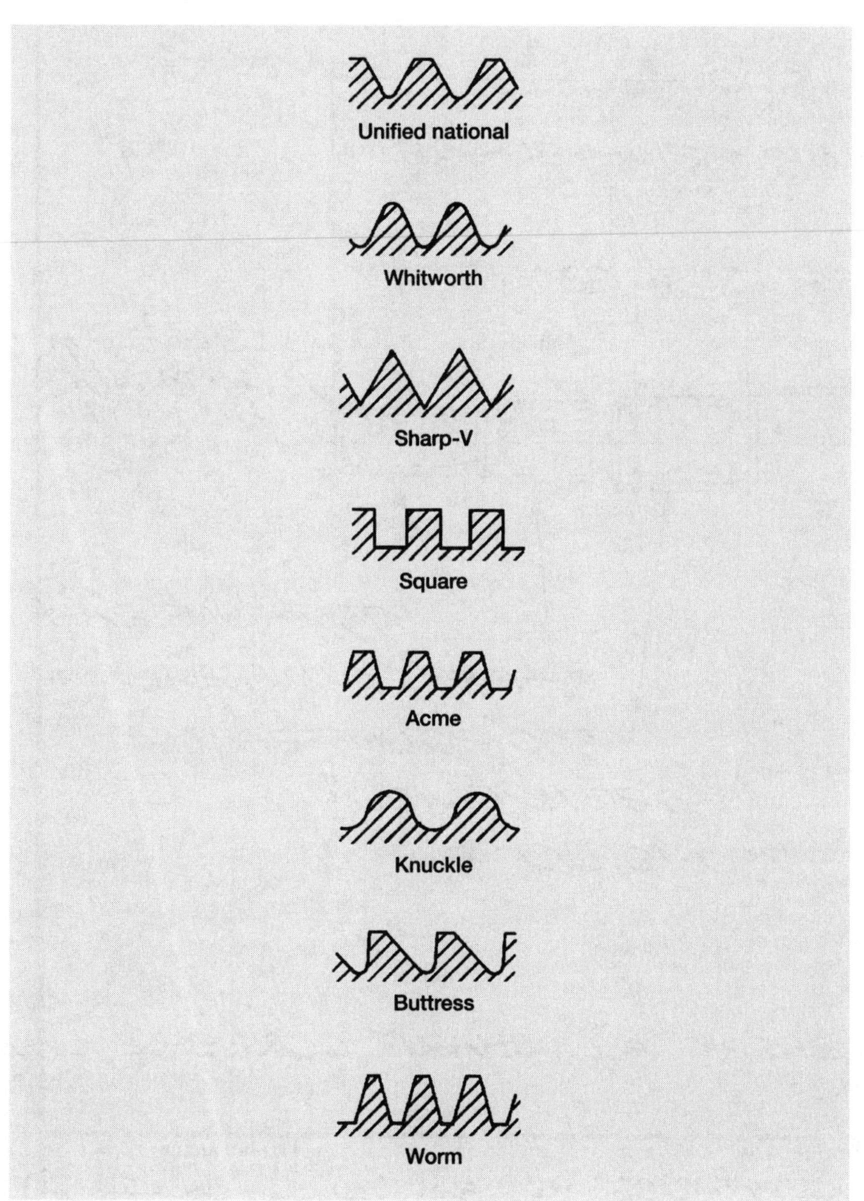

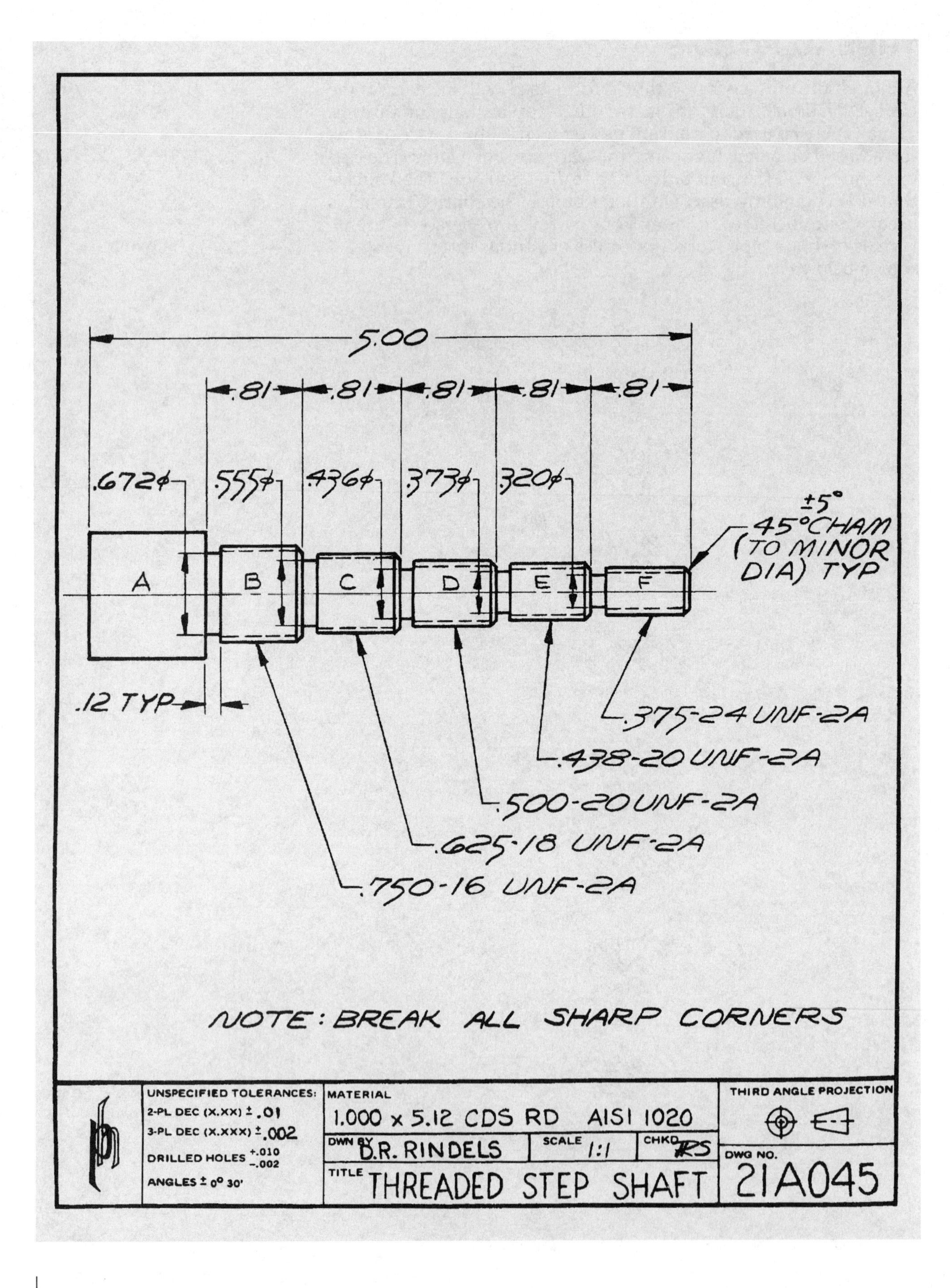
5.00
.81
.81
.81
.81
.81
.672⌀
.555⌀
.436⌀
.373⌀
.320⌀
±5°
45°CHAM
(TO MINOR
DIA) TYP
A
B
C
D
E
F
.12 TYP
.375-24 UNF-2A
.438-20 UNF-2A
.500-20 UNF-2A
.625-18 UNF-2A
.750-16 UNF-2A
NOTE: BREAK ALL SHARP CORNERS
UNSPECIFIED TOLERANCES:
2-PL DEC (X.XX) ± .01
3-PL DEC (X.XXX) ± .002
DRILLED HOLES +.010 −.002
ANGLES ± 0° 30'
MATERIAL
1.000 x 5.12 CDS RD AISI 1020
DWN BY D.R. RINDELS
SCALE 1:1
CHKD RS
TITLE THREADED STEP SHAFT
THIRD ANGLE PROJECTION
DWG NO.
21A045

INSTRUCTIONS: Refer to drawing 21A045 to answer the following questions.

1. Is the material considered to be low-, medium-, or high-carbon steel?	1. ______
2. What stock length is specified? (Blank length)	2. ______
3. What is the diameter of step A? (Blank diameter)	3. ______
4. How many chamfers does the shaft contain?	4. ______
5. How many necks are to be machined into the shaft?	5. ______
6. What method of dimensioning was used on this drawing? (Chain or broken-chain?)	6. ______
7. How much tolerance accumulates on the length of step A?	7. ______
8. What is the tolerance on the neck width dimension?	8. ______
9. What is the tolerance on the diameters of the necks?	9. ______
10. What is the tolerance on the chamfer angle?	10. ______
11. What method of thread representation was used?	11. ______
12. What thread *form* is specified?	12. ______
13. What thread *series* is specified?	13. ______
14. What class of fit is specified?	14. ______
15. What does the letter A after each thread specification designate?	15. ______
16. Are the threads right-hand or left-hand?	16. ______
17. How many full threads does step B contain? (Disregard the chamfer.)	17. ______
18. How many full threads does step C contain? (Disregard the chamfer.)	18. ______
19. How many full threads does step D contain? (Disregard the chamfer.)	19. ______
20. How many full threads does step F contain? (Disregard the chamfer.)	20. ______

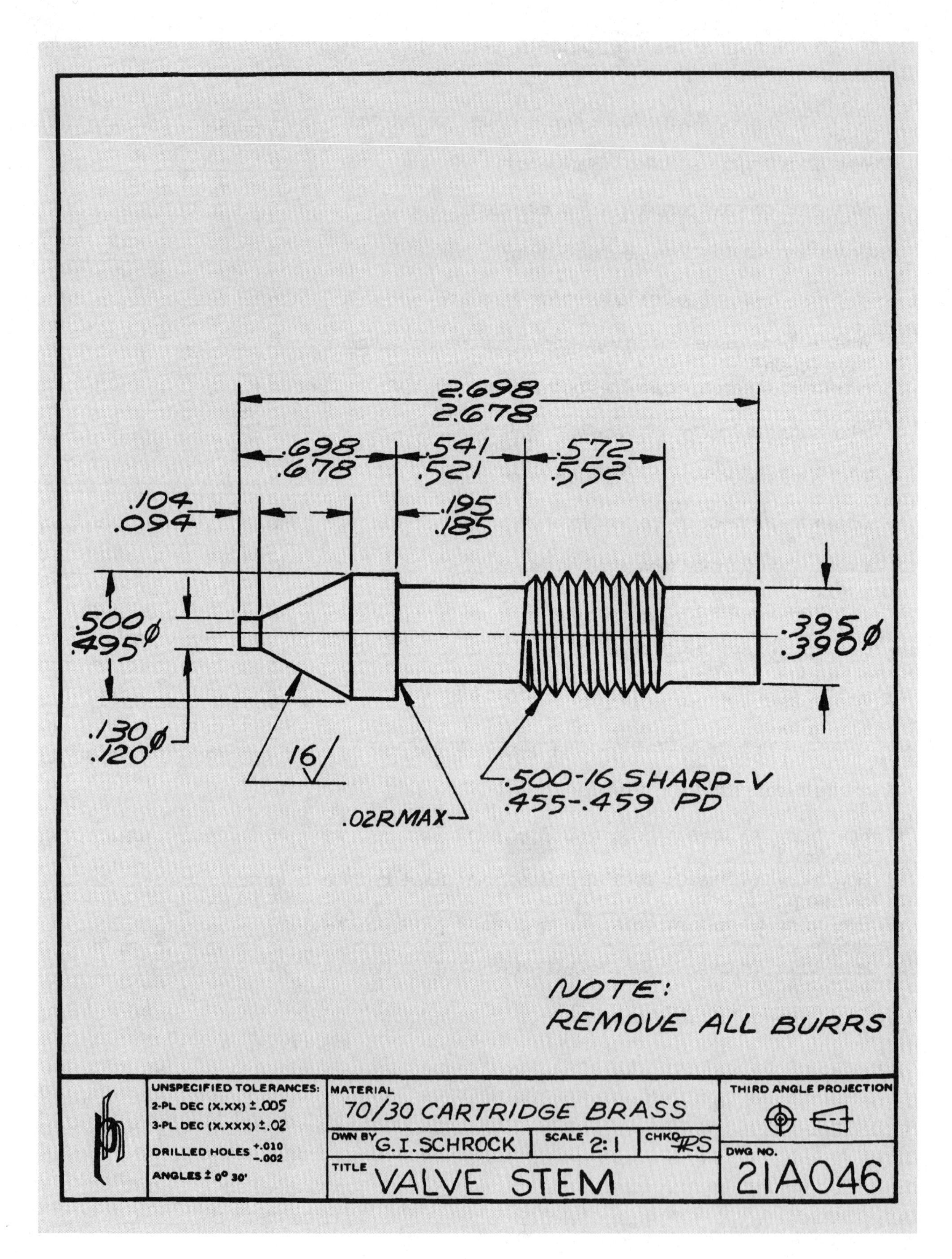
2.698
2.678
.698
.678
.541
.521
.572
.552
.104
.094
.195
.185
.500
.495 ø
.395 ø
.390
.130 ø
.120
16
.500-16 SHARP-V
.455-.459 PD
.02R MAX
NOTE:
REMOVE ALL BURRS
UNSPECIFIED TOLERANCES:
2-PL DEC (X.XX) ±.005
3-PL DEC (X.XXX) ±.02
DRILLED HOLES +.010 −.002
ANGLES ± 0° 30'
MATERIAL
70/30 CARTRIDGE BRASS
DWN BY G. I. SCHROCK
SCALE 2:1
CHKD
TITLE VALVE STEM
THIRD ANGLE PROJECTION
DWG NO.
21A046

INSTRUCTIONS: Refer to drawing 21A046 to answer the following questions.

1. Is the object drawn to half-scale or double-scale? 1. ____________
2. Is a ferrous or a nonferrous material specified? 2. ____________
3. What method of assigning tolerance was used on this drawing? 3. ____________
4. How much tolerance is allowed on the overall length? 4. ____________
5. Convert the overall length to a dimension with a bilateral tolerance. 5. ____________
6. What "single limit" dimension was used? 6. ____________
7. Show the decimal equivalent of the surface roughness height. 7. ____________
8. What is the maximum stem diameter adjacent to the threads? 8. ____________
9. What is the major diameter of the threads? 9. ____________
10. What method of thread representation is used? 10. ____________
11. What thread form is specified? 11. ____________
12. Are the threads left-hand or right-hand? 12. ____________
13. What do the letters PD in the thread specification abbreviate? 13. ____________
14. What is the decimal pitch of the threads? 14. ____________
15. What is the minimum length of the threaded portion? 15. ____________
16. Calculate the maximum number of full threads possible. 16. ____________
17. What is the minimum length of the tapered portion? 17. ____________
18. What is the maximum length of the tapered portion? 18. ____________
19. What is the minimum distance between the threaded portion and the right end? 19. ____________
20. What is the maximum distance between the threaded portion and the right end? 20. ____________

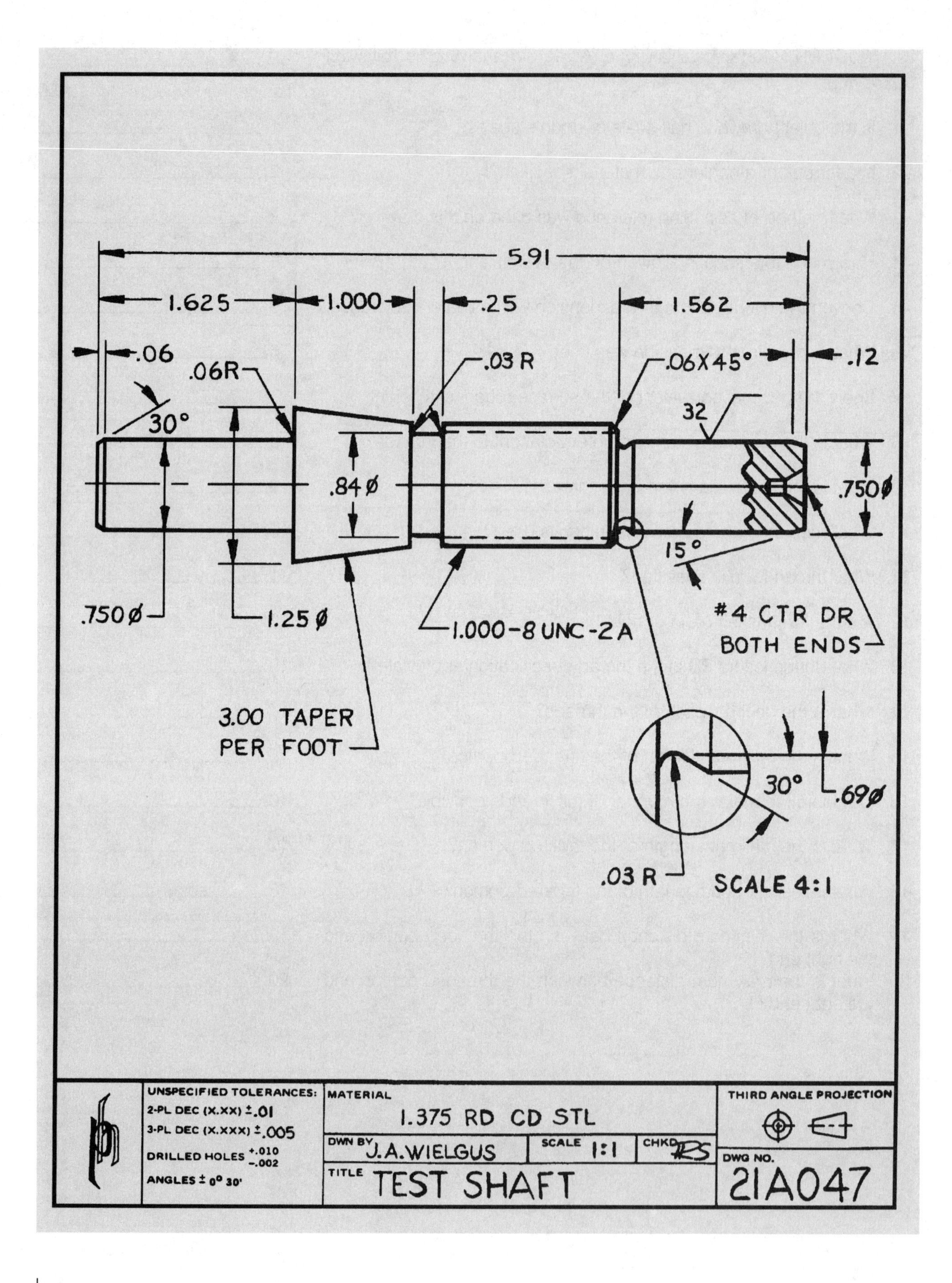
5.91
1.625
1.000
.25
1.562
.06
.06R
.03 R
.06X45°
.12
30°
32
.84 ∅
.750 ∅
15°
.750 ∅
1.25 ∅
1.000-8 UNC-2A
#4 CTR DR
BOTH ENDS
3.00 TAPER
PER FOOT
30°
.69 ∅
.03 R
SCALE 4:1
UNSPECIFIED TOLERANCES:
2-PL DEC (X.XX) ±.01
3-PL DEC (X.XXX) ±.005
DRILLED HOLES +.010 −.002
ANGLES ± 0° 30'
MATERIAL
1.375 RD CD STL
DWN BY J.A.WIELGUS
SCALE 1:1
CHKD
TITLE TEST SHAFT
THIRD ANGLE PROJECTION
DWG NO.
21A047

INSTRUCTIONS: Refer to drawing 21A047 to answer the following questions.

1. What is the diameter of the material specified? 1. ______
2. What do the letters CD in the material specs abbreviate? 2. ______
3. What center drill number is specified? 3. ______
4. How many chamfers does the shaft contain? 4. ______
5. What neck diameter is required adjacent to the thread chamfer? 5. ______
6. What is the diameter of the annular groove between the threaded portion and tapered portion? 6. ______
7. What is the length of the threaded portion? (Include the chamfer.) 7. ______
8. How many full threads will the threaded portion contain? 8. ______
9. What is the decimal pitch of the specified thread? 9. ______
10. What is the decimal equivalent of the surface roughness height? 10. ______
11. Calculate the small diameter of the tapered portion. 11. ______
12. What type of section was used to show the center drill? 12. ______
13. Was the chain method or the broken-chain method of dimensioning used? 13. ______
14. What method of thread representation was used? 14. ______
15. What thread class of fit is specified? 15. ______
16. What thread series is specified? 16. ______
17. How many TPI would a fine thread series contain for the same diameter? (Consult the table on page 206.) 17. ______
18. Show the limits of the overall length. 18. ______
19. Show the limits of the tapered length. 19. ______
20. What type of view was drawn to dimension the neck? 20. ______

TABULATED DIMENSIONS

Often, similar-shaped objects are combined on the same drawing. This can be accomplished by substituting letters in place of dimensions. The letters also appear in an accompanying table, assigning dimensional values to individual part numbers. Threaded fasteners lend themselves particularly well to this type of drawing. The example shown below combines five different hex head cap screws on the same drawing.

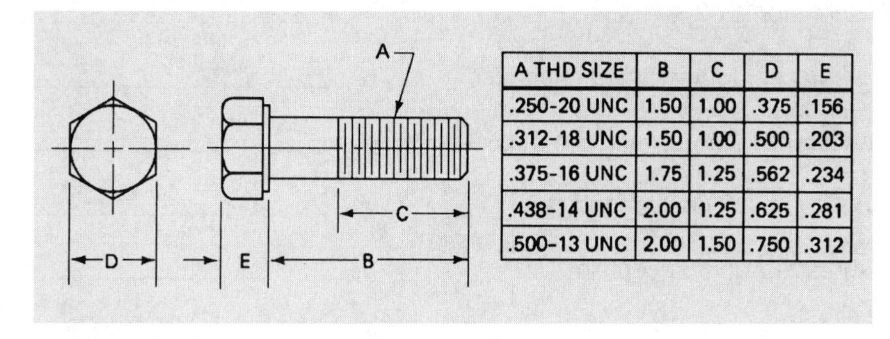

A THD SIZE	B	C	D	E
.250-20 UNC	1.50	1.00	.375	.156
.312-18 UNC	1.50	1.00	.500	.203
.375-16 UNC	1.75	1.25	.562	.234
.438-14 UNC	2.00	1.25	.625	.281
.500-13 UNC	2.00	1.50	.750	.312

MULTIPLE THREADS

Most threads are single-lead threads, which means that one revolution (360°) will cause a threaded member to advance a distance of one thread (equal to the pitch). However, some applications require a faster rate of advancement, so multiple threads may be specified. The specification (double, triple, etc.) will appear at the end of the thread callout on a drawing. A thread may be assumed to be single-lead if it is not specified otherwise. A double-lead thread will advance a distance of two threads per revolution (twice the pitch), a triple-lead thread will advance three threads, etc. Multiple threads may be recognized by observing more than one thread beginning at the end of a shaft or hole. However, on a drawing no attempt may be made to illustrate multiple threads differently than single-lead threads. (See the examples shown below.)

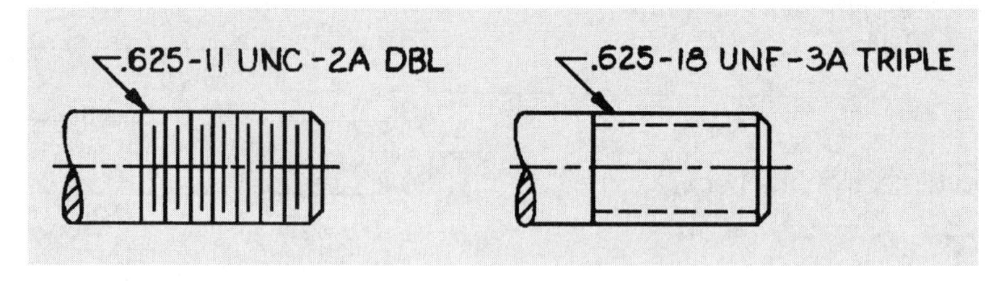

PIPE THREADS

Whereas all other thread specifications begin with the major thread diameter, pipe threads do not. Instead, the first figure represents the nominal pipe size, which is considerably smaller than the major thread diameter, since pipe is specified by ID and their threads occur on their OD. The abbreviation NPT (National pipe, tapered) or NPS (National pipe, straight) will appear in a pipe thread specification. Only one pitch size is available for each pipe size, whether it is a tapered or straight thread. Consult the table below for pipe thread standards. Tapered pipe threads may be illustrated as straight threads on a drawing. (See the example below.)

(Actual size shown)

PIPE THREADS

Nominal Size	*OD*	*TPI*
$\frac{1}{8}$	.405	27
$\frac{1}{4}$	.540	18
$\frac{3}{8}$	.675	18
$\frac{1}{2}$	.840	14
$\frac{3}{4}$	1.050	14
1	1.315	$11\frac{1}{2}$
$1\frac{1}{4}$	1.660	$11\frac{1}{2}$
$1\frac{1}{2}$	1.900	$11\frac{1}{2}$
2	2.375	$11\frac{1}{2}$
$2\frac{1}{2}$	2.875	8
3	3.500	8
$3\frac{1}{2}$	4.000	8
4	4.500	8
5	5.563	8
6	6.625	8
8	8.625	8
10	10.750	8
12	12.750	8

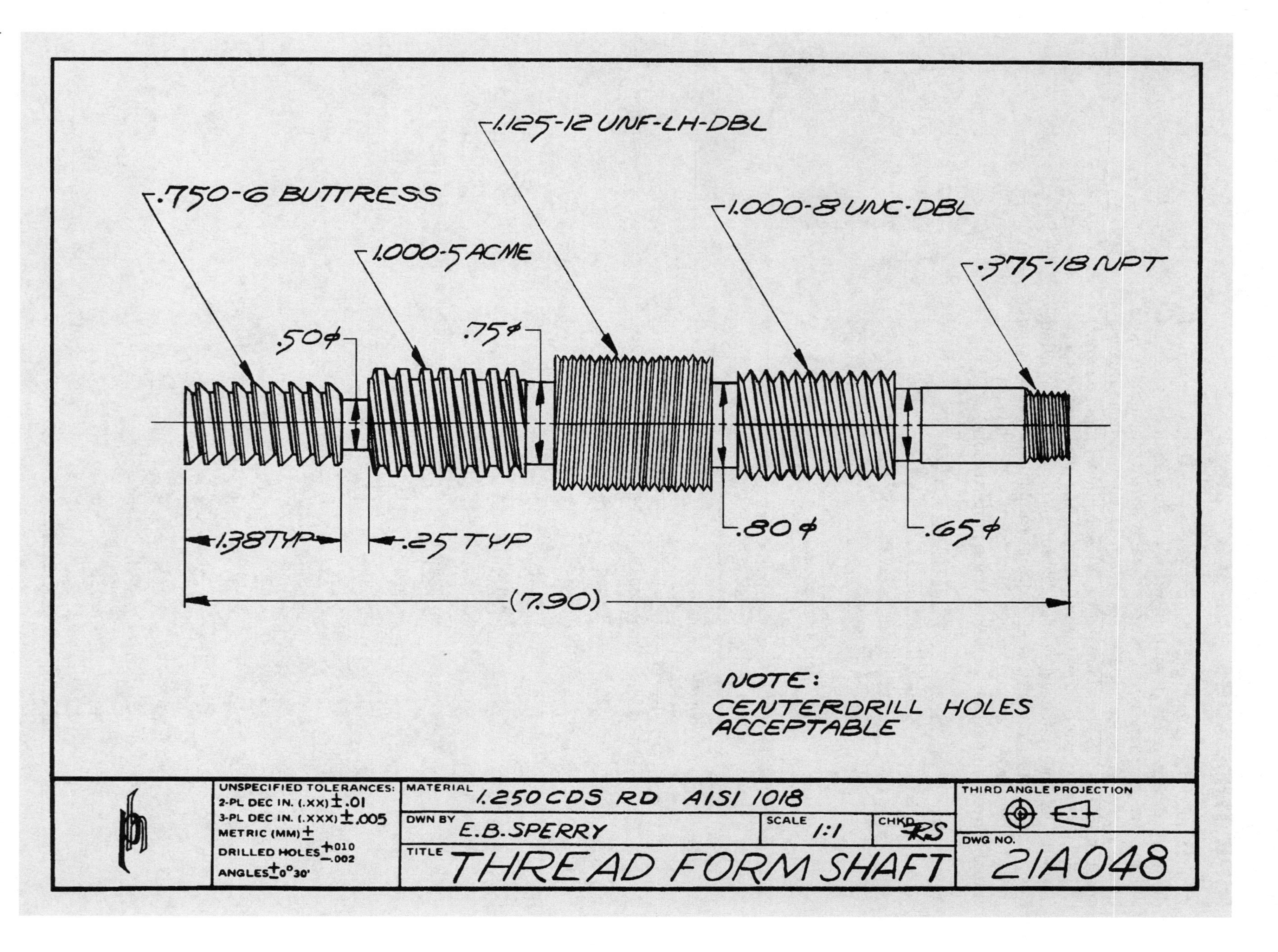

.750-6 BUTTRESS
1.000-5 ACME
1.125-12 UNF-LH-DBL
1.000-8 UNC-DBL
.375-18 NPT
.50⌀
.75⌀
.80⌀
.65⌀
1.38 TYP
.25 TYP
(7.90)
NOTE:
CENTERDRILL HOLES
ACCEPTABLE
UNSPECIFIED TOLERANCES:
2-PL DEC IN. (.XX) ±.01
3-PL DEC IN. (.XXX) ±.005
METRIC (MM) ±
DRILLED HOLES +.010 −.002
ANGLES ±0°30'
MATERIAL 1.250 CDS RD AISI 1018
DWN BY E.B. SPERRY
SCALE 1:1
CHKD RS
TITLE THREAD FORM SHAFT
THIRD ANGLE PROJECTION
DWG NO. 21A048

INSTRUCTIONS: Refer to drawing 21A048 to answer the following questions.

1. What method of thread representation was used on this drawing? 1. ____________
2. How many of the threaded portions contain multiple-lead threads? 2. ____________
3. How many of the threaded portions contain single-lead threads? 3. ____________
4. How many of the threaded portions contain left-hand threads? 4. ____________
5. How many full threads will the Buttress portion contain? 5. ____________
6. What is the decimal pitch of the Buttress thread? (Three decimal places.) 6. ____________
7. How many full threads will the Acme portion contain? 7. ____________
8. Figure the distance decimally that the Acme thread would advance in one revolution. 8. ____________
9. Figure the distance decimally that the UNC thread would advance in one revolution. 9. ____________
10. Are the National Pipe threads straight or are they tapered? 10. ____________
11. What is the nominal size of the pipe thread? 11. ____________
12. What is the actual major thread diameter of the pipe thread? (Consult the table on page 227.) 12. ____________
13. What is the decimal pitch of the pipe thread? (Three decimal places.) 13. ____________
14. How much tolerance applies to the length of each threaded portion? 14. ____________
15. How much tolerance could accumulate on the overall length dimension? 15. ____________

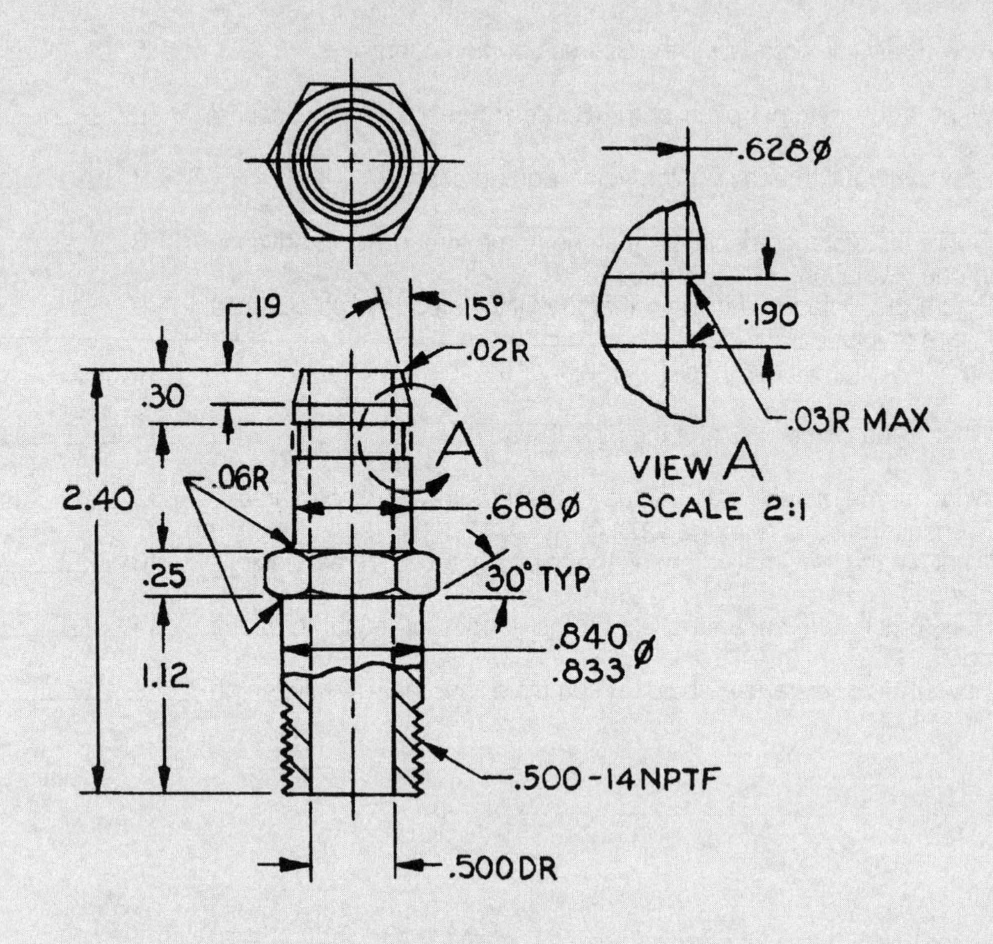

UNSPECIFIED TOLERANCES:	MATERIAL			THIRD ANGLE PROJECTION
2-PL DEC (X.XX) ± .01	70/30 BRASS .875 HEX BAR STOCK			
3-PL DEC (X.XXX) ± .005	DWN BY S. D. GRAHAM	SCALE 1:1	CHKD RS	
DRILLED HOLES +.010 −.002	TITLE QUICK CONNECTOR			DWG NO. 21A049
ANGLES ± 0° 30'				

DRYSEAL PIPE THREADS

Dryseal pipe threads are based upon the USA (American) pipe thread. They differ, however, in that they are designed to seal pressure-tight joints without the necessity of using sealing compounds. The following dryseal pipe thread abbreviations may appear on drawings: NPTF—Dryseal USA Standard Taper Pipe Thread; NPSF—Dryseal USA Standard Fuel Internal Straight Pipe Thread; NPSI—Dryseal USA Standard Intermediate Internal Straight Pipe Thread. A dryseal pipe thread appears on drawing 21A049.

HEX BAR STOCK

Hex bar stock may be purchased by shape, just as round or square bar stock. Whereas round bar stock is always specified by diameter, square and hex stock is always specified by the dimension "across flats." Also, hex shapes are normally illustrated "across corners" on their side views, so as not to be mistaken for square shapes. Observe this practice used on drawing 21A049.

INSTRUCTIONS: Refer to drawing 21A049 to answer the following questions.

1. What type of material is specified? 1. ____________
2. What type of section is used to dimension the ID? 2. ____________
3. What type of view is drawn to dimension the groove? 3. ____________
4. What is the distance across flats of the hexagonal portion? 4. ____________
5. Are the pipe threads left-hand or right-hand? 5. ____________
6. Are the pipe threads tapered or straight? 6. ____________
7. What is the decimal pitch of the threads? (Three decimal places.) 7. ____________
8. What is the nominal size of the pipe thread? 8. ____________
9. What is the actual major thread diameter? (Consult the table on page 227.) 9. ____________
10. Convert the overall length dimension into limits. 10. ____________
11. How deep is the annular groove? 11. ____________
12. Calculate the minimum wall thickness dimension at the annular groove. 12. ____________
13. How much of the .688∅ remains between the groove and the chamfer? 13. ____________
14. What is the length of the .688∅ between the hex portion and the groove? (Disregard the radius.) 14. ____________
15. What is the accumulated tolerance on the answer to question 14? 15. ____________

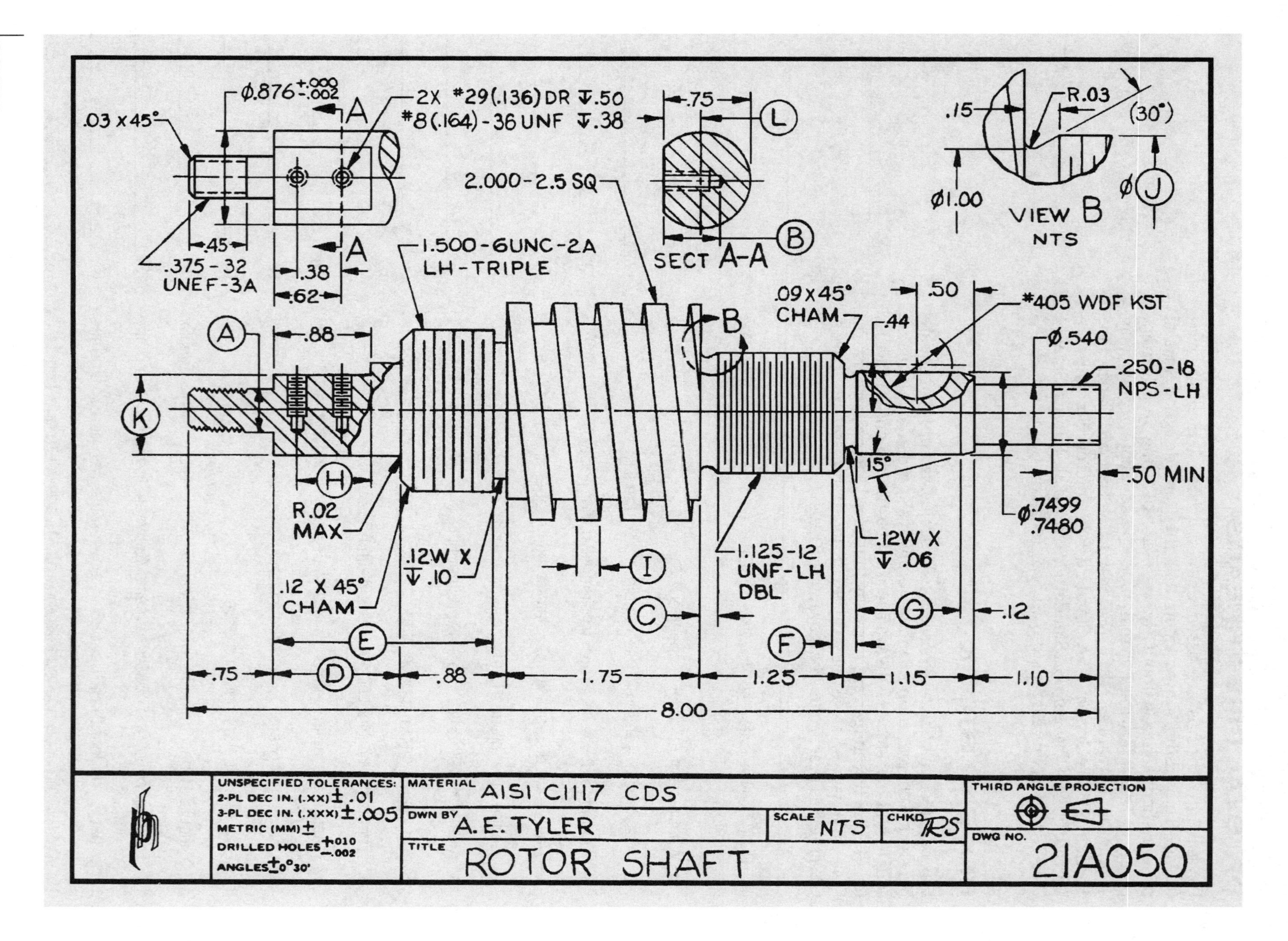
.03 X 45°
ϕ.876 +.000 -.002
2X #29(.136) DR ↧.50
#8(.164)-36 UNF ↧.38
A
A
.45
.375-32
UNEF-3A
.38
.62
.75
L
B
SECT A-A
2.000-2.5 SQ
1.500-6UNC-2A
LH-TRIPLE
.15
R.03
(30°)
ϕ1.00
VIEW B
NTS
ϕ J
A
.88
K
H
R.02
MAX
.12 X 45°
CHAM
.12W X
↧.10
I
C
E
D
B
.09 X 45°
CHAM
.50
.44
#405 WDF KST
ϕ.540
.250-18
NPS-LH
15°
.50 MIN
ϕ.7499
.7480
.12W X
↧.06
1.125-12
UNF-LH
DBL
G
.12
F
.75
.88
1.75
1.25
1.15
1.10
8.00
UNSPECIFIED TOLERANCES:
2-PL DEC IN. (.XX) ± .01
3-PL DEC IN. (.XXX) ± .005
METRIC (MM) ±
DRILLED HOLES +.010 -.002
ANGLES ± 0°30'
MATERIAL AISI C1117 CDS
DWN BY A. E. TYLER
SCALE NTS
CHKD RS
TITLE ROTOR SHAFT
THIRD ANGLE PROJECTION
DWG NO. 21A050

INSTRUCTIONS: Refer to drawing 21A050 to answer the following questions.

1. How many different thread series are specified for the Unified threads? 1. ____________
2. How many full threads will the UNEF portion contain after chamfering? 2. ____________
3. How many single limit dimensions appear on this drawing? 3. ____________
4. Show the decimal dimension of the tap drill diameter. 4. ____________
5. Show the assumed class of fit, followed by the appropriate letter, for the tapped holes. 5. ____________
6. Show the major diameter of the pipe threads. 6. ____________
7. Show the MMC of the diameter with the keyseat. 7. ____________
8. How far would the shaft advance with one revolution engaged in the square threads? 8. ____________
9. Show the minimum neck diameter between the UNF threads and the diameter with the keyseat. 9. ____________
10. Show the minimum neck diameter between the UNF threads and the square threaded portion. 10. ____________
11. Show the lead (decimally) of the UNC threads. 11. ____________
12. How far is the center of the Woodruff keyseat from the center of the *left* tapped hole? 12. ____________
13. How much tolerance can accumulate between their centers? (Question 12) 13. ____________

INSTRUCTIONS: Enter the dimensions for the following letters.

Ⓐ ____________ Ⓖ ____________

Ⓑ ____________ Ⓗ ____________

Ⓒ ____________ Ⓘ ____________

Ⓓ ____________ Ⓙ ____________

Ⓔ ____________ Ⓚ ____________

Ⓕ ____________ Ⓛ ____________

TYPICAL DIMENSIONS

To avoid repetitious dimensions on a drawing, the abbreviation TYP (for typical) is used. When it appears after a radial dimension, all undimensioned radii of similar size are intended to be the same radius as the one designated TYP. When it appears after a location dimension, it is intended that undimensioned locations are the same as the one designated. For example, a perforated pattern need only show typical dimensions if the same pattern is repeated.

RIGHT TRIANGLES

When a right triangle is created on a drawing due to an inclined line or plane, it may be dimensioned either by two sides (legs) or one side and one angle. The other sides or angles can then be calculated by trigonometry. However, as stated before, when the angle is 45°, both sides are obviously equal. Also, whenever the angle is 45°, the hypotenuse of the triangle will equal a distance of 1.414 times the side distance. For example, a right triangle of 45° with a side measuring 1 in. would have a hypotenuse measuring 1.414 in. But remember, this formula may only be used for right triangles with a 45° angle. (See the example shown below.)

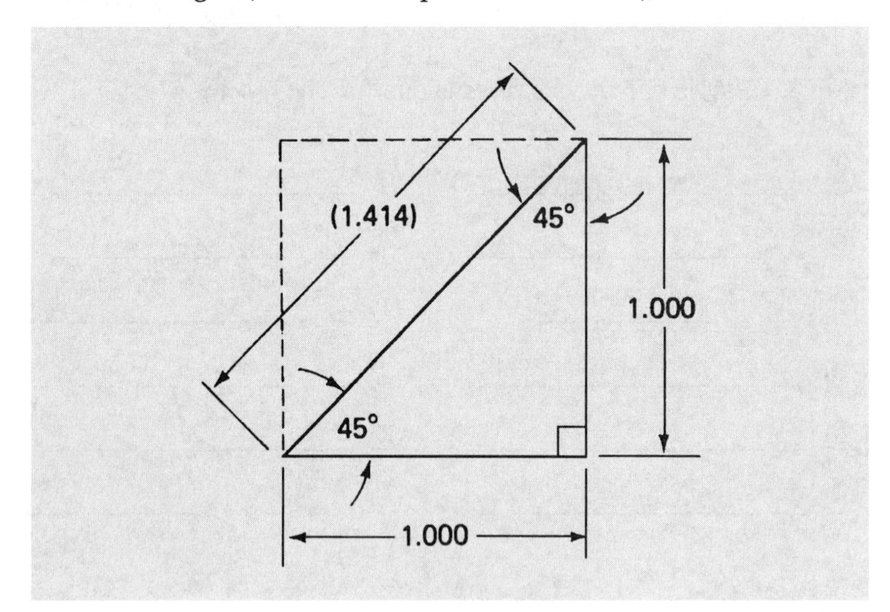

REPETITIVE FEATURES

Repetitive features or dimensions may be specified using an "X" after the quantity (to designate "times") followed by the dimension, when applicable. (See the examples shown below.)

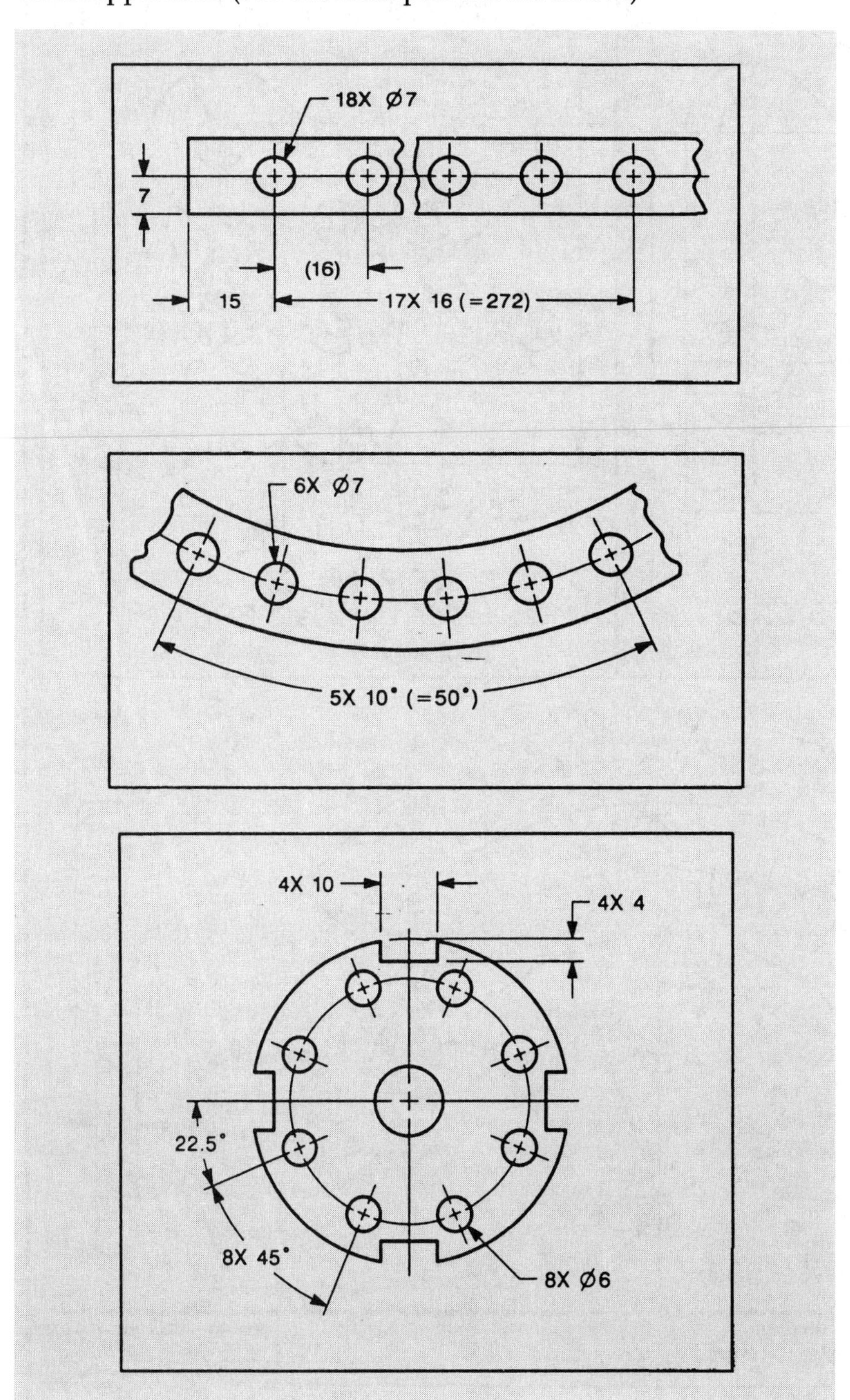

Reprinted from ASME Y14.5M–1994, by permission of The American Society of Mechanical Engineers.

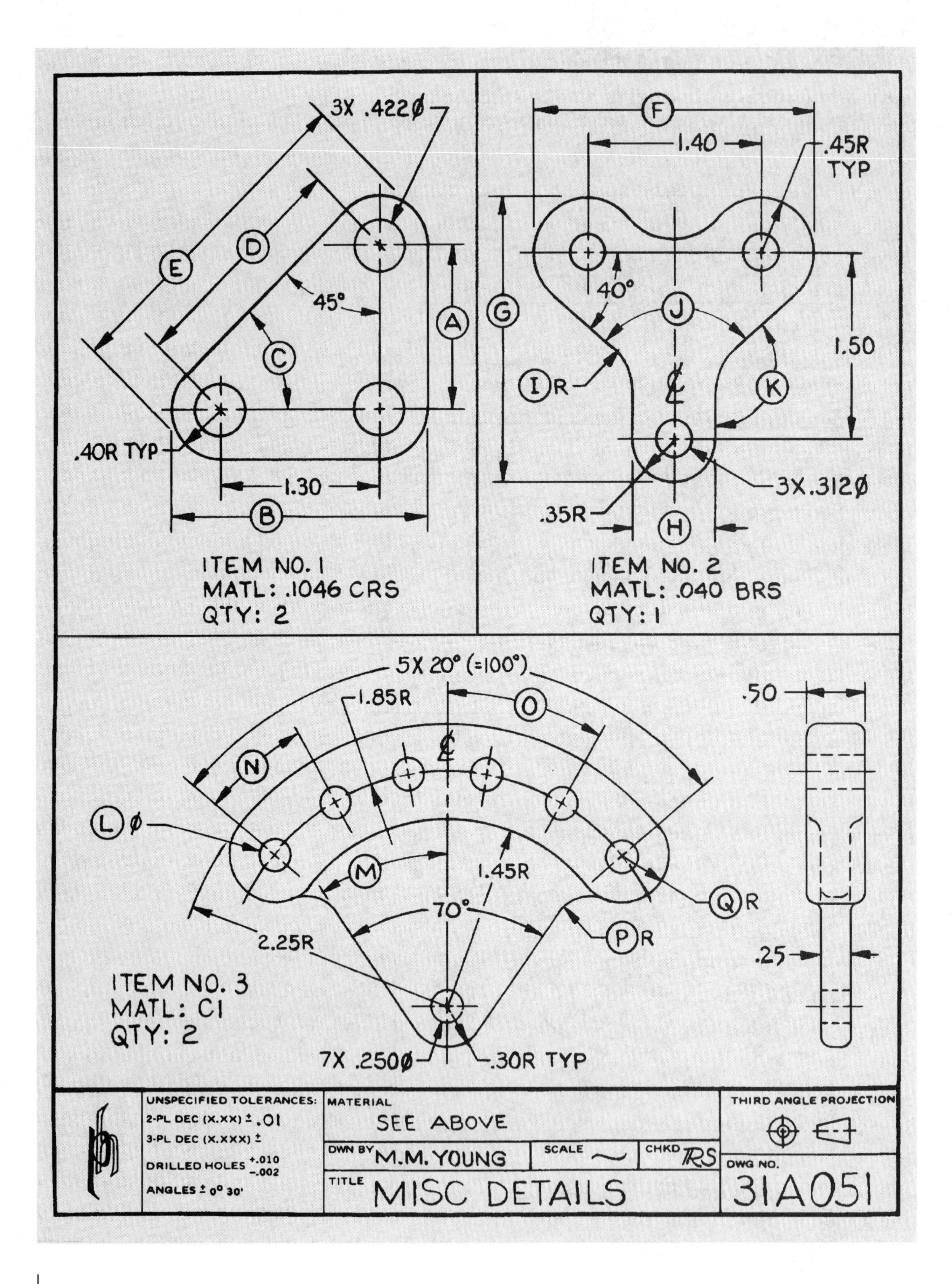
3X .422Ø
E
D
45°
A
C
.40R TYP
1.30
B
ITEM NO. 1
MATL: .1046 CRS
QTY: 2
F
1.40
.45R
TYP
G
40°
J
1.50
I R
K
3X.312Ø
.35R
H
ITEM NO. 2
MATL: .040 BRS
QTY: 1
5X 20° (=100°)
1.85R
O
.50
N
L Ø
M
1.45R
Q R
70°
P R
2.25R
.25
ITEM NO. 3
MATL: C1
QTY: 2
7X .250Ø
.30R TYP
UNSPECIFIED TOLERANCES:
2-PL DEC (X.XX) ± .01
3-PL DEC (X.XXX) ±
DRILLED HOLES +.010 −.002
ANGLES ± 0° 30'
MATERIAL
SEE ABOVE
DWN BY M.M. YOUNG
SCALE ~
CHKD RS
TITLE MISC DETAILS
THIRD ANGLE PROJECTION
DWG NO.
31A051

DETAIL DRAWINGS

Usually, detail drawings are confined to one object per sheet to avoid confusion and to control the part numbers. Occasionally, more than one detail may appear on a single drawing if the production is limited to small quantities. Such is the case for the items described on drawing 31A051.

INSTRUCTIONS: Refer to drawing 31A051 to answer the following questions.

1. How much material remains between a hole and the nearest outside edge of item 1? 1. ____________
2. What is the specified material and thickness of item 2? 2. ____________
3. What is the total quantity of the three items shown? 3. ____________

INSTRUCTIONS: Calculate the dimensions for the following letters.

Ⓐ ____________

Ⓑ ____________

Ⓒ ____________

Ⓓ ____________

Ⓔ ____________

Ⓕ ____________

Ⓖ ____________

Ⓗ ____________

Ⓘ ____________

Ⓙ ____________

Ⓚ ____________

Ⓛ ____________

Ⓜ ____________

Ⓝ ____________

Ⓞ ____________

Ⓟ ____________

Ⓠ ____________

Optional Math Exercises:

1a. Using only the dimensions given, calculate the X & Y coordinates for each of the six .250∅ holes in relation to the single .250∅ hole.

2a. Using the dimensions given, calculate the MOP (Measurement Over Pins) of adjacent .250∅ holes in Item No. 3.

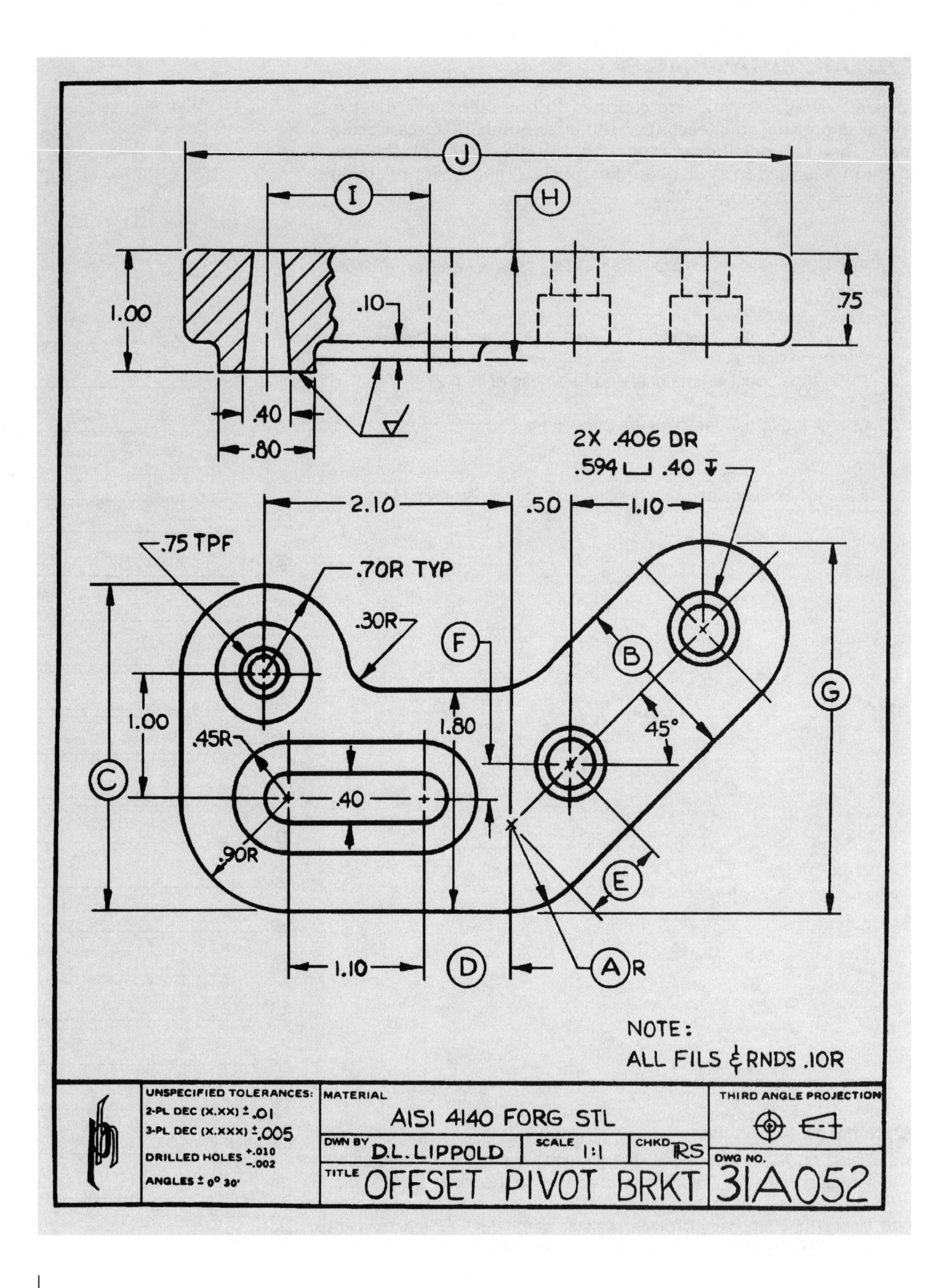
J
I
H
1.00
.75
.10
.40
.80
2X .406 DR
.594 ⌴ .40 ↧
2.10
.50
1.10
.75 TPF
.70R TYP
.30R
F
B
G
1.00
1.80
45°
.45R
C
.40
.90R
E
1.10
D
A R
NOTE:
ALL FILS & RNDS .10R
UNSPECIFIED TOLERANCES:
2-PL DEC (X.XX) ± .01
3-PL DEC (X.XXX) ± .005
DRILLED HOLES +.010 -.002
ANGLES ± 0° 30'
MATERIAL
AISI 4140 FORG STL
DWN BY D.L. LIPPOLD
SCALE 1:1
CHKD RS
TITLE OFFSET PIVOT BRKT
THIRD ANGLE PROJECTION
DWG NO.
31A052

INSTRUCTIONS: Refer to drawing 31A052 to answer the following questions.

1. What type of sectional view appears on the drawing? 1. ____________
2. Is the material (a) nickel, (b) molybdenum, or (c) chromium alloy steel? (See page 165.) 2. ____________
3. Is the material (a) low-, (b) medium-, or (c) high-carbon steel? (See page 167.) 3. ____________
4. How far does the boss extend above its surrounding surface? 4. ____________
5. How far does the pad extend above its surrounding surface? 5. ____________
6. What is the small diameter of the tapered hole? 6. ____________
7. What is the maximum width of the slotted hole? 7. ____________
8. What is the minimum length of the slotted hole? 8. ____________
9. What is the minimum width of the pad? 9. ____________
10. What is the maximum length of the pad? 10. ____________

INSTRUCTIONS: Calculate the dimensions for the following letters (disregard tolerances).

Ⓐ ____________

Ⓑ ____________

Ⓒ ____________

Ⓓ ____________

Ⓔ ____________

Ⓕ ____________

Ⓖ ____________

Ⓗ ____________

Ⓘ ____________

Ⓙ ____________

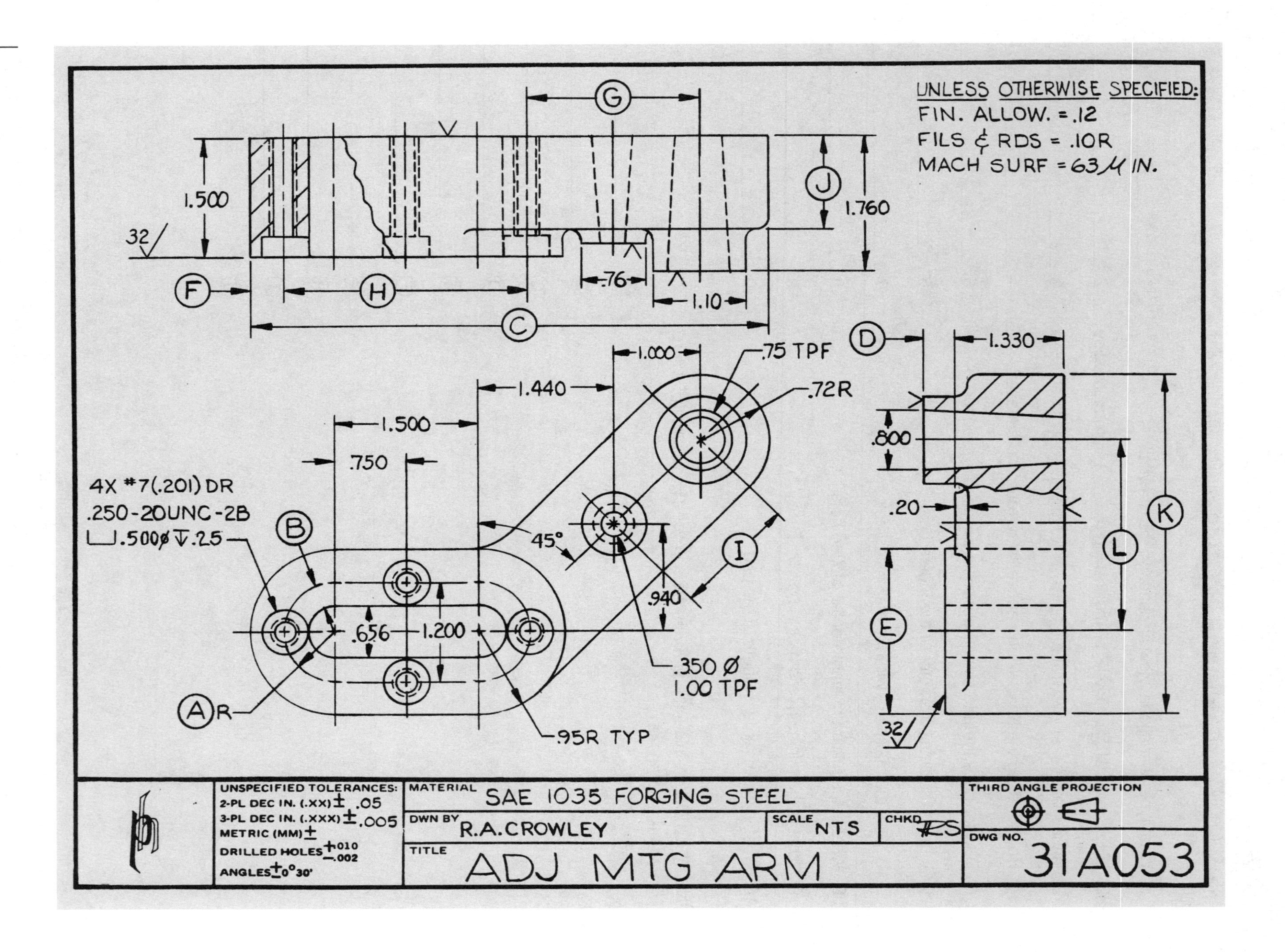
UNLESS OTHERWISE SPECIFIED:
FIN. ALLOW. = .12
FILS & RDS = .10R
MACH SURF = 63μ IN.
G
1.500
32
J
1.760
F
H
.76
1.10
C
D
1.330
1.000
.75 TPF
1.440
.72R
1.500
.750
.800
4X #7(.201) DR
.250-20UNC-2B
.500Ø ⊽.25
B
45°
.20
K
I
L
.940
.656
1.200
E
.350 Ø
1.00 TPF
A R
.95R TYP
32
UNSPECIFIED TOLERANCES:
2-PL DEC IN. (.XX) ± .05
3-PL DEC IN. (.XXX) ± .005
METRIC (MM) ±
DRILLED HOLES +.010 -.002
ANGLES ± 0°30'
MATERIAL SAE 1035 FORGING STEEL
DWN BY R.A.CROWLEY
SCALE NTS
CHKD
THIRD ANGLE PROJECTION
TITLE ADJ MTG ARM
DWG NO. 31A053

INSTRUCTIONS: Refer to drawing 31A053 to answer the following questions.

1. What is the pitch (decimally) of the threads? 1. ______
2. What thread series is specified? 2. ______
3. What does the B in the thread specification designate? 3. ______
4. How much surface roughness height is permitted on the pad face? 4. ______
5. How much surface roughness height is permitted on the boss faces? 5. ______
6. What was the overall thickness dimension of the part before machining? 6. ______
7. How much tolerance does the overall length dimension accumulate? 7. ______
8. Calculate the small diameter of the large tapered hole. 8. ______
9. Calculate the large diameter of the small tapered hole. (Three decimal places) 9. ______
10. How many full threads will each tapped hole contain? 10. ______
11. Are the counterbores tangent with the slotted hole? If not, show the dimensional difference. 11. ______
12. What is the maximum length of the slotted hole? 12. ______
13. What is the maximum length of the pad? 13. ______

INSTRUCTIONS: Enter the dimensions for the following letters.

Ⓐ ______

Ⓑ ______

Ⓒ ______

Ⓓ ______

Ⓔ ______

Ⓕ ______

Ⓖ ______

Ⓗ ______

Ⓘ ______

Ⓙ ______

Ⓚ ______

Ⓛ ______

DRAWING REVISIONS

Once an engineering drawing has been assigned a permanent part number and is released for production, all further changes or alterations must be recorded. Normally, a revision block is provided somewhere on the drawing for this purpose. A brief description of each change and a small balloon adjacent to the affected area or dimension are usually sufficient on small, simple drawings. On larger, more complex drawings it may be necessary to use zoning, as on road maps, to help find the area of change.

When a revision causes a noticeable dimensional change, the drafter usually places a heavy line underneath the revised dimension to alert the blueprint reader of the not-to-scale condition. Earlier practices called for the abbreviation NTS (not to scale) adjacent to the dimension, or a wavy line beneath it.

You should be aware that drawings may be out-of-scale due to reasons other than revisions, and they may or may not contain lines beneath the affected dimensions.

A typical revision with not-to-scale dimensions is shown below.

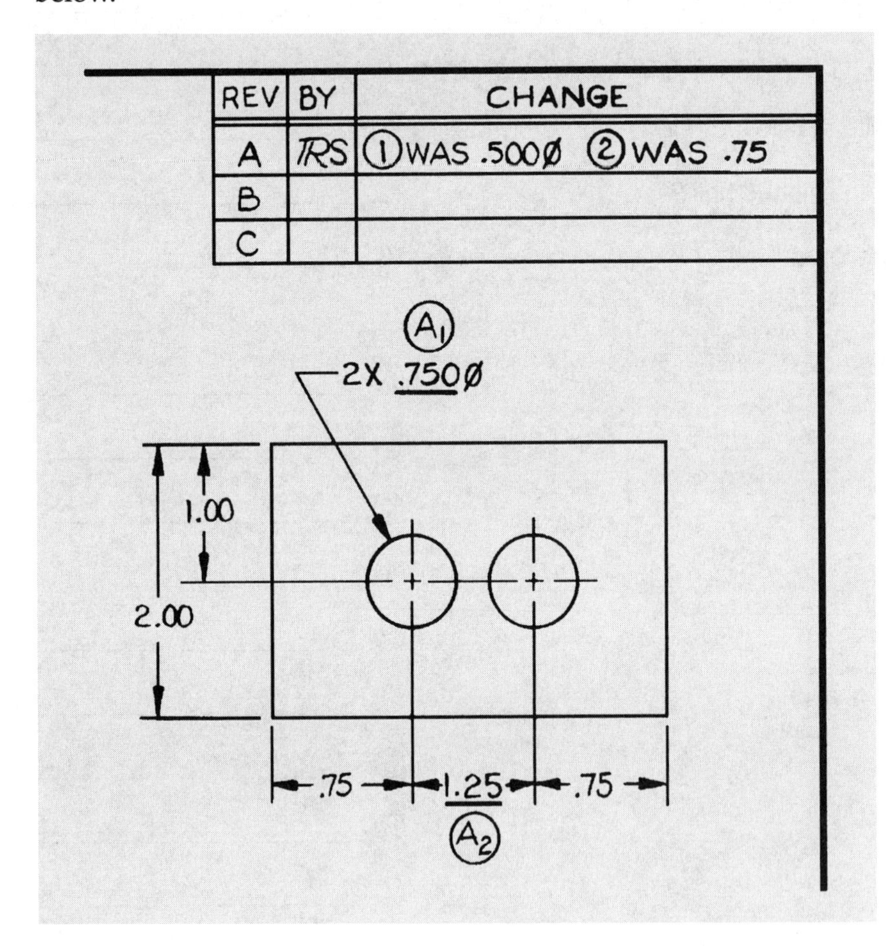

CONICAL TAPERS

Previous drawings in this book specified conical tapers by TPF (taper per foot). Another method involves specifying the ratio. A ratio of .2:1 designates a reduction of 0.20 inch on the diameter for every 1.00 inch of length. In the example shown below, the small end of the tapered portion would measure .60 diameter. Observe the use of the new symbol for conical taper on the illustration below and on drawing 31A054 on page 244.

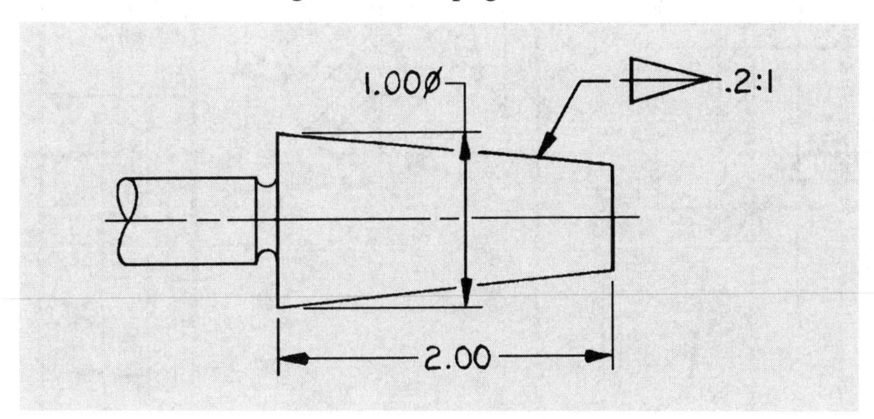

ROCKWELL HARDNESS TEST

The Rockwell hardness test measures hardness by measuring the depth of penetration into the specimen. The penetrator may be either a steel ball or a diamond sphero-conical penetrator. The hardness number is related to the depth of indentation; the harder the material, the higher the number. The various Rockwell scales and their applications are shown in the table below.

Scale	Testing Application
A	For tungsten carbide and other extremely hard materials. Also for thin, hard sheets.
B	For materials of medium hardness such as low and medium carbon steels in the annealed condition.
C	For materials harder than Rockwell B–100.
D	Where somewhat lighter load is desired than on C scale, as on case hardened pieces.
E	For very soft materials such as bearing metals.
F	Same as E scale but using 1/16-inch ball.
G	For metals harder than tested on B scale.
H & K	For softer metals.
15–N; 30–N; 45–N	Where shallow impression or small area is desired. For hardened steel and hard alloys.
15–T; 30–T; 45–T	Where shallow impression or small area is desired for materials softer than hardened steel.

(Courtesy Machinery's Handbook, *24th ed. Published by Industrial Press, New York)*

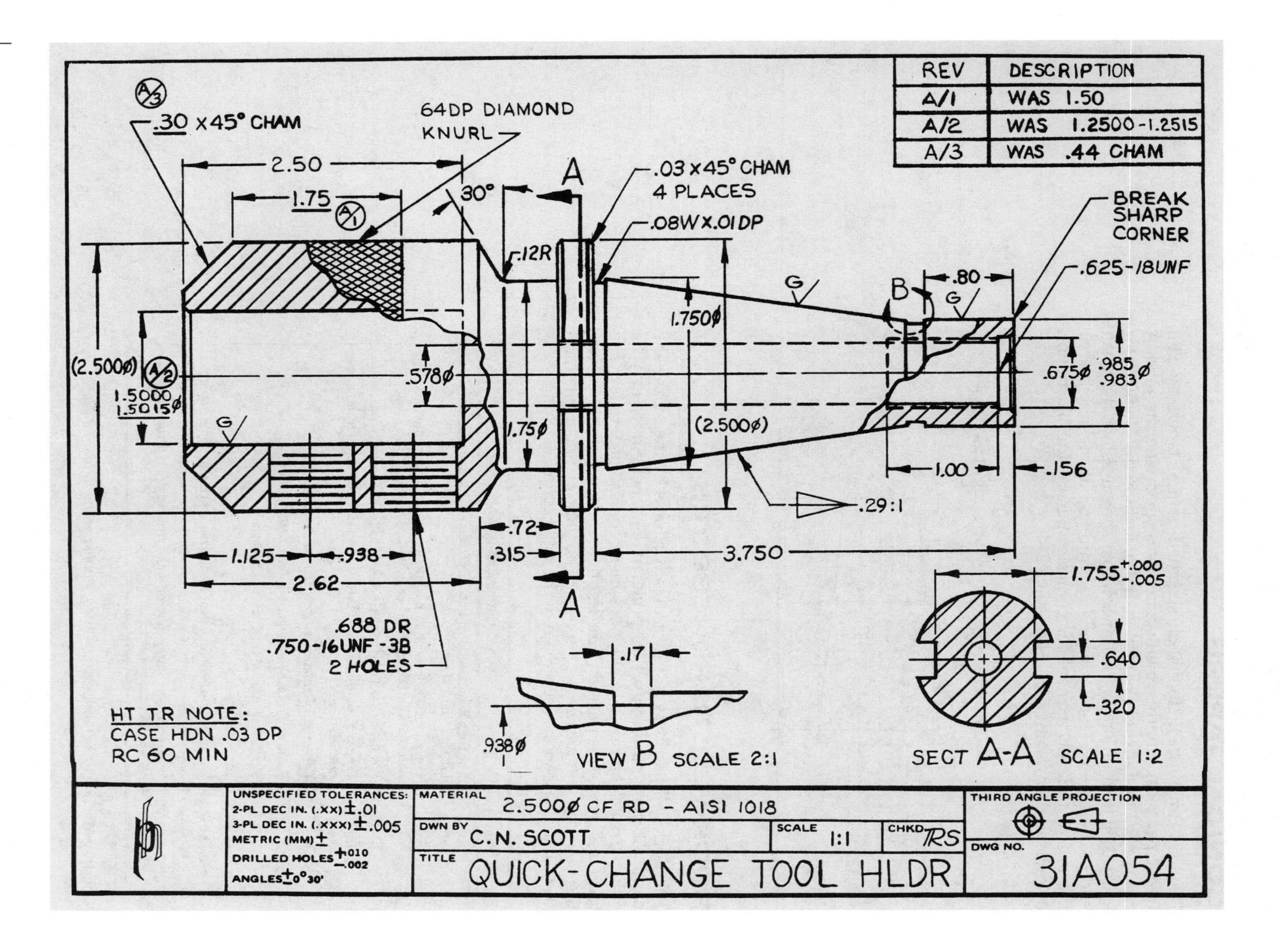

REV
DESCRIPTION
A/1
WAS 1.50
A/2
WAS 1.2500-1.2515
A/3
WAS .44 CHAM
A/3
.30 X 45° CHAM
64DP DIAMOND KNURL
2.50
1.75 A/1
30°
A
.03 X 45° CHAM 4 PLACES
.08W X .01DP
BREAK SHARP CORNER
.625-18UNF
.12R
.80
B
G
1.750Ø
(2.500Ø)
A/2
1.5000
1.5015Ø
.578Ø
1.75Ø
(2.500Ø)
.675Ø
.985
.983Ø
1.00
.156
.29:1
.72
.315
1.125
.938
2.62
3.750
A
1.755 +.000 -.005
.688 DR
.750-16UNF-3B
2 HOLES
.17
.640
.320
HT TR NOTE:
CASE HDN .03 DP
RC 60 MIN
.938Ø
VIEW B SCALE 2:1
SECT A-A SCALE 1:2
UNSPECIFIED TOLERANCES:
2-PL DEC IN. (.XX) ±.01
3-PL DEC IN. (.XXX) ±.005
METRIC (MM) ±
DRILLED HOLES +.010 -.002
ANGLES ±0°30'
MATERIAL
2.500Ø CF RD - AISI 1018
DWN BY
C.N. SCOTT
SCALE
1:1
CHKD
RS
THIRD ANGLE PROJECTION
DWG NO.
31A054
TITLE
QUICK-CHANGE TOOL HLDR

THREAD SPECIFICATION PLACEMENT

The circular view of threaded holes is the preferred position for placement of thread specifications. However, when no circular view appears, such as on drawing 31A054, the leader arrowhead will touch the axis centerline at the first thread.

INSTRUCTIONS: Refer to drawing 31A054 to answer the following questions.

1. What style of knurling is specified? 1. ____________
2. What type of section view is section A-A? 2. ____________
3. What type of view is view B? 3. ____________
4. What type of section was drawn to expose the threads? 4. ____________
5. What method of representation was used to illustrate the .750 holes? 5. ____________
6. What method of representation was used to illustrate the .625 holes? 6. ____________
7. What was the former dimension of the knurled portion? 7. ____________
8. Was the chamfer (a) increased or (b) decreased by revision A? 8. ____________
9. What does the heavy line underneath the revised dimension figures designate? 9. ____________
10. What is the OD at section A-A? 10. ____________
11. What is the MMC of the 1.50 counterbore diameter? 11. ____________
12. What is the MMC of the OD at the right end? 12. ____________
13. What is the neck diameter if the large end of the taper measures 1.750? 13. ____________
14. What is the groove diameter at the small end of the taper? 14. ____________
15. What is the small diameter of the taper? (Calculate) 15. ____________
16. What thread series is specified? 16. ____________
17. How many threads will each .750 hole contain? 17. ____________
18. What is the decimal pitch of the .625 thread? (Three decimal places) 18. ____________
19. What is the tap drill diameter for the .625 threads? 19. ____________
20. How many threads will the .625 tapped hole contain? 20. ____________

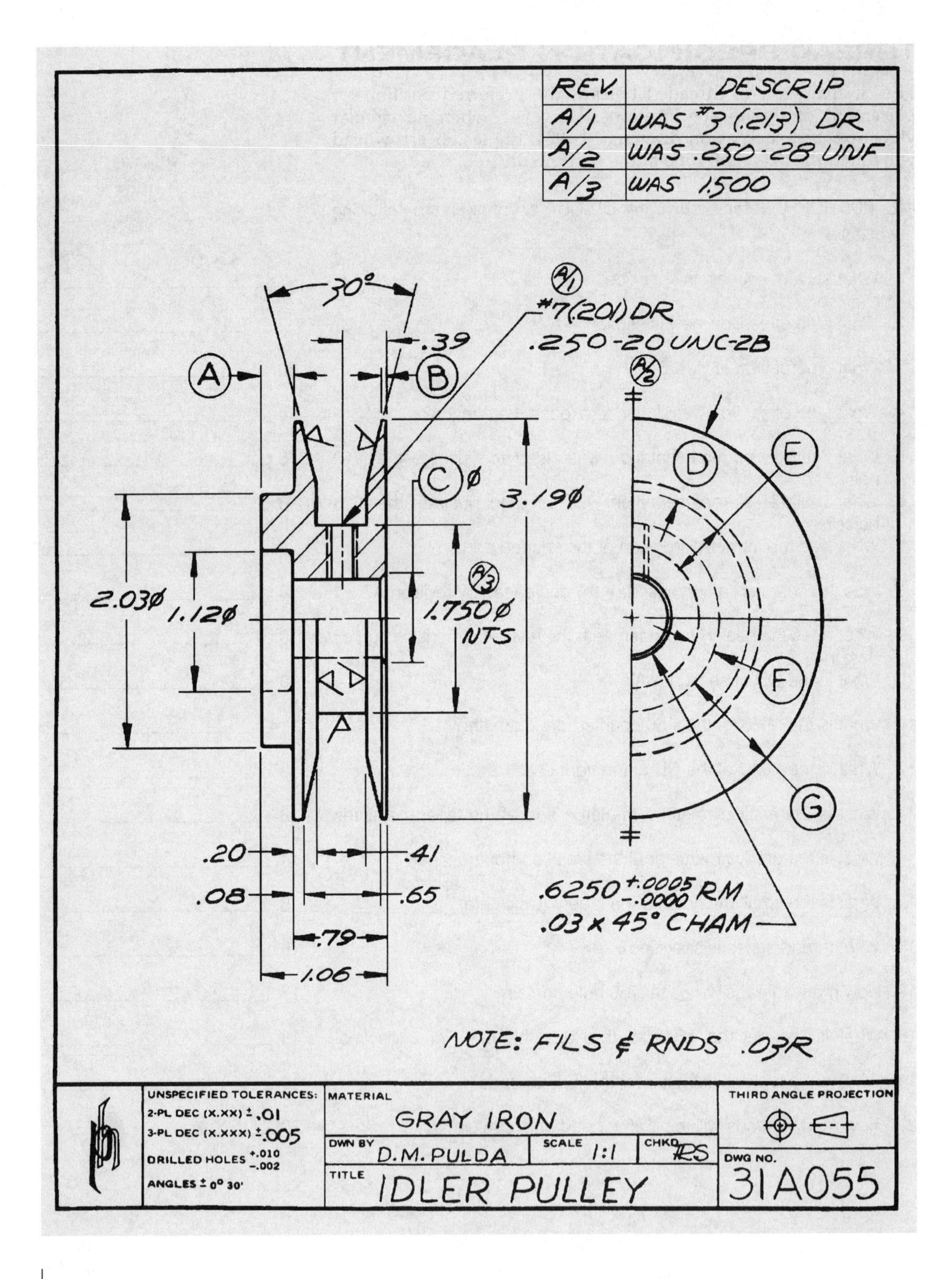

REV.
DESCRIP
A/1
WAS #3 (.213) DR
A/2
WAS .250-28 UNF
A/3
WAS 1.500
30°
#7(.201) DR
.250-20 UNC-2B
.39
A
B
C Ø
D
E
F
G
3.19Ø
2.03Ø
1.12Ø
1.750Ø
NTS
.20
.41
.08
.65
.79
1.06
.6250 +.0005 -.0000 RM
.03 x 45° CHAM
NOTE: FILS & RNDS .03R
UNSPECIFIED TOLERANCES:
2-PL DEC (X.XX) ± .01
3-PL DEC (X.XXX) ± .005
DRILLED HOLES +.010 -.002
ANGLES ± 0° 30'
MATERIAL
GRAY IRON
DWN BY
D.M. PULDA
SCALE
1:1
CHKD
RS
TITLE
IDLER PULLEY
THIRD ANGLE PROJECTION
DWG NO.
31A055

INSTRUCTIONS: *Refer to drawing 31A055 to answer the following questions.*

1. How many changes have been made since the drawing was released? 1. ____________
2. What thread series was specified prior to revision A/2? 2. ____________
3. Is the present tap drill size larger or smaller than the original size? 3. ____________
4. How much tolerance is allowed on the reamed hole size? 4. ____________
5. How many threads per inch are specified for the tapped hole? 5. ____________
6. How many full threads does the tapped hole contain? 6. ____________
7. What method of thread representation was used in the section view? 7. ____________
8. What thread class of fit is specified? 8. ____________
9. What type of section view is the front view? 9. ____________
10. What type of view is the right-side view? 10. ____________
11. What does the abbreviation NTS represent? 11. ____________
12. How many visible lines would a left-side view contain? 12. ____________
13. How many finished surfaces does the pulley contain? (Disregard holes) 13. ____________

INSTRUCTIONS: *Calculate the dimensions for the following letters.*

Ⓐ ____________

Ⓑ ____________

Ⓒ ____________

Ⓓ ____________

Ⓔ ____________

Ⓕ ____________

Ⓖ ____________

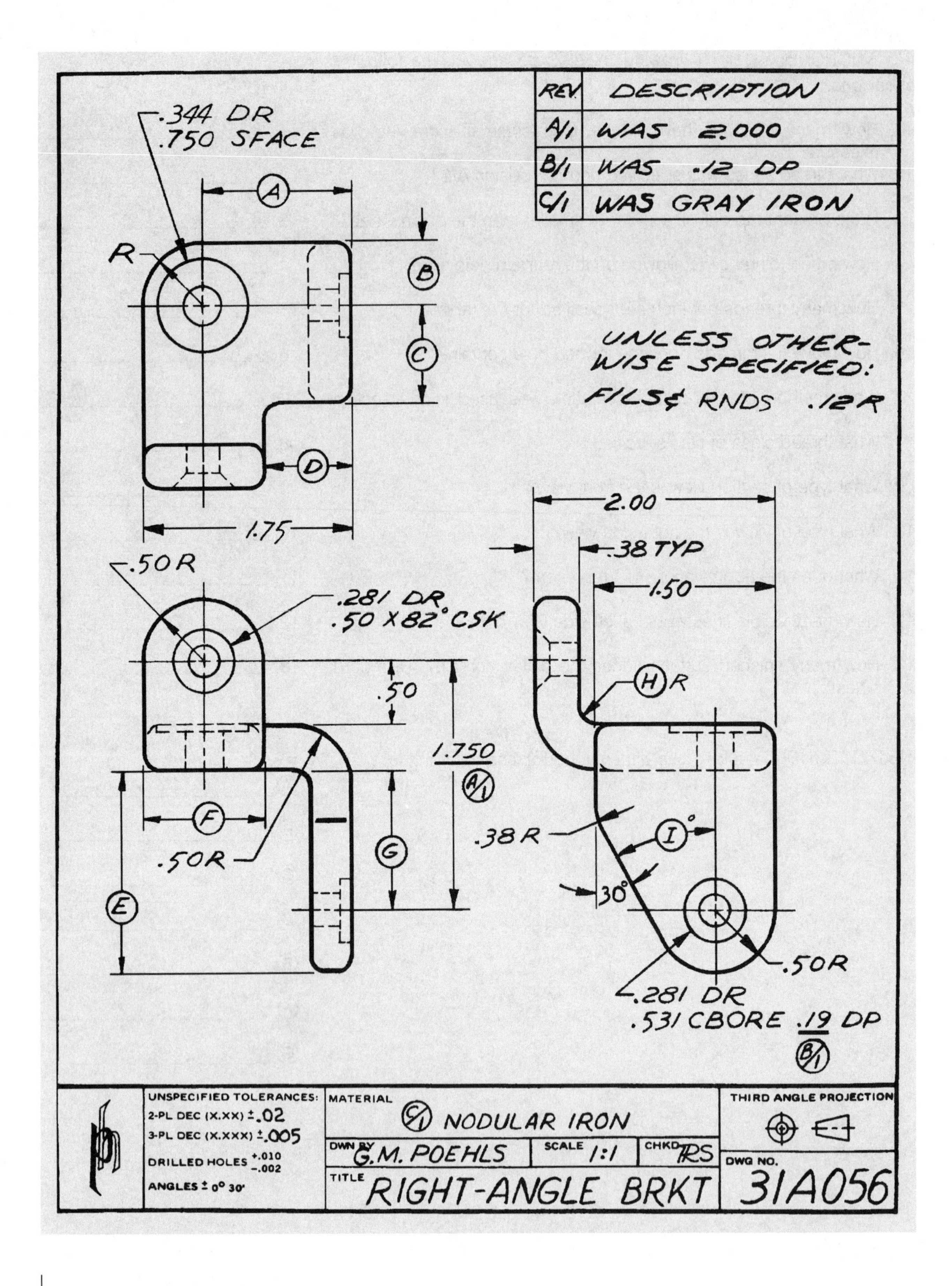

REV.
DESCRIPTION
A/1
WAS 2.000
B/1
WAS .12 DP
C/1
WAS GRAY IRON
.344 DR
.750 SFACE
A
R
B
C
D
1.75
UNLESS OTHER-WISE SPECIFIED:
FILS & RNDS .12R
2.00
.38 TYP
1.50
.50R
.281 DR
.50 X 82° CSK
.50
1.750
A/1
H R
F
.50R
G
E
.38R
I°
30°
.50R
.281 DR
.531 CBORE .19 DP
B/1
UNSPECIFIED TOLERANCES:
2-PL DEC (X.XX) ±.02
3-PL DEC (X.XXX) ±.005
DRILLED HOLES +.010 −.002
ANGLES ± 0° 30'
MATERIAL
C/1 NODULAR IRON
DWN BY G.M. POEHLS
SCALE 1:1
CHKD RS
TITLE RIGHT-ANGLE BRKT
THIRD ANGLE PROJECTION
DWG NO.
31A056

INSTRUCTIONS: *Refer to drawing 31A056 to answer the following questions.*

1. What is the overall height of the object? 1. ______
2. What was the former height before revision? 2. ______
3. How many revisions have been made to the drawing? 3. ______
4. What did revision C affect? 4. ______
5. What was the former depth of the counterbore? 5. ______
6. What does the heavy line underneath the revised dimensions designate? 6. ______
7. What type of iron is presently specified? 7. ______
8. What size are the fillets and rounds? 8. ______
9. What is the tolerance on the fillets and rounds? 9. ______
10. What is the maximum permissible spacing (front view) between countersunk and counterbored holes? (C to C) 10. ______
11. What is the dimension of the large radius shown on the top view? 11. ______
12. How many machined surfaces (excluding holes) does the object contain? 12. ______
13. How many different drills are required? 13. ______
14. What type of fastener head is intended for use in the countersunk hole? 14. ______
15. What size fastener is intended for use in the spotfaced hole? 15. ______
16. What size fastener is intended for use in the counterbored hole? 16. ______

INSTRUCTIONS: *Calculate the dimensions for the following letters.*

Ⓐ ______

Ⓑ ______

Ⓒ ______

Ⓓ ______

Ⓔ ______

Ⓕ ______

Ⓖ ______

Ⓗ ______

Ⓘ ______

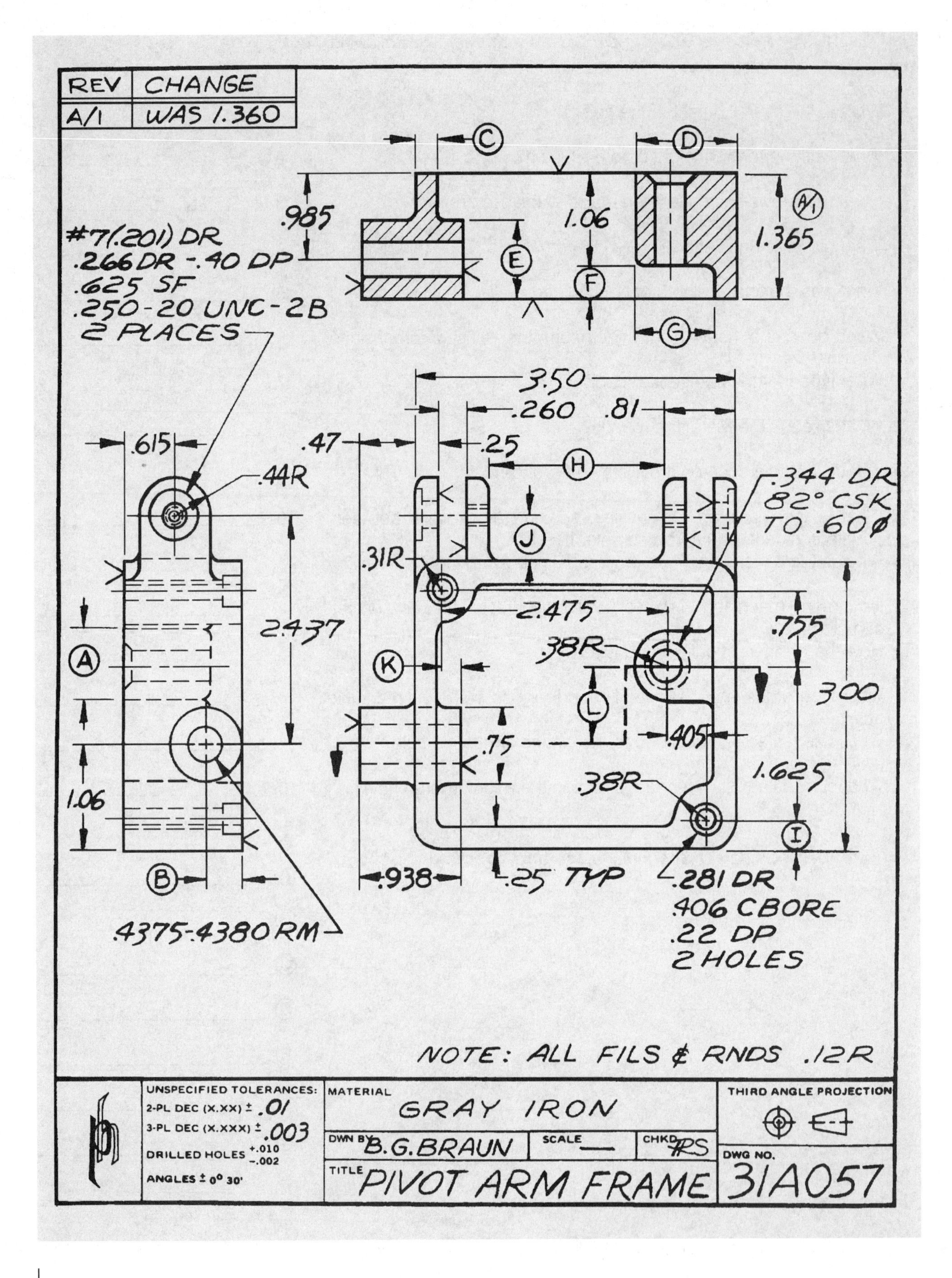

REV
CHANGE
A/1
WAS 1.360
#7(.201) DR
.266 DR -.40 DP
.625 SF
.250-20 UNC-2B
2 PLACES
.985
1.06
1.365
3.50
.260
.81
.615
.47
.25
.44R
.344 DR
82° CSK
TO .60⌀
.31R
2.437
2.475
.755
.38R
300
.405
.75
1.06
1.625
.938
.25 TYP
.281 DR
.406 CBORE
.22 DP
2 HOLES
.4375-.4380 RM
NOTE: ALL FILS & RNDS .12R
UNSPECIFIED TOLERANCES:
2-PL DEC (X.XX) ± .01
3-PL DEC (X.XXX) ± .003
DRILLED HOLES +.010 -.002
ANGLES ± 0° 30'
MATERIAL
GRAY IRON
DWN BY B.G.BRAUN
SCALE —
CHKD RS
THIRD ANGLE PROJECTION
DWG NO.
TITLE PIVOT ARM FRAME
31A057

INSTRUCTIONS: Refer to drawing 31A057 to answer the following questions.

1. What type of section view is shown? 1. ______
2. How many tapped holes does the part contain? 2. ______
3. What is the major thread diameter of the tapped holes? 3. ______
4. What is the tap drill diameter? 4. ______
5. How many threads will each tapped hole contain? 5. ______
6. How many counterbored holes does the part contain? 6. ______
7. How deep are the counterbores? 7. ______
8. What is the diameter of the counterbores? 8. ______
9. How many countersunk holes does the part contain? 9. ______
10. What is the diameter of the countersink? 10. ______
11. What is the diameter of the spotface? 11. ______
12. How many reamed holes does the part contain? 12. ______
13. Was the frame thickness (a) increased or (b) decreased by revision A? 13. ______

INSTRUCTIONS: Calculate the dimensions for the following letters.

Ⓐ ______

Ⓑ ______

Ⓒ ______

Ⓓ ______

Ⓔ ______

Ⓕ ______

Ⓖ ______

Ⓗ ______

Ⓘ ______

Ⓙ ______

Ⓚ ______

Ⓛ ______

COMPUTER-AIDED DRAFTING

Many drawings today are being created by means of computers with specialized drafting software and plotters. Commonly referred to as CAD (for computer-aided drafting), it has increased in usage at a tremendous rate in recent years. Once a CAD operator is well trained, the amount of time it takes to create a drawing can be reduced substantially. Also, changes to drawings created by CAD can be completed quickly, with little effort.

The hardware used in CAD is divided into five categories: a CPU (central processing unit), display devices, input devices, output devices, and memory devices. A brief description of each category follows.

Central Processing Unit

The CPU is generally regarded as the brain of the computer. It accepts a software program from a floppy or hard disk, then prompts the operator to provide necessary information. The CPU processes that information into a form that can be used to produce a drawing.

Graphics Display Device

The graphics display device is usually a CRT (cathode ray tube), sometimes called a monitor, and is similar to a TV screen. It displays the data on which the CAD operator is working, allowing the user to view a picture of the design as it is being entered into the system.

Input Devices

One input device used to establish or maintain communications between the operator and the CPU is the keyboard. It is used to enter text, system commands, X-Y coordinates, and any other type of alphanumeric information. Other peripheral input devices may include digitizers (graphics tablets), cursors, light pens, joysticks, or scanners.

Output Devices

The output device produces hard copy data, duplicating the image appearing on the CRT. A pen plotter is used when quality reproduction is required. Plotters are available as flatbed, drum, or electrostatic models. A printer may be used when speed of reproduction is more important than quality, such as for check prints.

Memory Devices

The memory device is used to store and retrieve programs and other data. Information needed to program the CPU or to create the drawing is stored on disks or magnetic tape. Permanent memory is referred to as read-only memory (ROM), while temporary memory is referred to as random-access memory (RAM).

Computervision's UNIX-based CADDStation systems. (Courtesy of Computervision Corporation)

Digitizers (Courtesy Houston Instrument)

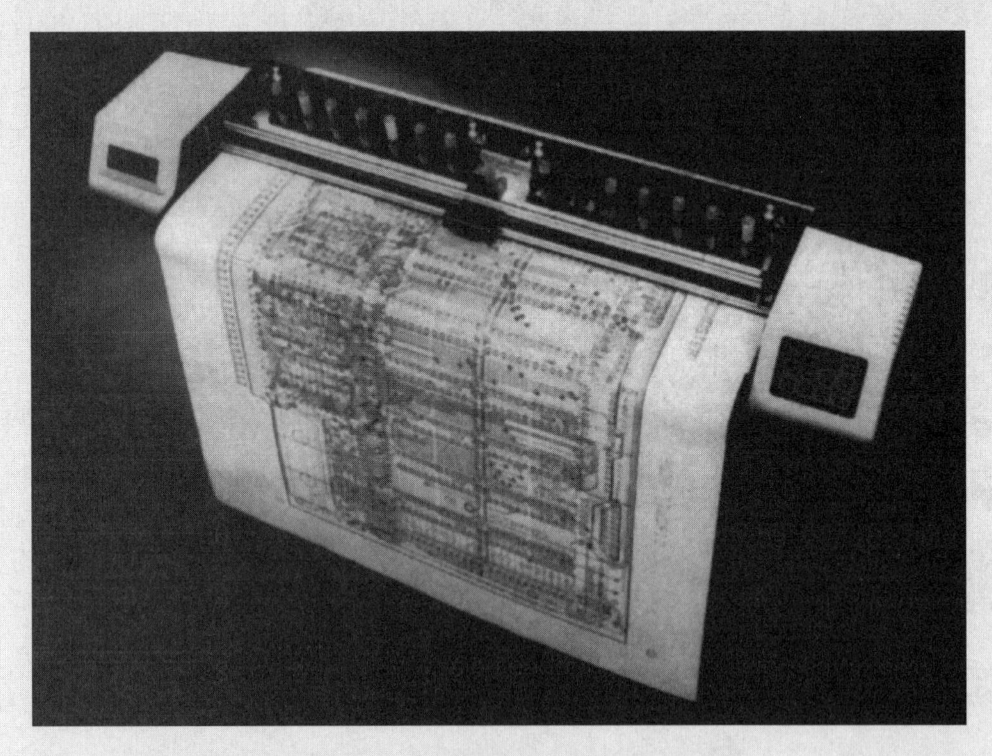

14-pen plotter (Courtesy Houston Instrument)

ADVANTAGES OF CAD

Computer-aided drafting stations are rapidly replacing the conventional manual drafting tables. In addition to the drawing speed advantage, CAD offers advantages in accuracy, neatness, consistency, revisions, and drawing storage problems. CAD hardware is consistently accurate to .001 inch or better. Line widths and lettering are always uniform, eliminating the skill factor required with manual drafting. Revisions, including corrections and changes, are much easier and faster to accomplish on CAD drawings. Large drawing storage files can be replaced with disk containers, and back-up disks can be made quickly, easily, and inexpensively.

The next drawing to appear in this text, 31A058 on page 256, was created on a CAD system. Uniformity of the numbers and letters on CAD drawings is usually the most noticeable difference from hand-drawn drawings. The print reader should be able to distinguish between a CAD print and a hand-drawn print. Turn the page and see if you can notice the difference.

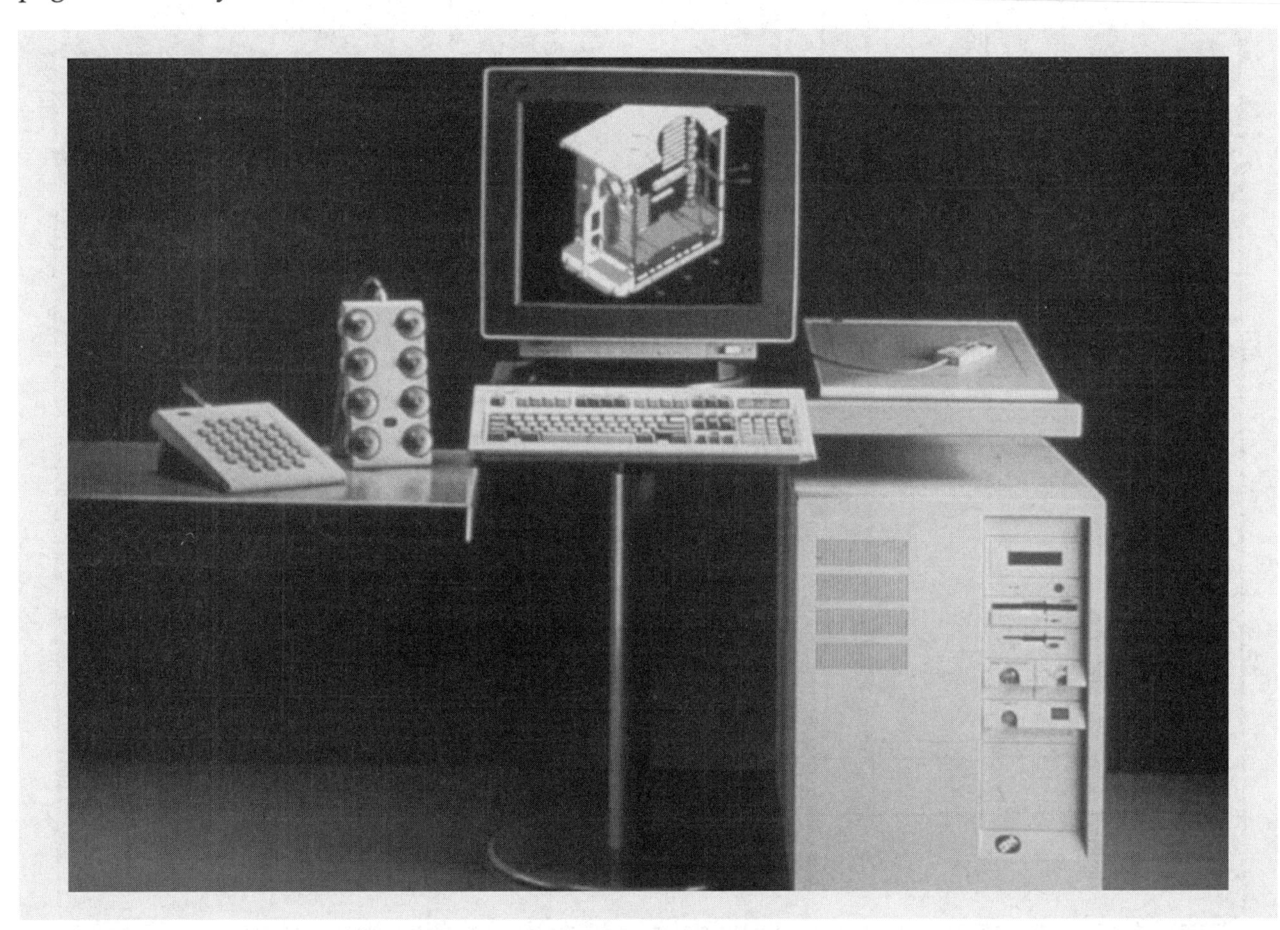

RISC System/6000 (Courtesy of International Business Machines Corporation)

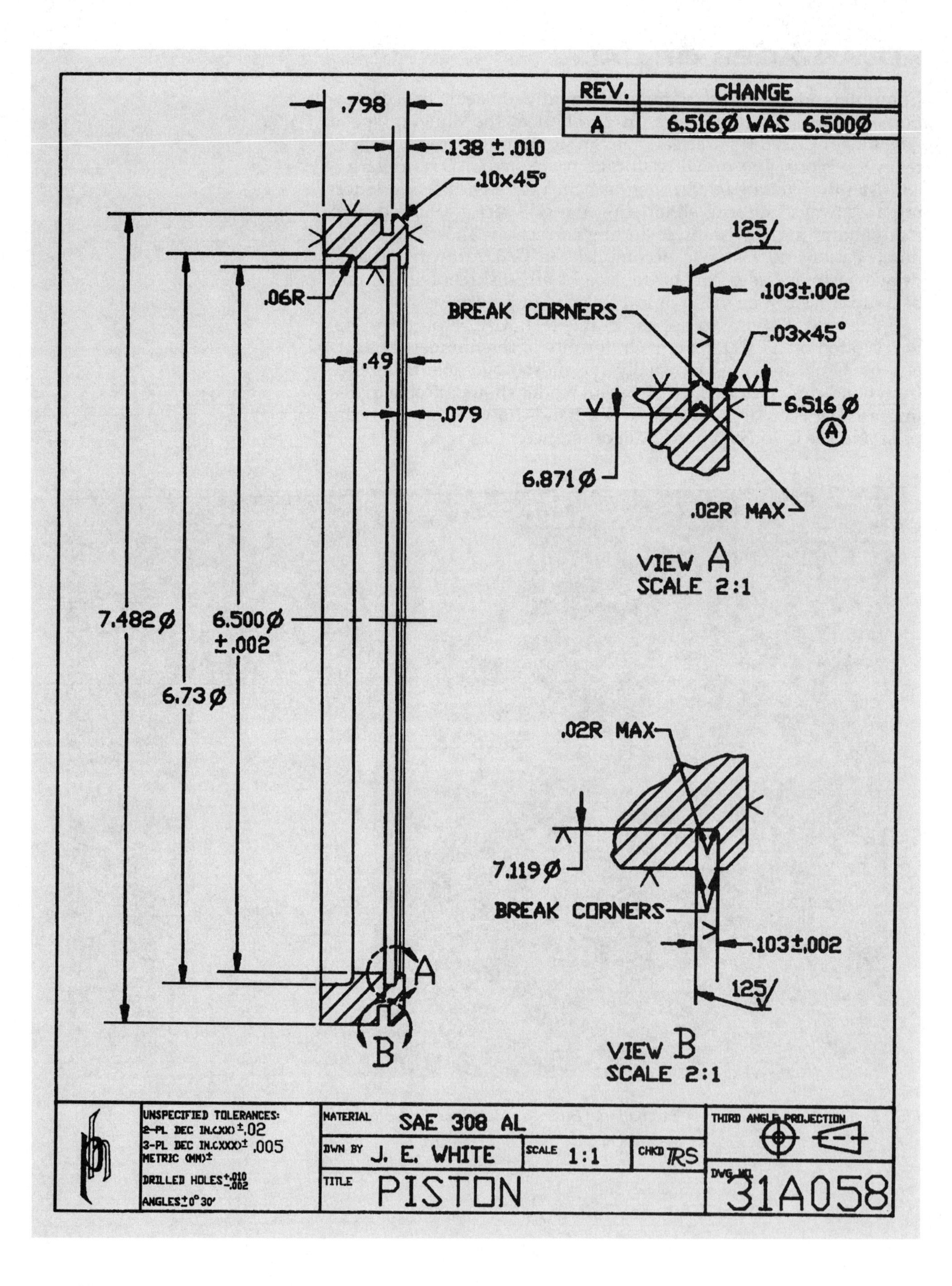
REV.
CHANGE
A
6.516Ø WAS 6.500Ø
.798
.138 ± .010
.10×45°
.06R
.49
.079
7.482Ø
6.500Ø
±.002
6.73Ø
A
B
125/
.103±.002
BREAK CORNERS
.03×45°
6.516 Ø
A
6.871Ø
.02R MAX
VIEW A
SCALE 2:1
.02R MAX
7.119Ø
BREAK CORNERS
.103±.002
125/
VIEW B
SCALE 2:1
UNSPECIFIED TOLERANCES:
2-PL DEC IN.(XX) ±.02
3-PL DEC IN.(XXX)± .005
METRIC (MM)±
DRILLED HOLES +.010 -.002
ANGLES ± 0° 30'
MATERIAL SAE 308 AL
DWN BY J. E. WHITE
SCALE 1:1
CHKD TRS
TITLE PISTON
THIRD ANGLE PROJECTION
DWG NO. 31A058

INSTRUCTIONS: Refer to drawing 31A058 to answer the following questions.

1. Was the drawing hand-drawn or CAD-drawn? 1. ____________
2. Did revision A increase or decrease the diameter? 2. ____________
3. What does AL in the material specification abbreviate? 3. ____________
4. What "single limit" dimension is used on this drawing? 4. ____________
5. What is the tolerance on the outside diameter? 5. ____________
6. What is the tolerance on the smallest inside diameter? 6. ____________
7. What is the tolerance on the counterbore diameter? 7. ____________

INSTRUCTIONS: Calculate the dimensions for the following letters.

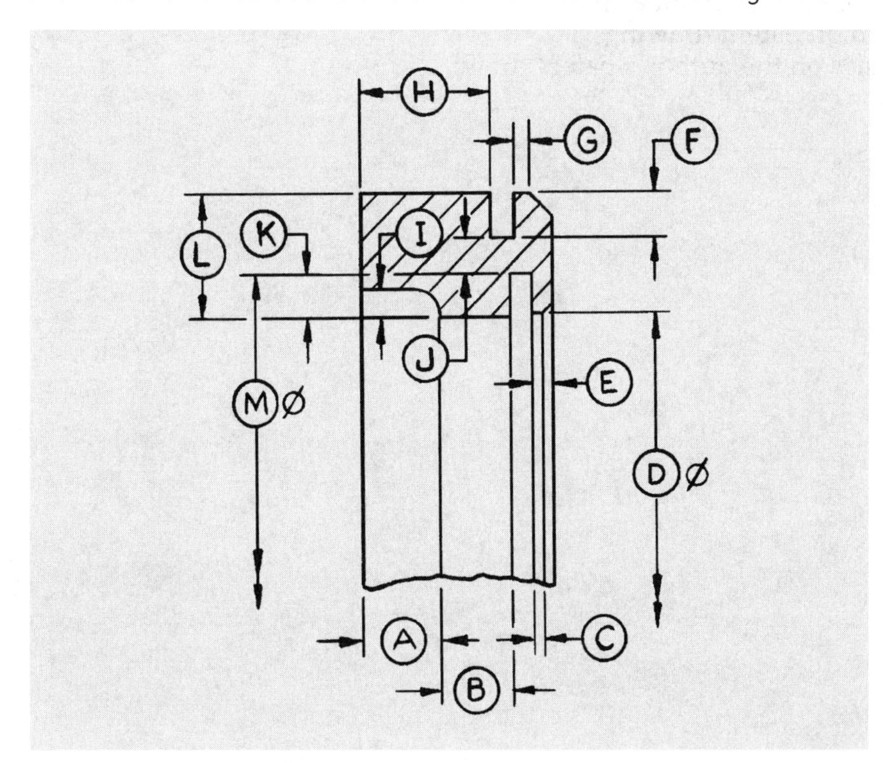

Ⓐ ____________
Ⓑ ____________
Ⓒ ____________
Ⓓ ____________
Ⓔ ____________
Ⓕ ____________
Ⓖ ____________
Ⓗ ____________
Ⓘ ____________
Ⓙ ____________
Ⓚ ____________
Ⓛ ____________
Ⓜ ____________

PATENT DRAWINGS

A patent when granted excludes anyone else from manufacturing, using, or selling the device covered by that patent for a period of seventeen years. The invention will be fully described in a patent document that usually includes a patent drawing.

Patent drawings must be prepared in accordance with U.S. Patent Office rules, and must be submitted to that office along with a written description and a petition. Special drafters, trained specifically to produce well-executed patent drawings that conform to those special rules, are usually employed by, and work closely with, patent attorneys.

The drawing must show every feature of the invention specified in the claims. Unlike the detail drawings shown throughout this text, patent drawings will include line shading wherever surface delineation occurs. The different views should be consecutively numbered and identified as figures. All figures must be separately identified with plain, legible, carefully formed numbers. The same part of an invention appearing in more than one figure must be designated by the same character number in all views. Descriptive matter is not permitted on patent drawings.

Observe the conformity to rules on the author's patent drawing illustrated on page 259.

Aug. 3, 1965 R. R. SCHULTZ 3,198,462

ELECTRIC LIGHT FIXTURE

Filed May 28, 1962

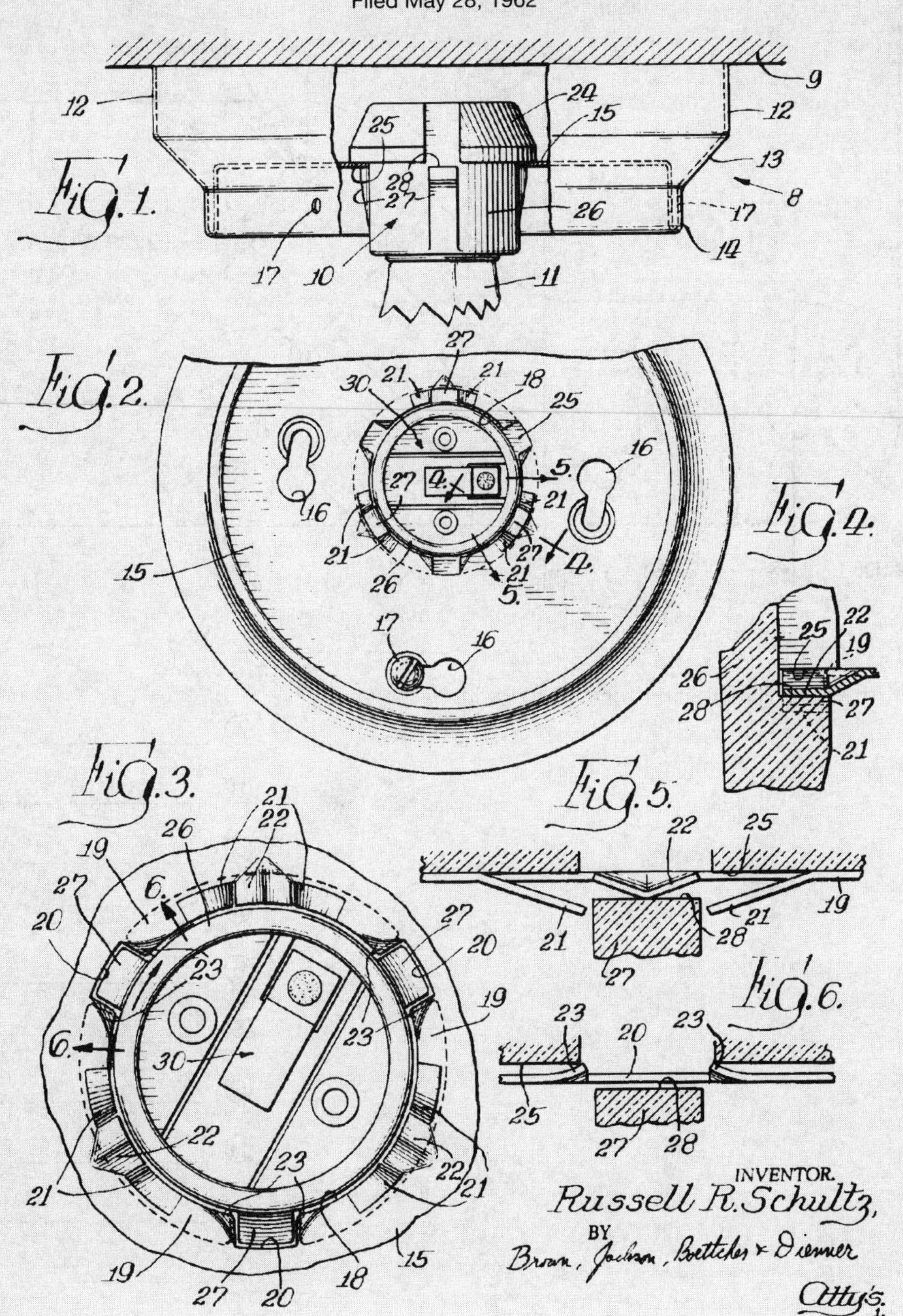

Review Exercise

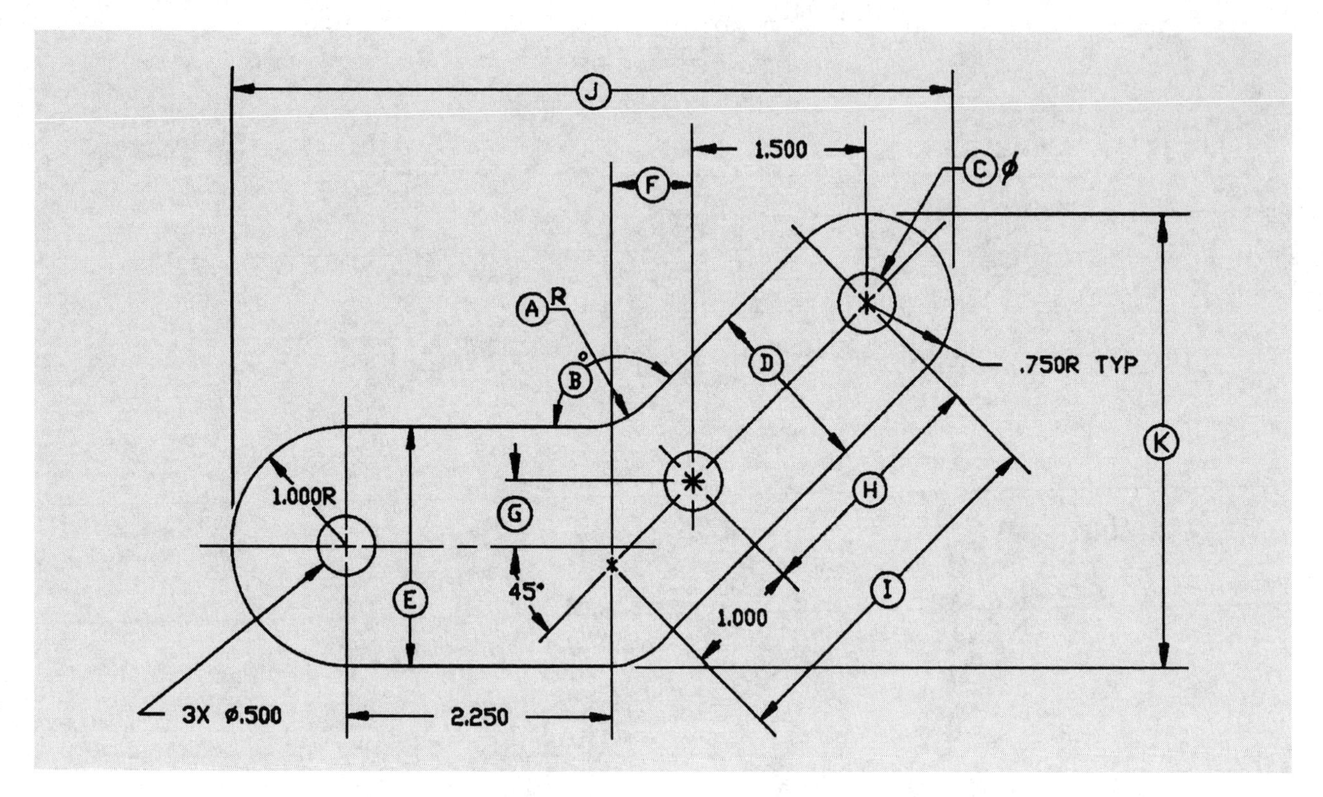

INSTRUCTIONS: Enter the dimensions for the following letters.

Ⓐ ______________________

Ⓑ ______________________

Ⓒ ______________________

Ⓓ ______________________

Ⓔ ______________________

Ⓕ ______________________

Ⓖ ______________________

Ⓗ ______________________

Ⓘ ______________________

Ⓙ ______________________

Ⓚ ______________________

UNIT 11

METRIC DRAWINGS

More and more companies are producing metric drawings today. They may be recognized by a block enclosing the word METRIC in large letters, or a note stating that all dimensions are in millimeters. When a new part is designed metrically, an effort is usually made to use whole millimeters or half-millimeters, except where standard tools are involved (drills, reamers, counterbores, etc.). Unlike the inch system of dimensioning, metric dimensions do not show insignificant zeros to the right of the decimal. Thus, a whole millimeter would not even include a decimal point and a half-millimeter would include only one digit to the right of the decimal (for example: 31.5). Also, a dimension of less than 1 millimeter will always include a zero to the left of the decimal (for example: 0.5).

Metric dimensions may be converted to inches by multiplying millimeters by .03937, but it is preferred to have the machinist "think metric" by using metric measuring instruments. Some metric drawings include the inch equivalents of all dimensions in a printed chart on the left side of a drawing. Other metric drawings are dual-dimensioned, with the inch equivalent adjacent each metric dimension.

Metric drawings were the first to have symbols precede the dimension figures (see examples below), while inch-dimensioned drawings continued to have the symbol follow the figures. However, the trend today is for both systems to uniformly place the symbol in front.

DIMENSIONING SYMBOLS

A major move toward international standardization of drawings has been to adopt symbols in place of words and abbreviations. Below are dimensioning symbols that you must be able to recognize for the correct interpretation of the most recently drawn blueprints.

SYMBOL	INTERPRETATION	EXAMPLE
Ø	DIAMETER	Ø25
R	RADIUS	R12.5
SØ	SPHERICAL DIAMETER	SØ25
SR	SPHERICAL RADIUS	SR12.5
( )	REFERENCE DIMENSION	(25)
___	DIMENSION NOT-TO-SCALE	50 (underlined)
X	NO. OF TIMES / PLACES	6X
⌒	ARC LENGTH	⌒75
□	SQUARE (shape)	□25
↧	DEPTH / DEEP	↧25
⌴	COUNTERBORE / SPOTFACE	⌴Ø25
⌵	COUNTERSINK	⌵Ø25
⌲	CONICAL TAPER	⌲ 0.2:1
⌳	SLOPE	⌳ 0.1:1
⌖→	DIMENSION ORIGIN	⌖ 25 →

METRIC THREADS

Metric threads are specified on engineering drawings in a system similar to Unified threads. That is, the major thread diameter is listed first (after the prefix M, for ISO metric), then the pitch, followed by the tolerance class designation. Whereas the inch system specified the threads per inch, requiring the pitch dimension to be calculated, the metric system specifies the pitch dimension (for example: M10 × 1.5). Only two series of threads are standard in the metric system: coarse, the most common, falls between our Unified coarse and Unified fine series, and fine, which falls between our Unified fine and Unified extra-fine series. When the pitch does not appear in the thread specifications, it may be assumed that the thread series is coarse. The fine series is not available for threads smaller than 8 mm. See the table on page 263 for standard sizes and tap drills.

Tolerance Class

Numbers and letters following the pitch in a thread specification designate the tolerance for the pitch and major diameters of external threads and the pitch and minor diameters of internal threads. The numbers indicate the amount of tolerance permissible; the smaller the number, the smaller the tolerance. The letters indicate the position of the thread tolerance in relation to its basic diameter. Lowercase letters are used for external threads, with *e* indicating a large allowance, *g* a small allowance, and *h* no allowance. Uppercase letters are used for internal threads, with G indicating a small allowance and H indicating no allowance. For general-purpose applications, tolerance classes 6H/6g are usually assigned. They are comparable to the U.S. 2A/2B classes of fit, and may be assumed when omitted from the specifications. A class 4H5H/4h6h is approximately equal to the U.S. class 3A/3B.

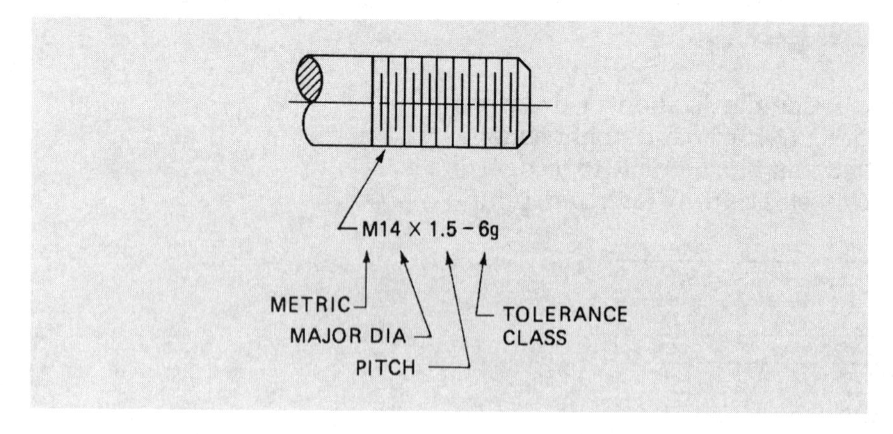

METRIC THREAD TABLE

Recommended Metric Standard			Closest Inch Standard	
Tap Size	Drill Size	Inch Equivalent	Drill Size	Inch Equivalent
M1.6 × 0.35	1.25	.0492		
M1.8 × 0.35	1.45	.0571		
M2 × 0.4	1.60	.0630	#52	.0635
M2.2 × 0.45	1.75	.0689		
M2.5 × 0.45	2.05	.0807	#46	.0810
M3 × 0.5	2.50	.0984	#40	.0980
M3.5 × 0.6	2.90	.1142	#33	.1130
M4 × 0.7	3.30	.1299	#30	.1285
M4.5 × 0.75	3.70	.1457	#26	.1470
M5 × 0.8	4.20	.1654	#19	.1660
M6 × 1	5.00	.1968	#9	.1960
M7 × 1	6.00	.2362	15/64	.2344
M8 × 1.25	6.70	.2638	17/64	.2656
M8 × 1	7.00	.2756	J	.2770
M10 × 1.5	8.50	.3346	Q	.3320
M10 × 1.25	8.70	.3425	11/32	.3438
M12 × 1.75	10.20	.4016	Y	.4040
M12 × 1.25	10.80	.4252	27/64	.4219
M14 × 2	12.00	.4724	15/32	.4688
M14 × 1.5	12.50	.4921		
M16 × 2	14.00	.5512	35/64	.5469
M16 × 1.5	14.50	.5709		
M18 × 2.5	15.50	.6102	39/64	.6094
M18 × 1.5	16.50	.6496		
M20 × 2.5	17.50	.6890	11/16	.6875
M20 × 1.5	18.50	.7283		
M22 × 2.5	19.50	.7677	49/64	.7656
M22 × 1.5	20.50	.8071		
M24 × 3	21.00	.8268	53/64	.8281
M24 × 2	22.00	.8661		
M27 × 3	24.00	.9449	15/16	.9375
M27 × 2	25.00	.9843	63/64	.9844

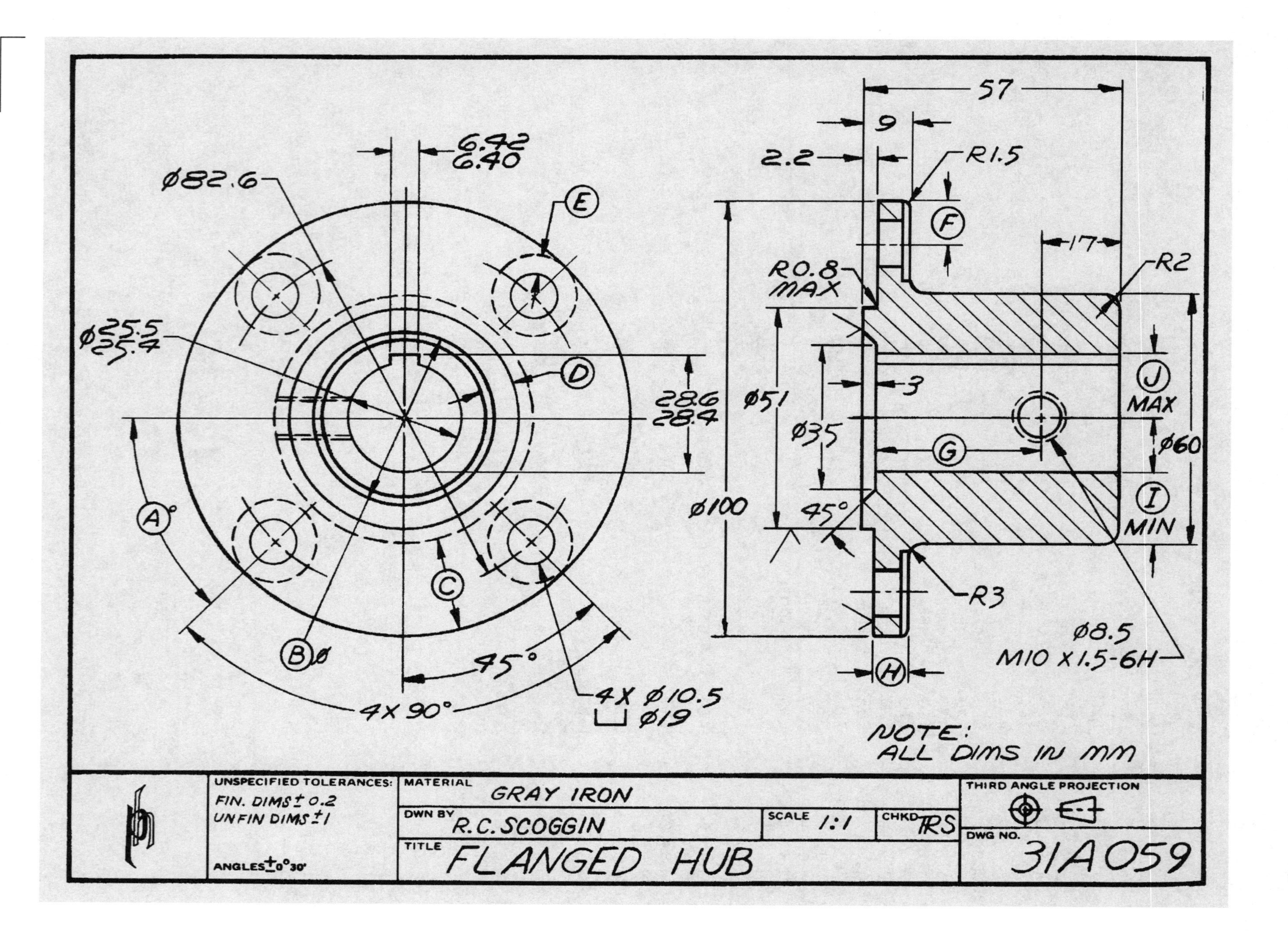

Ø82.6
6.42
6.40
Ø25.5
Ø25.4
28.6
28.4
A°
B Ø
4X 90°
45°
4X Ø10.5
Ø19
57
9
2.2
R1.5
R0.8 MAX
R2
17
3
Ø51
Ø35
Ø100
45°
R3
Ø60
J MAX
I MIN
Ø8.5
M10 X 1.5-6H
NOTE:
ALL DIMS IN MM
UNSPECIFIED TOLERANCES:
FIN. DIMS ± 0.2
UNFIN DIMS ± 1
ANGLES ± 0° 30'
MATERIAL
GRAY IRON
DWN BY R.C. SCOGGIN
SCALE 1:1
CHKD TRS
TITLE
FLANGED HUB
THIRD ANGLE PROJECTION
DWG NO.
31A059

INSTRUCTIONS: *Refer to drawing 31A059 to answer the following questions.*

1. Are the units of measurement in (a) inches, (b) millimeters, or (c) centimeters? — 1. ______
2. Is the section view a (a) half-section or a (b) full section? — 2. ______
3. What tolerance applies to the flange diameter? — 3. ______
4. What is the MMC of the bore diameter? — 4. ______
5. Write the flange diameter in limit form. — 5. ______
6. Write the bore diameter with bilateral tolerance. — 6. ______
7. Convert the flange diameter to its decimal inch equivalent. — 7. ______
8. What is the diameter of the bolt circle? — 8. ______
9. What is the diameter of the spotface? — 9. ______
10. Does the spotface run tangent with the diameter of the flange? (Calculate) — 10. ______
11. What is the major diameter of the threads? — 11. ______
12. What is the pitch of the threads? — 12. ______
13. What is the tolerance class of the threads? — 13. ______
14. Are the threads coarse or fine? (Consult the table on page 263.) — 14. ______
15. How many full threads will the tapped hole contain? — 15. ______

INSTRUCTIONS: *Calculate the dimensions for the following letters.*

Ⓐ ______

Ⓑ ______

Ⓒ ______

Ⓓ ______

Ⓔ ______

Ⓕ ______

Ⓖ ______

Ⓗ ______

Ⓘ MIN: ______

Ⓙ MAX: ______

Optional Math Exercises:

1a. Using only the dimensions given, calculate the X & Y coordinates of the four 10.5mm holes.

2a. Using the dimensions given, calculate the MOP (Measurement Over Pins) of adjacent 10.5∅ holes.

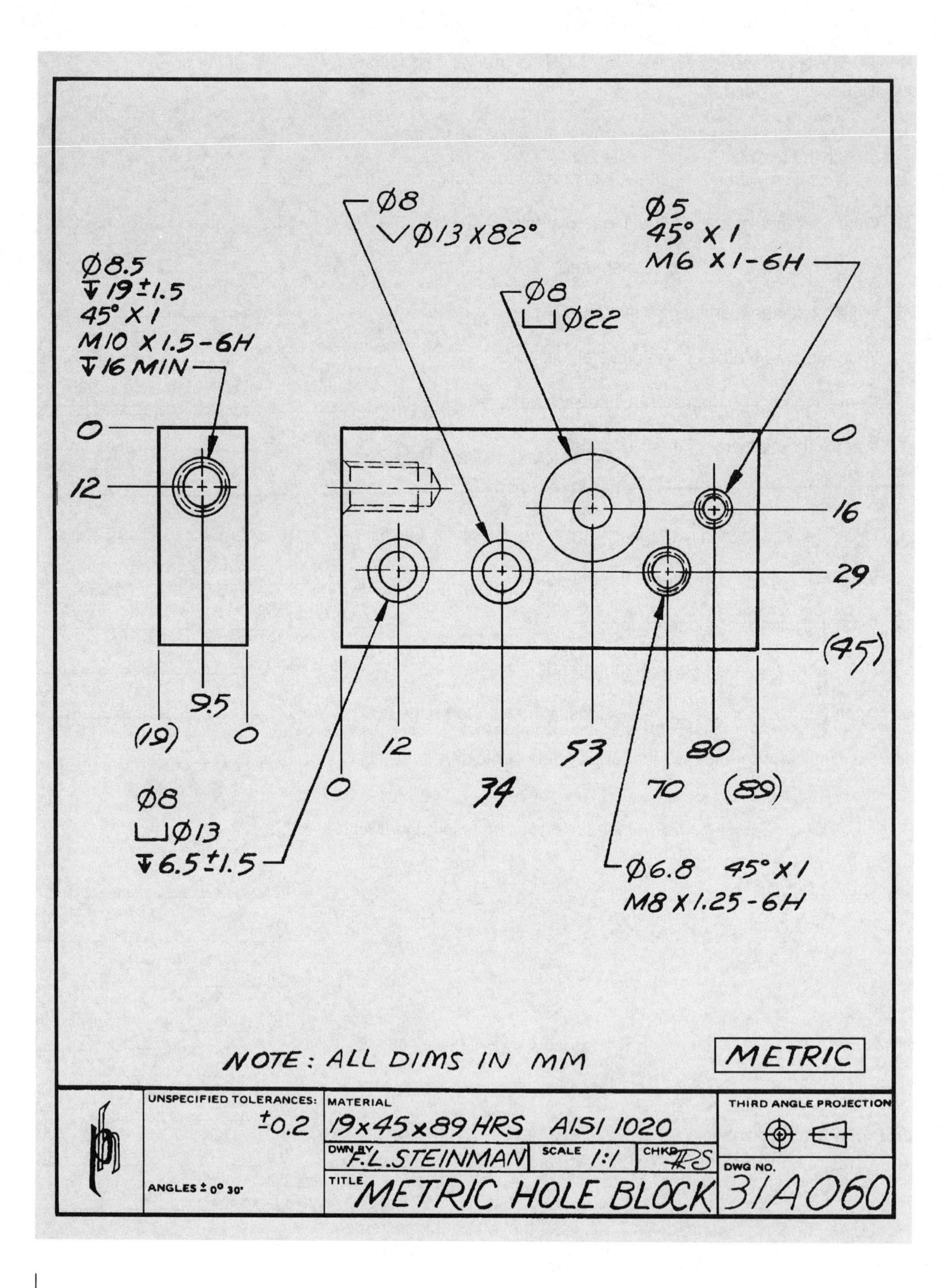
Ø8.5
↧19±1.5
45° X 1
M10 X 1.5-6H
↧16 MIN
Ø8
⌵Ø13 X 82°
Ø8
⌴Ø22
Ø5
45° X 1
M6 X 1-6H
0
12
9.5
(19)
0
0
16
29
(45)
0
12
34
53
70
80
(89)
Ø8
⌴Ø13
↧6.5±1.5
Ø6.8 45° X 1
M8 X 1.25-6H
NOTE: ALL DIMS IN MM
METRIC
UNSPECIFIED TOLERANCES:
±0.2
ANGLES ± 0° 30'
MATERIAL
19x45x89 HRS AISI 1020
DWN BY F.L.STEINMAN
SCALE 1:1
CHKD RS
TITLE METRIC HOLE BLOCK
THIRD ANGLE PROJECTION
DWG NO.
31A060

INSTRUCTIONS: Refer to drawing 31A060 to answer the following questions.

1. What units of measurement are used for dimensioning? 1. ____________
2. What do the parentheses enclosing some of the dimensions designate? 2. ____________
3. Are the holes located by chain dimensioning or datum dimensioning? 3. ____________
4. What is the maximum tolerance accumulation between any two hole locations? 4. ____________
5. Is the specified material an alloy steel or a carbon steel? 5. ____________
6. Is the specified material a high-, medium-, or low-carbon steel? 6. ____________
7. What tolerance class is specified for the threaded holes? 7. ____________
8. What is the major diameter of the smallest threaded hole? 8. ____________
9. Are the M8 threads coarse or fine? (Consult the table on page 263.) 9. ____________
10. What is the thread pitch of the blind hole? 10. ____________
11. What is the tolerance on the depth of the tap drill in the blind hole? 11. ____________
12. What is the chamfer angle on the threaded holes? 12. ____________
13. Calculate the minimum number of full threads possible in the blind hole. (Disregard chamfer.) 13. ____________
14. How far apart are the countersunk hole and the counterbored hole? (C to C) 14. ____________
15. How far apart are the spotfaced hole and the M6 threaded hole? (C to C) 15. ____________
16. What is the diameter of the spotface? 16. ____________
17. What is the approximate depth of the spotface? 17. ____________
18. What is the diameter of the counterbore? 18. ____________
19. Show the limits of the counterbore depth. 19. ____________
20. What is the included angle of the countersink? 20. ____________

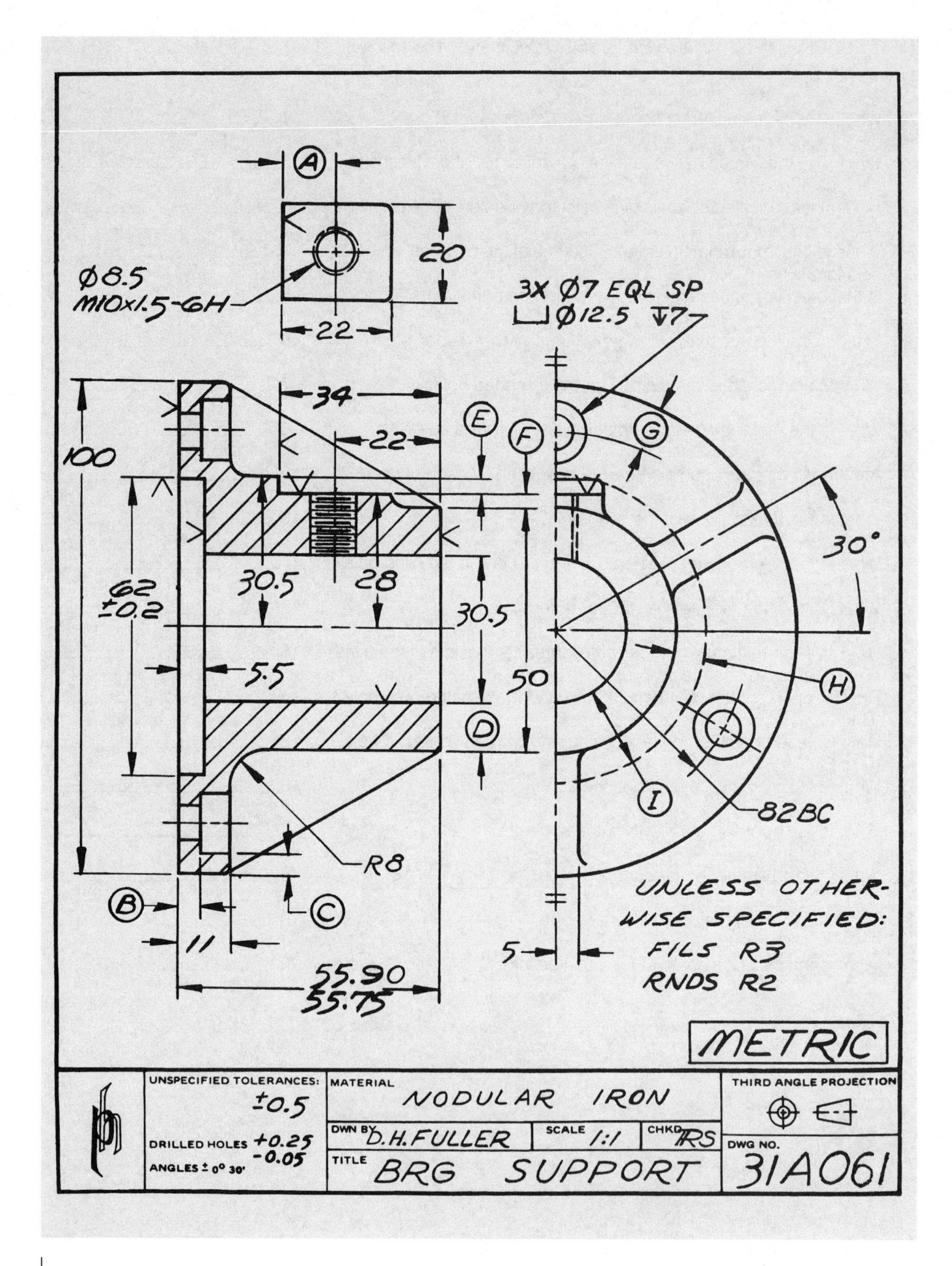

A
Ø8.5
M10x1.5-6H
20
22
3X Ø7 EQL SP
⌴ Ø12.5 ↧7
100
34
22
E
F
G
30°
62
±0.2
30.5
28
30.5
50
5.5
D
H
I
82BC
R8
B
C
11
5
UNLESS OTHER-
WISE SPECIFIED:
FILS R3
RNDS R2
55.90
55.75
METRIC
UNSPECIFIED TOLERANCES:
±0.5
DRILLED HOLES +0.25
-0.05
ANGLES ± 0° 30'
MATERIAL
NODULAR IRON
DWN BY D.H.FULLER
SCALE 1:1
CHKD TRS
TITLE BRG SUPPORT
THIRD ANGLE PROJECTION
DWG NO.
31A061

INSTRUCTIONS: Refer to drawing 31A061 to answer the following questions.

1. What unit of measurement was used for dimensioning? 1. ____________
2. Is the part symmetrical about its (a) vertical centerline, (b) horizontal centerline, or (c) both? 2. ____________
3. Is the section view considered (a) a full section or (b) a half-section? 3. ____________
4. Does the section view contain any rotated features? 4. ____________
5. Is the partial view (a) a projected top view or (b) a removed view? 5. ____________
6. What is the angular dimension between mounting holes? 6. ____________
7. What is the angular dimension between a mounting hole and a rib? 7. ____________
8. What is the thickness of each rib? 8. ____________
9. What is the fillet radius between the flange and the ribs? 9. ____________
10. What is the fillet radius between the flange and the hub? 10. ____________
11. What is the tolerance on the bolt circle? 11. ____________
12. What is the major thread diameter of the tapped hole? 12. ____________
13. What is the pitch of the threads? 13. ____________
14. What tap drill size is specified? 14. ____________
15. Are the threads coarse or fine? (Consult the table on page 263.) 15. ____________
16. How many full threads will the tapped hole contain? 16. ____________

INSTRUCTIONS: Enter the dimensions for the following letters.

Ⓐ ____________

Ⓑ ____________

Ⓒ ____________

Ⓓ ____________

Ⓔ ____________

Ⓕ ____________

Ⓖ ____________

Ⓗ ____________

Ⓘ ____________

Optional Math Exercises:

1a. Using only the dimensions given, calculate the X & Y coordinates of the three 7mm∅ holes.

2a. Using the dimensions given, calculate the MOP (Measurement Over Pins) of adjacent 7mm∅ holes.

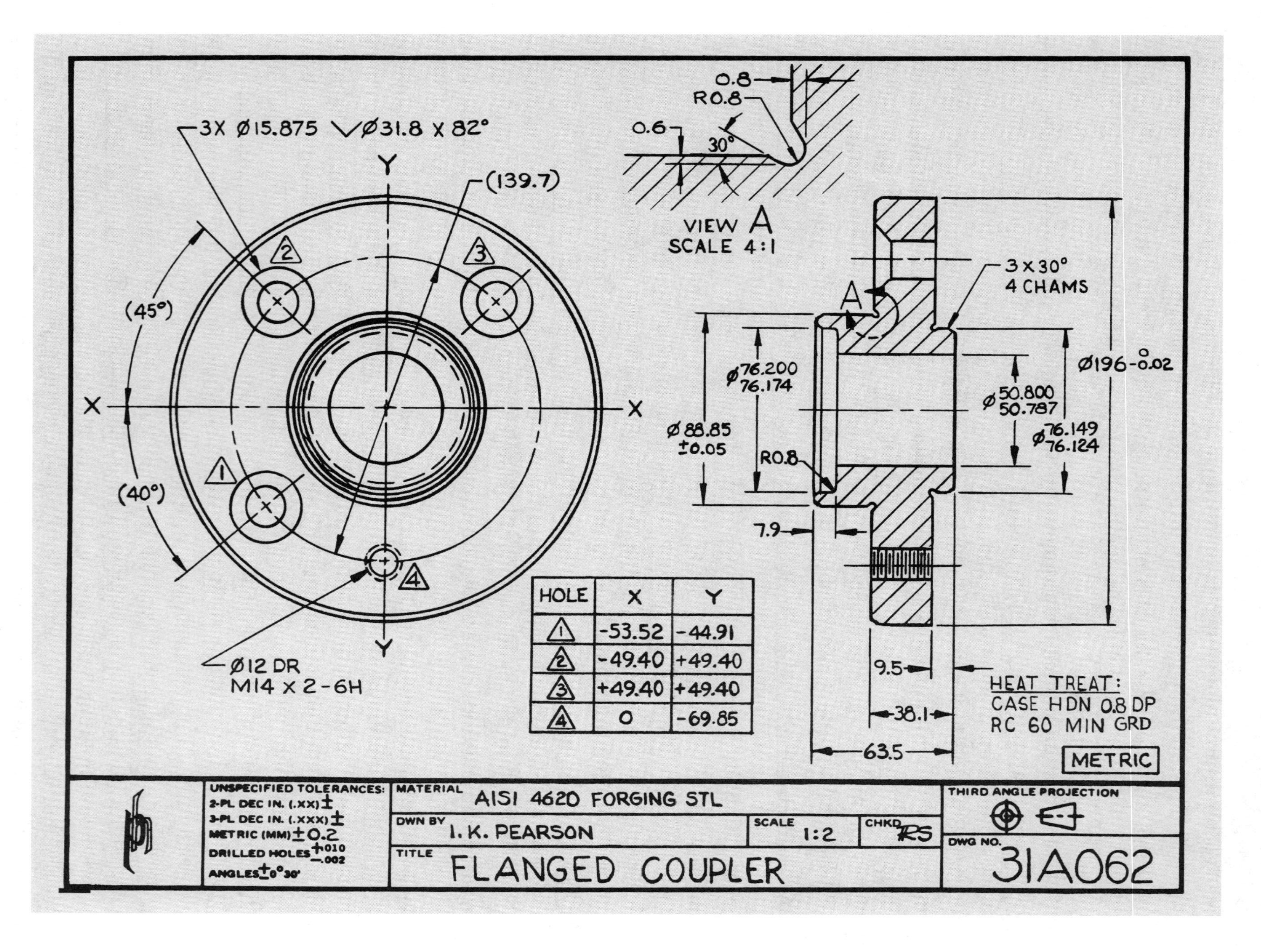

3X Ø15.875 ⌵Ø31.8 X 82°
(139.7)
(45°)
(40°)
Ø12 DR
M14 X 2-6H
X
Y
HOLE X Y
1 -53.52 -44.91
2 -49.40 +49.40
3 +49.40 +49.40
4 0 -69.85
0.8
R0.8
0.6
30°
VIEW A
SCALE 4:1
A
3 X 30°
4 CHAMS
Ø196-0 -0.02
Ø50.800 50.787
Ø76.149 76.124
Ø76.200 76.174
R0.8
Ø88.85 ±0.05
7.9
9.5
38.1
63.5
HEAT TREAT:
CASE HDN 0.8 DP
RC 60 MIN GRD
METRIC
UNSPECIFIED TOLERANCES:
2-PL DEC IN. (.XX) ±
3-PL DEC IN. (.XXX) ±
METRIC (MM) ±0.2
DRILLED HOLES +.010 -.002
ANGLES ±0°30'
MATERIAL AISI 4620 FORGING STL
DWN BY I. K. PEARSON
SCALE 1:2
CHKD RS
THIRD ANGLE PROJECTION
TITLE FLANGED COUPLER
DWG NO. 31A062

COORDINATE DIMENSIONING

With the advent of numerical control (N/C) machining, coordinate dimensioning has become popular. Instead of using angular dimensions to locate holes or other features, they may be dimensioned horizontally and vertically from X and Y coordinates. When this method is used, datums X and Y must be properly labeled. As with drawing 31A062, the angular dimension may still appear as a reference on the drawing.

INSTRUCTIONS: Refer to drawing 31A062 to answer the following questions.

1. What new method of dimensioning was used to locate the holes? 1. ____________
2. Are the hole locations controlled by (a) angular tolerance or (b) linear tolerance? 2. ____________
3. Why are the angular dimensions and the bolt circle dimension in parentheses? 3. ____________
4. Are holes 1 and 3 diametrically opposite? (180°) 4. ____________
5. What type of view is view A? 5. ____________
6. Is the right-side view (a) a full section or (b) a half-section? 6. ____________
7. What practice was used to show the countersunk hole in the section view even though the hole is not on the vertical CL? 7. ____________
8. Show the limits of the flange diameter. 8. ____________
9. Show the limits of the flange thickness. 9. ____________
10. What is the total tolerance on the bored hole? 10. ____________
11. What is the MMC of the bored hole? 11. ____________
12. What is the MMC of the 88.85 diameter? 12. ____________
13. What is the minimum wall thickness at the counterbore? 13. ____________
14. What is the maximum wall thickness between the bore and the 76.149 diameter? 14. ____________
15. Is the specified material (a) a carbon steel or (b) an alloy steel? 15. ____________
16. What does the note FAO abbreviate? 16. ____________
17. What is the minimum acceptable Rockwell hardness? (Include the letters.) 17. ____________
18. What is the pitch of the threads? 18. ____________
19. Are the threads coarse or fine? (Consult the table on page 263.) 19. ____________
20. How many full threads will the tapped hole contain? 20. ____________

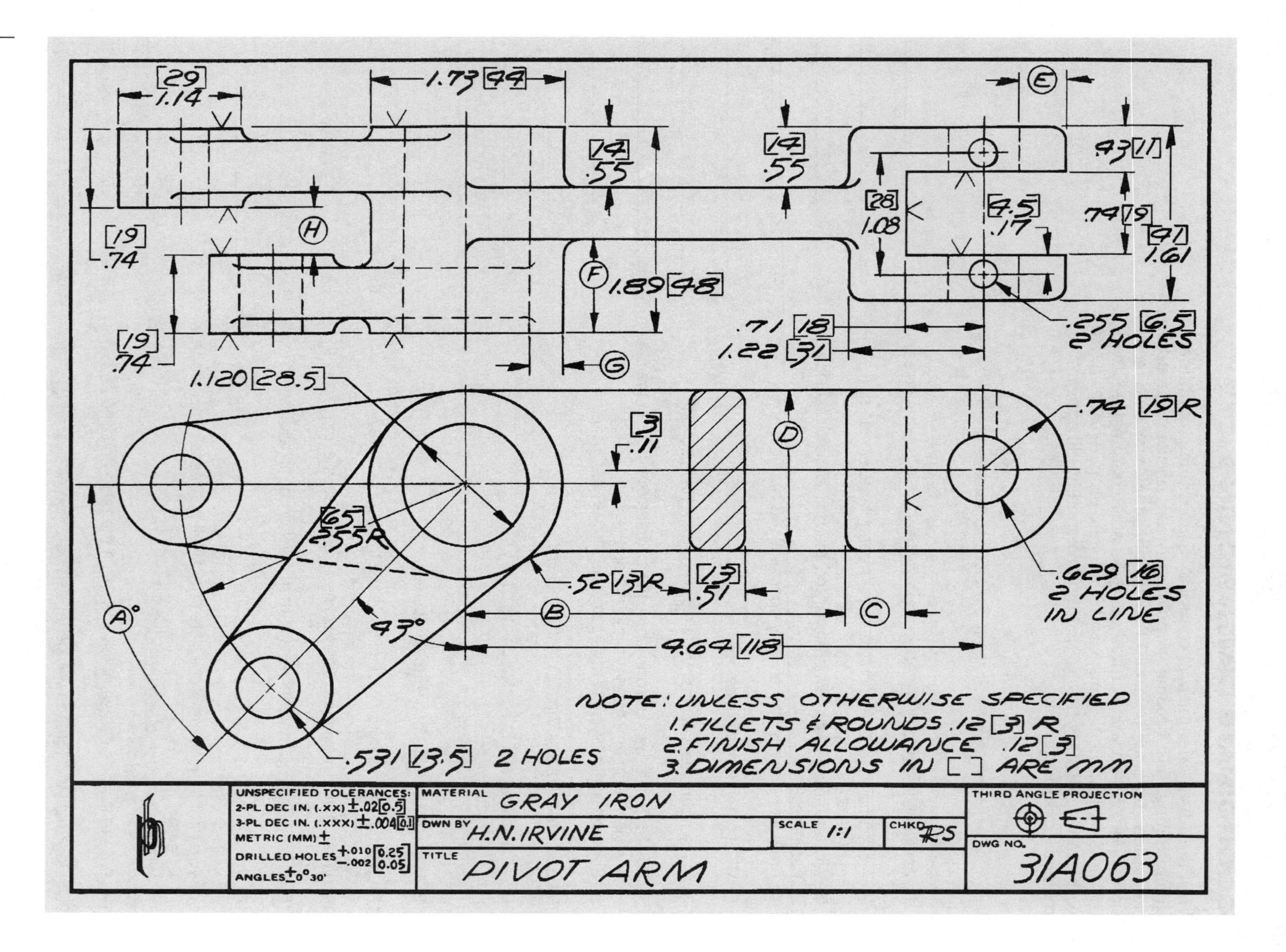

NOTE: UNLESS OTHERWISE SPECIFIED
1. FILLETS & ROUNDS .12 [3] R
2. FINISH ALLOWANCE .12 [3]
3. DIMENSIONS IN [ ] ARE mm
.531 [13.5] 2 HOLES
1.120 [28.5]
2.55R [65]
.52 [13] R
.629 [16] 2 HOLES IN LINE
.74 [19] R
.255 [6.5] 2 HOLES
4.64 [118]
1.89 [48]
1.73 [44]
1.14 [29]
.74 [19]
.55 [14]
1.08 [28]
.17 [4.5]
.43 [11]
.74 [19]
1.61 [41]
.71 [18]
1.22 [31]
.11 [3]
.51 [13]
43°
UNSPECIFIED TOLERANCES:
2-PL DEC IN. (.XX) ±.02 [0.5]
3-PL DEC IN. (.XXX) ±.004 [0.1]
METRIC (MM) ±
DRILLED HOLES +.010 -.002 [0.25] [0.05]
ANGLES ±0°30'
MATERIAL GRAY IRON
DWN BY H.N.IRVINE
SCALE 1:1
CHKD RS
TITLE PIVOT ARM
THIRD ANGLE PROJECTION
DWG NO. 31A063

DUAL DIMENSIONING

In an effort to make the change from inch dimensioning to metric dimensioning gradual, some companies chose to include both units of measurement on the drawings during the transitional period. This method was referred to as dual dimensioning. To distinguish between the two, metric dimensions were enclosed between a pair of brackets. This practice was used on drawing 31A063. Observe the example at the right showing the difference between designing to whole inches and designing to whole millimeters.

DUAL DIMENSIONING

[25.4]	vs.	[25]
1.000		.984

Inch-Designed vs. Metric-Designed
(Whole Inch) (Whole Millimeter)

INSTRUCTIONS: Refer to drawing 31A063 to answer the following questions.

1. What method of dimensioning was used? 1. ____________________
2. What type of section view is shown? 2. ____________________
3. What is the metric tolerance on 2-place decimal inch dimensions? 3. ____________________
4. How many holes are specified? 4. ____________________
5. How many flat finished surfaces does the part contain? 5. ____________________
6. How much material (metric) is provided for each machined surface? (See finish allowance note.) 6. ____________________
7. How much material (metric) remains between a 6.5 hole and the adjacent finished surface? 7. ____________________
8. How far (metric) are the 13.5 holes from the 28.5 hole? (C to C) 8. ____________________

INSTRUCTIONS: Calculate the dimensions for the following letters in the units specified. Do not convert your first answers for letters F, G, and H. Recalculate them in the units specified.

Ⓐ ____________________ °

Ⓑ ____________________ in.

Ⓒ ____________________ mm

Ⓓ ____________________ in.

Ⓔ ____________________ mm

Ⓕ ____________________ mm

Ⓕ ____________________ in. (max)

Ⓖ ____________________ mm

Ⓖ ____________________ in. (min)

Ⓗ ____________________ in.

Ⓗ ____________________ mm (max)

Ⓗ ____________________ mm (min)

DIMENSION ORIGIN

In some cases it is important to designate that a dimension between two features must originate from one of these features, but not the other. Such is the case in the example illustrated below, where a part having two parallel surfaces of unequal length is to be mounted on the shorter surface. In this example the dimension origin symbol is used to designate that the dimension originates from the shorter surface. Without the use of this symbol, the longer surface could have been selected as the origin, thereby permitting a much greater angular variation between surfaces.

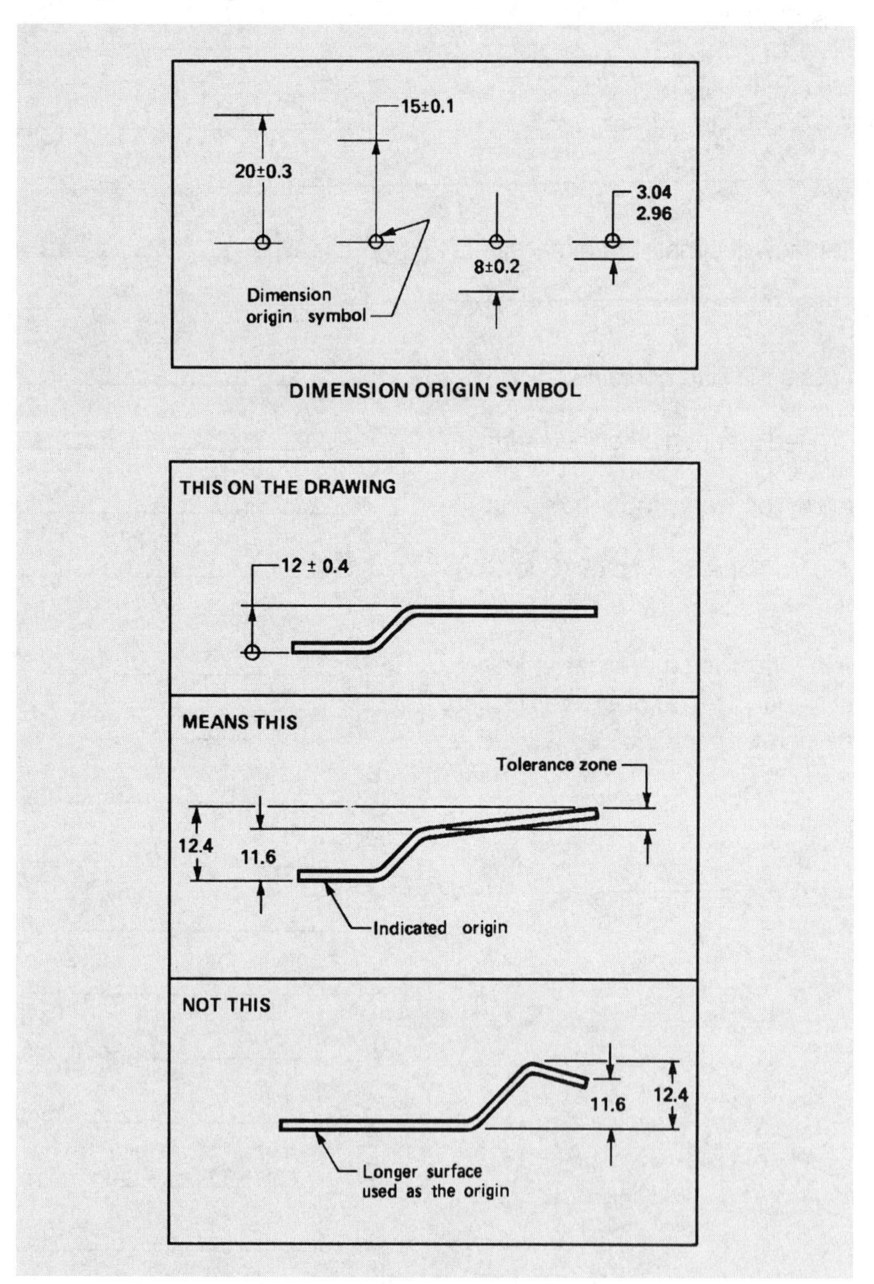

Reprinted from ASME Y14.5M–1994, by permission of The American Society of Mechanical Engineers.

Symbol Quiz 1

INSTRUCTIONS: Shown below are examples using symbols introduced to you earlier. Select the correct interpretation from the right-hand column and write it alongside each corresponding symbol.

Ø25	Arc length
R12.5	Conical taper
SØ25	Counterbore/Spotface
SR12.5	Countersink
(25)	Depth/Deep
50 (underlined)	Diameter
6X	Dimension not-to-scale
⌒75	Dimension origin
□25	Metric dimension with dual dimensioning
↧25	No. of times/places
⌴25	Radius
⌵25	Reference dimension
▷ 0.2:1	Slope
◺ 0.1:1	Spherical diameter
⌀— 25 →	Spherical radius
[25]	Square (shape)

UNIT 12

AUXILIARY VIEWS

Thus far we have concentrated our efforts on reading drawings that contain only principal views (views projected vertically and horizontally from one another). However, many objects contain inclined or oblique surfaces, and when holes or other features occur on those surfaces, an auxiliary view is required to show their true shape for dimensioning. An auxiliary view does not necessarily create an additional view on a drawing because it can often be used to replace one of the principal views. (See the illustration shown below.)

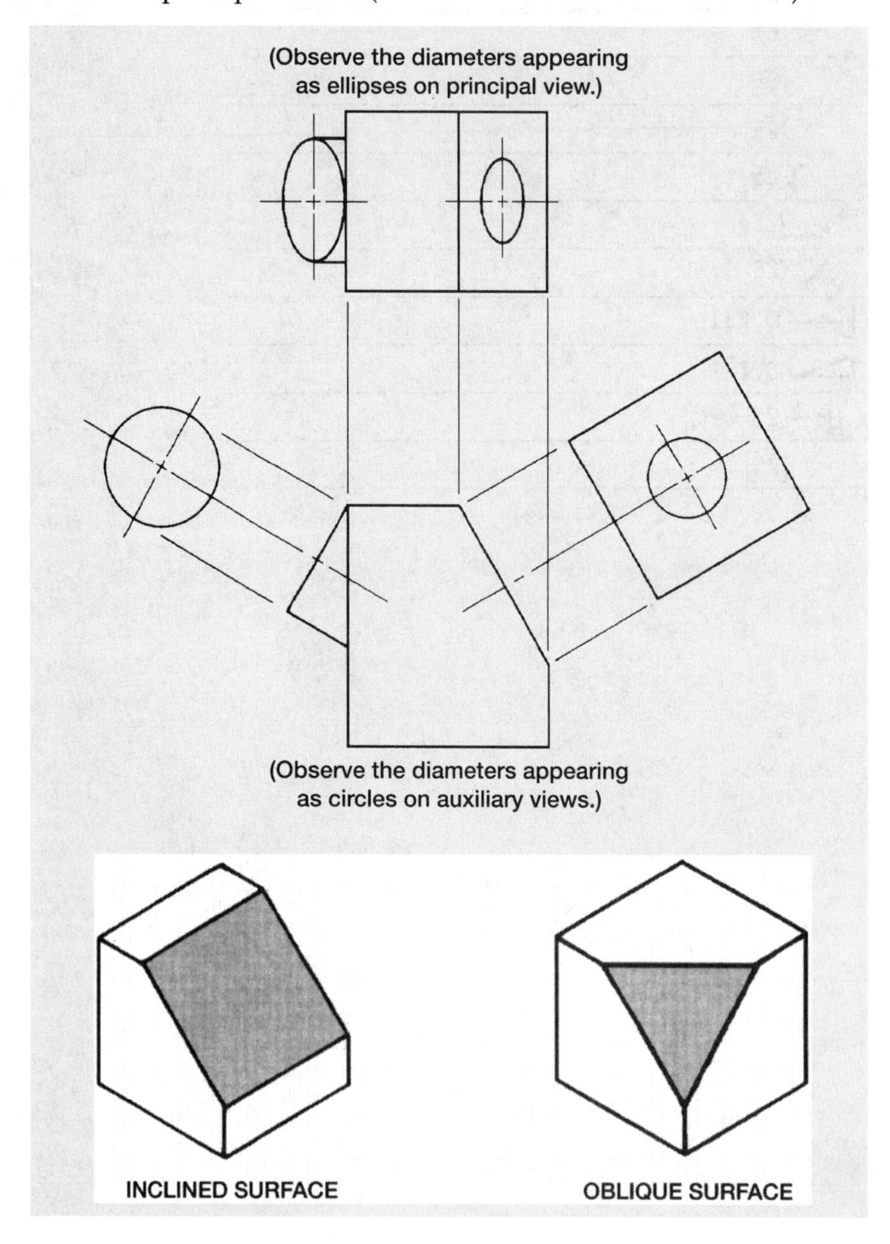

INCLINED PLANES

An inclined plane will appear as an edge in a principal view. Its true shape may be shown by projecting a view perpendicular to it. This can be accomplished by drawing a primary auxiliary view. They are often drawn as partial views, exposing only the true shape of the inclined plane. Partial views may terminate with short break lines or use existing visible lines where applicable. (See the illustrations shown below.)

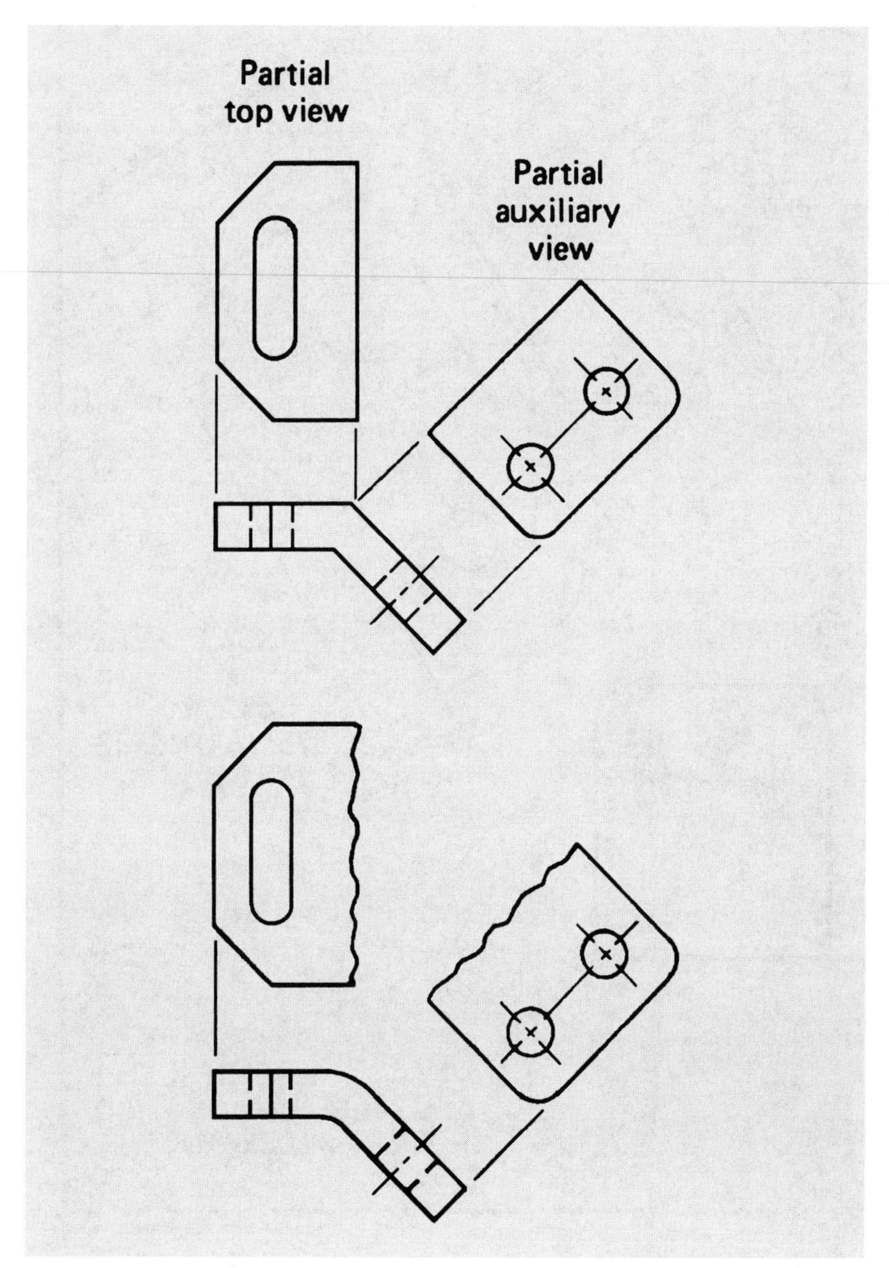

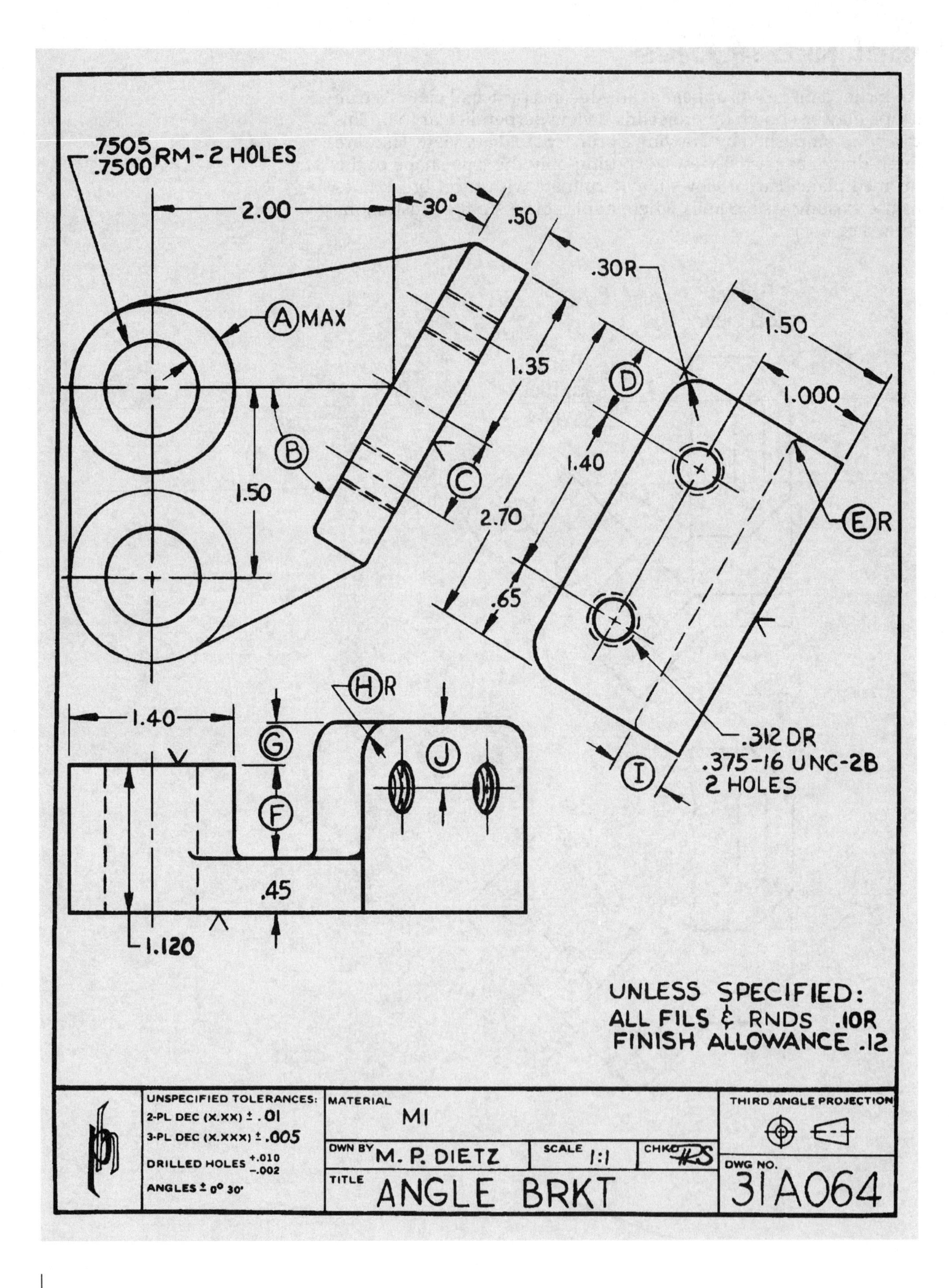

.7505
.7500 RM-2 HOLES
2.00
30°
.50
.30R
1.50
(A) MAX
1.35
(D)
1.000
(B)
1.40
(C)
1.50
2.70
(E) R
.65
(H) R
1.40
(G)
(J)
(F)
.312 DR
.375-16 UNC-2B
2 HOLES
(I)
.45
1.120
UNLESS SPECIFIED:
ALL FILS & RNDS .10R
FINISH ALLOWANCE .12
UNSPECIFIED TOLERANCES:
2-PL DEC (X.XX) ± .01
3-PL DEC (X.XXX) ± .005
DRILLED HOLES +.010 −.002
ANGLES ± 0° 30'
MATERIAL
MI
DWN BY M. P. DIETZ
SCALE 1:1
CHKD
TITLE ANGLE BRKT
THIRD ANGLE PROJECTION
DWG NO.
31A064

INSTRUCTIONS: Refer to drawing 31A064 to answer the following questions.

1. What type of iron is specified for the casting? (Do not abbreviate.) 1. ____________
2. What principal view did the auxiliary view replace? 2. ____________
3. Is the auxiliary view a partial view or is it complete? 3. ____________
4. What view shows the true shape of the reamed holes? 4. ____________
5. What view shows the true height of the bosses? 5. ____________
6. What view shows the true shape and spacing of the tapped holes? 6. ____________
7. What tap drill size is specified for the threaded holes? 7. ____________
8. How many threads does each tapped hole contain? 8. ____________
9. What is the total tolerance on the reamed holes? 9. ____________
10. What would be the overall height of the casting before machining? 10. ____________

INSTRUCTIONS: Enter the dimensions for the following letters.

Ⓐ MAX: ____________

Ⓑ ____________

Ⓒ ____________

Ⓓ ____________

Ⓔ ____________

Ⓕ ____________

Ⓖ ____________

Ⓗ ____________

Ⓘ ____________

Ⓙ ____________

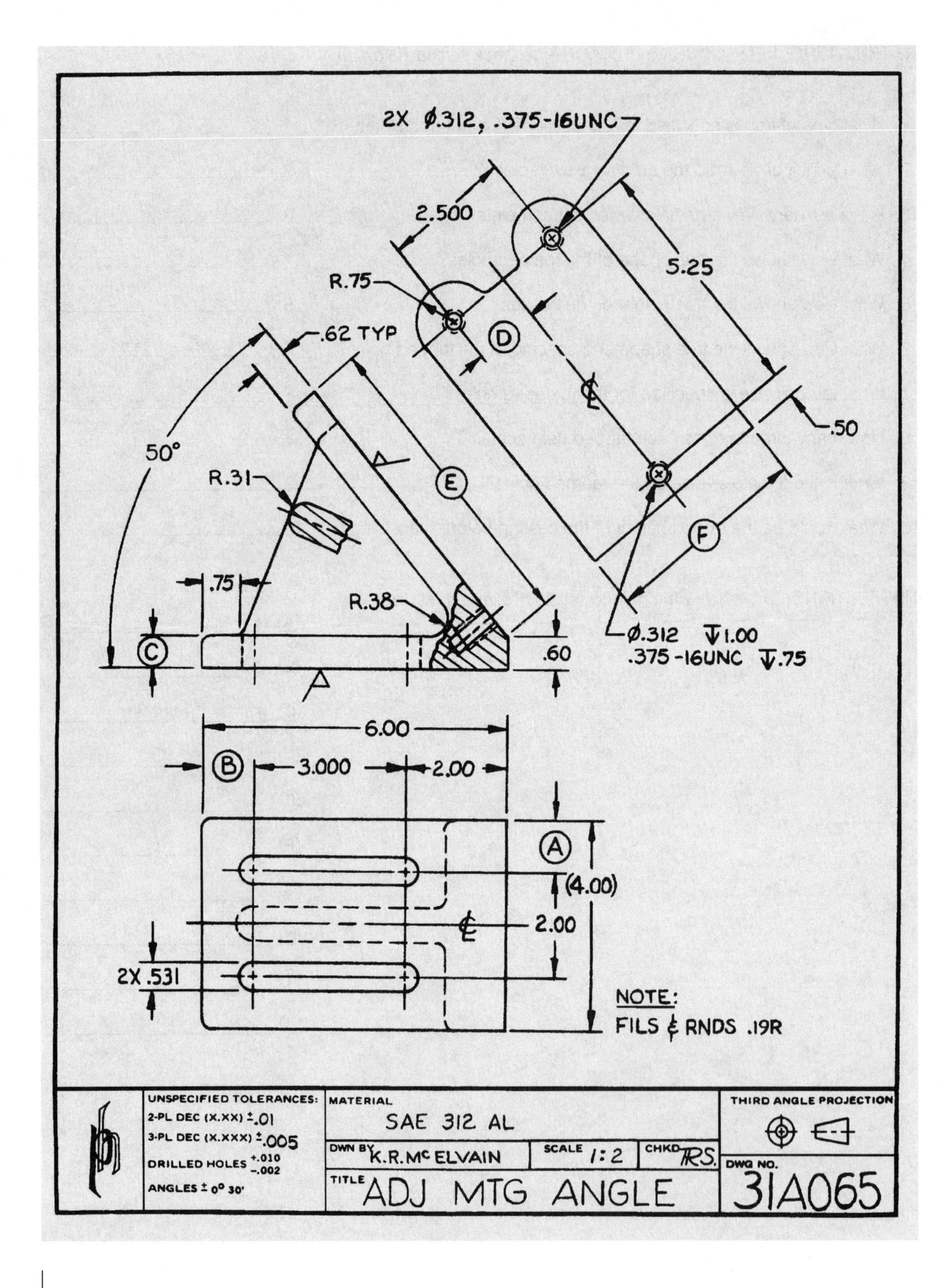
2X ∅.312, .375-16UNC
2.500
5.25
R.75
D
.62 TYP
.50
50°
R.31
E
F
.75
R.38
C
.60
∅.312 ↧1.00
.375-16UNC ↧.75
6.00
B
3.000
2.00
A
(4.00)
2.00
2X .531
NOTE:
FILS & RNDS .19R
UNSPECIFIED TOLERANCES:
2-PL DEC (X.XX) ±.01
3-PL DEC (X.XXX) ±.005
DRILLED HOLES +.010 −.002
ANGLES ± 0° 30'
MATERIAL
SAE 312 AL
DWN BY K.R.Mc ELVAIN
SCALE 1:2
CHKD RS.
TITLE ADJ MTG ANGLE
THIRD ANGLE PROJECTION
DWG NO.
31A065

INSTRUCTIONS: Refer to drawing 31A065 to answer the following questions.

1. Is the auxiliary view a complete view or a partial view? 1. ____________
2. Are the tapped holes true shape or elliptical on the auxiliary view? 2. ____________
3. Would the tapped holes be round or elliptical on a principal view? 3. ____________
4. What is the major thread diameter of the tapped holes? 4. ____________
5. What size tap drill is specified? 5. ____________
6. What thread series is specified? 6. ____________
7. What is the pitch of the threads? (Answer decimally.) 7. ____________
8. What class of fit may be assumed for the threaded holes? 8. ____________
9. How many full threads will the blind hole contain? 9. ____________
10. What type of section is used to expose the blind hole? 10. ____________
11. What type of section is used to show the radius on the rib? 11. ____________
12. How much material remains between the side of the slotted hole and the rib? 12. ____________
13. What is the overall length of the slotted hole? 13. ____________
14. What size fasteners are intended for use in the slotted holes? 14. ____________

INSTRUCTIONS: Calculate the dimensions for the following letters.

Ⓐ ____________

Ⓑ ____________

Ⓒ ____________

Ⓓ ____________

Ⓔ MAX: ____________

Ⓕ MAX: ____________

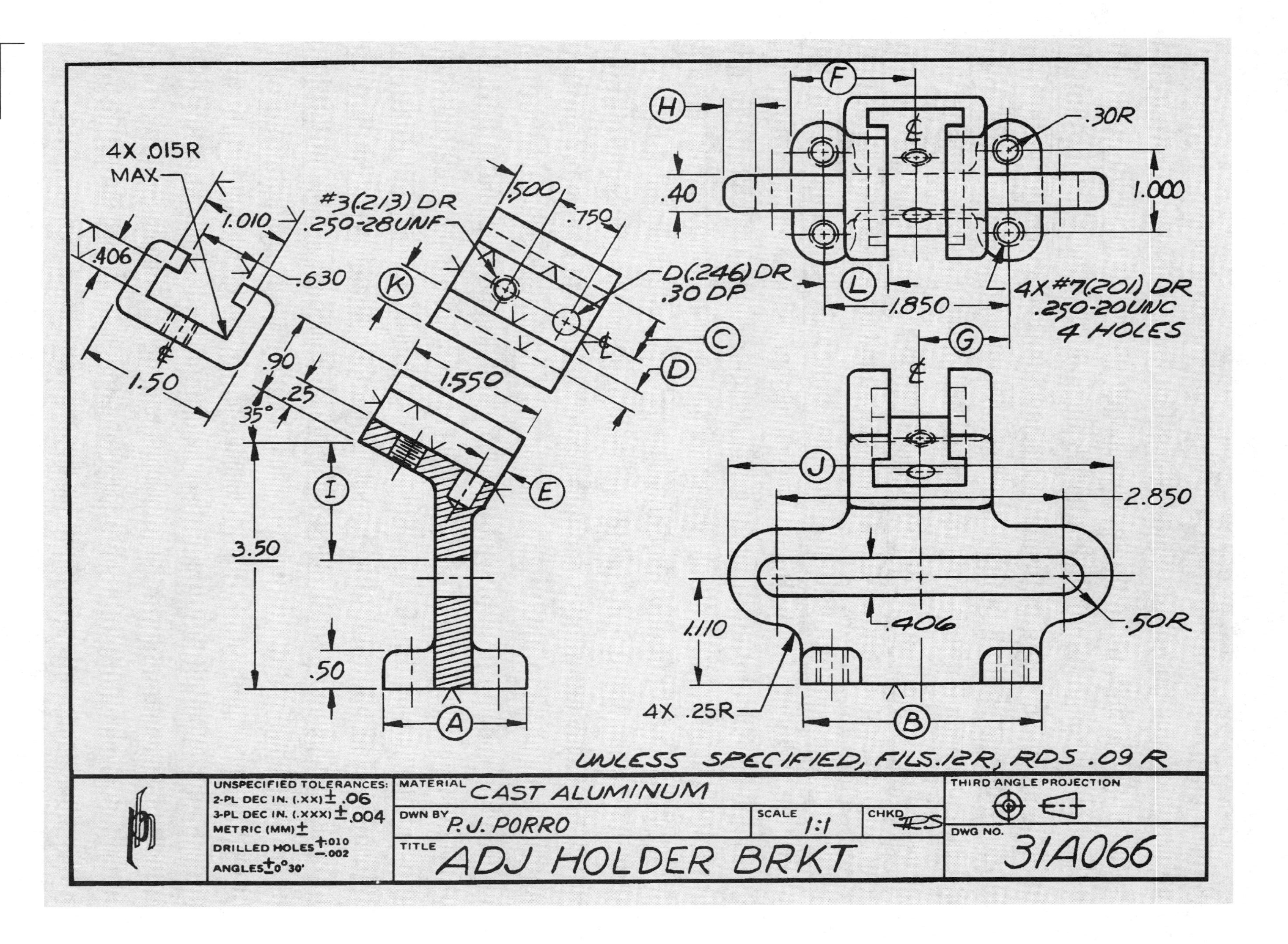

4X .015R MAX
#3(.213) DR .250-28UNF
1.010
.406
.630
.500
.750
D(.246) DR .30 DP
1.50
.90
.25
35°
1.550
3.50
.50
.40
1.000
.30R
1.850
4X #7(.201) DR .250-20UNC 4 HOLES
2.850
1.110
.406
.50R
4X .25R
A
B
C
D
E
F
G
H
I
J
K
L
UNLESS SPECIFIED, FILS.12R, RDS .09 R
UNSPECIFIED TOLERANCES:
2-PL DEC IN. (.XX) ± .06
3-PL DEC IN. (.XXX) ± .004
METRIC (MM) ±
DRILLED HOLES +.010 -.002
ANGLES ± 0°30'
MATERIAL CAST ALUMINUM
DWN BY P.J. PORRO
SCALE 1:1
CHKD
THIRD ANGLE PROJECTION
TITLE ADJ HOLDER BRKT
DWG NO. 31A066

INSTRUCTIONS: Refer to drawing 31A066 to answer the following questions.

1. What type of sectional view is the left-side view? 1. ____________
2. How many auxiliary views does the drawing contain? 2. ____________
3. What is the pitch of the threads tapped into the tee-slot? (Three decimal places.) 3. ____________
4. Do the coarse threads require a larger or a smaller size tap drill than the fine threads? 4. ____________
5. How many threads will the tapped hole in the tee-slot contain? 5. ____________
6. What is the MMC of the blind drilled hole diameter? 6. ____________
7. What is the maximum permissible length of the slotted hole? 7. ____________
8. Is the view showing the true shape of the holes in the tee-slot a *primary* auxiliary view? 8. ____________

INSTRUCTIONS: Enter the dimensions for the following letters.

Ⓐ ____________

Ⓑ ____________

Ⓒ ____________

Ⓓ ____________

Ⓔ ____________

Ⓕ ____________

Ⓖ ____________

Ⓗ ____________

Ⓘ ____________

Ⓙ MAX: ____________

Ⓚ MIN: ____________

Ⓛ MAX: ____________

OBLIQUE PLANES

Unlike an inclined surface that shows as an edge in one of the principal views, an oblique surface will never appear as an edge in any principal view. When it is necessary to show the true shape of an oblique surface, a primary auxiliary view must first be constructed to show the edge of that surface. Then, a secondary auxiliary view must be projected perpendicular to the edge, producing the true shape. This is only done when a feature, such as a hole, must be located on the oblique surface. This is the case with drawings 31A067 and 31A068 on pages 286 and 288 respectively.

Once again, principal views may be omitted if they are not needed for clarity. The auxiliary views may be partial views to simplify the drawing. Sometimes, light projection lines or continuous centerlines are shown on the drawing to help the reader visualize the relationship of views. (See the illustrations shown below and on the opposite page.)

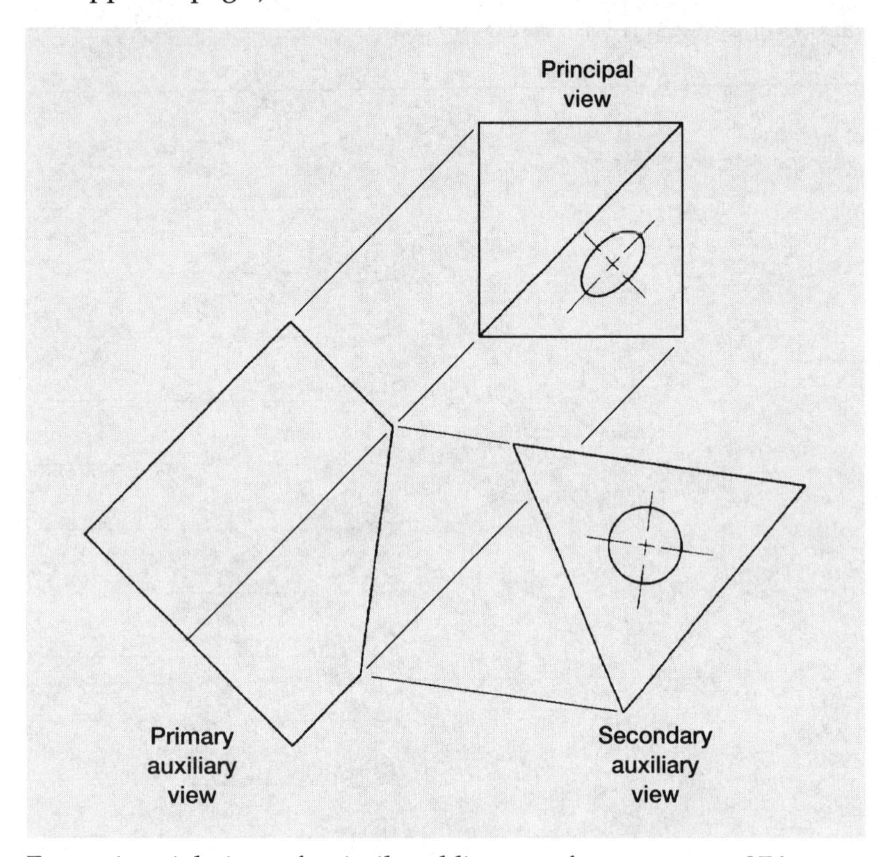

For a pictorial view of a similar oblique surface, see page 276.

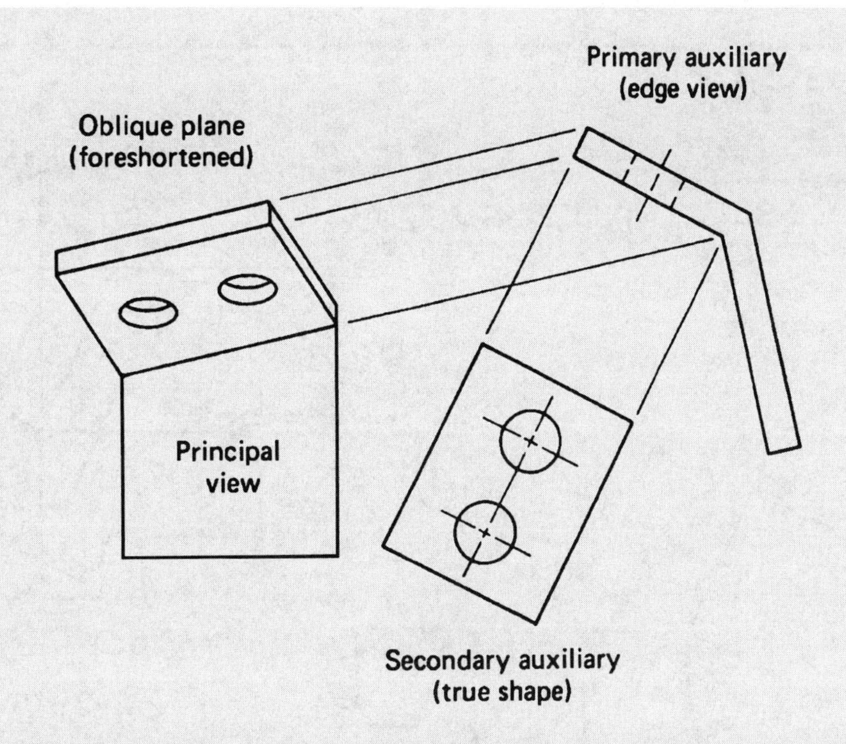
Primary auxiliary
(edge view)
Oblique plane
(foreshortened)
Principal
view
Secondary auxiliary
(true shape)

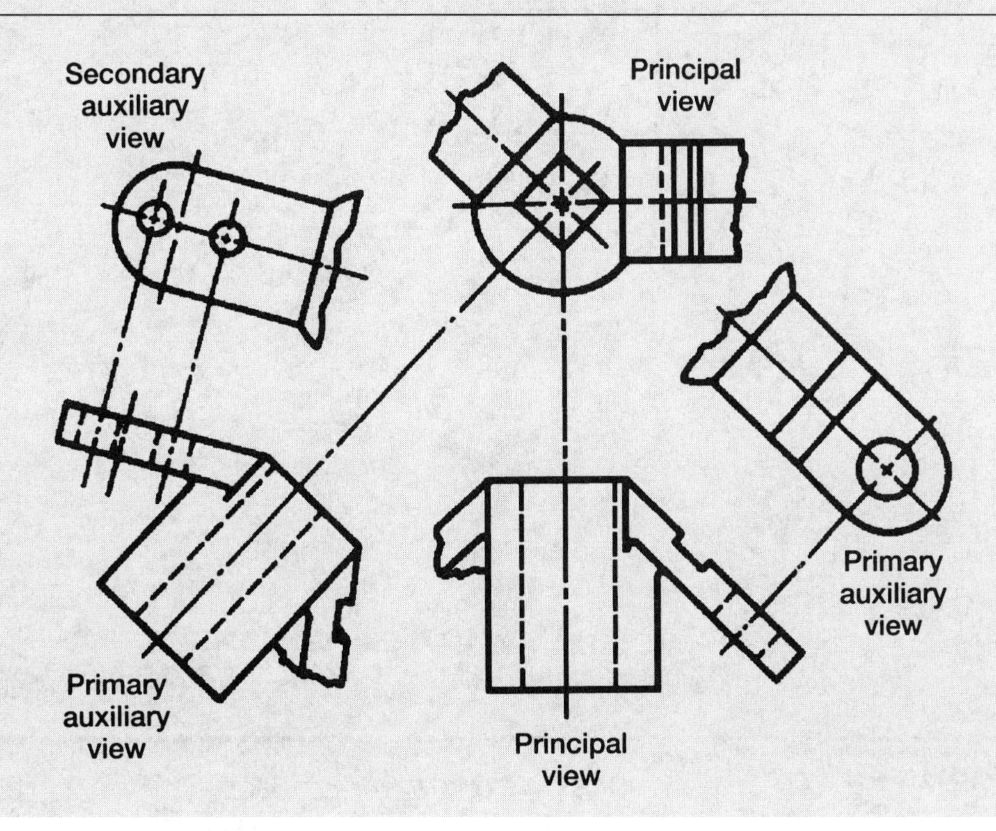
Secondary
auxiliary
view
Principal
view
Primary
auxiliary
view
Primary
auxiliary
view
Principal
view

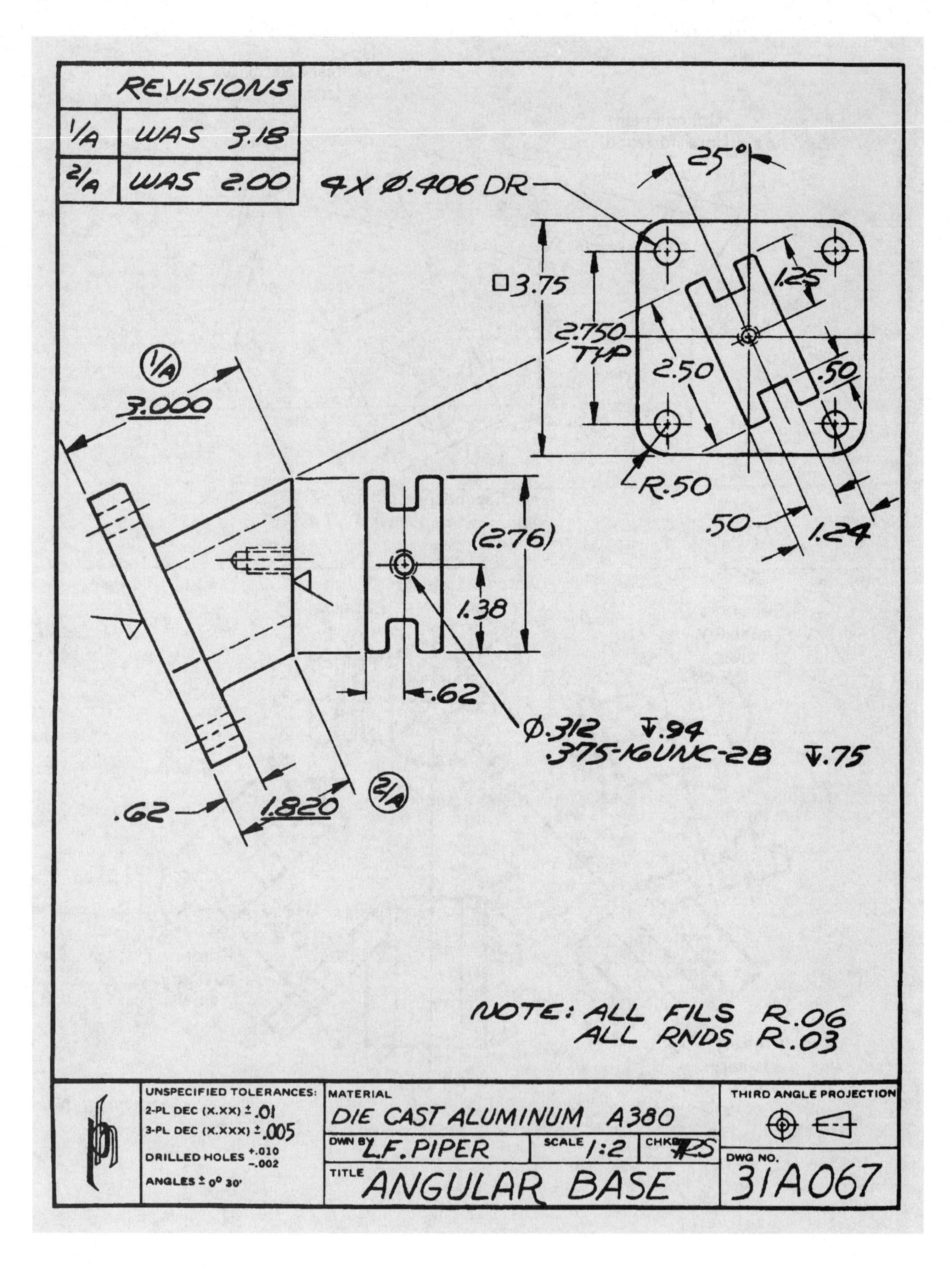
REVISIONS
1/A WAS 3.18
2/A WAS 2.00
4X Ø.406 DR
25°
□3.75
1.25
2.750 TYP
2.50
.50
R.50
.50
1.24
1/A
3.000
(2.76)
1.38
.62
Ø.312 ↧.94
.375-16UNC-2B ↧.75
.62
1.820
2/A
NOTE: ALL FILS R.06
ALL RNDS R.03
UNSPECIFIED TOLERANCES:
2-PL DEC (X.XX) ± .01
3-PL DEC (X.XXX) ± .005
DRILLED HOLES +.010 −.002
ANGLES ± 0° 30'
MATERIAL
DIE CAST ALUMINUM A380
DWN BY L.F. PIPER
SCALE 1:2
CHKBY
TITLE ANGULAR BASE
THIRD ANGLE PROJECTION
DWG NO.
31A067

INSTRUCTIONS: Refer to drawing 31A067 to answer the following questions.

1. How many principal views appear on this drawing? 1. ______
2. How many partial views appear on this drawing? 2. ______
3. What does the heavy line beneath the 3.000 dimension symbolize? 3. ______
4. What do the parentheses enclosing the 2.76 dimension symbolize? 4. ______
5. What does the box preceding the 3.75 dimension symbolize? 5. ______
6. What does the arrow preceding the .94 dimension symbolize? 6. ______
7. What size fasteners are intended for use in the four clearance holes? 7. ______
8. What is the MMC of the clearance holes? 8. ______
9. How many threads will the blind hole contain? 9. ______
10. How much deeper is the blind hole drilled than it is tapped? 10. ______
11. Is the surface with the tapped hole inclined or oblique? 11. ______
12. Is the view containing the thread specs a primary or a secondary auxiliary view? 12. ______
13. How much material remains between the side of a clearance hole and the radius around it? 13. ______
14. Did the height dimension increase or decrease as a result of the revision? 14. ______
15. Did the revision affect the size of the primary auxiliary view, the secondary auxiliary view, or both? 15. ______

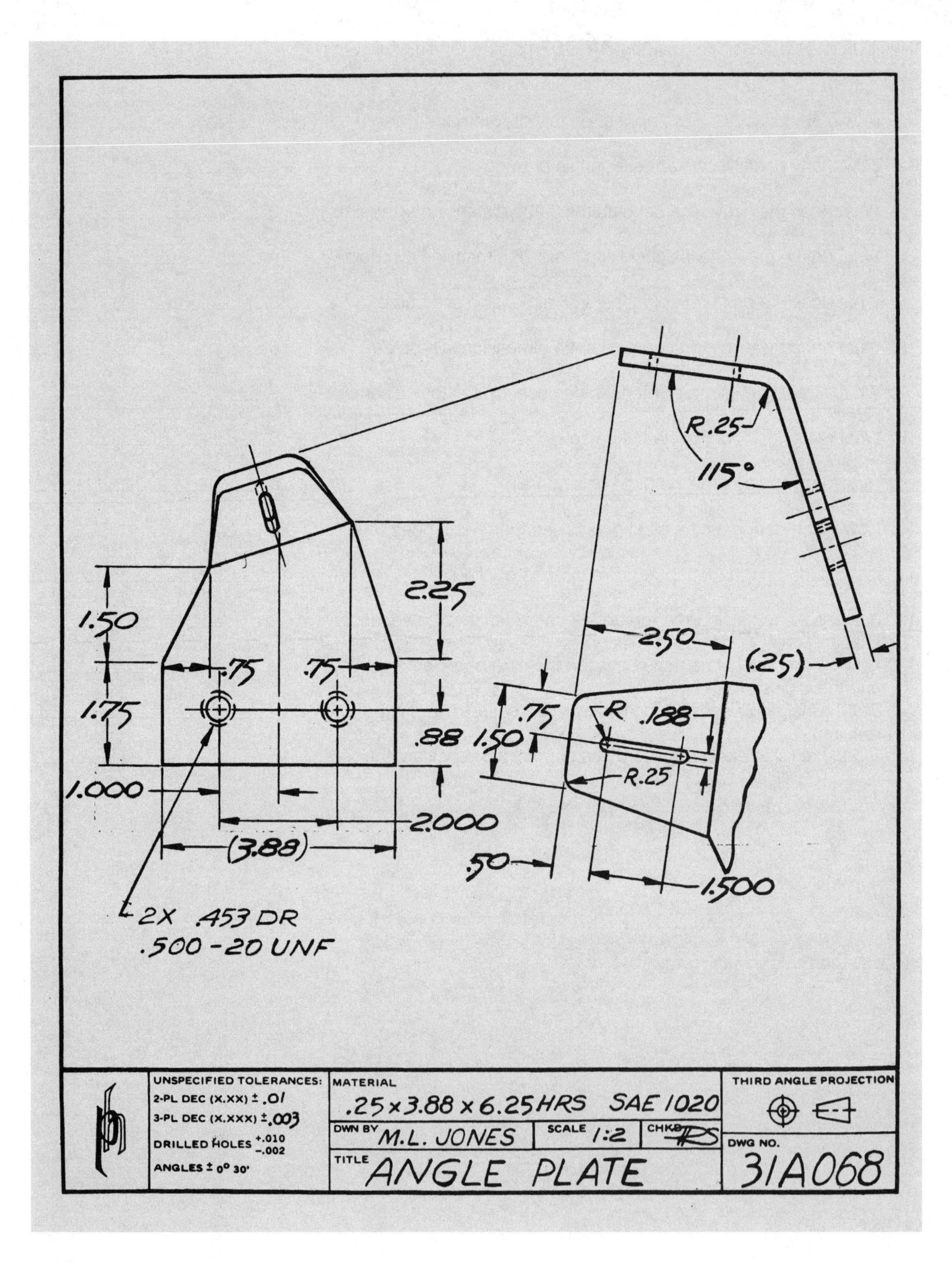

R.25
115°
2.25
1.50
1.75
.75
.75
2.50
(.25)
.75
R .188
.88 1.50
R.25
1.000
2.000
(3.88)
.50
1.500
2X .453 DR
.500 -20 UNF
UNSPECIFIED TOLERANCES:
2-PL DEC (X.XX) ± .01
3-PL DEC (X.XXX) ± .003
DRILLED HOLES +.010 −.002
ANGLES ± 0° 30'
MATERIAL
.25 x 3.88 x 6.25 HRS SAE 1020
DWN BY M.L. JONES
SCALE 1:2
CHKD
TITLE ANGLE PLATE
THIRD ANGLE PROJECTION
DWG NO.
31A068

INSTRUCTIONS: Refer to drawing 31A068 to answer the following questions.

1. How many principal views appear on this drawing? 1. ______
2. Which view (primary or secondary auxiliary) shows the bend angle? 2. ______
3. Which view (primary or secondary auxiliary) shows true shape of the slot? 3. ______
4. Which surface (tapped hole surface or slotted hole surface) is oblique? 4. ______
5. How many degrees is the angle plate bent from its flat stock position? 5. ______
6. What dimension is the inside bend radius? 6. ______
7. What would be the approximate outside bend radius? 7. ______
8. Is the material (a) low-, (b) medium-, or (c) high-carbon steel? 8. ______
9. Is the drawing (a) half-scale or (b) double-scale? 9. ______
10. Are the threads (a) left-hand or (b) right-hand? 10. ______
11. What class of fit are the threads? 11. ______
12. What thread form is specified? 12. ______
13. What thread series is specified? 13. ______
14. What is the thread pitch? (Answer decimally.) 14. ______
15. How many threads will each hole contain? 15. ______
16. What is the overall length of the slotted hole? 16. ______
17. What is the radius at the ends of the slotted hole? 17. ______
18. What do the parentheses enclosing the .25 dimension indicate? 18. ______
19. What is the minimum spacing permissible between threaded holes? (C to C) 19. ______
20. What is the maximum overall length permissible for the slotted hole? 20. ______

Optional Math Exercise:

1a. Calculate the angle of rotation between the principal surface and the oblique surface.

BEND ALLOWANCE

In sheet-metal dimensioning, allowance must be made for bends. The developed length (before bending) of the pattern equals the sum of the flat sides plus the distance around the bend measured along the neutral axis. The neutral axis is assumed to be approximately 0.44 of the metal thickness, measured from the inside surface. To calculate bend allowance (BA), use the following empirical formula:

$$BA = (.017453R + .0078T)\ N$$

where R = inside radius, T = metal thickness, and N = number of degrees of bend.

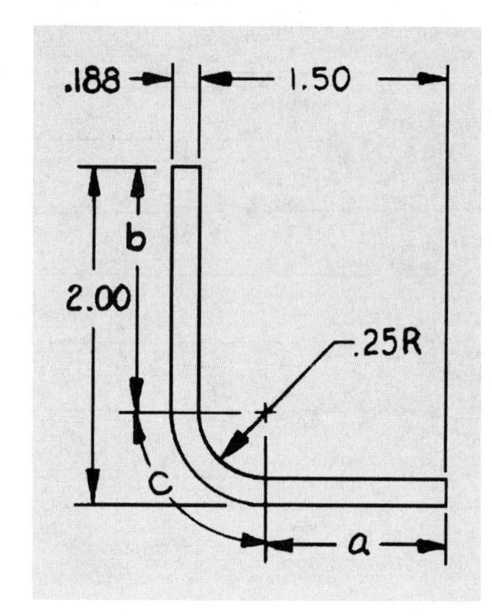

Example

$a = 1.50 - .25(R)$
$a = 1.25$

$b = 2.00 - .25(R) - .188(T)$
$b = 2.00 - .438$
$b = 1.562$

$c = (.017453R + .0078T)\ N$
$c = (.0043632 + .0014664)\ N$
$c = (.0058296)\ N$
$c = .0058296 \times 90°$
$c = .524664$

Developed blank = a + b + c
a = 1.25
b = 1.562
c = .525
DEV BLK = 3.337

Bend Allowance Calculations

INSTRUCTIONS: Calculate the developed blank for each of the shapes shown below. Use the empirical formula to determine the bend allowance. Round your answers to three decimal places.

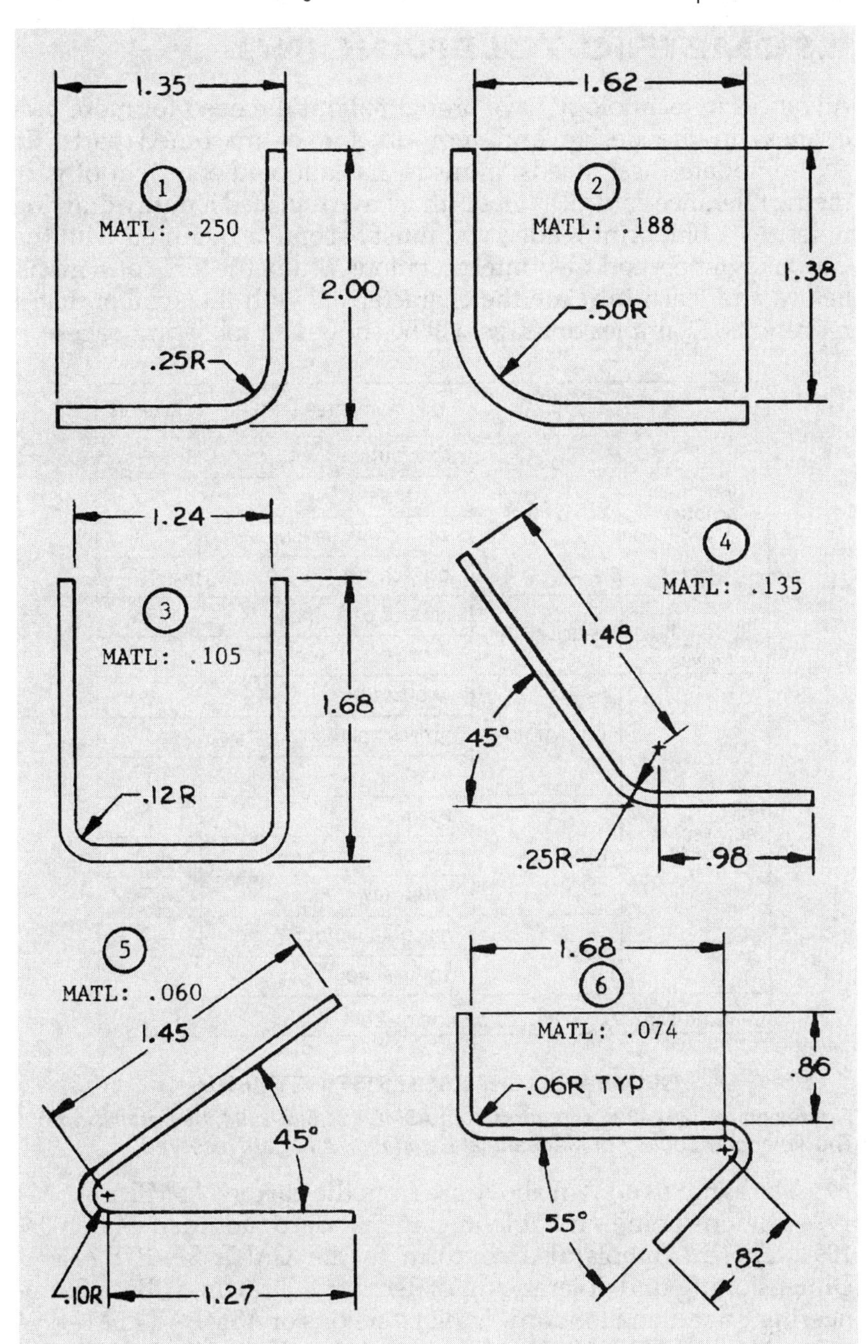

1. ______________

2. ______________

3. ______________

4. ______________

5. ______________

6. ______________

UNIT 13

GEOMETRIC TOLERANCING

Advances in technology have brought about the need for more preciseness in the design and reproduction of machined parts. To accommodate these needs, industry has adopted a system of geometric tolerance control symbols that have replaced lengthy drawing notes. As a blueprint reader, you must become acquainted with the various symbols and their interpretations. Study the table of symbols below, and learn to relate the characteristic with the symbol that it represents. Examples of usage will be shown on following pages.

	TYPE OF TOLERANCE	CHARACTERISTIC	SYMBOL
FOR INDIVIDUAL FEATURES	FORM	STRAIGHTNESS	—
		FLATNESS	⏥
		CIRCULARITY (ROUNDNESS)	○
		CYLINDRICITY	⌭
FOR INDIVIDUAL OR RELATED FEATURES	PROFILE	PROFILE OF A LINE	⌒
		PROFILE OF A SURFACE	⌓
FOR RELATED FEATURES	ORIENTATION	ANGULARITY	∠
		PERPENDICULARITY	⊥
		PARALLELISM	//
	LOCATION	POSITION	⌖
		CONCENTRICITY	◎
		SYMMETRY	⌯
	RUNOUT	CIRCULAR RUNOUT	↗•
		TOTAL RUNOUT	⌰•
• ARROWHEADS MAY BE FILLED OR NOT FILLED			

GEOMETRIC CHARACTERISTIC SYMBOLS

The symbols shown above are from the current ASME Y14.5M–1994 Dimensioning and Tolerancing Standard, adopted March 14, 1994. These symbols also conform to the CAN/CSA–B78.2–M91 Dimensioning and Tolerancing of Technical Drawings, Basic Engineering Canadian Standard. Earlier versions of ASME Y14.5M–1994 were: ANSI Y14.5M–1982, ANSI Y14.5–1973, USASI Y14.5–1966, and ASA Y14.5–1957. Other previous standards consolidated were the MIL–STD–8C and the SAE Automotive Aerospace Drawing Standards. Since many drawings in existence today were created when earlier standards were in effect, those drawings might still contain some of the old symbols. Therefore, it will be necessary for you to also become familiar with the former symbols shown near the end of Unit 14.

FEATURE CONTROL FRAME

The feature control frame is divided into separate compartments. The geometric characteristic symbol will always appear in the first compartment of the frame. The second compartment will contain the tolerance. Where applicable, the tolerance is preceded by the diameter symbol and/or followed by a material condition symbol. See the examples shown below. (*Note:* the examples in this unit are expressed in millimeters.)

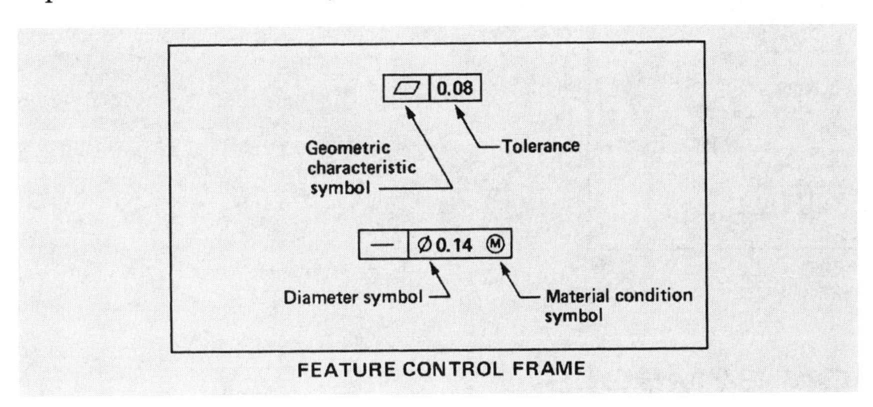

FEATURE CONTROL FRAME

Where a geometric tolerance is related to a datum, the datum reference will occupy the compartment following the tolerance, inside the feature control frame. Earlier versions had the datum reference appearing between the symbol and the tolerance. (See the example below.)

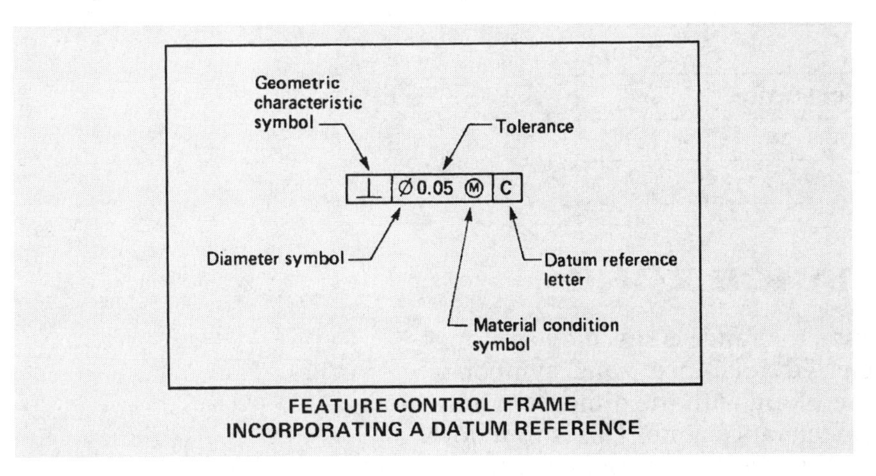

FEATURE CONTROL FRAME
INCORPORATING A DATUM REFERENCE

Where a datum is established by two datum features, such as an axis established by two datum diameters, both datum reference letters appear, separated by a dash. (See the example shown below.)

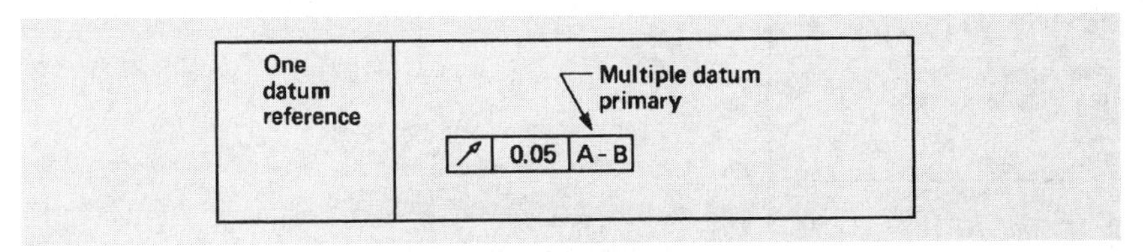

Where more than one datum is referenced, the letters (followed by a material condition symbol, where applicable) appear inside separate compartments in order of precedence, from left to right. (See the example below.)

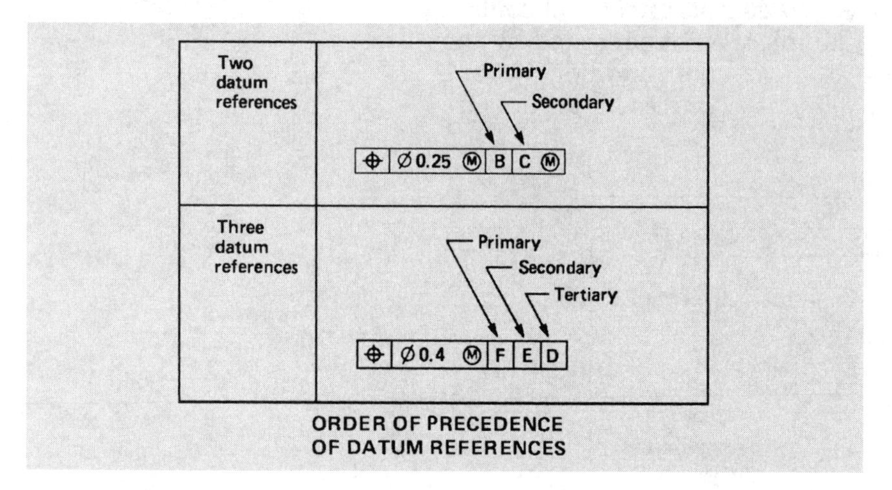

ORDER OF PRECEDENCE
OF DATUM REFERENCES

MATERIAL CONDITION SYMBOLS

The symbols used to indicate "at maximum material condition," and "at least material condition" are shown below. They may be used in the feature control frame to indicate how the tolerance or datum reference is applied. Examples on this and the preceding page include material condition symbols.

TERM	SYMBOL
AT MAXIMUM MATERIAL CONDITION	Ⓜ
AT LEAST MATERIAL CONDITION	Ⓛ
PROJECTED TOLERANCE ZONE	Ⓟ

PROJECTED TOLERANCE ZONE

Where a positional or an orientation tolerance is specified as a projected tolerance zone, the projected tolerance zone symbol is placed in the feature control frame, along with the dimension indicating the minimum height of the tolerance zone. This is to follow the stated tolerance and any modifier. (See the example below.)

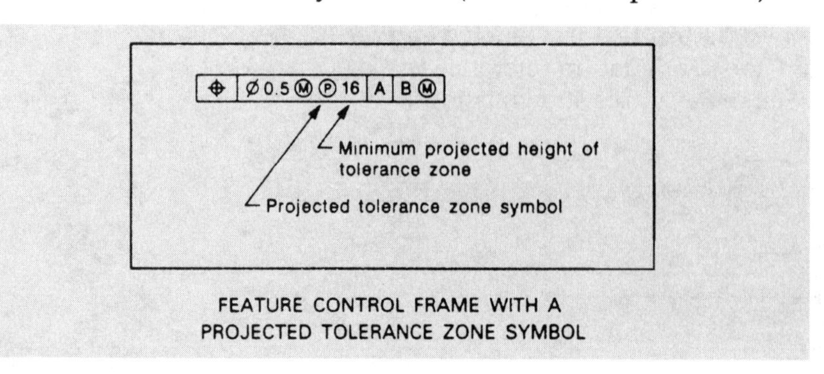

FEATURE CONTROL FRAME WITH A
PROJECTED TOLERANCE ZONE SYMBOL

BASIC DIMENSION SYMBOL

A basic dimension is a theoretically exact value, with no tolerance. It is used as the basis from which permissible variations are established by tolerances in the feature control frames. The symbol is to enclose the dimension figure in a rectangular frame, as the example shown below.

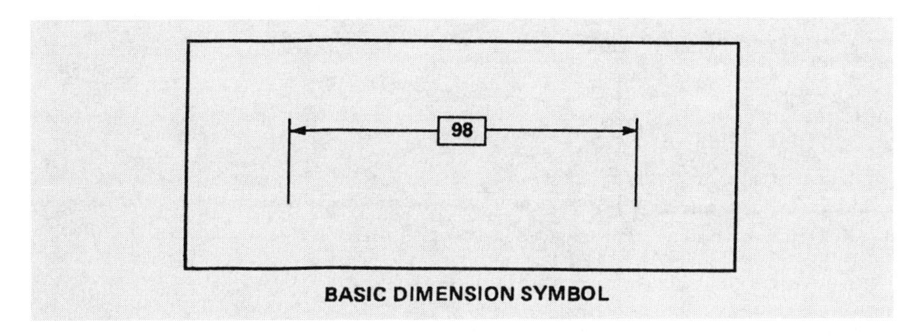

BASIC DIMENSION SYMBOL

DATUM FEATURE SYMBOL

A datum feature is a physical feature of a part used to establish a datum. The symbol consists of a square frame containing the datum identifying letter and a leader line extending from the frame to the feature, terminating with a triangle. (See the example below.)

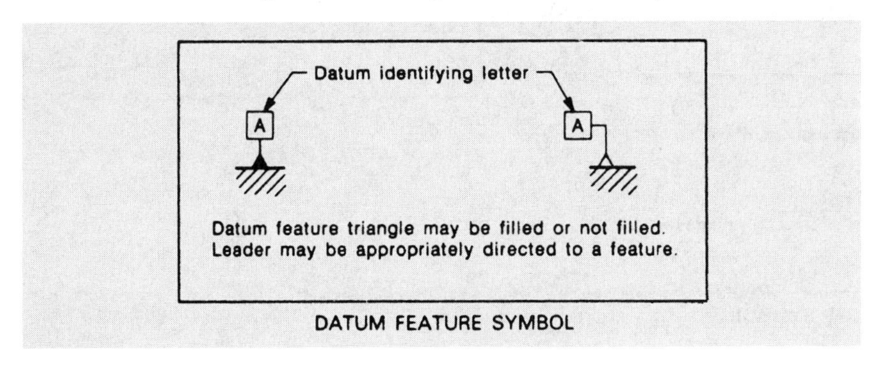

DATUM FEATURE SYMBOL

COMBINED SYMBOLS

When a feature that is controlled by a geometric tolerance also serves as a datum feature, the feature control frame and the datum feature symbol may be combined. In the example shown below, the feature is controlled for position in relation to both datum A and datum B, and is identified as datum feature C. Whenever datum C is referenced elsewhere on the drawing, the reference applies only to datum C, not to datum A or B.

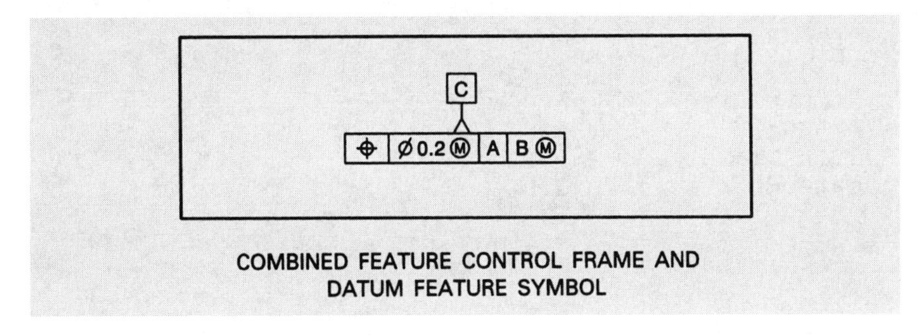

COMBINED FEATURE CONTROL FRAME AND DATUM FEATURE SYMBOL

COMPOSITE FRAMES

Where more than one tolerance is specified for the same geometric characteristic of a feature, the composite frame may be used. It will contain a single geometric characteristic symbol followed by each tolerance and datum requirement, one above the other. (See the example below.)

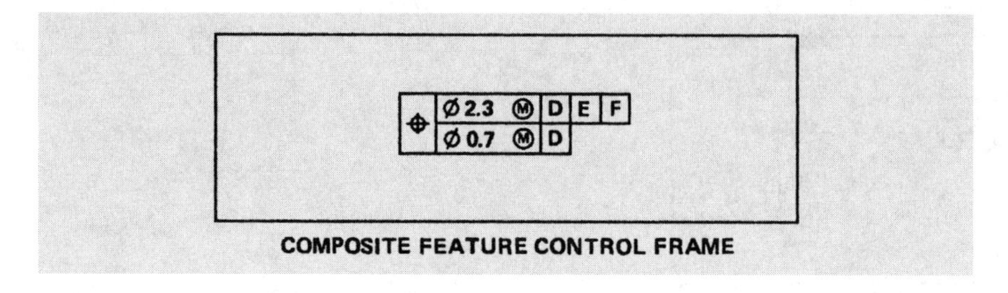

COMPOSITE FEATURE CONTROL FRAME

DATUM TARGET SYMBOL

A datum target is a specified point, line, or area used to establish datum points, lines, planes, or areas for manufacturing or inspecting purposes. The datum target symbol is a circle divided into halves. The bottom half contains the datum identification letter followed by the target number. If the datum target is an area, the size may appear in the top half of the symbol. (See the example below.)

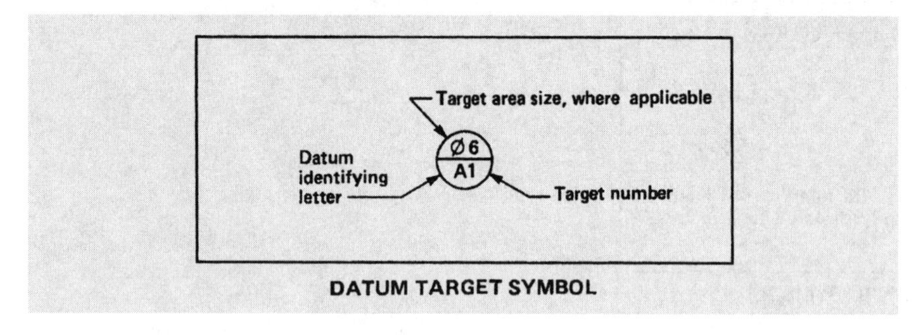

DATUM TARGET SYMBOL

ALL AROUND SYMBOL

A circle at the junction of the leader elbow is the symbol that indicates a profile tolerance applies to surfaces all around the part. (See the example below.)

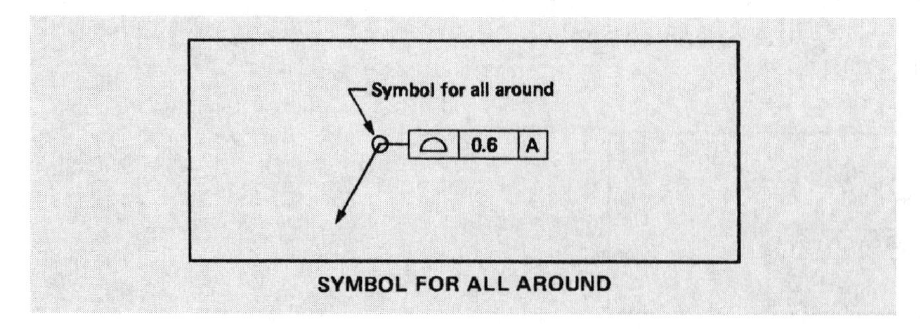

SYMBOL FOR ALL AROUND

Symbol Quiz 2

INSTRUCTIONS: Shown below are the various symbols introduced to you on the preceding pages. Without referring to those pages, try to match the symbols with the list of words, then write the correct words alongside each symbol.

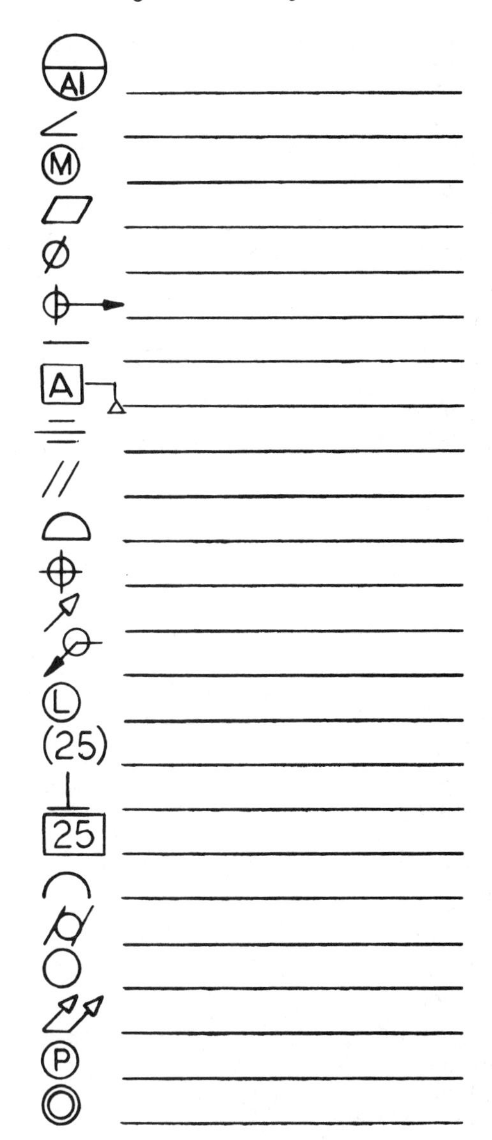

All around
Angularity
At least material condition
At maximum material condition
Basic dimension
Circular runout
Circularity
Concentricity
Cylindricity
Datum feature
Datum target
Diameter
Dimension origin
Flatness
Parallelism
Perpendicularity
Position
Profile of a line
Profile of a surface
Projected tolerance zone
Reference dimension
Straightness
Symmetry
Total runout

INSTRUCTIONS: Identify each component of the feature control frame drawn below. Write your answers on the lines provided.

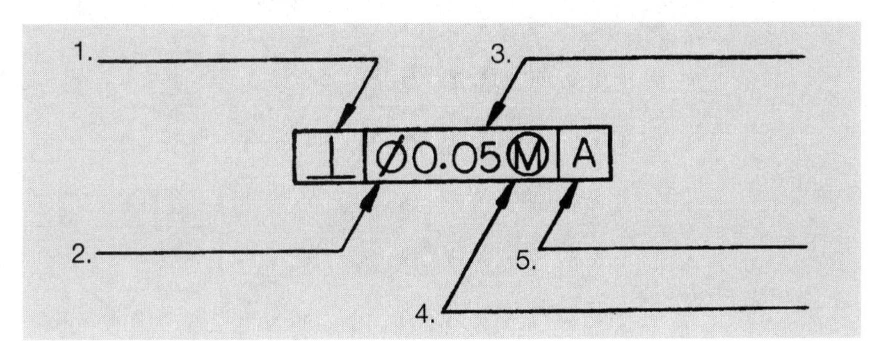

FORM TOLERANCES

The starting point for all geometric dimensioning and tolerancing (GD&T) is with Rule #1. This rule states that the form of an individual feature is controlled by its limits of size to the extent that the surfaces of a feature shall not extend beyond a boundary (envelope) of perfect form at MMC. This boundary is the true geometric form represented by the drawing. With few exceptions, no variation in form is permitted if the feature is produced at its MMC limit of size. Where the actual size of a feature has departed from MMC toward LMC, a variation in form is allowed equal to the amount of such departure. Therefore, if a feature has a large size range, it is possible for this same feature to have a large form error if only a size is given. To control the amount of form error that a feature might have, form controls are used. A form tolerance specifies a zone within which the considered feature, its line elements, its axis, or its centerplane must be contained. Where the tolerance value represents the diameter of a cylindrical zone, it is preceded by the diameter symbol. Form tolerances are applicable to individual features only. Therefore, feature control frames associated with form controls will not contain datum reference letters. All form controls applied to a surface apply RFS (regardless of feature size). Only straightness, when it is applied to a feature of size, can use the Ⓜ modifier. The four form tolerances are straightness, flatness, circularity, and cylindricity.

Straightness Tolerance

Straightness is a condition where an element of a surface or an axis is a straight line. A straightness tolerance specifies a tolerance zone. The line element or axis must lie within this zone. A straightness tolerance applies in the view where the elements to be controlled are shown as a straight line. Straightness can be applied to a surface. This results in a two dimensional tolerance zone (two parallel straight lines apart by the tolerance value). It can also be applied to a feature of size, resulting in a three dimensional tolerance zone that consists of either a cylinder (having the diameter of the specified tolerance zone), or two parallel planes apart by the tolerance value. All points of the derived median line must lie within the specified tolerance zone. Observe from the examples on page 300 how the tolerance zone increases as the feature size deviates from MMC when Ⓜ is included in the feature control frame.

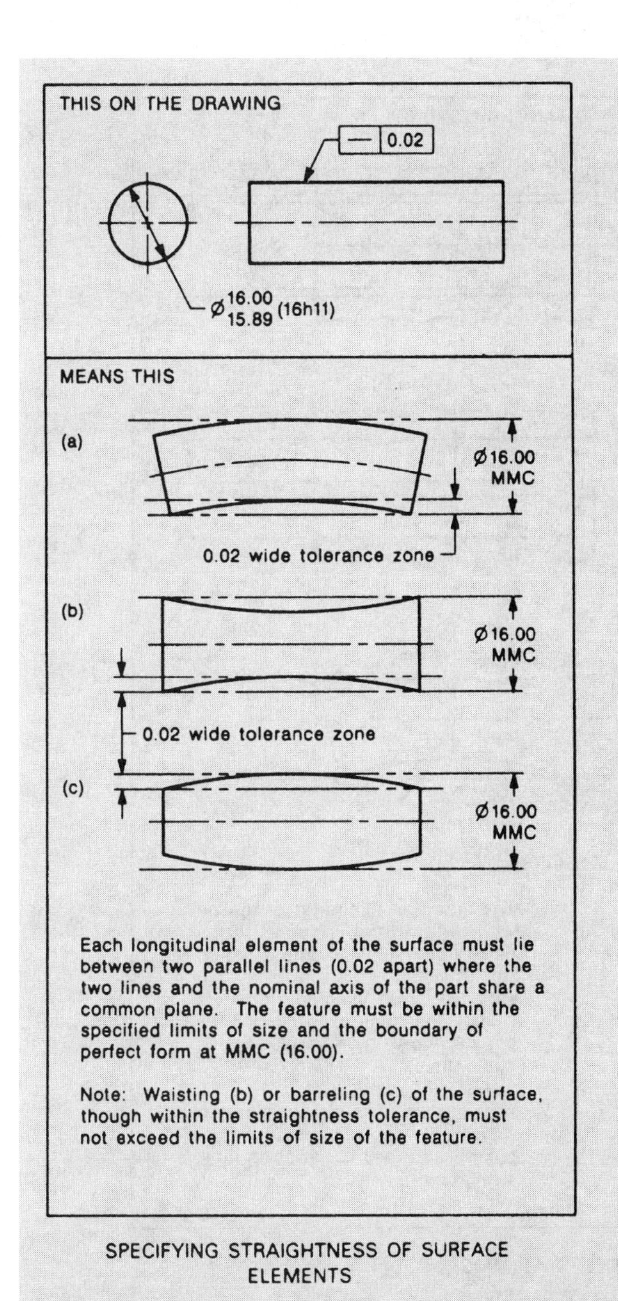

SPECIFYING STRAIGHTNESS OF SURFACE ELEMENTS

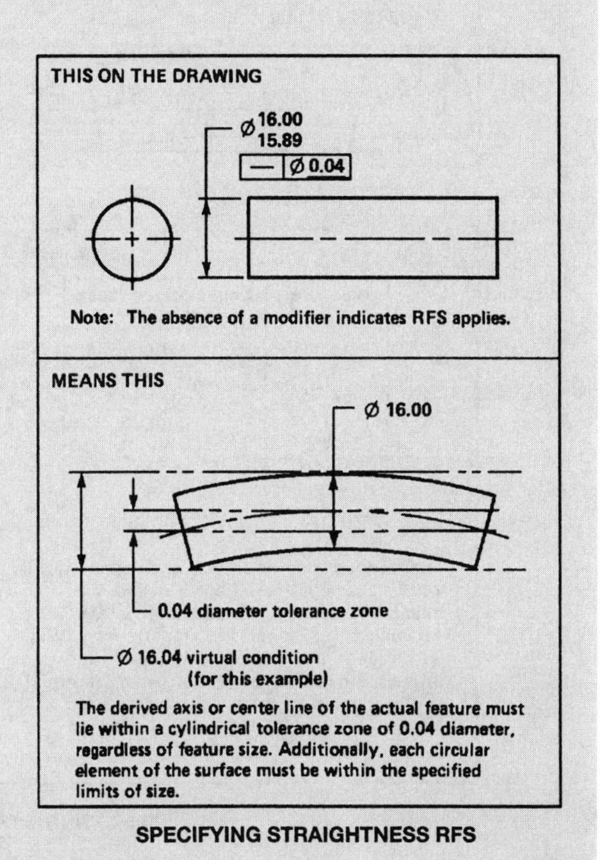

SPECIFYING STRAIGHTNESS RFS

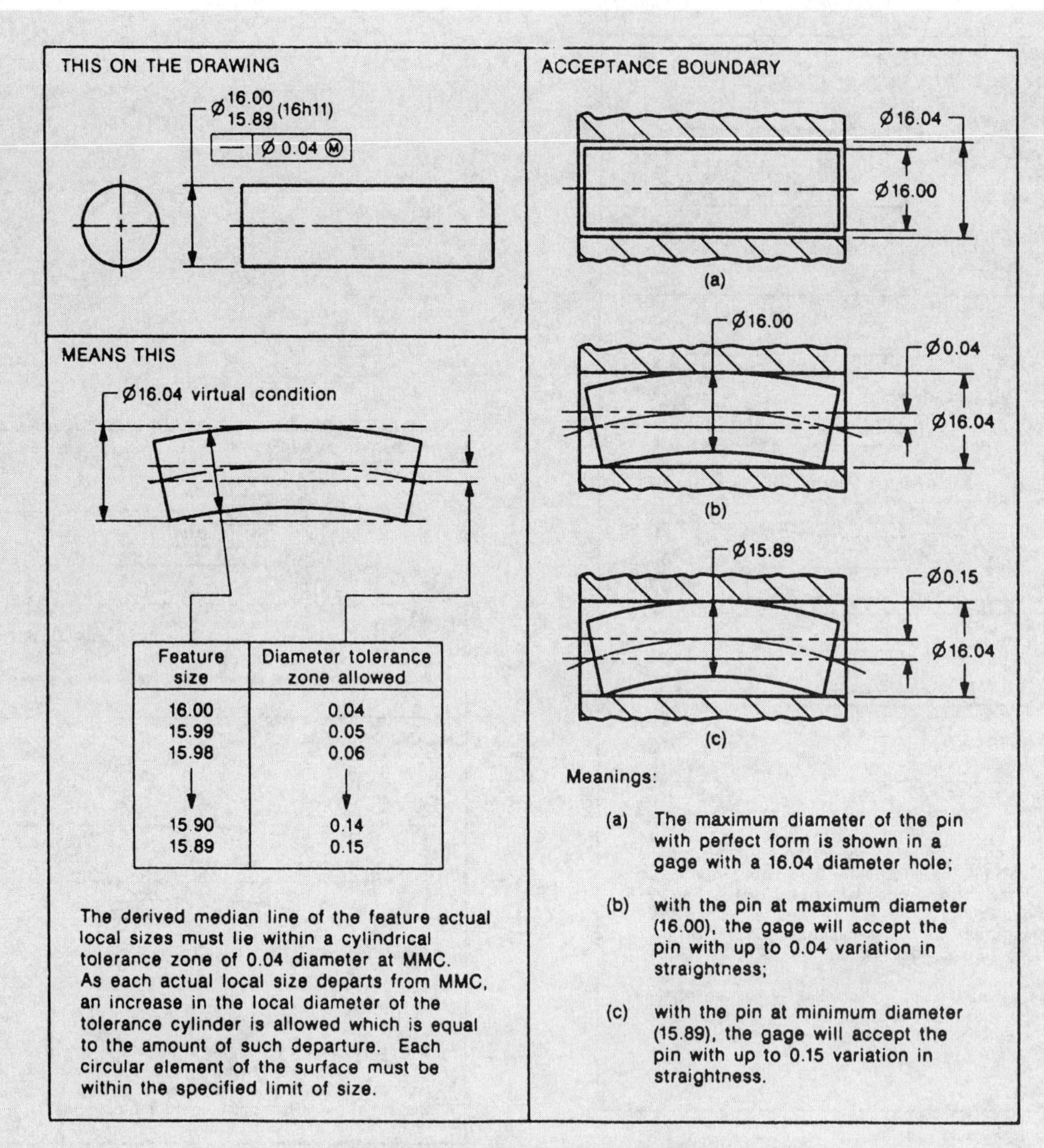

Feature size	Diameter tolerance zone allowed
16.00	0.04
15.99	0.05
15.98	0.06
↓	↓
15.90	0.14
15.89	0.15

The derived median line of the feature actual local sizes must lie within a cylindrical tolerance zone of 0.04 diameter at MMC. As each actual local size departs from MMC, an increase in the local diameter of the tolerance cylinder is allowed which is equal to the amount of such departure. Each circular element of the surface must be within the specified limit of size.

Meanings:

(a) The maximum diameter of the pin with perfect form is shown in a gage with a 16.04 diameter hole;

(b) with the pin at maximum diameter (16.00), the gage will accept the pin with up to 0.04 variation in straightness;

(c) with the pin at minimum diameter (15.89), the gage will accept the pin with up to 0.15 variation in straightness.

SPECIFYING STRAIGHTNESS AT MMC

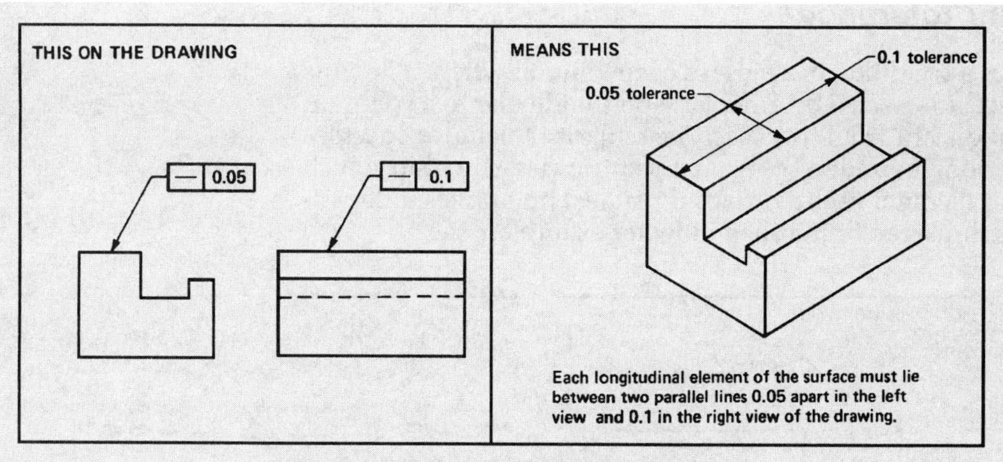

SPECIFYING STRAIGHTNESS OF FLAT SURFACES

Reprinted from ASME Y14.5M-1994, by permission of The American Society of Mechanical Engineers. All rights reserved.

Flatness Tolerance

Flatness is the condition of a surface having all elements in one plane. A flatness tolerance (which is a 3-dimensional tolerance zone) specifies a zone defined by two parallel planes within which the surface must lie. A flatness tolerance and its symbol will appear in the view where the surface elements to be controlled are represented by a line. Compare the example of flatness (3-dimensional) shown below with the application of straightness of flat surfaces (2-dimensional) shown in the example above.

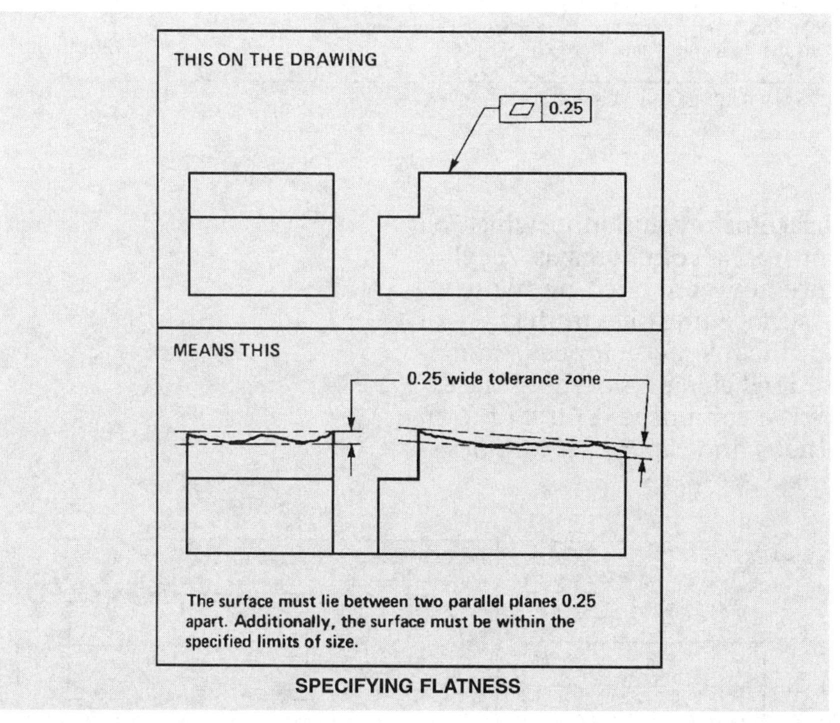

SPECIFYING FLATNESS

Reprinted from ASME Y14.5M-1994, by permission of The American Society of Mechanical Engineers. All rights reserved.

Circularity Tolerance

Circularity is a condition of a surface of revolution where all points of the surface intersected by any plane perpendicular to a common axis are equidistant from that axis. A circularity tolerance specifies a tolerance zone bounded by two concentric circles within which each circular element of the surface must lie, and applies independently at each plane. (Study the following example.)

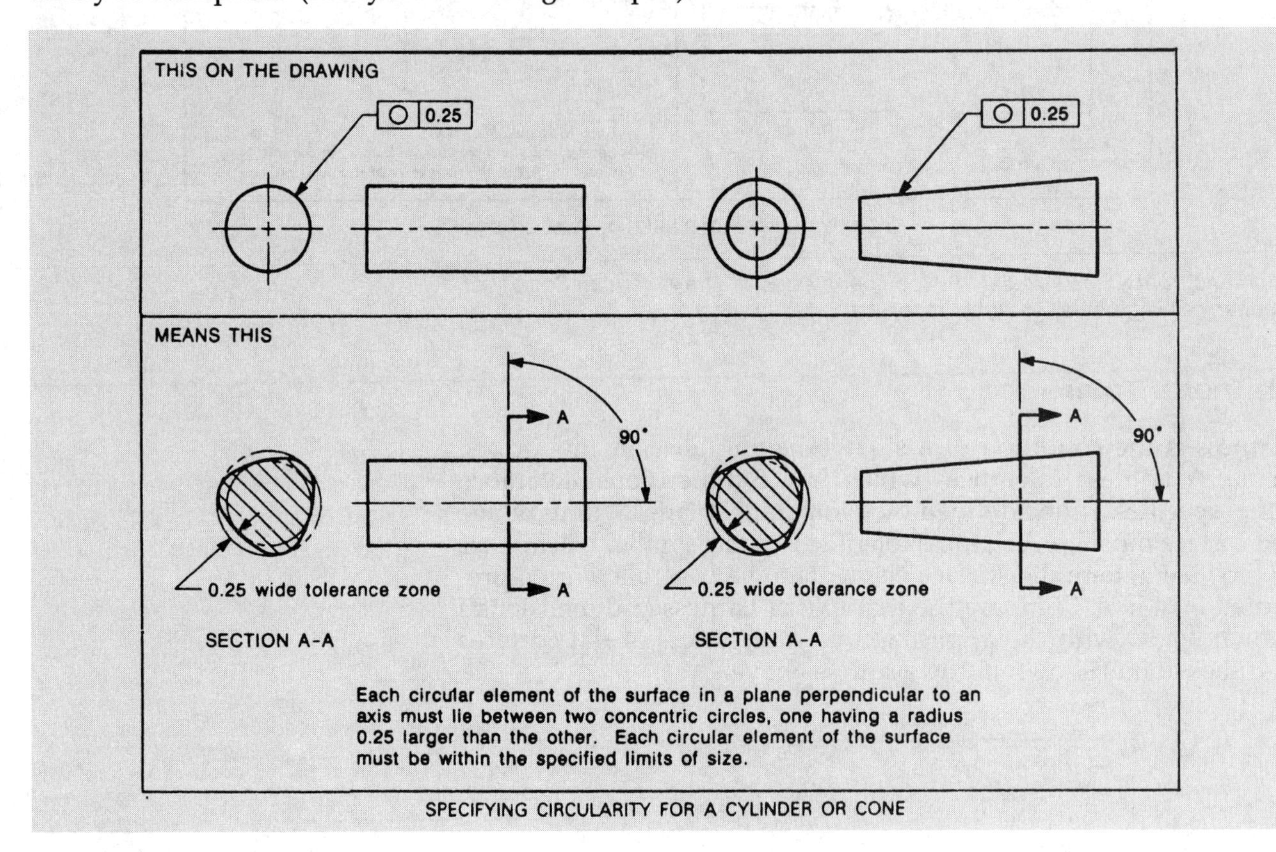

SPECIFYING CIRCULARITY FOR A CYLINDER OR CONE

Cylindricity Tolerance

Cylindricity is a condition of a surface of revolution in which all points of the surface are equidistant from a common axis. A cylindricity tolerance specifies a tolerance zone bounded by two concentric cylinders within which the surface must lie. In the case of cylindricity, unlike that of circularity, the tolerance applies simultaneously to both circular and longitudinal elements of the surface.

Note: The cylindricity tolerance is a composite control of form which includes circularity, straightness, and taper of a cylindrical feature. (Study the following example.)

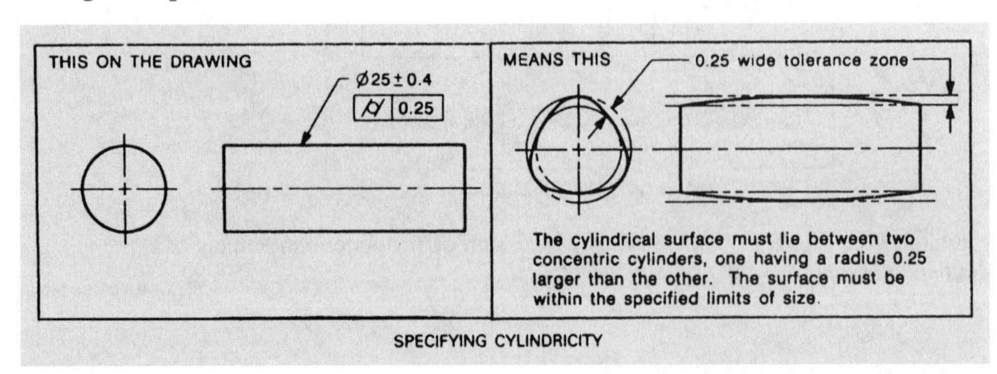

SPECIFYING CYLINDRICITY

Tolerance Calculations 3

INSTRUCTIONS: Calculate the maximum diameter tolerance zone for each of the feature sizes listed below. Refer to the example on page 300.

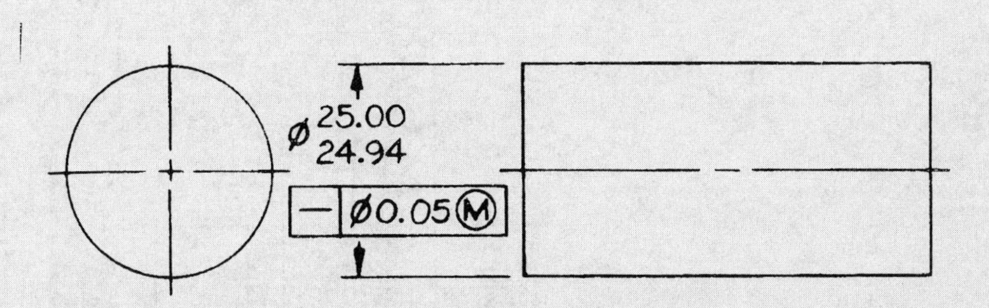

Feature Size	Diameter Tolerance Zone Allowed
25.00	
24.99	
24.98	
24.97	
24.96	
24.95	
24.94	

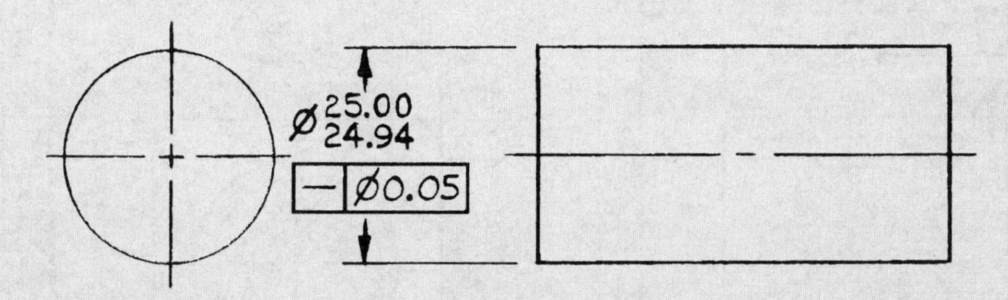

Feature Size	Diameter Tolerance Zone Allowed
25.00	
24.99	
24.98	
24.97	
24.96	
24.95	
24.94	

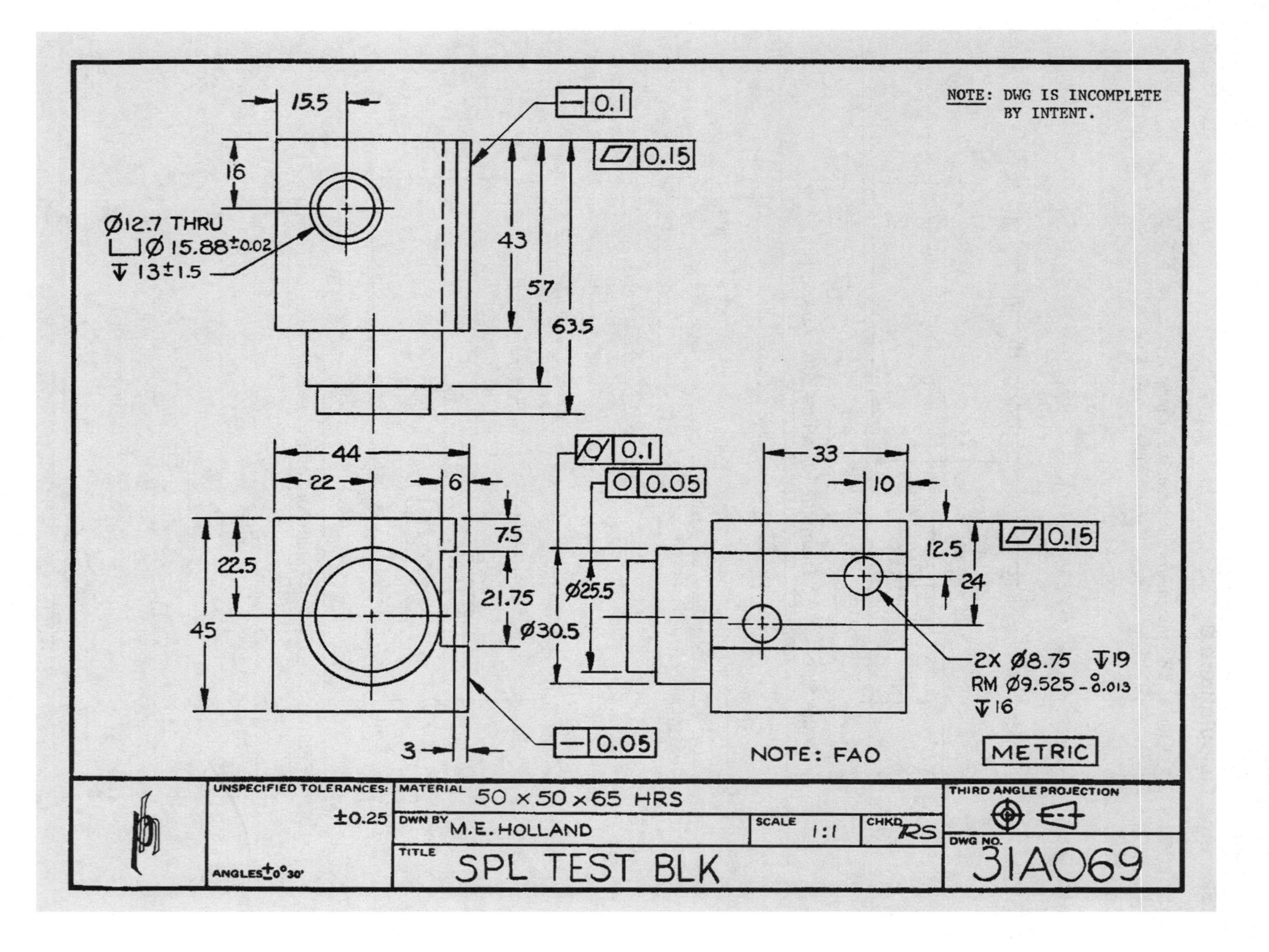
NOTE: DWG IS INCOMPLETE BY INTENT.
15.5
16
0.1
0.15
Ø12.7 THRU
⌴ Ø 15.88 ±0.02
↧ 13 ±1.5
43
57
63.5
44
22
6
0.1
0.05
33
10
7.5
22.5
21.75
Ø25.5
45
Ø30.5
12.5
0.15
24
2X Ø8.75 ↧19
RM Ø9.525 -0 -0.013
↧16
3
0.05
NOTE: FAO
METRIC
UNSPECIFIED TOLERANCES:
±0.25
ANGLES ±0°30'
MATERIAL 50 × 50 × 65 HRS
DWN BY M.E. HOLLAND
SCALE 1:1
CHKD RS
TITLE SPL TEST BLK
THIRD ANGLE PROJECTION
DWG NO. 31A069

INSTRUCTIONS: Refer to drawing 31A069 to answer the following questions.

1. How many geometric tolerances are specified? (Total) 1. ____________
2. How many *different* geometric characteristic symbols are shown? 2. ____________
3. Interpret the geometric characteristic symbol appearing on the front view. 3. ____________
4. Interpret the geometric characteristic symbol appearing on the large diameter. 4. ____________
5. Interpret the geometric characteristic symbol appearing on the small diameter. 5. ____________
6. Interpret the geometric characteristic symbol appearing on the top surface of the block. 6. ____________
7. How much geometric tolerance is applied to cylindricity? 7. ____________
8. How much geometric tolerance is applied to flatness? 8. ____________
9. Do the geometric tolerances apply at (a) MMC, (b) LMC, or (c) RFS (regardless of feature size)? 9. ____________
10. Are the geometric tolerances (a) form, (b) orientation, or (c) location? 10. ____________
11. Are the dimensions in the top view arranged by (a) chain, (b) broken-chain, or (c) datum method? 11. ____________
12. How far does the 25.5 diameter extend beyond the end of the 30.5 diameter? 12. ____________
13. What is the tolerance on the counterbore diameter? 13. ____________
14. What is the tolerance on the diameter of the reamed holes? 14. ____________
15. Will either of the blind holes intersect with the counterbored hole? (Calculate.) 15. ____________
16. How much tolerance accumulates (C to C) between the blind holes? 16. ____________
17. With the tolerances assigned, can either of the blind holes overlap the side of the slot? (Calculate.) 17. ____________
18. With the tolerances assigned, can the large diameter ever extend above the bottom of the slot? (Calculate.) 18. ____________
19. How much material is to be removed from the length of the blank? (Raw material size) 19. ____________
20. What does the note FAO abbreviate? 20. ____________

NOTE: Students may return to this drawing (after completing Unit 14) to add the proper positional tolerance controls for the (2) reamed holes and the counterbored hole.

PROFILE TOLERANCES

The profile tolerance specifies a uniform boundary along the true profile within which the elements of the surface must lie. It can be used to control the form, location, orientation, and size of a part feature, either singly or in a combination. Profile of a line or surface can be used to tolerance simple or complex shapes.

The true profile should be defined with basic dimensions, basic angular dimensions, and/or basic radii. The tolerance is assigned either bilaterally (equal or unequal) to both sides of the true profile, or unilaterally to either one side or the other. For unilateral tolerances, a short phantom line is drawn parallel to the true profile to indicate if the tolerance zone boundary is inside or outside of the true profile. Where an equally disposed bilateral tolerance zone is intended, no phantom lines are drawn. Observe the examples on the following pages.

Profile tolerances are applicable to individual features, or to related features that require datum reference letters. Profile tolerance must always be applied RFS. As far as the feature being toleranced, however, the datum reference can be applied at RFS or at MMC.

Profile of a Line

The tolerance zone established by the profile of a line tolerance is *two dimensional,* extending along the length of the considered feature. This applies to the profiles of parts having a varying cross section (such as the tapered wing of an aircraft), or to random cross sections of parts (as in the following illustration, where it is not desired to control the entire surface of the feature as a single entity).

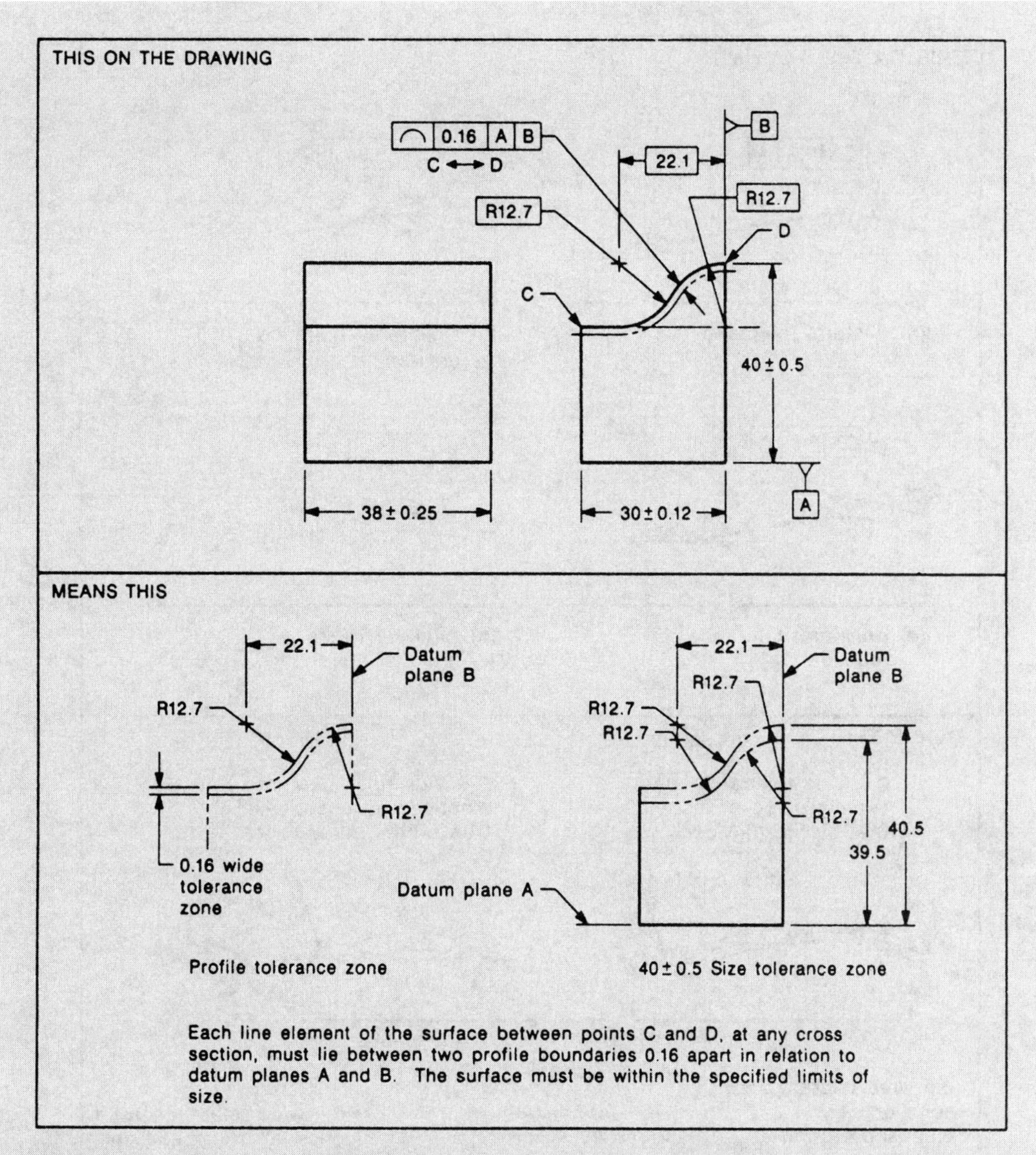

PROFILE OF A LINE AND SIZE CONTROL

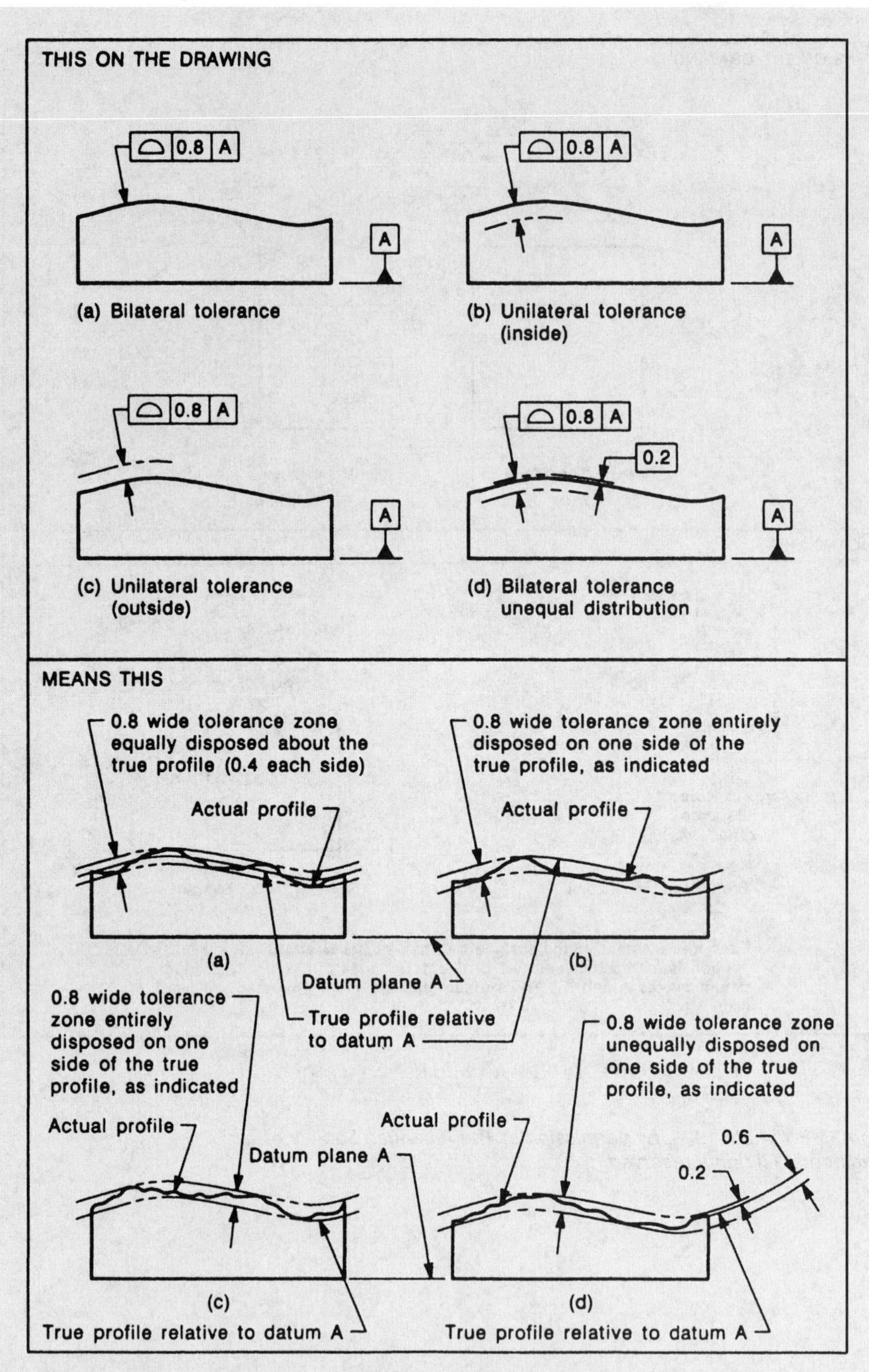

APPLICATION OF PROFILE OF A SURFACE TOLERANCE TO A BASIC CONTOUR

Profile of a Surface

The tolerance zone established by the profile of a surface tolerance is *three dimensional,* extending along the length and width of the considered feature or features. This applies to parts having a constant cross section, to parts having a surface of revolution, or to parts defined by profile tolerances applying "ALL OVER" indicated below the feature control frame, such as castings. Where a profile tolerance applies all around the profile of a part, the symbol used to designate "all around" will appear on the leader from the feature control frame. (See the example below.)

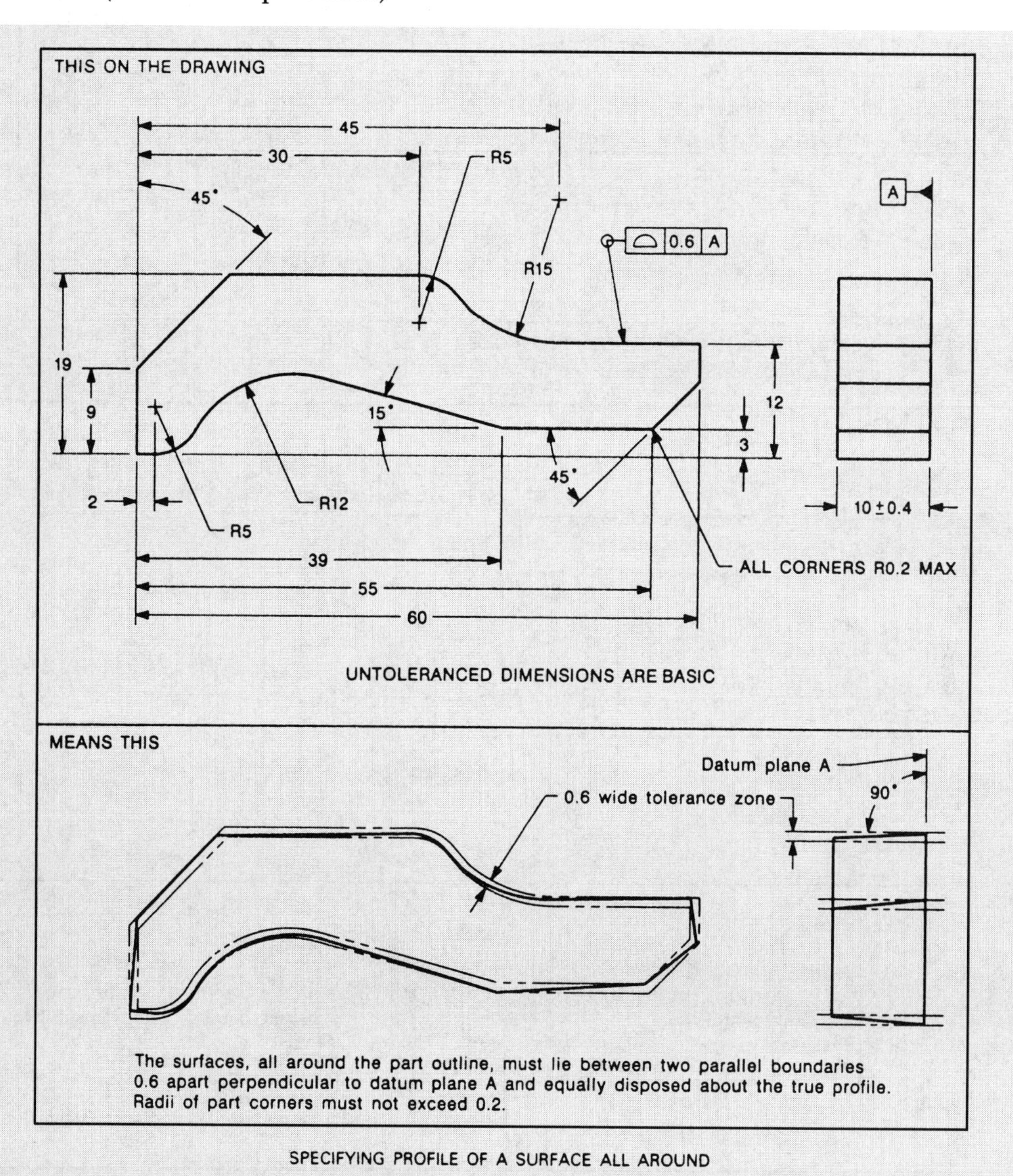

SPECIFYING PROFILE OF A SURFACE ALL AROUND

Reprinted from ASME Y14.5M-1994, by permission of The American Society of Mechanical Engineers. All rights reserved.

Using the "between" symbol underneath the feature control frame with the profile tolerancing control enables the designer to place different amounts of profile control to different segments of the same part. (See the example below.)

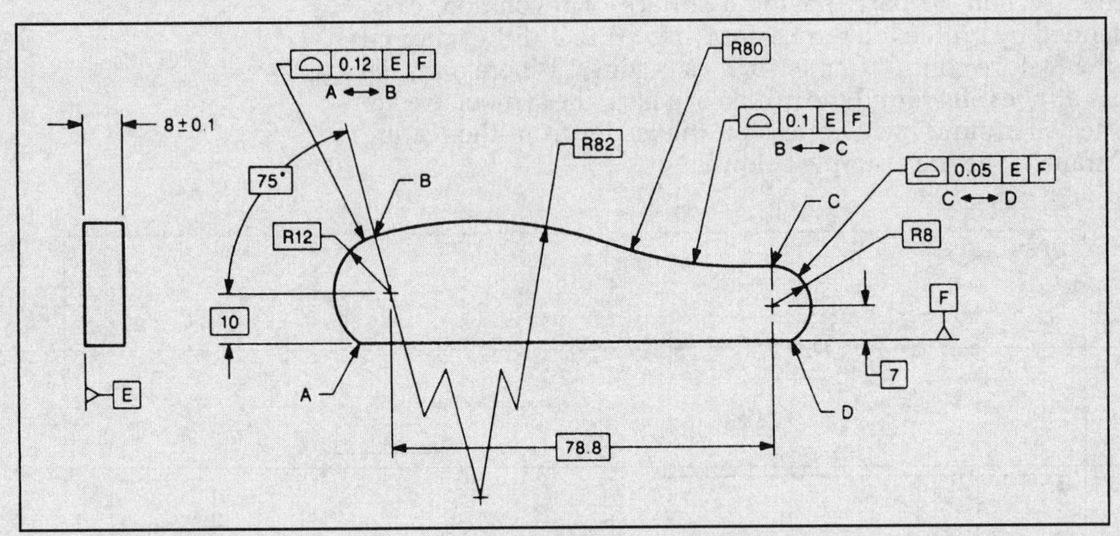

SPECIFYING DIFFERENT PROFILE TOLERANCES ON SEGMENTS OF A PROFILE

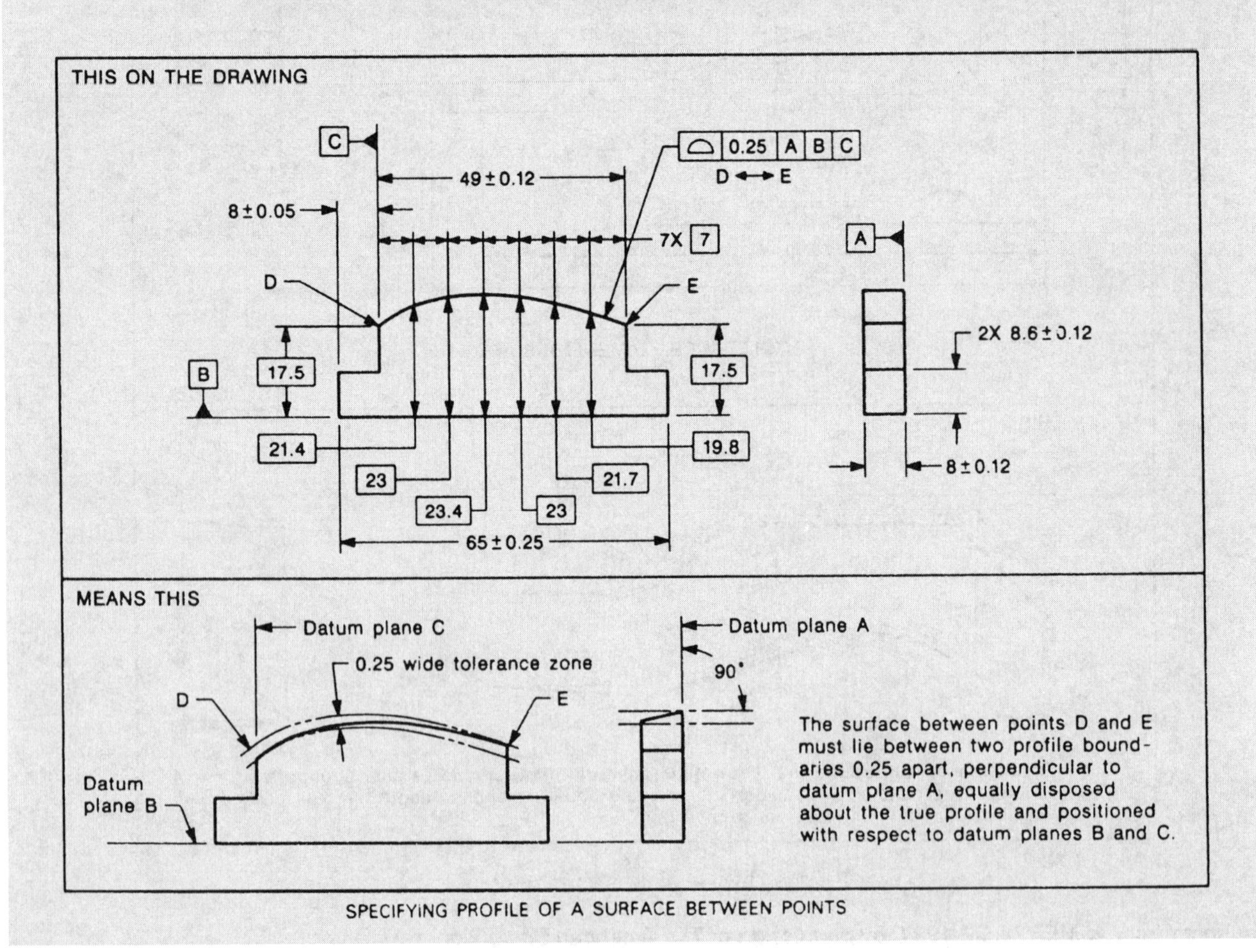

SPECIFYING PROFILE OF A SURFACE BETWEEN POINTS

Figures above reprinted from ASME, Y14.5M-1994, by permission of The American Society of Mechanical Engineers.

Profile of a surface control may be combined with other controls in order to provide specific types of control for a given feature. (See the example below.)

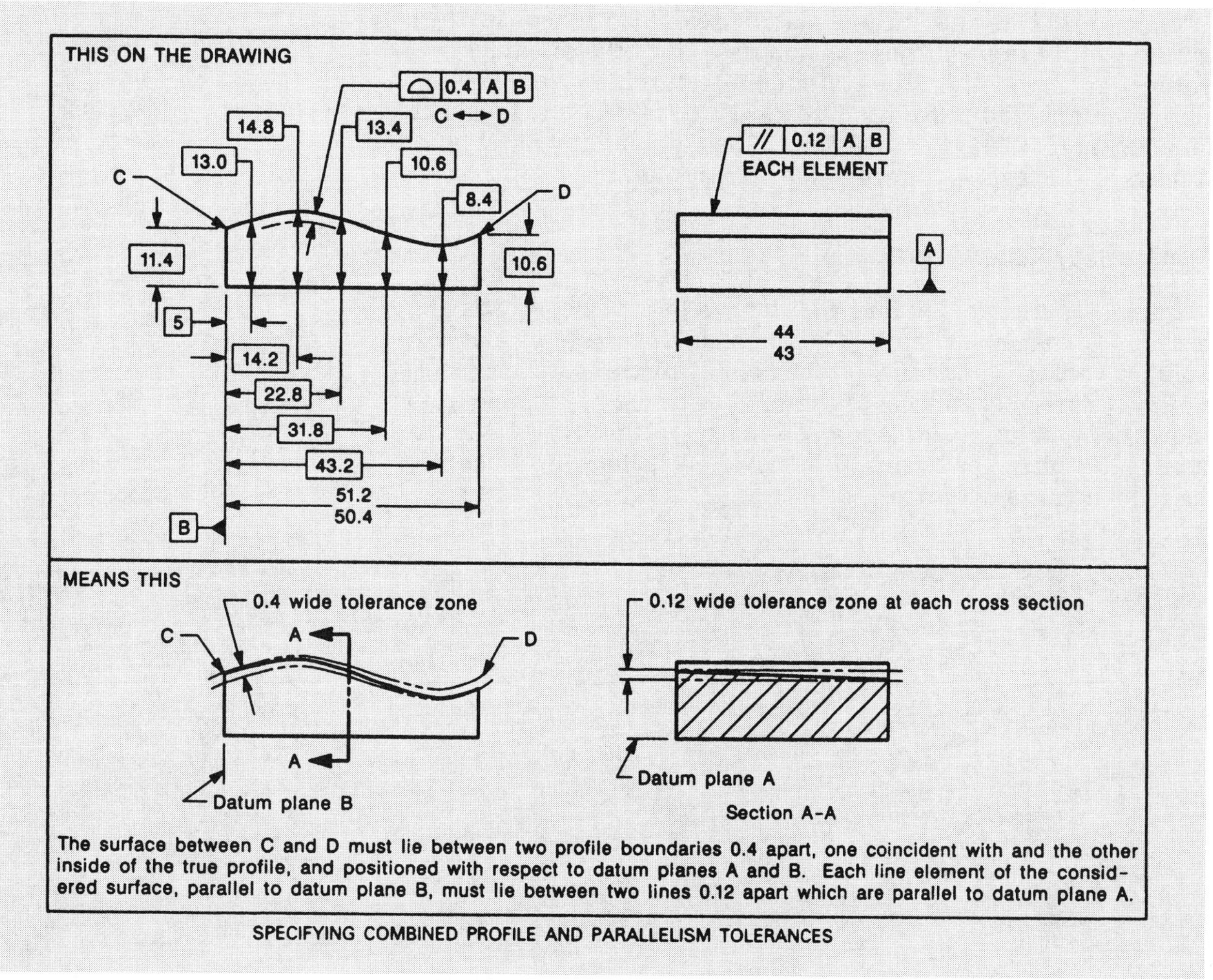

The surface between C and D must lie between two profile boundaries 0.4 apart, one coincident with and the other inside of the true profile, and positioned with respect to datum planes A and B. Each line element of the considered surface, parallel to datum plane B, must lie between two lines 0.12 apart which are parallel to datum plane A.

SPECIFYING COMBINED PROFILE AND PARALLELISM TOLERANCES

Reprinted from ASME Y14.5M-1994, by permission of The American Society of Mechanical Engineers.

ORIENTATION TOLERANCES

Angularity, parallelism, perpendicularity, and in some instances profile, are orientation tolerances. All are applicable to related features, and will always include datum reference letters in their feature control frames. These tolerances control the orientation of features to one another. The considered feature may be related to more than one datum feature if required to stabilize the tolerance zone in more than one direction. When no modifying symbol appears in the feature control frame, RFS applies.

Angularity Tolerance

Angularity is the condition of a surface, center plane, or axis at a specified angle (other than 90°) from a datum plane or axis. An angularity tolerance specifies a zone defined by two parallel planes. It also specifies a cylindrical tolerance zone at the basic angle from one or more datum planes, or a datum axis, within which the surface, center plane, or axis of the considered feature must lie. (See the following examples.)

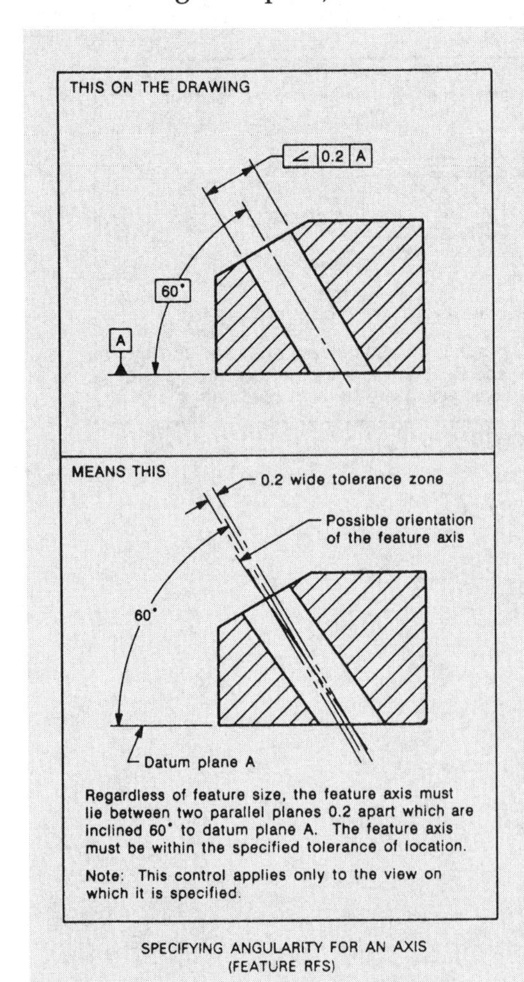

SPECIFYING ANGULARITY FOR AN AXIS (FEATURE RFS)

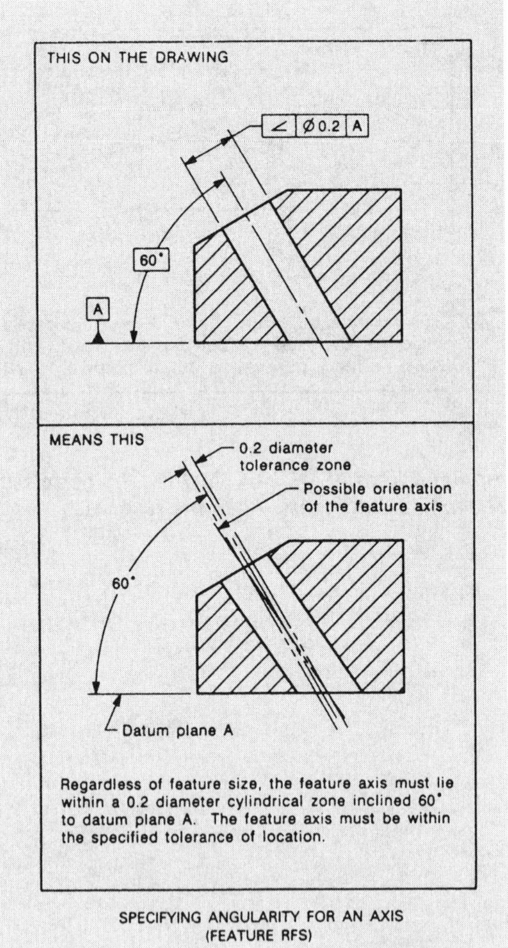

SPECIFYING ANGULARITY FOR AN AXIS (FEATURE RFS)

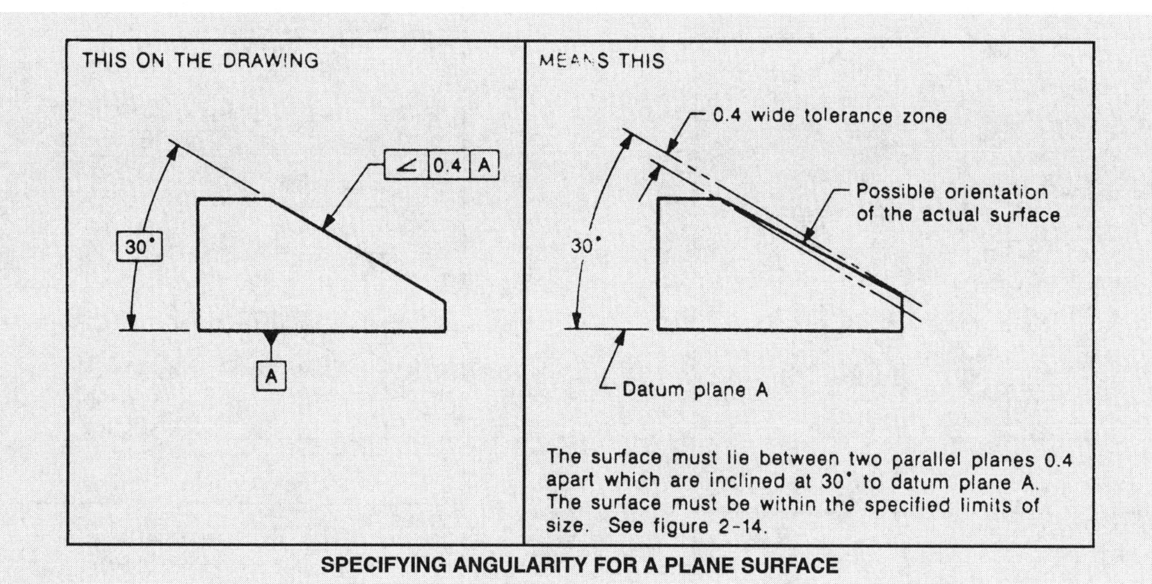

SPECIFYING ANGULARITY FOR A PLANE SURFACE

Parallelism Tolerance

Parallelism is the condition of a surface or center plane equidistant at all points from a datum plane. It also describes an axis, equidistant along its length from one or more datum planes or a datum axis. A parallelism tolerance specifies a tolerance zone defined by two planes parallel to a datum plane (or axis), within which the surface or center plane of the considered feature must lie. It may also establish a cylindrical tolerance zone whose axis is parallel to a datum axis within which the axis of the feature must lie. Observe the examples shown below and on the following page.

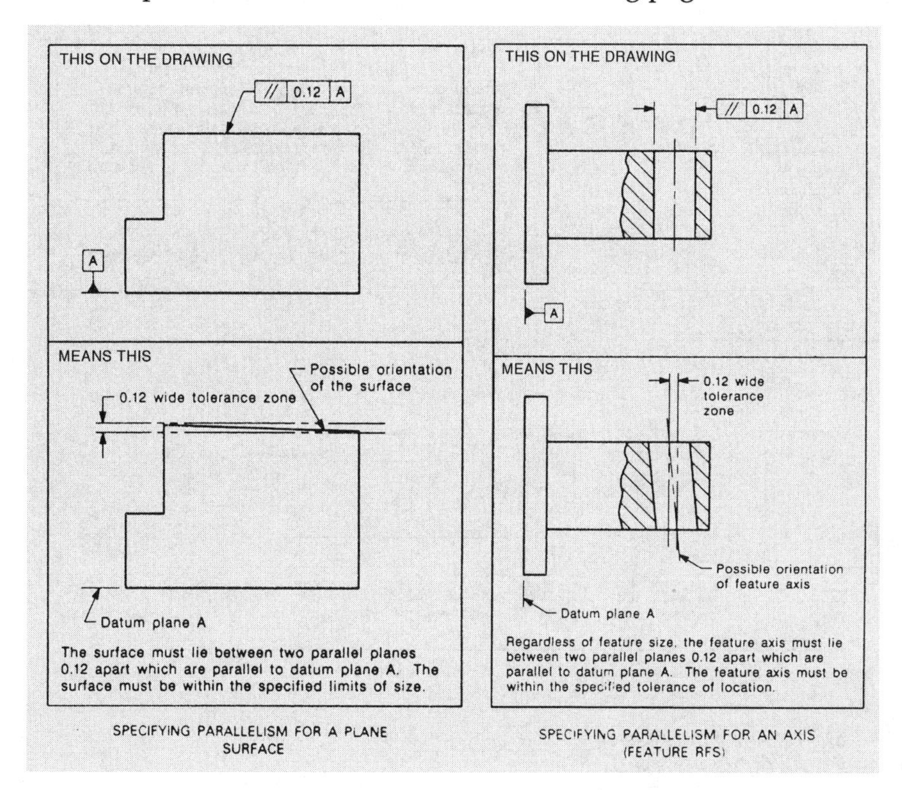

SPECIFYING PARALLELISM FOR A PLANE SURFACE

SPECIFYING PARALLELISM FOR AN AXIS (FEATURE RFS)

Figures on this page reprinted from ASME Y14.5M- 1994, by permission of The American Society of Mechanical Engineers.

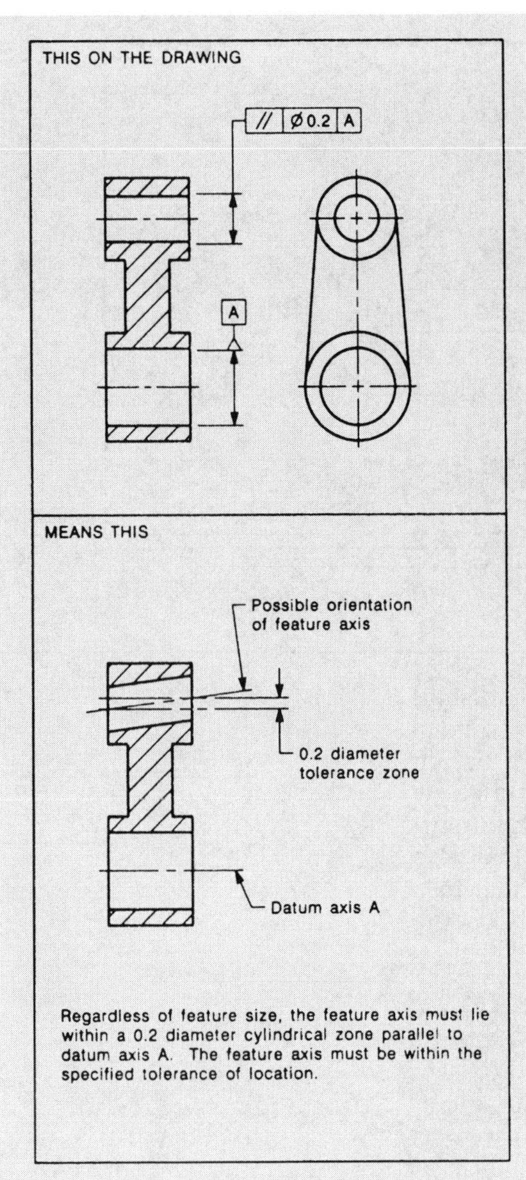

SPECIFYING PARALLELISM FOR AN AXIS
(BOTH FEATURE AND DATUM FEATURE RFS)

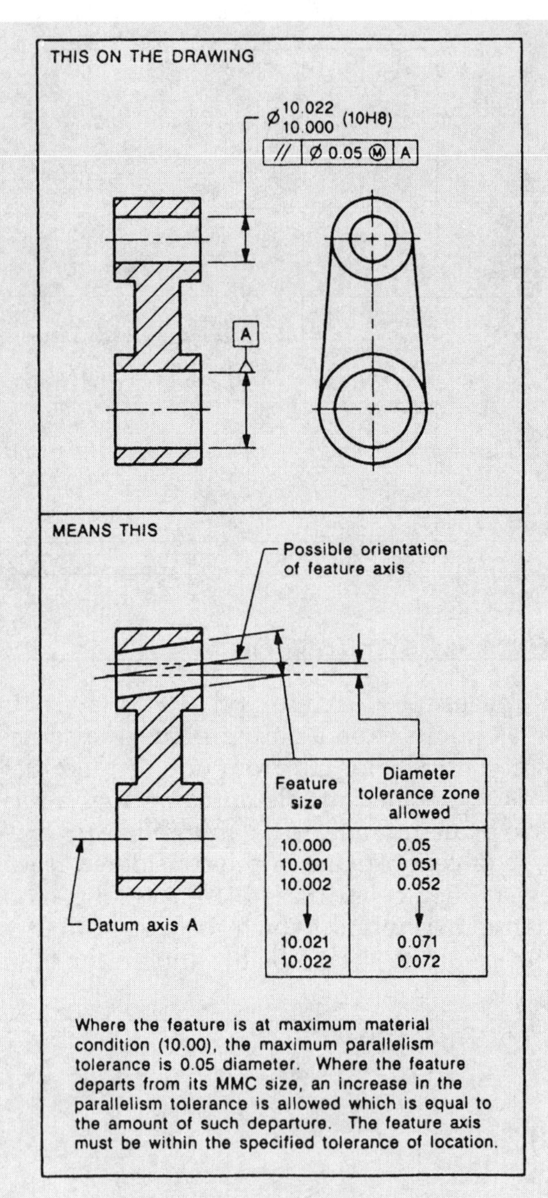

Feature size	Diameter tolerance zone allowed
10.000	0.05
10.001	0.051
10.002	0.052
↓	↓
10.021	0.071
10.022	0.072

SPECIFYING PARALLELISM FOR AN AXIS
(FEATURE AT MMC AND DATUM FEATURE RFS)

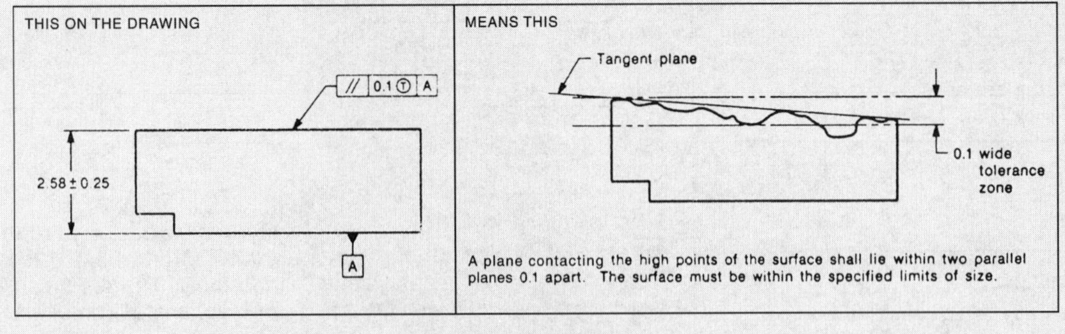

SPECIFYING A TANGENT PLANE

Perpendicularity Tolerance

Perpendicularity is a condition of a surface, center plane, or axis at a right angle to a datum plane or axis. It may specify a tolerance zone defined by two parallel planes perpendicular to a datum plane or axis within which the surface, center plane, or axis of the considered feature must lie; or it may establish a cylindrical tolerance zone perpendicular to a datum plane within which the axis of the considered feature must lie. Observe the examples below and on the following pages.

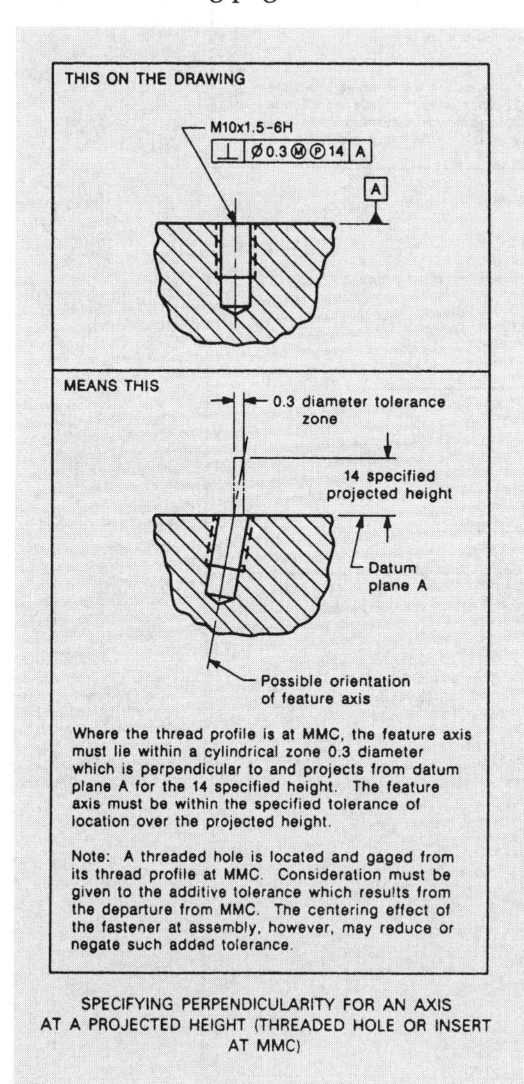

SPECIFYING PERPENDICULARITY FOR AN AXIS AT A PROJECTED HEIGHT (THREADED HOLE OR INSERT AT MMC)

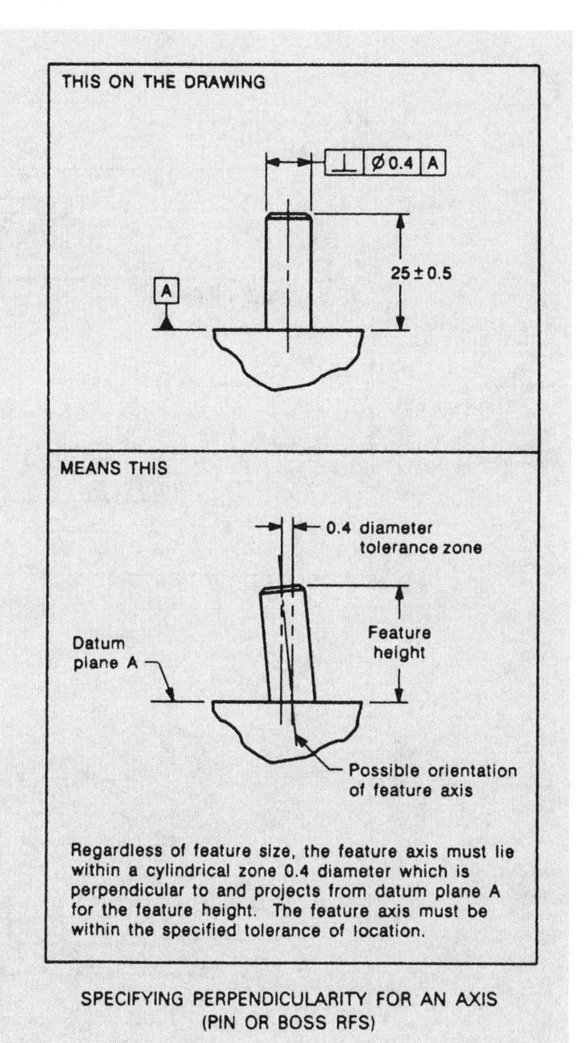

SPECIFYING PERPENDICULARITY FOR AN AXIS (PIN OR BOSS RFS)

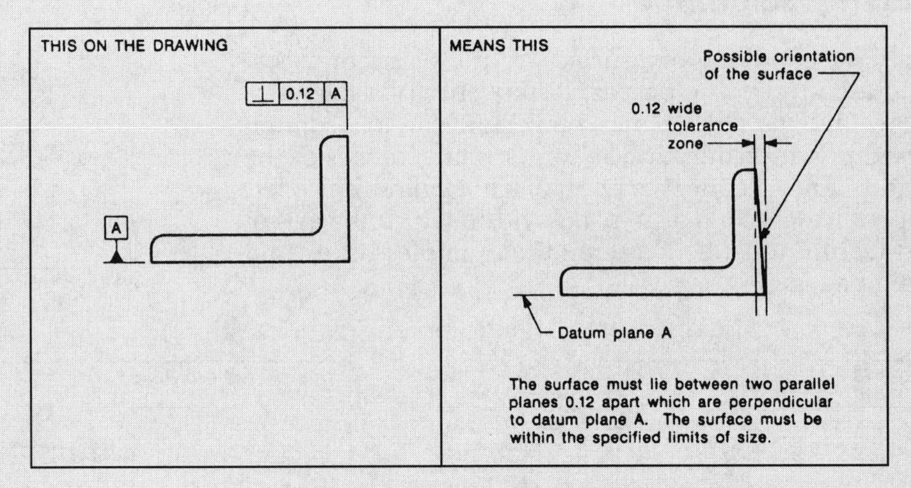

SPECIFYING PERPENDICULARITY FOR A PLANE SURFACE

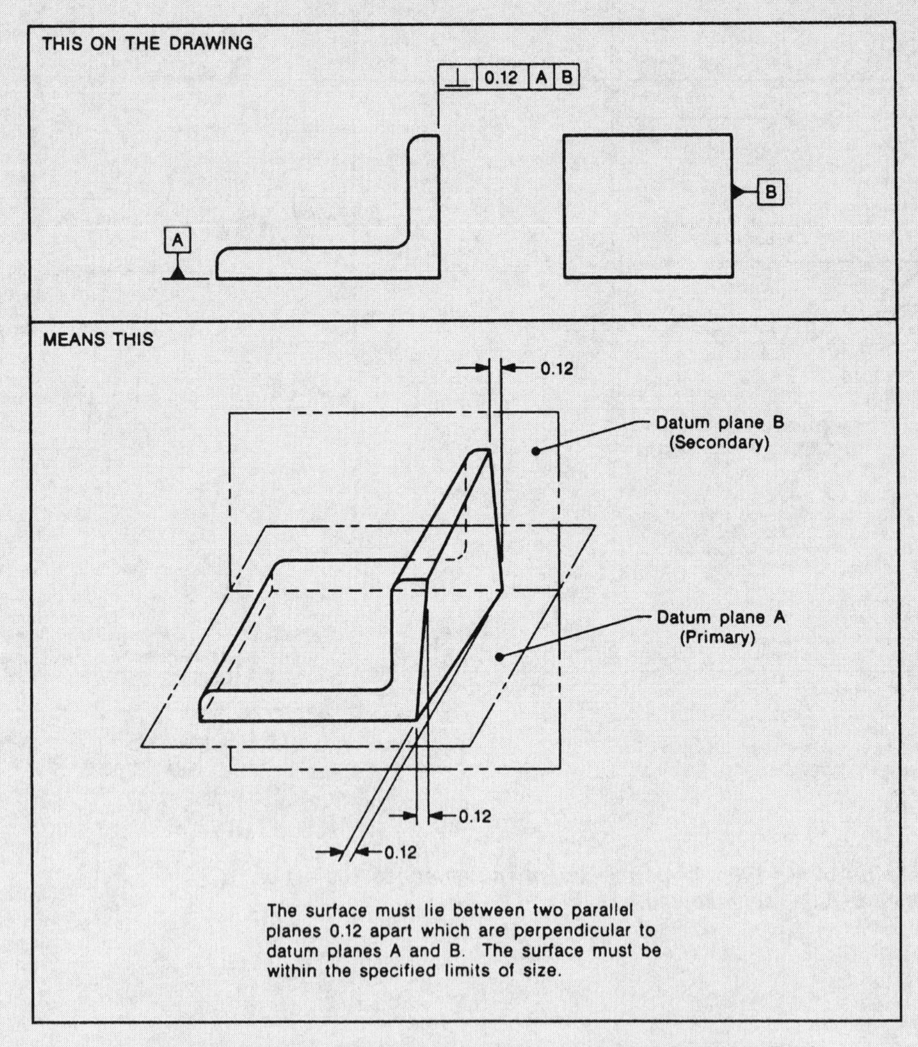

SPECIFYING PERPENDICULARITY FOR A PLANE SURFACE RELATIVE TO TWO DATUMS

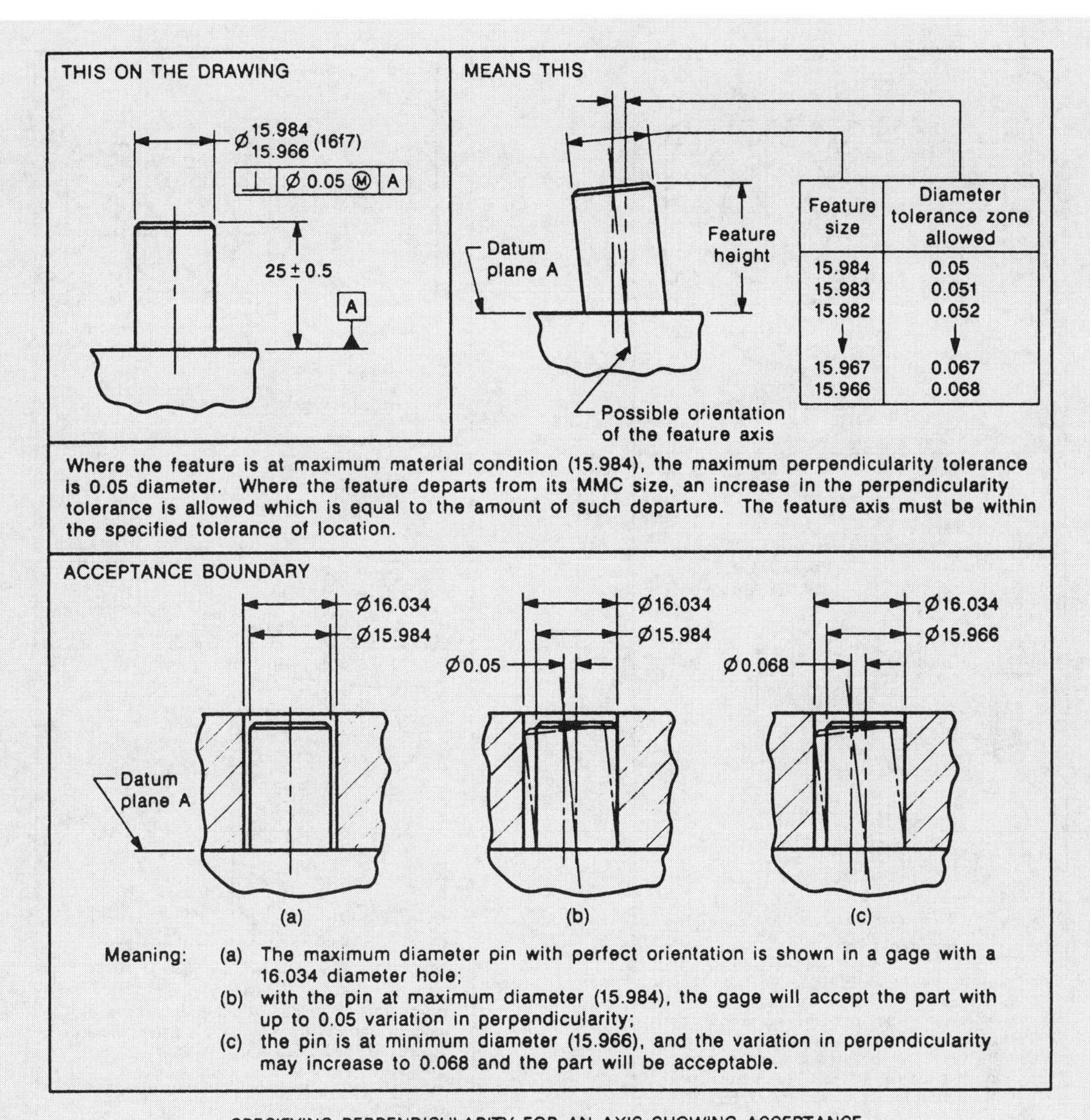

SPECIFYING PERPENDICULARITY FOR AN AXIS SHOWING ACCEPTANCE BOUNDARY (PIN OR BOSS AT MMC)

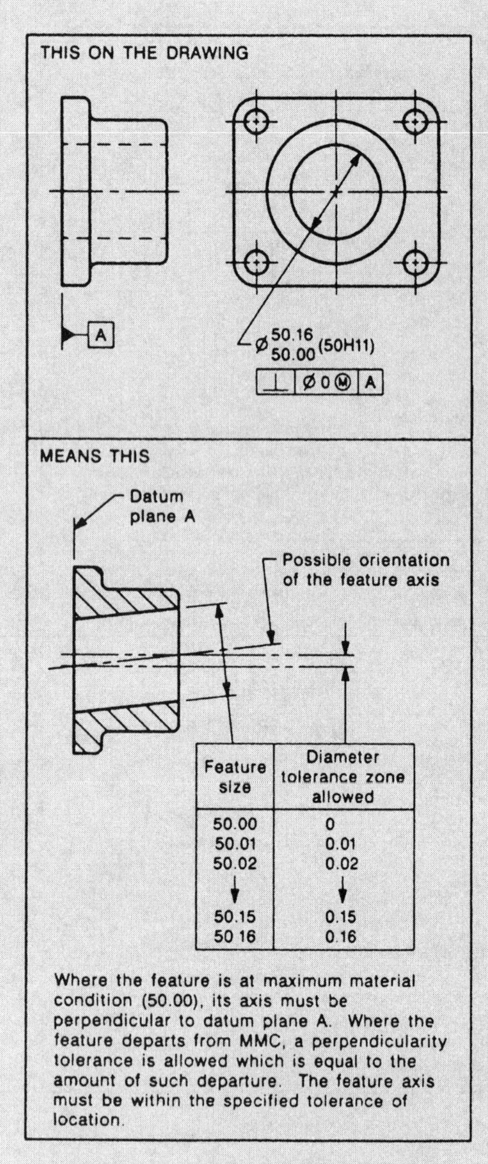

SPECIFYING PERPENDICULARITY FOR AN AXIS (ZERO TOLERANCE AT MMC)

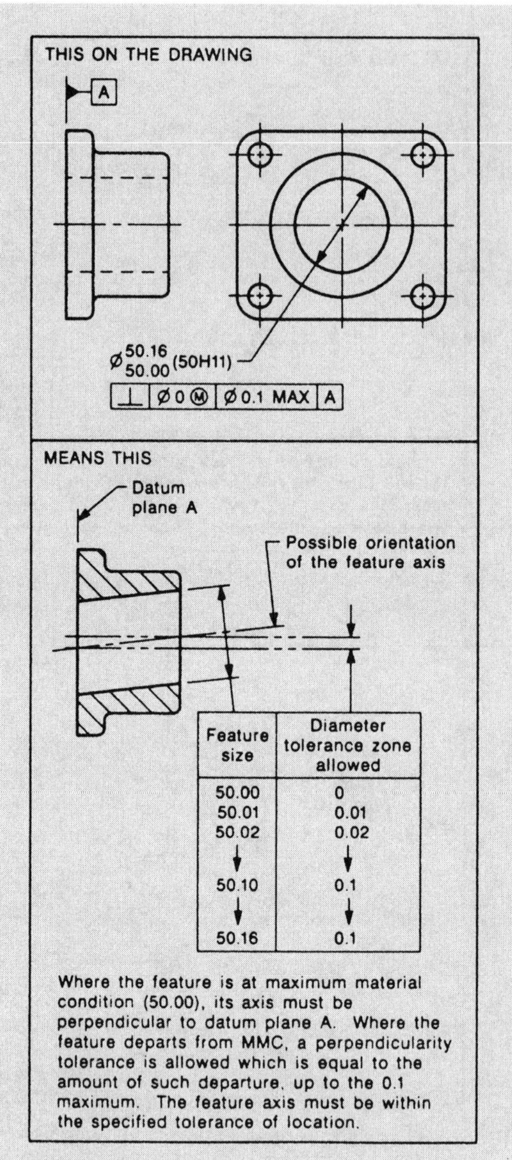

SPECIFYING PERPENDICULARITY FOR AN AXIS (ZERO TOLERANCE AT MMC WITH A MAXIMUM SPECIFIED)

Tolerance Calculations 4

INSTRUCTIONS: Calculate the maximum diameter tolerance zone for each of the feature sizes listed below. Refer to the examples shown on page 318.

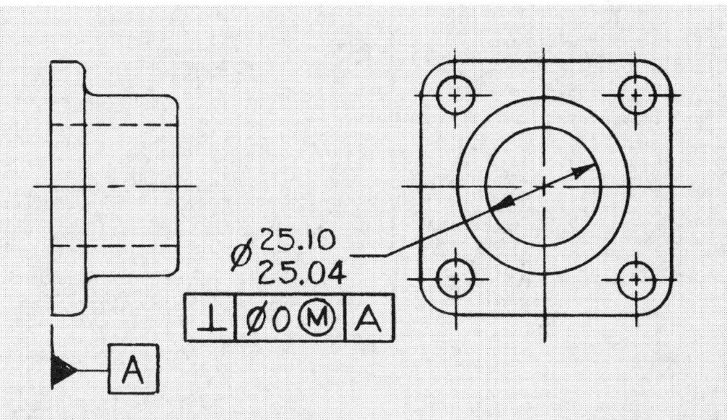

FEATURE SIZE	DIAMETER TOLERANCE ZONE ALLOWED
25.04	
25.05	
25.06	
25.07	
25.08	
25.09	
25.10	

⊥ | Ø0 Ⓜ | Ø0.02 MAX | A

FEATURE SIZE	DIAMETER TOLERANCE ZONE ALLOWED
25.04	
25.05	
25.06	
25.07	
25.08	
25.09	
25.10	

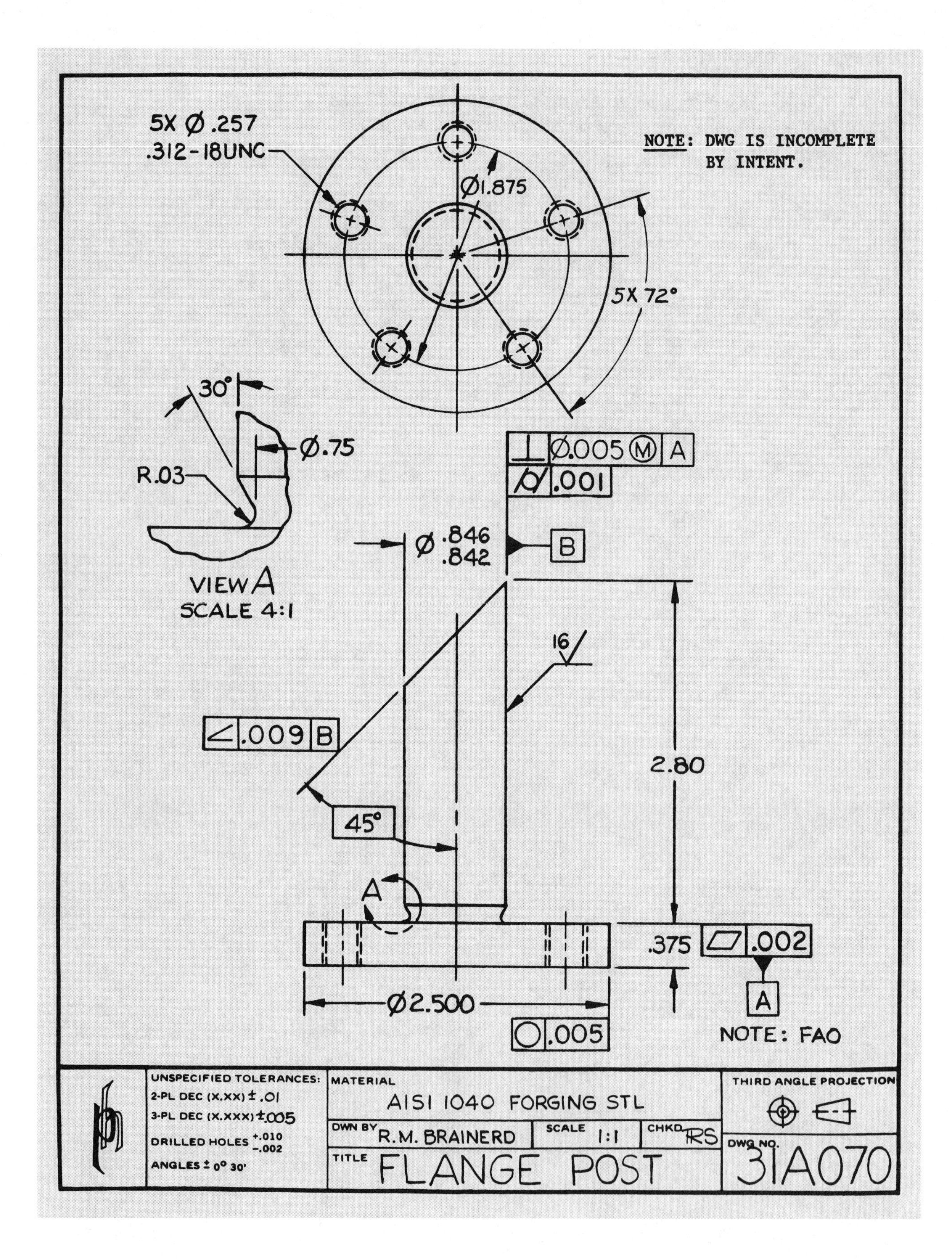
5X Ø.257
.312-18UNC
NOTE: DWG IS INCOMPLETE BY INTENT.
Ø1.875
5X 72°
30°
Ø.75
R.03
VIEW A
SCALE 4:1
Ø.005 Ⓜ A
.001
Ø .846 .842
B
16
.009 B
2.80
45°
A
.375
.002
A
Ø2.500
.005
NOTE: FAO
UNSPECIFIED TOLERANCES:
2-PL DEC (X.XX) ± .01
3-PL DEC (X.XXX) ± .005
DRILLED HOLES +.010 −.002
ANGLES ± 0° 30'
MATERIAL
AISI 1040 FORGING STL
DWN BY R.M. BRAINERD
SCALE 1:1
CHKD RS
TITLE FLANGE POST
THIRD ANGLE PROJECTION
DWG NO. 31A070

INSTRUCTIONS: Refer to drawing 31A070 to answer the following questions.

1. How many geometric tolerances are specified? 1. ______
2. How many surfaces serve as datum features? 2. ______
3. What does the geometric symbol on the flange diameter represent? 3. ______
4. What does the geometric symbol on the flange face represent? 4. ______
5. What does the upper geometric symbol on the post diameter represent? 5. ______
6. What does the lower geometric symbol on the post diameter represent? 6. ______
7. What does the geometric symbol on the post face represent? 7. ______
8. What does the rectangle enclosing the 45° dimension represent? 8. ______
9. What does the modifier on the perpendicularity tolerance represent? 9. ______
10. What is the flatness tolerance? 10. ______
11. What is the circularity (roundness) tolerance? 11. ______
12. What is the cylindricity tolerance? 12. ______
13. What is the angularity tolerance? 13. ______
14. What is the datum reference for the perpendicularity tolerance? 14. ______
15. What is the perpendicularity tolerance when the post diameter is .846? 15. ______
16. What is the perpendicularity tolerance when the post diameter is .842? 16. ______
17. What is the diameter of the bolt circle? 17. ______
18. What is the diameter of the neck? 18. ______
19. How many full threads does each hole contain? 19. ______
20. What is the accumulated tolerance on the overall height? 20. ______

Optional Exercise Using Tolerance of Position:

1a. After completing Unit 14, students may return to this drawing and create a feature control frame for the 5 × .312-18UNC threaded holes that states the following conditions:

The axis of each threaded hole is to be located within a .010 diameter cylindrical tolerance zone at RFS that is oriented perpendicular to datum A, and located relative to datum B at MMC. Designate the proper dimensions as basic by enclosing them inside a rectangular frame.

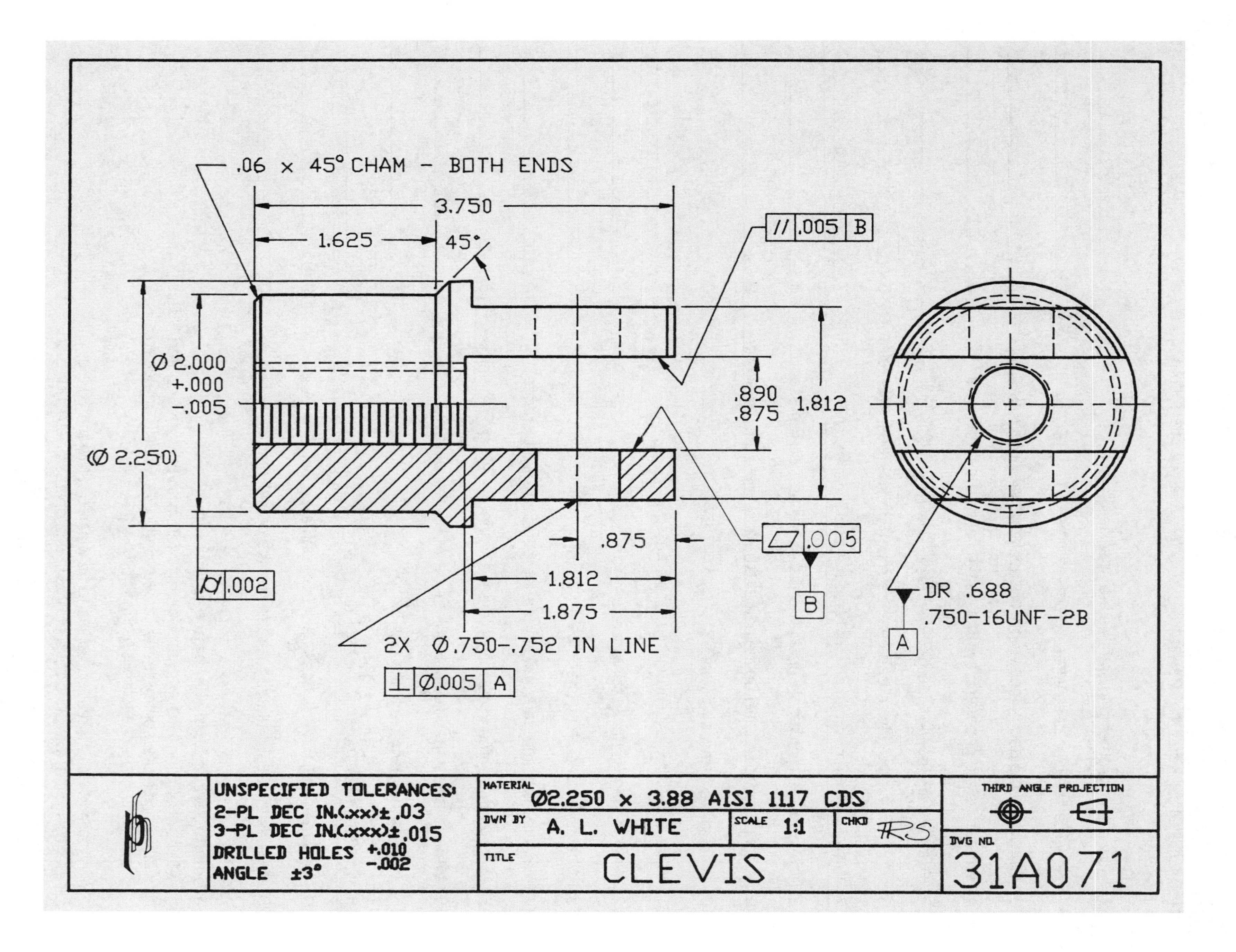

.06 × 45° CHAM - BOTH ENDS
3.750
1.625
45°
// .005 B
Ø 2.000
+.000
-.005
(Ø 2.250)
.890
.875
1.812
.875
1.812
1.875
⏥ .002
▱ .005
B
2X Ø.750-.752 IN LINE
⊥ Ø.005 A
DR .688
.750-16UNF-2B
A
UNSPECIFIED TOLERANCES:
2-PL DEC IN.(.xx)± .03
3-PL DEC IN.(.xxx)± .015
DRILLED HOLES +.010 -.002
ANGLE ±3°
MATERIAL
Ø2.250 × 3.88 AISI 1117 CDS
DWN BY A. L. WHITE
SCALE 1:1
CHKD
TITLE
CLEVIS
THIRD ANGLE PROJECTION
DWG NO.
31A071

INSTRUCTIONS: Refer to drawing 31A071 to answer the following questions.

1. How many geometric tolerances are specified? 1. ____________
2. How many of the geometric tolerances are considered an orientation type of tolerance? 2. ____________
3. Do the geometric tolerances apply at MMC or RFS? 3. ____________
4. What do the letters at the ends of the feature control frames represent? 4. ____________
5. How many surfaces serve as datum features? 5. ____________
6. Interpret the geometric characteristic symbol appearing on the 2.000 diameter. 6. ____________
7. Interpret the geometric characteristic symbol appearing on the .750 clearance holes. 7. ____________
8. Interpret the geometric characteristic symbols appearing on the flat surfaces. 8. ____________
9. How much geometric tolerance is assigned to the 2.000 diameter? 9. ____________
10. How much size tolerance is assigned to the 2.000 diameter? 10. ____________
11. What is the size tolerance on the .750 clearance holes? 11. ____________
12. What is the maximum metal thickness where each .750 hole passes through? 12. ____________
13. What type of section view is drawn? 13. ____________
14. What do the parentheses around the large-diameter dimension represent? 14. ____________
15. Is the clevis symmetrical? 15. ____________
16. How much material is removed from the blank length to produce the clevis? 16. ____________
17. Is the specified material a low- or a medium-carbon steel? 17. ____________
18. Is the specified material a sulfurized steel? 18. ____________
19. What size tap drill is specified? 19. ____________
20. How many full threads will the tapped hole contain? 20. ____________

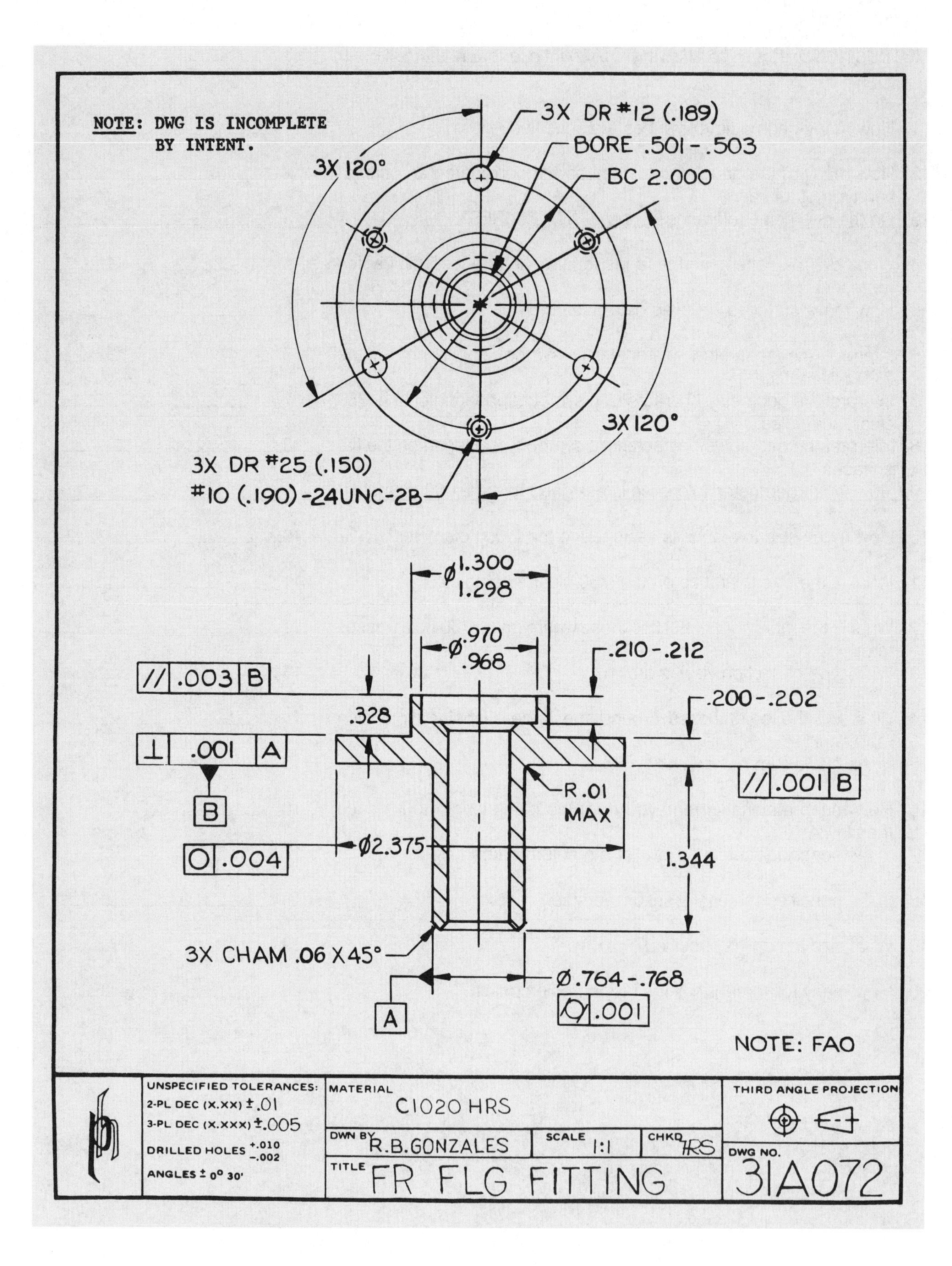
NOTE: DWG IS INCOMPLETE BY INTENT.
3X DR #12 (.189)
BORE .501 - .503
BC 2.000
3X 120°
3X 120°
3X DR #25 (.150)
#10 (.190) - 24UNC-2B
Ø 1.300 1.298
Ø .970 .968
.210 - .212
.200 - .202
// .003 B
.328
⊥ .001 A
B
R .01 MAX
// .001 B
O .004
Ø2.375
1.344
3X CHAM .06 X 45°
Ø .764 - .768
.001
A
NOTE: FAO
UNSPECIFIED TOLERANCES:
2-PL DEC (X.XX) ±.01
3-PL DEC (X.XXX) ±.005
DRILLED HOLES +.010 −.002
ANGLES ± 0° 30'
MATERIAL
C1020 HRS
DWN BY R.B.GONZALES
SCALE 1:1
CHKD HRS
TITLE FR FLG FITTING
THIRD ANGLE PROJECTION
DWG NO.
31A072

INSTRUCTIONS: Refer to drawing 31A072 to answer the following questions.

1. Interpret the geometric characteristic symbol on datum A. 1. ______
2. How much geometric tolerance is assigned to datum A? 2. ______
3. How much geometric tolerance is assigned to the flange diameter? 3. ______
4. Show the major diameter (decimally) of the threads in the tapped holes. 4. ______
5. Show the thread pitch. (Three decimal places.) 5. ______
6. How many *full* threads will each threaded hole contain? 6. ______
7. What is the maximum permissible overall length of the fitting? 7. ______
8. What is the MMC of datum A? 8. ______
9. What is the MMC of the bore diameter? 9. ______
10. Show the upper and lower limits of the flange diameter. 10. ______
11. Show the percentage of carbon content in the steel specified. 11. ______
12. Is the parallelism tolerance assigned at MMC, LMC, or RFS? 12. ______
13. Is the perpendicularity tolerance assigned at MMC, LMC, or RFS? 13. ______
14. How much size tolerance is permitted on the unthreaded holes in the flange? 14. ______
15. What is the minimum wall thickness at the bore? 15. ______
16. What is the minimum wall thickness at the counterbore? 16. ______
17. What is the total size tolerance permissible on datum A? 17. ______
18. How many of the geometric tolerances do not involve a datum reference? 18. ______
19. How many of the feature control frames include an orientation tolerance? 19. ______
20. What is the minimum amount of material permissible between an unthreaded hole and the OD of the flange? 20. ______

NOTE: Students may return to this drawing (after completing Unit 14) to add the proper positional tolerance controls for the three (3) clearance holes and the three (3) threaded holes.

UNIT 14

LOCATION TOLERANCES

Location tolerances include tolerance of symmetry, concentricity, and position. They are used to control the center distances, location, coaxiality, and concentricity or symmetry between features such as holes, slots, bosses, and tabs.

Symmetry Tolerance

Symmetry is the condition where the median points of all opposed elements of two or more feature surfaces are located within a tolerance zone of two parallel planes equally disposed on either side of the axis or centerplane of the datum reference. The symmetry tolerance and the datum reference can only be applied at RFS. (See the example below.)

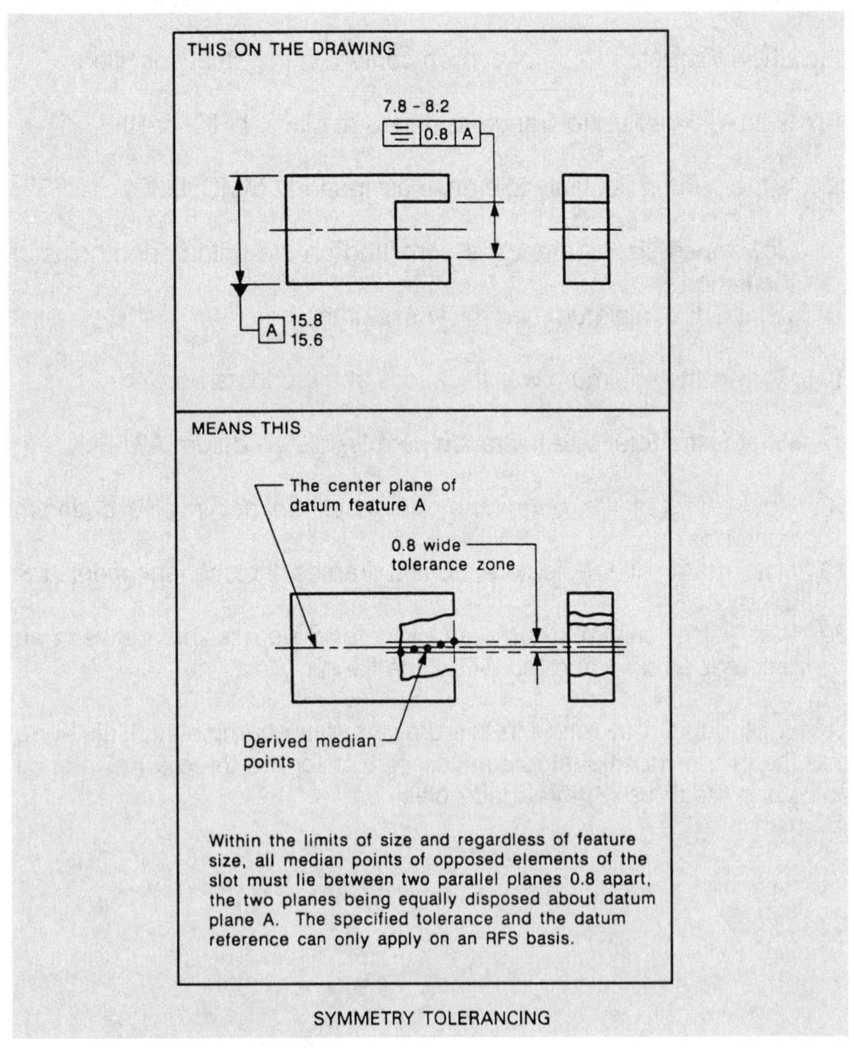

SYMMETRY TOLERANCING

Concentricity Tolerance

Concentricity is the condition where the median points of all diametrically opposed elements of a cylindrical feature are located within a cylindrical tolerance zone equally disposed around the axis of the datum reference. The concentricity tolerance and the datum reference can only be applied at RFS.

Both concentricity and symmetry are very expensive controls to use. Therefore, it is preferred to use other controls such as position, runout, or profile in place of concentricity, and to use position in place of symmetry.

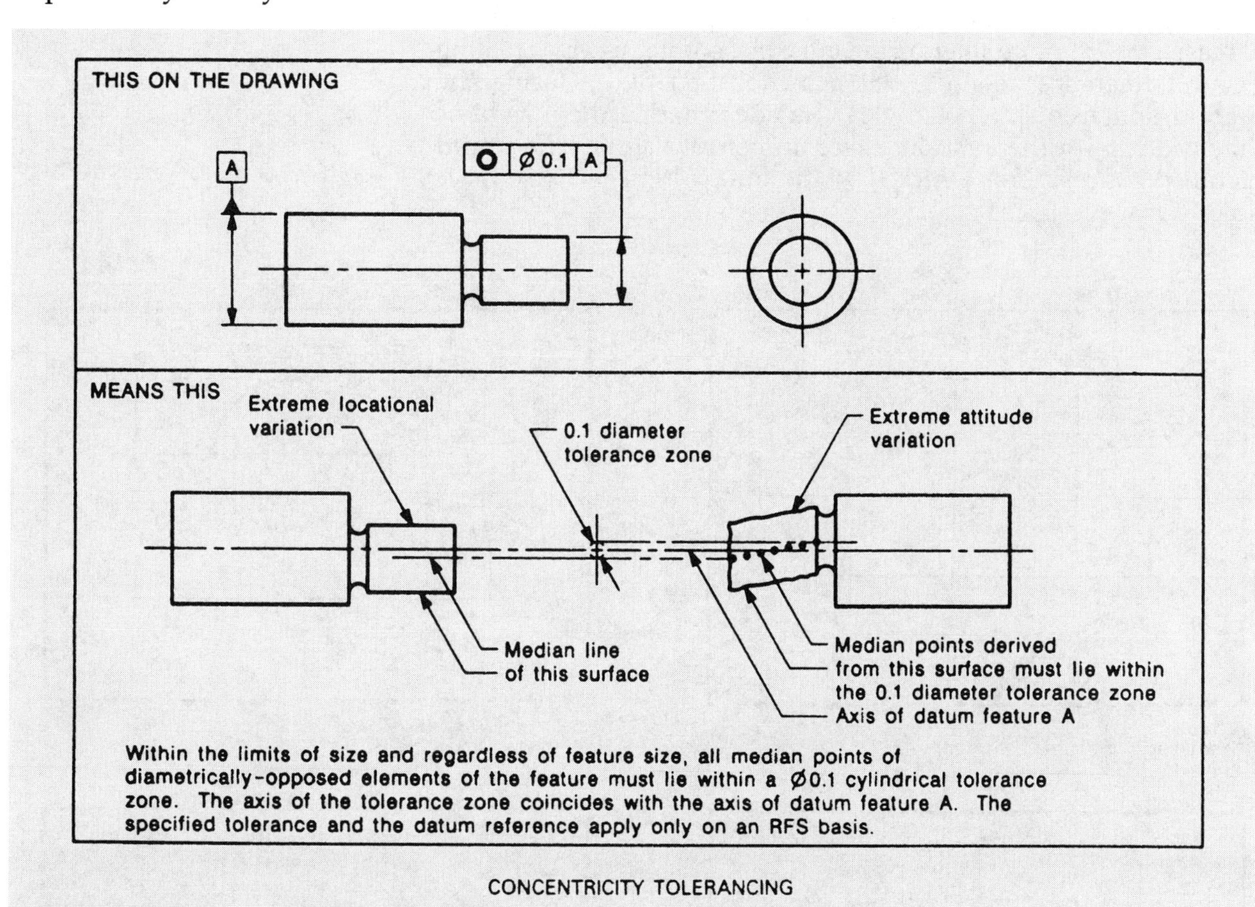

CONCENTRICITY TOLERANCING

Reprinted from ASME Y14.5M-1994, by permission of The American Society of Mechanical Engineers.

Positional Tolerance

A positional tolerance defines a zone within which the center, axis, or center plane of a feature of size is permitted to vary from true (theoretically exact) position. Basic dimensions establish the true position from specified datum features and between interrelated features.

The advantages of positional tolerancing can be clearly seen in the examples that follow. Observe the .010 square tolerance zone that results from coordinate plus and minus tolerancing. This tolerancing method permits a location of .007 from the true center if it occurs in the corners of the .010 square zone (1.4 × .005). An increase of 57% more tolerance zone can be obtained by creating a round tolerance zone. This is accomplished by enclosing the location dimensions in rectangles, thus designating them as basic dimensions with theoretically exact numerical values. The round tolerance zone is then centered at the intersection of these basic dimensions.

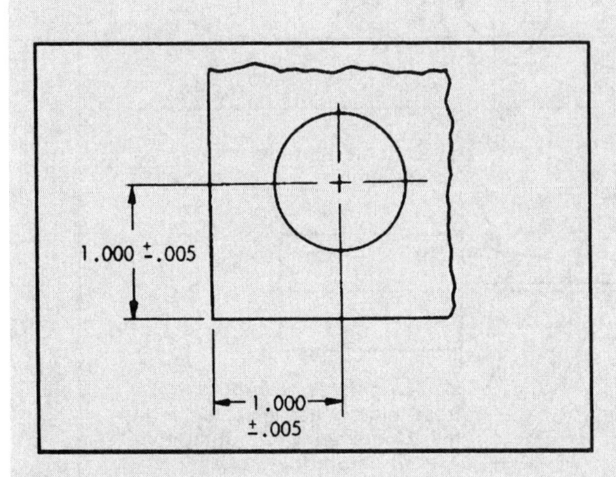

Coordinate tolerancing.

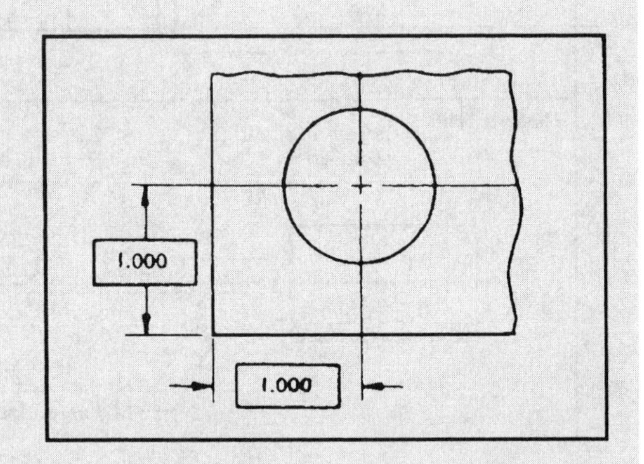

Positional tolerancing.

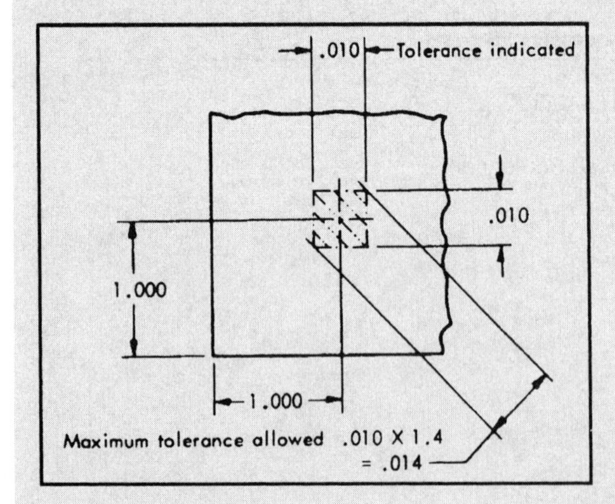

Coordinate tolerance zone (square).

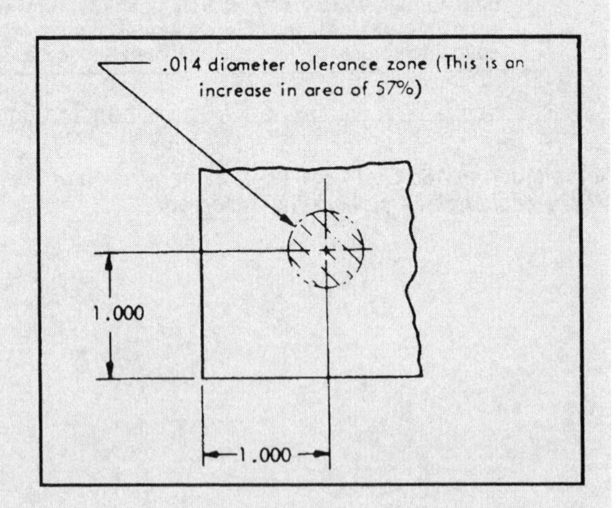

Positional tolerance zone (round).

Positional Tolerance Modifiers

With the publication of the ASME Y14.5M-1994 standard, a new method of applying positional tolerancing was introduced. The previous ANSI Y14.5M standard stated that all position controls had to have the modifier MMC, LMC, or RFS in either the tolerance portion or the datum reference portion of the feature control frame. Now, in the ASME Y14.5M-1994 standard, when no material condition modifier appears (either Ⓜ or Ⓛ), then the control automatically applies at RFS. RFS is referred to as the default condition for all tolerance controls or datum references. (Rule #2)

Maximum material condition means that internal features such as holes and slots would be at their minimum allowable size, whereas external features such as shafts would be at their maximum allowable size. For example, MMC of a .500-.505 diameter hole is at .500, while MMC of a .500-.505 diameter shaft is at .505 diameter. When positional tolerance is specified at MMC, perfect form is required (Rule #1) and the tolerance zone size is dependent on the size of the considered feature. Where the actual size of the feature has departed from MMC, an increase in the tolerance is allowed equal to the amount of the departure. For example, if a .014 dia. tolerance zone is applied to a .500-.505 hole at MMC, the tolerance zone would be .014 diameter, if the hole measured exactly .500 diameter. If the hole measured .501, the tolerance would increase to .015 dia., .016 dia. for .502, .017 dia. for .503, and on up to .019 dia. tolerance zone for a hole measuring .505 diameter.

Specifying positional tolerance at LMC (least material condition) would mean that the stated positional tolerance applies when the feature contains the least amount of material permitted by its toleranced size dimension. Specification of positional tolerance at LMC requires that the feature have perfect form at LMC, but perfect form at MMC is not required. Where the feature departs from its LMC size, an increase in positional tolerance is allowed equal to the amount of the departure. Using the .500-.505 dia. hole as an example, when the hole is at LMC (.505), the positional tolerance zone is .014 dia. If the hole measured .504 dia., the tolerance would increase to .015 dia., .016 dia. for .503, and on up to .019 dia. for a hole measuring .500 (MMC).

When a feature is referenced at RFS (no Ⓜ or Ⓛ with the datum letters), then the tolerance zone remains the same as long as the feature size falls within the acceptable size range.

When using positional tolerancing, the tolerance zone may be of any convenient size, as long as it contains an Ⓜ or an Ⓛ material condition modifier that will allow the tolerance zone to grow as the feature size changes. Observe the examples on the following pages.

Zero Positional Tolerance at MMC

In applications where it is necessary to provide greater than normal tolerance within functional limits, the principle of positional tolerancing at MMC may be extended. This is accomplished by adjusting the minimum size of a hole to the absolute minimum required for insertion of a fastener located at true position, and specifying a zero positional tolerance at MMC.

Figures 1 and 2 below illustrate the same part, one with zero positional tolerance at MMC, the other with conventional positional tolerance at MMC. Note that the maximum size limit of the clearance hole remains the same, but the minimum was adjusted to correspond with a 14mm diameter fastener. This results in an increase in the size tolerance for the clearance holes, the increase being equal to the conventional positional tolerance specified in Fig. 2. Although the positional tolerance specified in Fig. 1 is zero at MMC, the positional tolerance allowed is in direct proportion to the actual clearance hole size as shown in the table that follows.

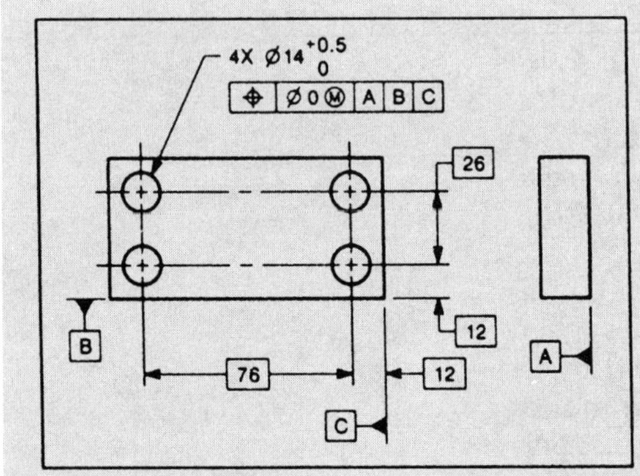

FIG. 1 ZERO POSITIONAL TOLERANCING AT MMC

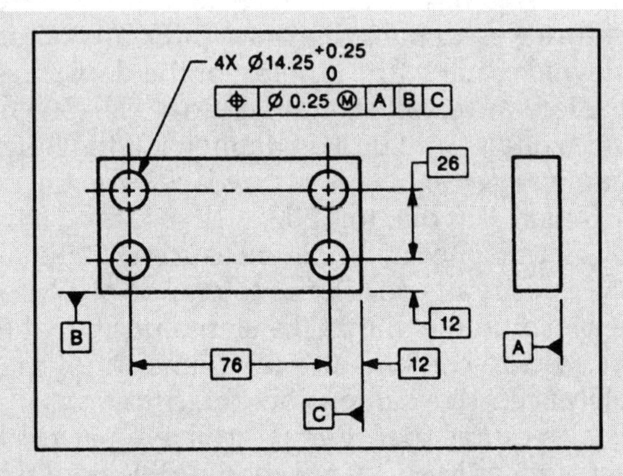

FIG. 2 CONVENTIONAL POSITIONAL TOLERANCING AT MMC

Clearance Hole Diameter (Feature Actual Mating Size)	Positional Tolerance Diameter Allowed (Zero Positional Tolerancing at MMC)	Positional Tolerance Diameter Allowed (Conventional Positional Tolerancing at MMC)
14	0	Part would function, but
14.1	0.1	would have to be rejected
14.2	0.2	because of size violation.
14.25	0.25	0.25
14.3	0.3	0.3
14.4	0.4	0.4
14.5	0.5	0.5

Zero positional tolerancing allows for all functional parts to be accepted.

Positional Tolerancing for Symmetrical Relationships

Positional tolerancing for symmetrical relationships is that condition where the center plane of the actual mating envelope of one or more features is congruent with the axis or center plane of a datum feature within specified limits. MMC, LMC, or RFS may be specified to apply to both the tolerance and the datum feature.

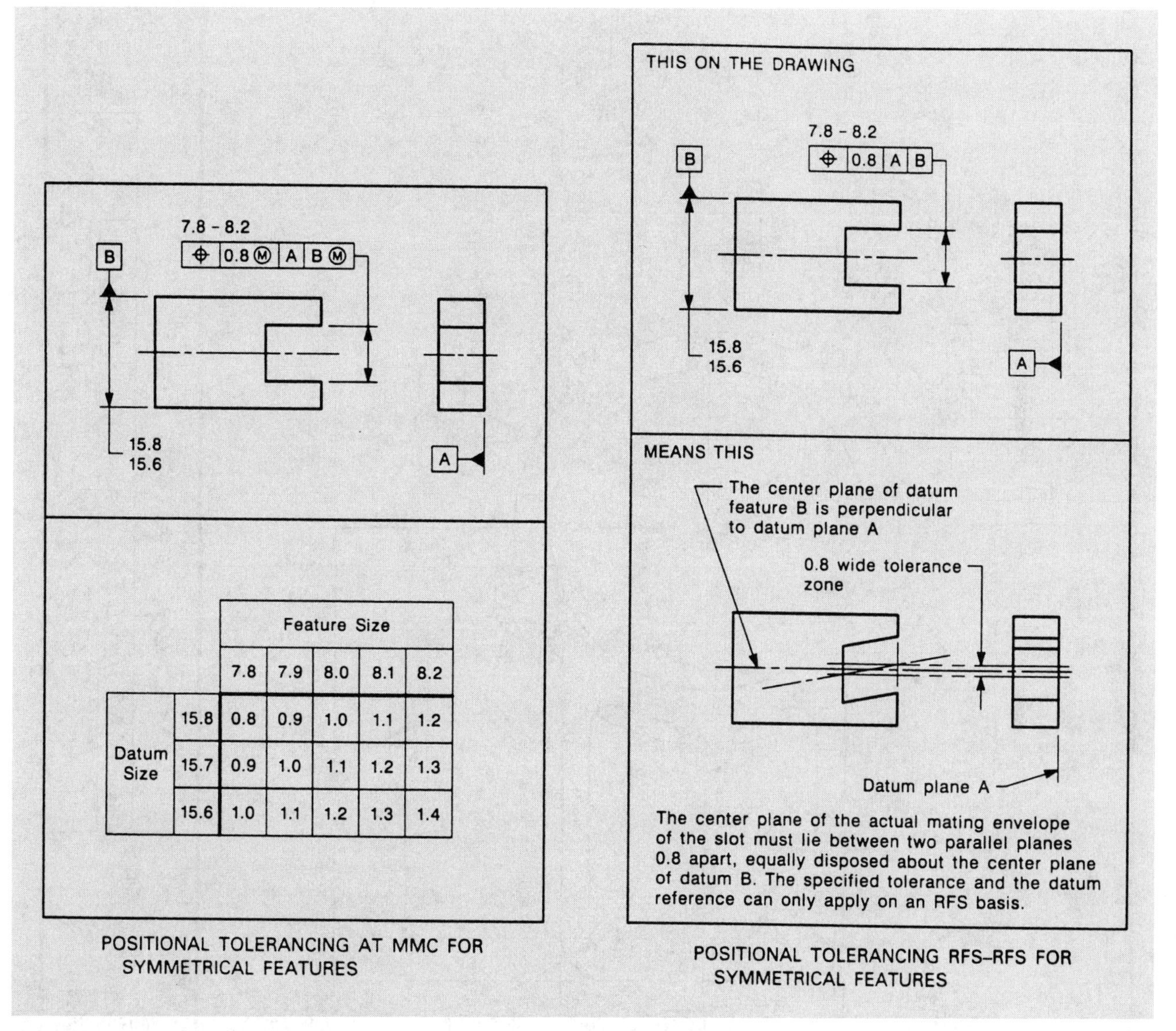

		Feature Size				
		7.8	7.9	8.0	8.1	8.2
Datum Size	15.8	0.8	0.9	1.0	1.1	1.2
	15.7	0.9	1.0	1.1	1.2	1.3
	15.6	1.0	1.1	1.2	1.3	1.4

POSITIONAL TOLERANCING AT MMC FOR SYMMETRICAL FEATURES

POSITIONAL TOLERANCING RFS–RFS FOR SYMMETRICAL FEATURES

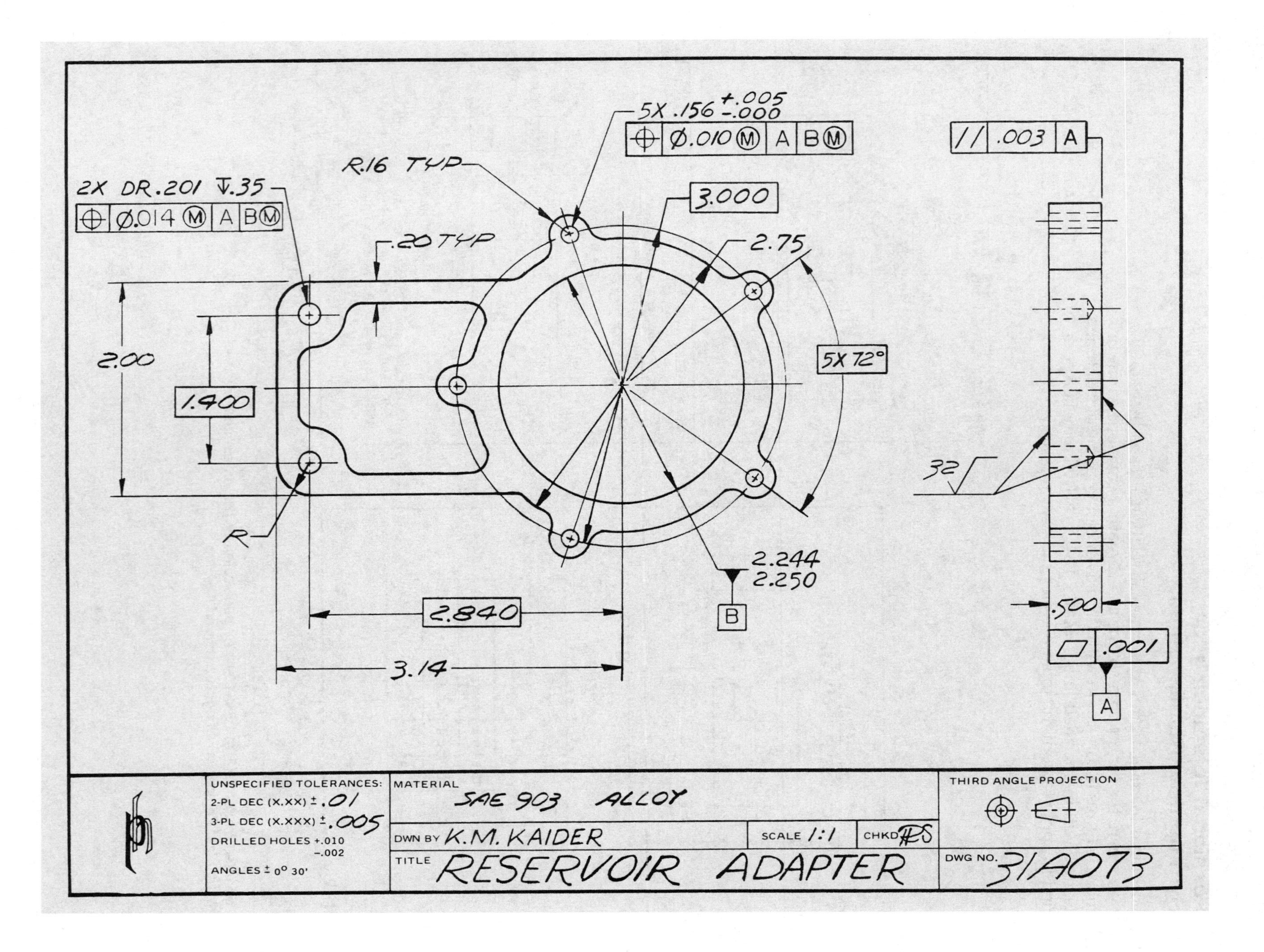
5X .156 +.005 -.000
⌖ Ø.010 Ⓜ A BⓂ
// .003 A
R.16 TYP
2X DR.201 ↧.35
⌖ Ø.014 Ⓜ A BⓂ
3.000
.20 TYP
2.75
5X 72°
2.00
1.400
32
R
2.244
2.250
B
2.840
.500
▱ .001
3.14
A
UNSPECIFIED TOLERANCES:
2-PL DEC (X.XX) ± .01
3-PL DEC (X.XXX) ± .005
DRILLED HOLES +.010 -.002
ANGLES ± 0° 30'
MATERIAL SAE 903 ALLOY
DWN BY K.M. KAIDER
SCALE 1:1
CHKD
THIRD ANGLE PROJECTION
TITLE RESERVOIR ADAPTER
DWG NO. 31A073

INSTRUCTIONS: Refer to drawing 31A073 to answer the following questions.

1. How many datum feature symbols appear on the drawing? 1. ____________
2. How many basic dimensions are designated? 2. ____________
3. How many geometric tolerances are specified? 3. ____________
4. How many (question 3) are locational tolerances? 4. ____________
5. What characteristic do the locational tolerance symbols represent? 5. ____________
6. What characteristic does the form tolerance symbol represent? 6. ____________
7. What characteristic does the orientation tolerance symbol represent? 7. ____________
8. What is the total tolerance on the surface of datum A? 8. ____________
9. What is the total tolerance on the diameter of datum B? 9. ____________
10. What is the positional tolerance on the .156 diameter holes at MMC? 10. ____________
11. What is the positional tolerance on the .201 diameter holes at MMC? 11. ____________
12. What is the size tolerance on the .201 diameter holes? 12. ____________
13. What size is the positional tolerance zone if the five holes measure .156? 13. ____________
14. What size is the positional tolerance zone if the five holes measure .157? 14. ____________
15. What size is the positional tolerance zone if the two holes measure .201? 15. ____________
16. Is the tolerance zone for the hole locations (a) square-shaped or (b) round-shaped? 16. ____________
17. What roughness height is specified for the two flat surfaces? 17. ____________
18. What size bolt circle is specified? 18. ____________
19. How far apart angularly are the .156 diameter holes? 19. ____________
20. What is the overall length measured along the horizontal centerline? 20. ____________

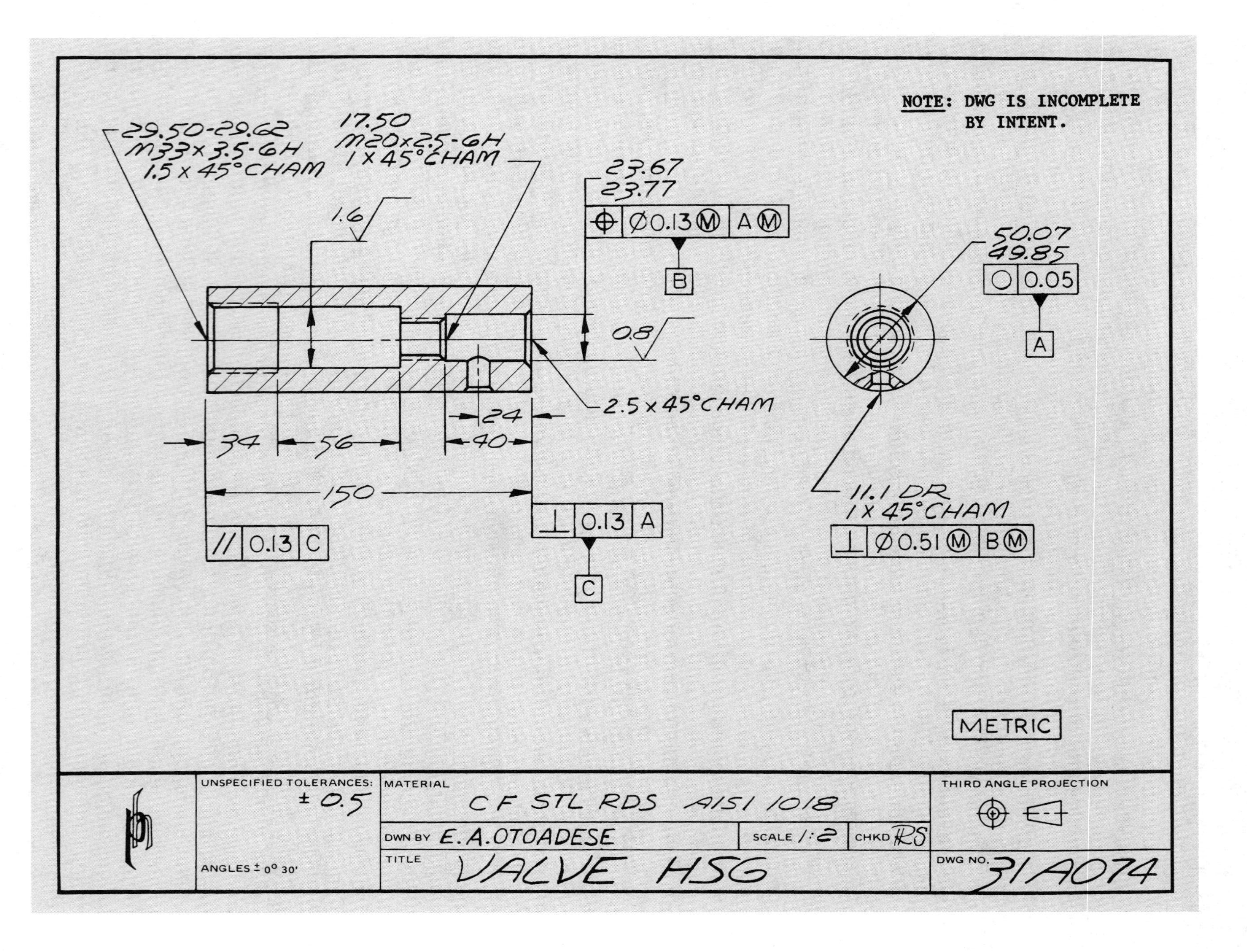
NOTE: DWG IS INCOMPLETE BY INTENT.
29.50-29.62
M33 x 3.5-6H
1.5 x 45°CHAM
17.50
M20 x 2.5-6H
1 x 45°CHAM
23.67
23.77
Ø0.13 Ⓜ A Ⓜ
B
1.6
0.8
2.5 x 45°CHAM
50.07
49.85
0.05
A
24
34
56
40
150
0.13 A
C
// 0.13 C
11.1 DR
1 x 45°CHAM
Ø0.51 Ⓜ B Ⓜ
METRIC
UNSPECIFIED TOLERANCES: ± 0.5
ANGLES ± 0° 30'
MATERIAL C F STL RDS AISI 1018
DWN BY E. A. OTOADESE
SCALE 1:2
CHKD RS
TITLE VALVE HSG
THIRD ANGLE PROJECTION
DWG NO. 31A074

INSTRUCTIONS: Refer to drawing 31A074 to answer the following questions.

1. What measurement units were used to dimension the drawing? 1. ____________
2. How many chamfers does the housing contain? 2. ____________
3. What tolerance is specified for the thread class of fit? 3. ____________
4. How many datum feature symbols appear on the drawing? 4. ____________
5. How many geometric tolerances are specified? 5. ____________
6. What characteristic does the tolerance symbol on datum A represent? 6. ____________
7. What characteristic does the tolerance symbol on datum B represent? 7. ____________
8. What characteristic does the tolerance symbol on datum C represent? 8. ____________
9. What does the modifier Ⓜ in the feature control frames represent? 9. ____________
10. How much roughness height tolerance is assigned to datum B? 10. ____________
11. How much geometric tolerance is assigned to datum B? 11. ____________
12. How much size tolerance is assigned to datum B? 12. ____________
13. How much size tolerance is assigned to datum A? 13. ____________
14. How much geometric tolerance is assigned to datum A? 14. ____________
15. What is the major diameter of the large threads? 15. ____________
16. What is the pitch of the small threads? 16. ____________
17. What is the length of the small threaded section? (Include the chamfer.) 17. ____________
18. What is the unthreaded diameter of the large hole? 18. ____________
19. What is the diameter of the cross-drilled hole? 19. ____________
20. What is the maximum wall thickness between datum B and the OD? 20. ____________

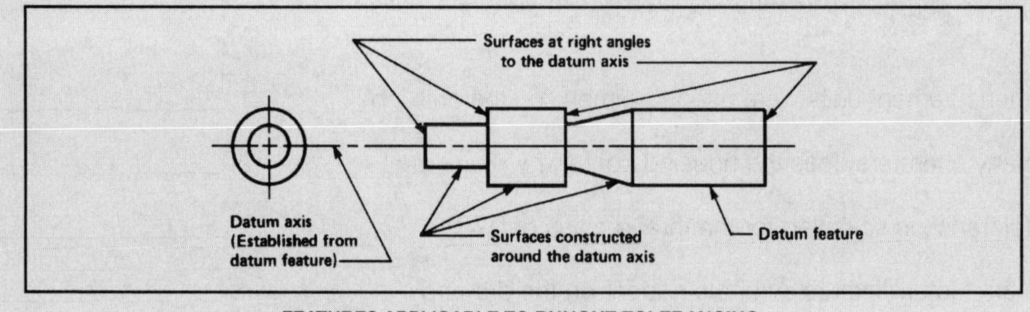

FEATURES APPLICABLE TO RUNOUT TOLERANCING

Reprinted from ASME Y14.5M-1994, by permission of The American Society of Mechanical Engineers.

RUNOUT TOLERANCES

Runout is a composite tolerance used to control the functional relationship of one or more features of a part to a datum axis. The types of features controlled by runout tolerances include those surfaces constructed around a datum axis and those constructed at right angles to a datum axis. (See the illustration above.)

The datum axis for a runout control is established in one of three ways. First, a datum axis can be established by a diameter of sufficient length to be repeatable. Second, two diameters having sufficient axial separation can be used. Third, a diameter and a face at right angles to it can be used and the datums are specified separately to indicate datum preference. In other words, it could be a face and a diameter, or a diameter and a face.

Each considered feature must be within its runout tolerance when rotated about the datum axis. The tolerance specified is the total tolerance, or full indicator movement (FIM). There are two types of runout control, circular runout and total runout.

Circular Runout

Circular runout provides control of circular elements of a surface. The tolerance is applied independently at any circular measuring position as the part is rotated 360°. Where applied to surfaces constructed around a datum axis, circular runout may be used to control the cumulative variations of circularity and coaxiality. Where applied to surfaces constructed at right angles to the datum axis, circular runout controls circular elements of a plane surface (wobble).

Total Runout

Total runout provides composite control of all surface elements. The tolerance is applied simultaneously to all circular and profile measuring positions as the part is rotated 360°. Where applied to surfaces constructed around a datum axis, total runout is used to control cumulative variations of circularity, straightness, coaxiality, angularity, taper, and profile of a surface. Where applied to surfaces constructed at right angles to a datum axis, total runout controls cumulative variations of perpendicularity (to detect wobble) and flatness (to detect concavity or convexity).

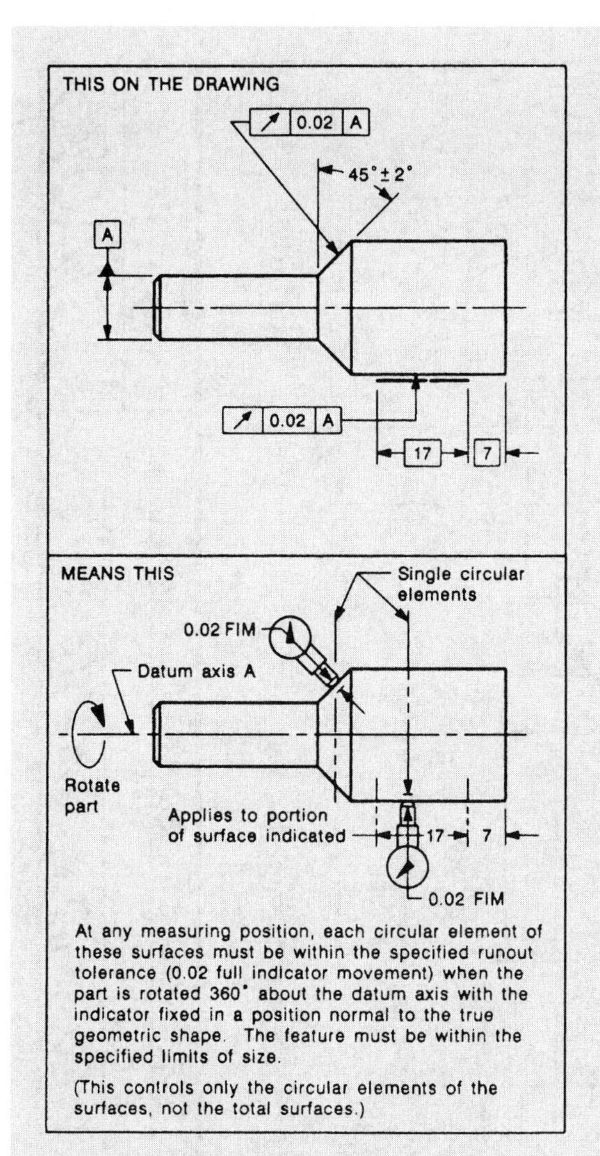

SPECIFYING CIRCULAR RUNOUT RELATIVE TO A DATUM DIAMETER

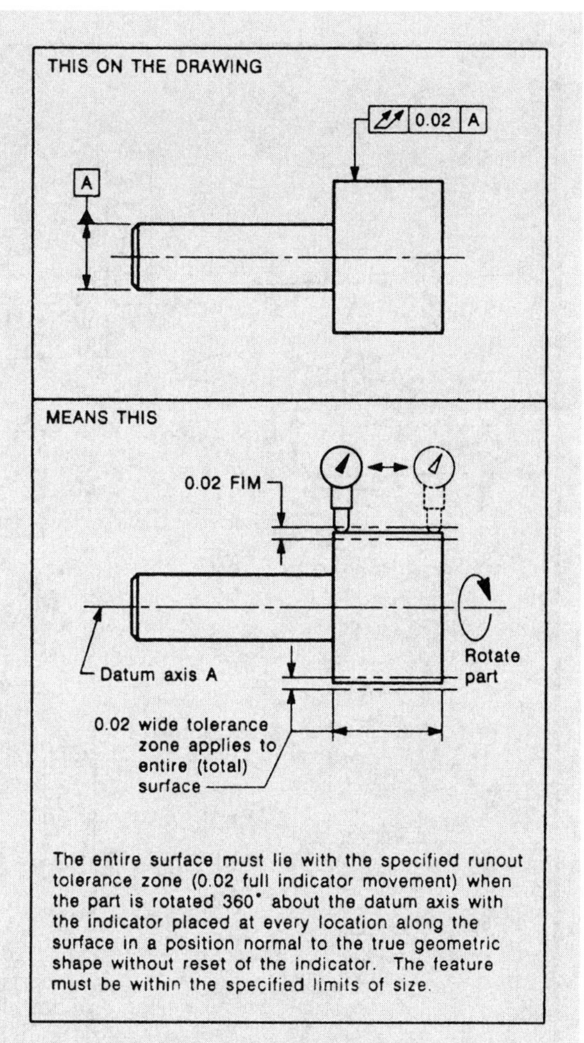

SPECIFYING TOTAL RUNOUT RELATIVE TO A DATUM DIAMETER

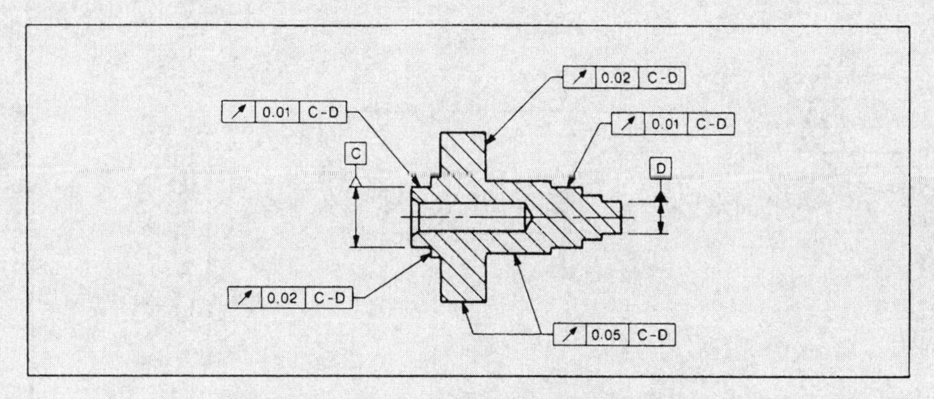

SPECIFYING RUNOUT RELATIVE TO TWO DATUM DIAMETERS

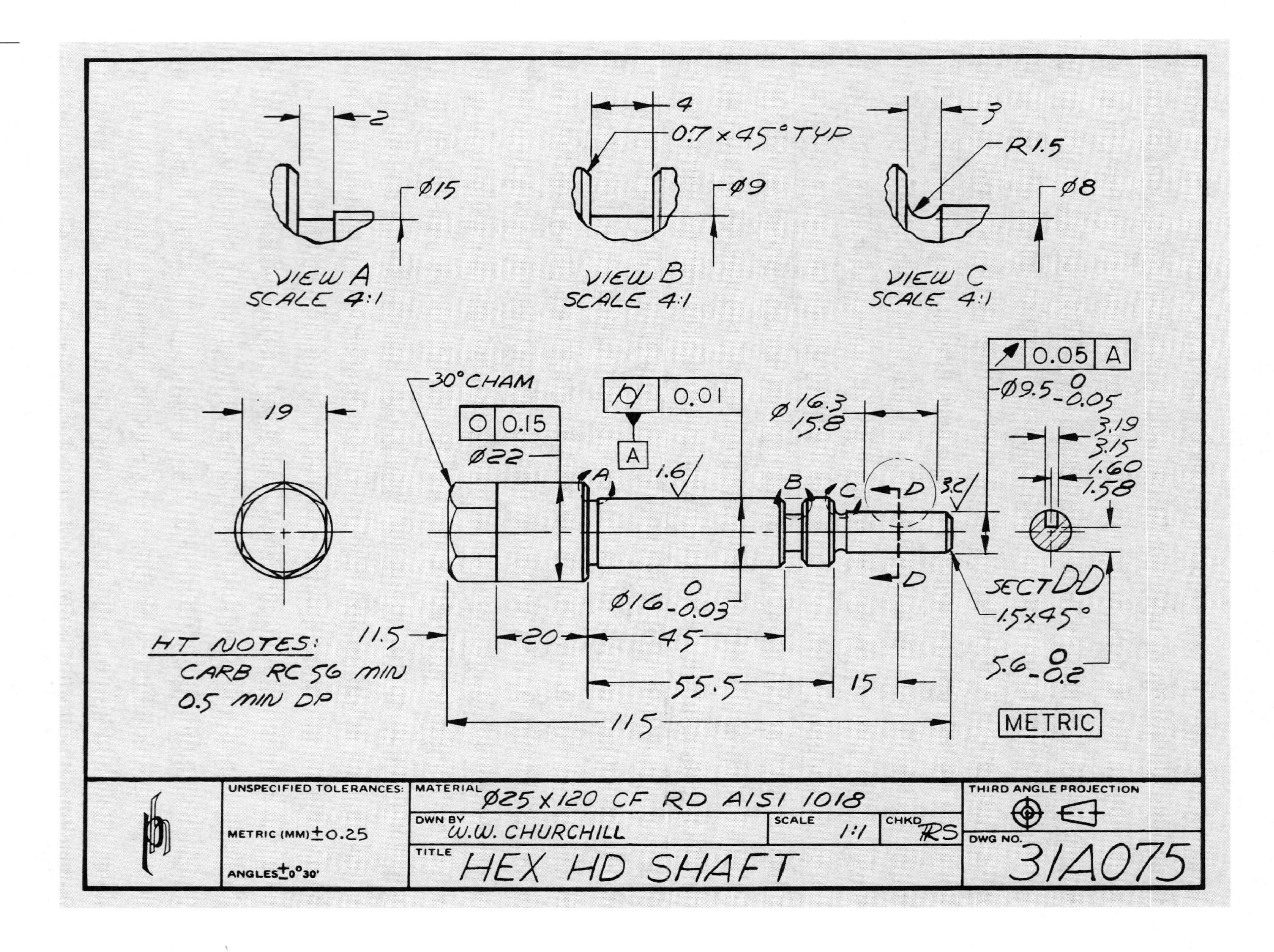

VIEW A
SCALE 4:1
2
Ø15
VIEW B
SCALE 4:1
4
0.7 × 45° TYP
Ø9
VIEW C
SCALE 4:1
3
R1.5
Ø8
19
30° CHAM
0.15
Ø22
0.01
A
1.6
B
C
D
3.2
Ø 16.3 15.8
0.05 A
Ø9.5 0 -0.05
3.19 3.15
1.60 1.58
SECT DD
1.5×45°
5.6 0 -0.2
Ø16 0 -0.03
11.5
20
45
55.5
15
115
HT NOTES:
CARB RC 56 MIN
0.5 MIN DP
METRIC
UNSPECIFIED TOLERANCES:
METRIC (MM) ±0.25
ANGLES ±0°30'
MATERIAL Ø25 X 120 CF RD AISI 1018
DWN BY W.W. CHURCHILL
SCALE 1:1
CHKD RS
TITLE HEX HD SHAFT
THIRD ANGLE PROJECTION
DWG NO. 31A075

INSTRUCTIONS: Refer to drawing 31A075 to answer the following questions.

1. What metric units do the dimensional tolerances represent? 1. ____
2. What metric units do the surface roughness tolerances represent? 2. ____
3. Which diameter (∅16 or ∅9.5) requires the smoothest surface? 3. ____
4. What is the minimum width of the keyseat? 4. ____
5. What is the maximum diameter of the keyseat cutter? 5. ____
6. What is the minimum Rockwell hardness permissible? 6. ____
7. What is the dimension across the flat sides of the hexagon? 7. ____
8. How many chamfers does the shaft contain? 8. ____
9. What characteristic does the tolerance symbol on the ∅22 represent? 9. ____
10. What characteristic does the tolerance symbol on the ∅16 represent? 10. ____
11. What characteristic does the tolerance symbol on the ∅9.5 represent? 11. ____
12. Which one of the above diameters is identified as datum A? 12. ____
13. Which one of the diameters uses datum A as a reference? 13. ____
14. How deep is the annular groove in datum A? 14. ____
15. What is the width of the annular groove in datum A? 15. ____
16. What is the width of the neck at the left side of datum A? 16. ____
17. What is the width of the neck at the left side of ∅9.5? 17. ____
18. What is the length of the ∅9.5 after necking and chamfering? 18. ____
19. How much size tolerance applies to ∅22? 19. ____
20. How much geometric tolerance applies to ∅22? 20. ____

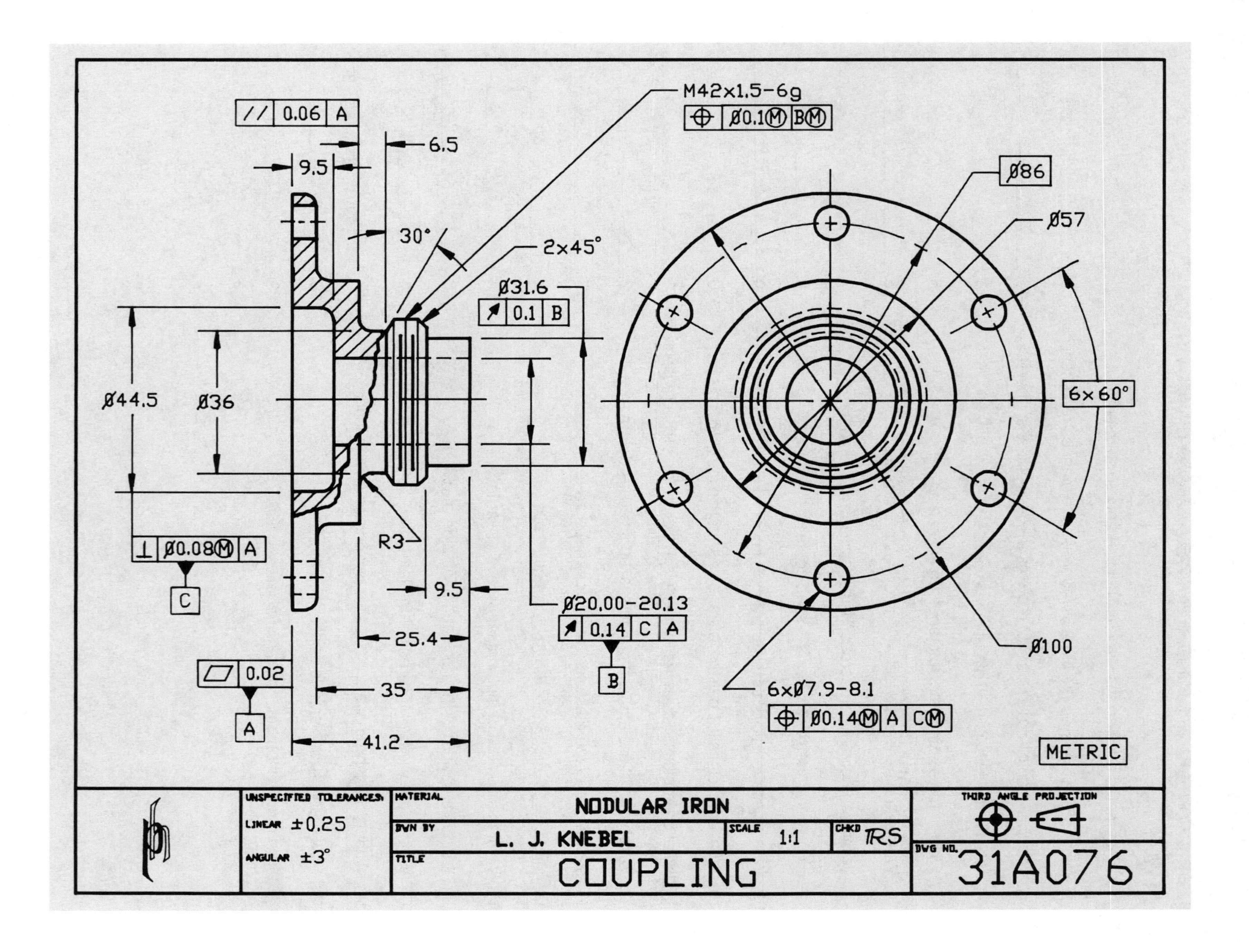
// 0.06 A
9.5
6.5
30°
M42x1.5-6g
⌖ Ø0.1Ⓜ BⓂ
2x45°
Ø86
Ø57
Ø31.6
↗ 0.1 B
Ø44.5
Ø36
6x60°
⊥ Ø0.08Ⓜ A
C
R3
9.5
Ø20.00-20.13
↗ 0.14 C A
25.4
B
Ø100
0.02
A
35
6xØ7.9-8.1
⌖ Ø0.14Ⓜ A CⓂ
41.2
METRIC
UNSPECIFIED TOLERANCES
LINEAR ±0.25
ANGULAR ±3°
MATERIAL
NODULAR IRON
DWN BY
L. J. KNEBEL
SCALE 1:1
CHKD TRS
TITLE
COUPLING
THIRD ANGLE PROJECTION
DWG NO.
31A076

INSTRUCTIONS: Refer to drawing 31A076 to answer the following questions.

1. How many feature control frames appear on the drawing? 1. ______
2. How many of them (question 1) contain at least one datum reference? 2. ______
3. How many of them (question 1) contain at least one material condition symbol? 3. ______
4. How many *different* geometric characteristic symbols appear? 4. ______
5. How many datum feature symbols appear? 5. ______
6. How many basic dimension symbols appear? 6. ______
7. Interpret the geometric characteristic symbol that appears on datum A. 7. ______
8. Interpret the geometric characteristic symbol that appears on datum B. 8. ______
9. Interpret the geometric characteristic symbol that appears on datum C. 9. ______
10. Interpret the geometric characteristic symbol that locates the mounting holes. 10. ______
11. Does the parallelism tolerance apply at MMC or RFS? 11. ______
12. Does the positional tolerance apply at MMC or RFS? 12. ______
13. How much size tolerance applies to datum B? 13. ______
14. How much geometric tolerance applies to datum B? 14. ______
15. How much size tolerance applies to datum C? 15. ______
16. How much geometric tolerance applies to datum C? 16. ______
17. How much size tolerance applies to the mounting holes? 17. ______
18. How much locational tolerance applies to the mounting holes? 18. ______
19. What size is the tolerance zone on the mounting hole locations if the hole size is exactly 7.9? 19. ______
20. What size is the tolerance zone on the mounting hole locations if the hole size is exactly 8.1? 20. ______

Optional Exercise Using Tolerance of Position:

1a. Return to Drawing 31A072 (p. 324) and create a feature control frame for the 3 × #10-24UNC threaded holes that states the following conditions:

The axis of each of the threaded holes is located within a .005 diameter cylindrical tolerance zone at RFS that is oriented perpendicular to datum B, and located relative to datum C at MMC. Designate datum C as the 1.300/1.298 diameter. Designate the proper dimensions as basic by enclosing them inside a rectangular frame.

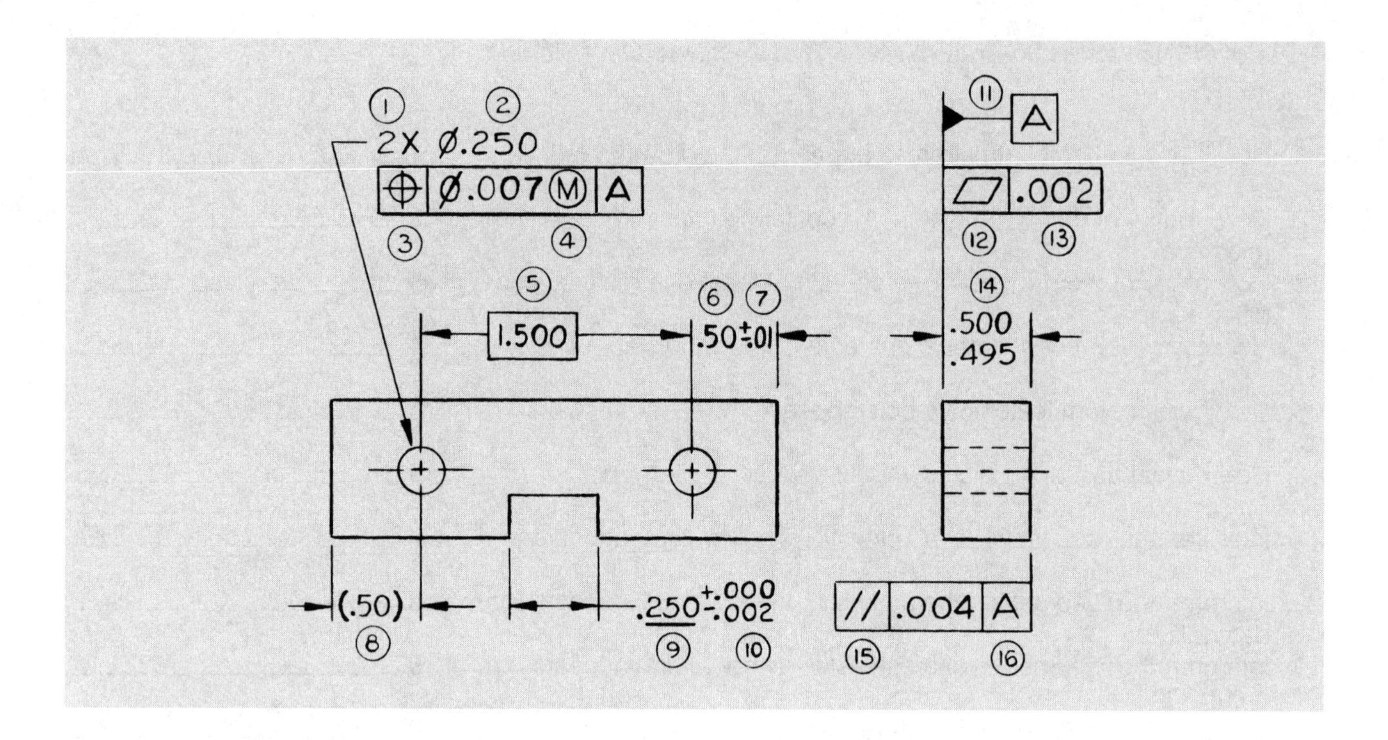

Symbol Quiz 3

INSTRUCTIONS: Match the descriptions listed below with the proper symbols and tolerances shown in the illustration above. Select the correct answers from the encircled numbers adjacent to the features.

1. Bilateral tolerance 1. ____________
2. Unilateral tolerance 2. ____________
3. Geometric tolerance 3. ____________
4. Datum reference 4. ____________
5. Datum feature symbol 5. ____________
6. Basic dimension symbol 6. ____________
7. Material condition symbol 7. ____________
8. Form tolerance symbol 8. ____________
9. Orientation tolerance symbol 9. ____________
10. Location tolerance symbol 10. ____________

PROJECTED TOLERANCE ZONE

This concept can be applied where the variation in perpendicularity of threaded or press-fit holes could cause fasteners such as screws, studs, or pins to interfere with mating parts. See the examples shown below. Note that it is the variation in perpendicularity of the portion of the fastener passing through the mating part that is significant. The location and perpendicularity of the threaded hole is of importance only insofar as it affects the extended portion of the engaging fastener.

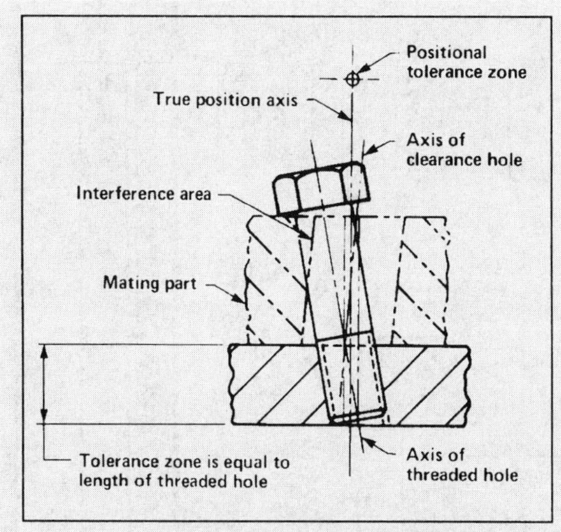

INTERFERENCE DIAGRAM, FASTENER AND HOLE

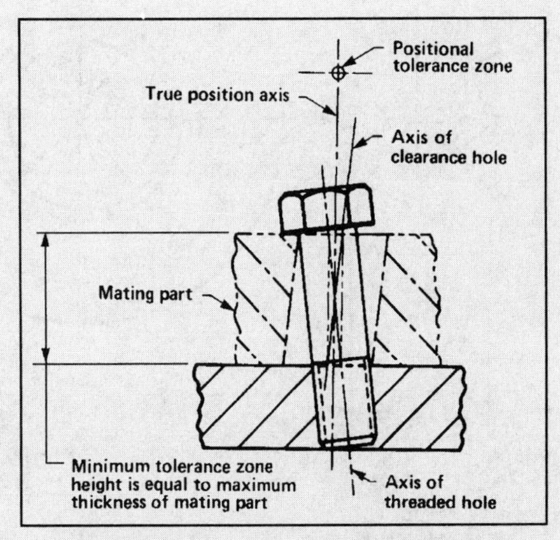

BASIS FOR PROJECTED TOLERANCE ZONE

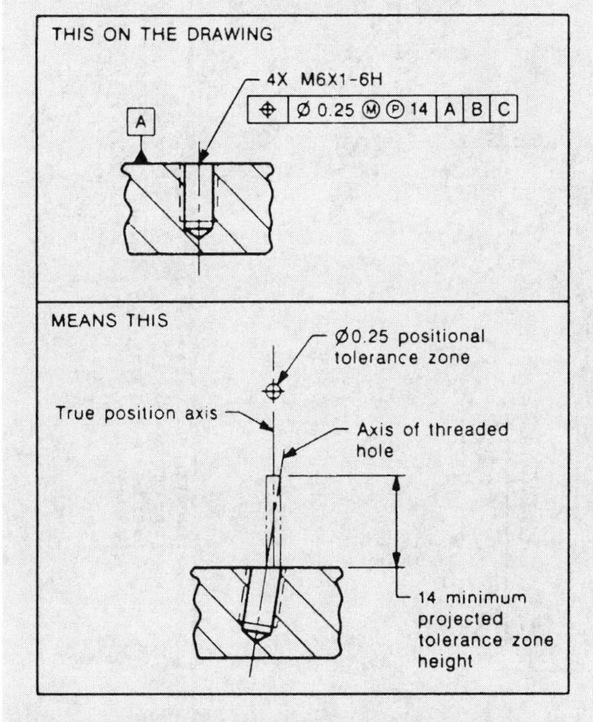

PROJECTED TOLERANCE ZONE SPECIFIED

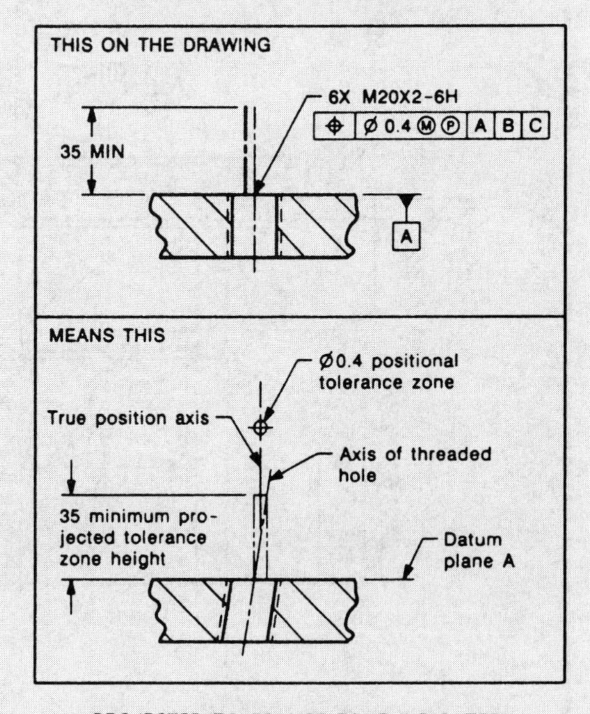

PROJECTED TOLERANCE ZONE INDICATED WITH CHAIN LINE

*Reprinted from ASME Y14.5M-1994, by permission of The American Society of Mechanical Engineers. *

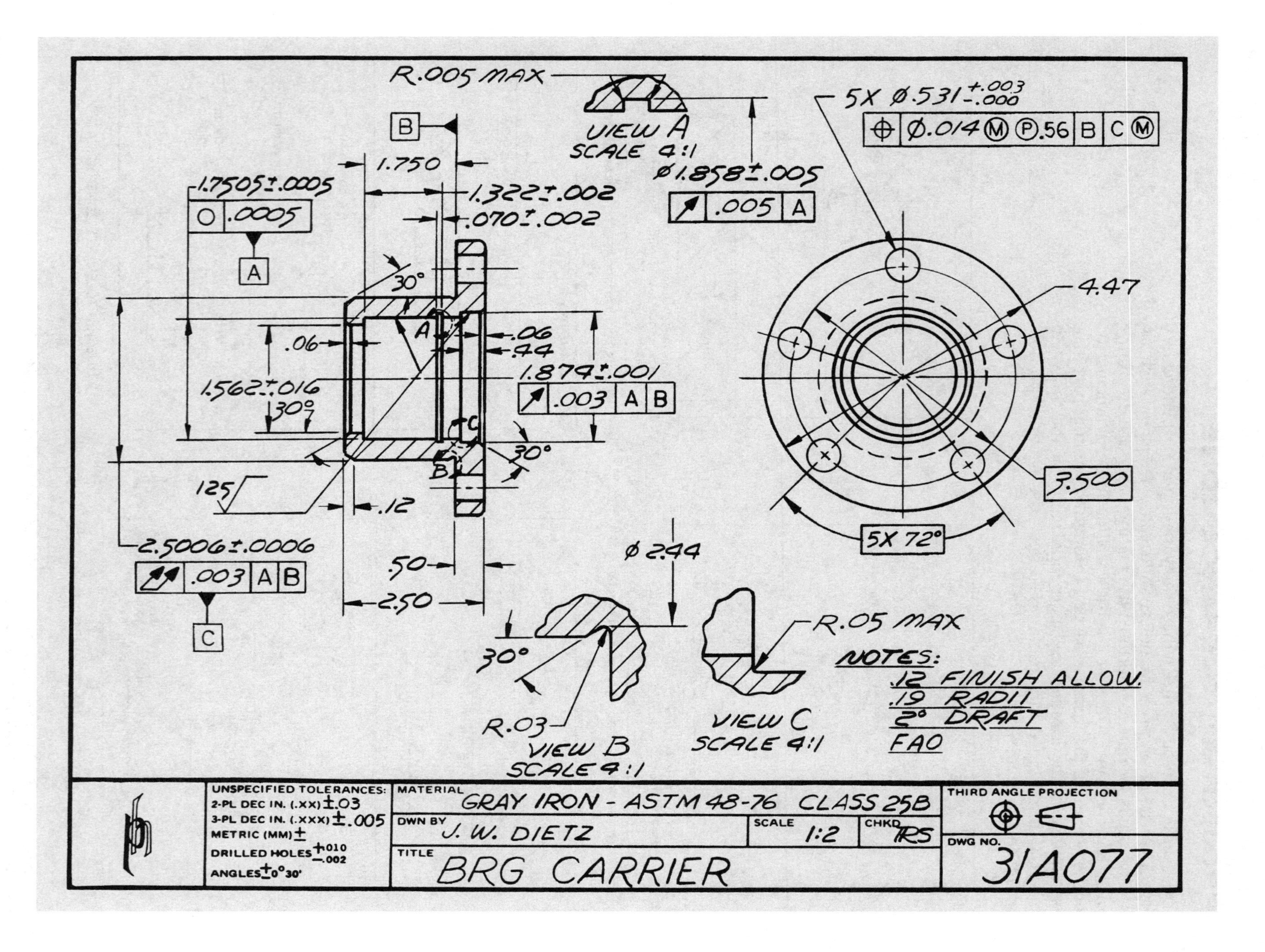

R.005 MAX
VIEW A
SCALE 4:1
Ø1.858±.005
.005 A
5X Ø.531 +.003 -.000
Ø.014 Ⓜ Ⓟ.56 B C Ⓜ
4.47
3.500
5X 72°
B
1.750
1.7505±.0005
.0005
A
1.322±.002
.070±.002
30°
.06
.06
.44
1.879±.001
.003 A B
1.562±.016
30°
A
B
C
125
.12
2.5006±.0006
.003 A B
C
.50
2.50
Ø2.44
30°
R.03
VIEW B
SCALE 4:1
R.05 MAX
VIEW C
SCALE 4:1
NOTES:
.12 FINISH ALLOW.
.19 RADII
2° DRAFT
FAO
UNSPECIFIED TOLERANCES:
2-PL DEC IN. (.XX) ±.03
3-PL DEC IN. (.XXX) ±.005
METRIC (MM) ±
DRILLED HOLES +.010 -.002
ANGLES ±0°30'
MATERIAL GRAY IRON - ASTM 48-76 CLASS 25B
DWN BY J. W. DIETZ
SCALE 1:2
CHKD RS
THIRD ANGLE PROJECTION
TITLE BRG CARRIER
DWG NO. 31A077

INSTRUCTIONS: Refer to drawing 31A077 to answer the following questions.

1. Interpret the geometric characteristic symbol on datum A. 1. ______
2. Interpret the geometric characteristic symbol on datum C. 2. ______
3. What is the size tolerance on datum C? 3. ______
4. What is the geometric tolerance on datum A? 4. ______
5. Does the geometric tolerance on datum A apply at MMC or RFS? 5. ______
6. Does the geometric tolerance on the location of the five holes apply at MMC or RFS? 6. ______
7. What would the material condition symbol Ⓛ signify if it appeared in a feature control frame? 7. ______
8. What is the shape of the tolerance zone for the location of the mounting holes? 8. ______
9. What two basic dimensions establish the location of the tolerance zones for the mounting holes? 9. ______
10. What size is the tolerance zone for the location of the mounting holes if their exact size is .531? 10. ______
11. What size is the tolerance zone for the location of the mounting holes if their exact size is .533? 11. ______
12. What is the minimum projected height of the positional tolerance zone for the mounting hole locations? 12. ______
13. What is the MMC of datum C? 13. ______
14. What is the MMC of the (.070) inside groove diameter? 14. ______
15. Is the geometric tolerance on the inside groove diameter total runout or circular runout? 15. ______
16. What is the diameter of the neck between datums C and B? 16. ______
17. Calculate the minimum wall thickness between datums C and A. 17. ______
18. Calculate the maximum wall thickness between datums C and A. 18. ______
19. Calculate the minimum distance between the left side of the 1.874 counterbore and the right side of the inside groove. 19. ______
20. Taking *all* tolerances into consideration, calculate the minimum possible web between the side of a mounting hole and the OD of the flange. 20. ______

Optional Exercise Using Tolerance of Position:

1a. Return to Drawing 31A072 (p.324) and create a feature control frame for the 3 × #12 drilled holes that states the following conditions:

The axis of each of the drilled holes is located within a .010 diameter cylindrical tolerance zone at MMC that is perpendicular to datum D and located relative to datum A at MMC. Designate datum D as the opposite side from datum B. Designate datum A as the .766–.764 diameter. Designate the proper dimensions as basic by enclosing them inside a rectangular frame.

GEOMETRIC CHARACTERISTIC SYMBOLS

The current dimensioning and tolerancing standards, ASME Y14.5M-1994 and CAN/CSA-B78.2-M91 were followed exclusively throughout this textbook. However, many older drawings still in use today contain geometric characteristic symbols from earlier standards. Therefore, it is imperative that you learn the old as well as the new symbols. Shown below are the symbols from the ANSI Y14.5-1973 standard. Study them, along with their footnotes.

		CHARACTERISTIC	SYMBOL	NOTES
INDIVIDUAL FEATURES	FORM TOLERANCES	STRAIGHTNESS	—	1
		FLATNESS	▱	1
		ROUNDNESS (CIRCULARITY)	○	
		CYLINDRICITY	⌭	
INDIVIDUAL OR RELATED FEATURES		PROFILE OF A LINE	⌒	2
		PROFILE OF A SURFACE	⌓	2
RELATED FEATURES		ANGULARITY	∠	
		PERPENDICULARITY (SQUARENESS)	⊥	
		PARALLELISM	//	3
	LOCATION TOLERANCES	POSITION	⌖	
		CONCENTRICITY	◎	3,7
		SYMMETRY	⌯	5
	RUNOUT TOLERANCES	CIRCULAR	↗	4
		TOTAL	↗	4,6

Note: 1) The symbol ∼ formerly denoted flatness.
The symbol ⌒ or — formerly denoted flatness and straightness.
2) Considered "related" features where datums are specified.
3) The symbol || and ◉ formerly denoted parallelism and concentricity, respectively.
4) The symbol ↗ without the qualifier "CIRCULAR" formerly denoted total runout.
5) Where symmetry applies, it is preferred that the position symbol be used.
6) "TOTAL" must be specified under the feature control symbol.
7) Consider the use of position or runout.

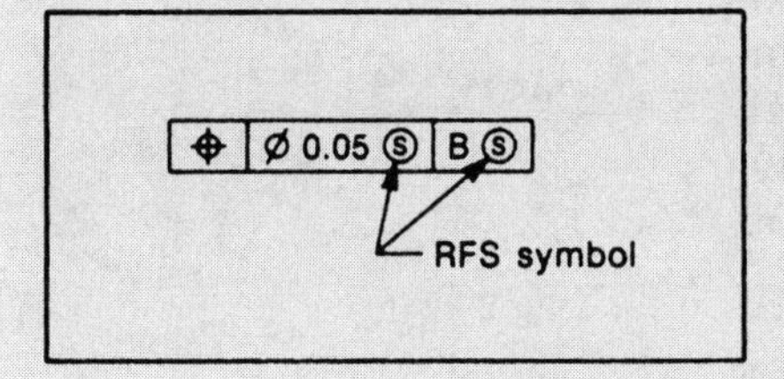

FORMER RFS SYMBOL APPLIED TO A FEATURE AND DATUM

FORMER DATUM FEATURE SYMBOL

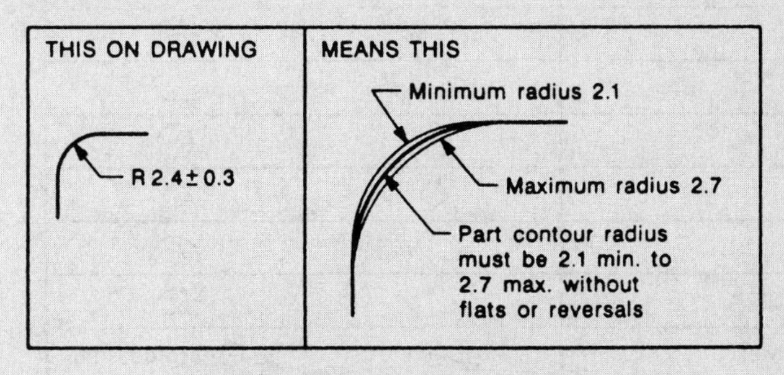

FORMER INTERPRETATION OF THE TOLERANCE ZONE CREATED BY THE SYMBOL R

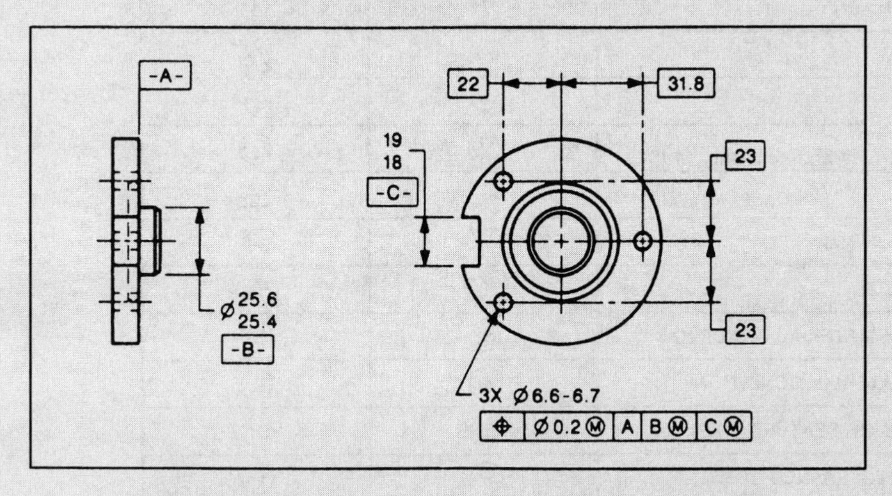

EXAMPLE OF FORMER DATUM FEATURE SYMBOL APPLICATIONS

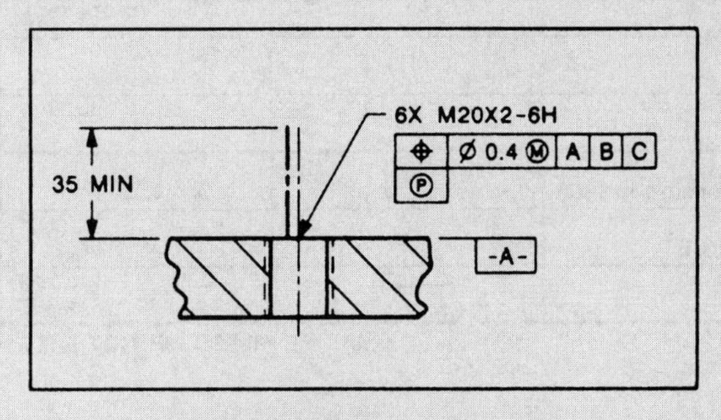

FORMER METHOD OF INDICATING A PROJECTED TOLERANCE ZONE

COMPARISON OF SYMBOLS

With the advent of global manufacturing, it is important that people involved in the manufacturing sector the world over have similar standards with which to work. A comparison chart of ASME Y14.5M-1994 and ISO symbols is shown below and on the following page. You will notice that the two systems of tolerancing are very similar. Some day, perhaps both standards will be the same.

SYMBOL FOR:	ASME Y14.5M	ISO
STRAIGHTNESS	⏤	⏤
FLATNESS	⏥	⏥
CIRCULARITY	○	○
CYLINDRICITY	⌭	⌭
PROFILE OF A LINE	⌒	⌒
PROFILE OF A SURFACE	⌓	⌓
ALL AROUND	↙⌀	↙⌀ (proposed)
ANGULARITY	∠	∠
PERPENDICULARITY	⊥	⊥
PARALLELISM	//	//
POSITION	⌖	⌖
CONCENTRICITY (concentricity and coaxiality in ISO)	◎	◎
SYMMETRY	⌯	⌯
CIRCULAR RUNOUT	•↗	↗
TOTAL RUNOUT	•⌰	⌰
AT MAXIMUM MATERIAL CONDITION	Ⓜ	Ⓜ
AT LEAST MATERIAL CONDITION	Ⓛ	Ⓛ
REGARDLESS OF FEATURE SIZE	NONE	NONE
PROJECTED TOLERANCE ZONE	Ⓟ	Ⓟ
TANGENT PLANE	Ⓣ	Ⓣ (proposed)
FREE STATE	Ⓕ	Ⓕ
DIAMETER	Ø	Ø
BASIC DIMENSION (theoretically exact dimension in ISO)	[50]	[50]
REFERENCE DIMENSION (auxiliary dimension in ISO)	(50)	(50)
DATUM FEATURE	•▲—[A]	•▲— or •▲—[A]

• MAY BE FILLED OR NOT FILLED

SYMBOL FOR:	ASME Y14.5M	ISO
DIMENSION ORIGIN	⌯→	⌯→
FEATURE CONTROL FRAME	⌖ | Ø0.5 Ⓜ | A | B | C	⌖ | Ø0.5 Ⓜ | A | B | C
CONICAL TAPER	⌲	⌲
SLOPE	⌳	⌳
COUNTERBORE/SPOTFACE	⌴	⌴ (proposed)
COUNTERSINK	⌵	⌵ (proposed)
DEPTH/DEEP	↧	↧ (proposed)
SQUARE	□	□
DIMENSION NOT TO SCALE	15 (underlined)	15 (underlined)
NUMBER OF PLACES	8X	8X
ARC LENGTH	⌒105	⌒105
RADIUS	R	R
SPHERICAL RADIUS	SR	SR
SPHERICAL DIAMETER	SØ	SØ
CONTROLLED RADIUS	CR	NONE
BETWEEN	•↔	NONE
STATISTICAL TOLERANCE	(ST)	NONE
DATUM TARGET	Ø6 / A1 or A1 —Ø6	Ø6 / A1 or A1 —Ø6
TARGET POINT	X	X

• MAY BE FILLED OR NOT FILLED

UNIT 15

SPUR GEAR TERMINOLOGY

ADDENDUM (A): radial distance on a tooth from the pitch circle to the outside circle.

CENTER DISTANCE: center-to-center distance between the axes of two meshing gears.

CHORDAL ADDENDUM: radial distance from the top of the tooth to the chord of the pitch circle.

CHORDAL THICKNESS: length of the chord along the pitch circle between the two sides of the tooth.

CIRCULAR PITCH: length of the pitch circle arc between corresponding points of adjacent teeth.

CIRCULAR THICKNESS: thickness of the tooth measured along a chord of the pitch circle.

CLEARANCE (C): distance allowed between the top of the tooth and the bottom of the mating tooth space, to avoid interference when meshing.

DEDENDUM (D): radial distance on a tooth from the pitch circle to the root circle.

DIAMETRAL PITCH (P): ratio of the number of teeth to the number of inches of pitch diameter (number of teeth per inch of pitch diameter). Mating gears must have the same diametral pitch.

NUMBER OF TEETH (N): total quantity of teeth formed along the pitch circle.

OUTSIDE DIAMETER (OD): maximum diameter (top of tooth).

PITCH DIAMETER (PD): diameter of the pitch circle. If two spur gears are in mesh, their pitch circles are tangent to each other. The center distance between two spur gears is the sum of the radii of the pitch circles.

PRESSURE ANGLE: standard angle at which pressure from the tooth of one gear is passed to the tooth of another gear. For involute gears, the pressure angle is usually $14\frac{1}{2}°$ or 20°.

ROOT DIAMETER (RD): diameter of the root circle (bottom of tooth spaces).

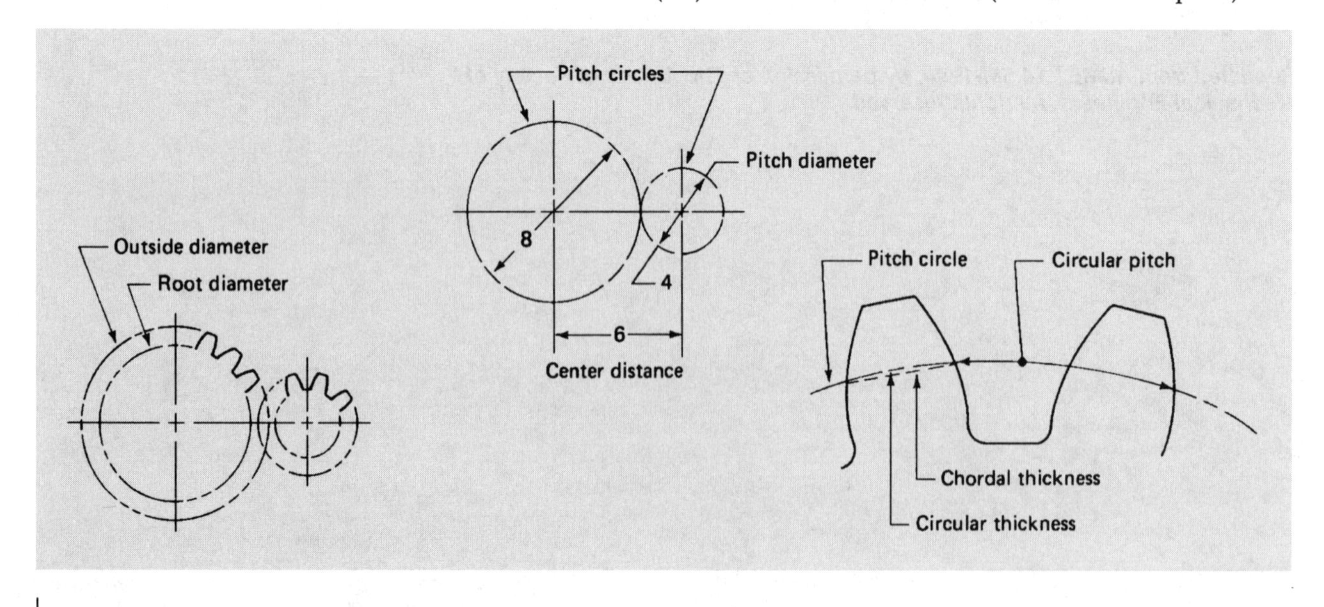

SPUR GEAR FORMULAS

$$\text{Center distance} = \frac{\text{PD of pinion} + \text{PD of gear}}{2}$$

$$\text{Chordal addendum} = A + \frac{PD}{2}\left[1.00 - \cos\left(\frac{90^\circ}{N}\right)\right]$$

$$\text{Chordal thickness} = PD \times \sin\left(\frac{90^\circ}{N}\right)$$

$$\text{Circular pitch} = \frac{\pi}{P}$$

$$\text{Circular thickness} = \frac{\pi \times PD}{2N}$$

$$\text{Working depth} = \text{A of pinion} + \text{A of gear}$$

Formulas		*Abbreviations*
$A = \frac{1.00}{P}$	$P = \frac{N}{PD}$ or $\frac{N+2}{OD}$	A = Addendum
$C = D - A$	$PD = \frac{N}{P}$	C = Clearance
$D = \frac{1.157}{P}$	$RD = PD - 2D$	D = Dedendum
$N = P \times PD$	$WD = A + D$	N = Number of Teeth
$OD = PD + 2A$ or $\frac{N+2}{P}$		OD = Outside Diameter
		P = Diametral Pitch
		PD = Pitch Diameter
		RD = Root Diameter
		WD = Whole Depth

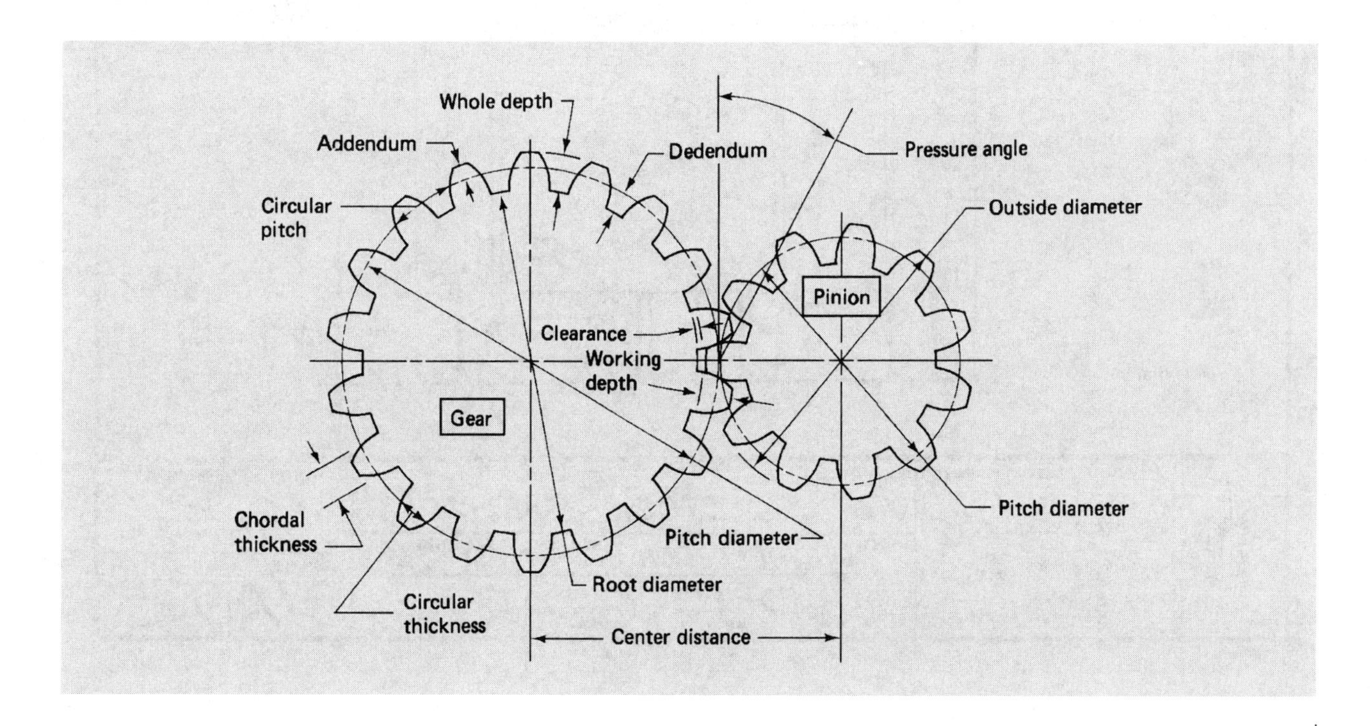

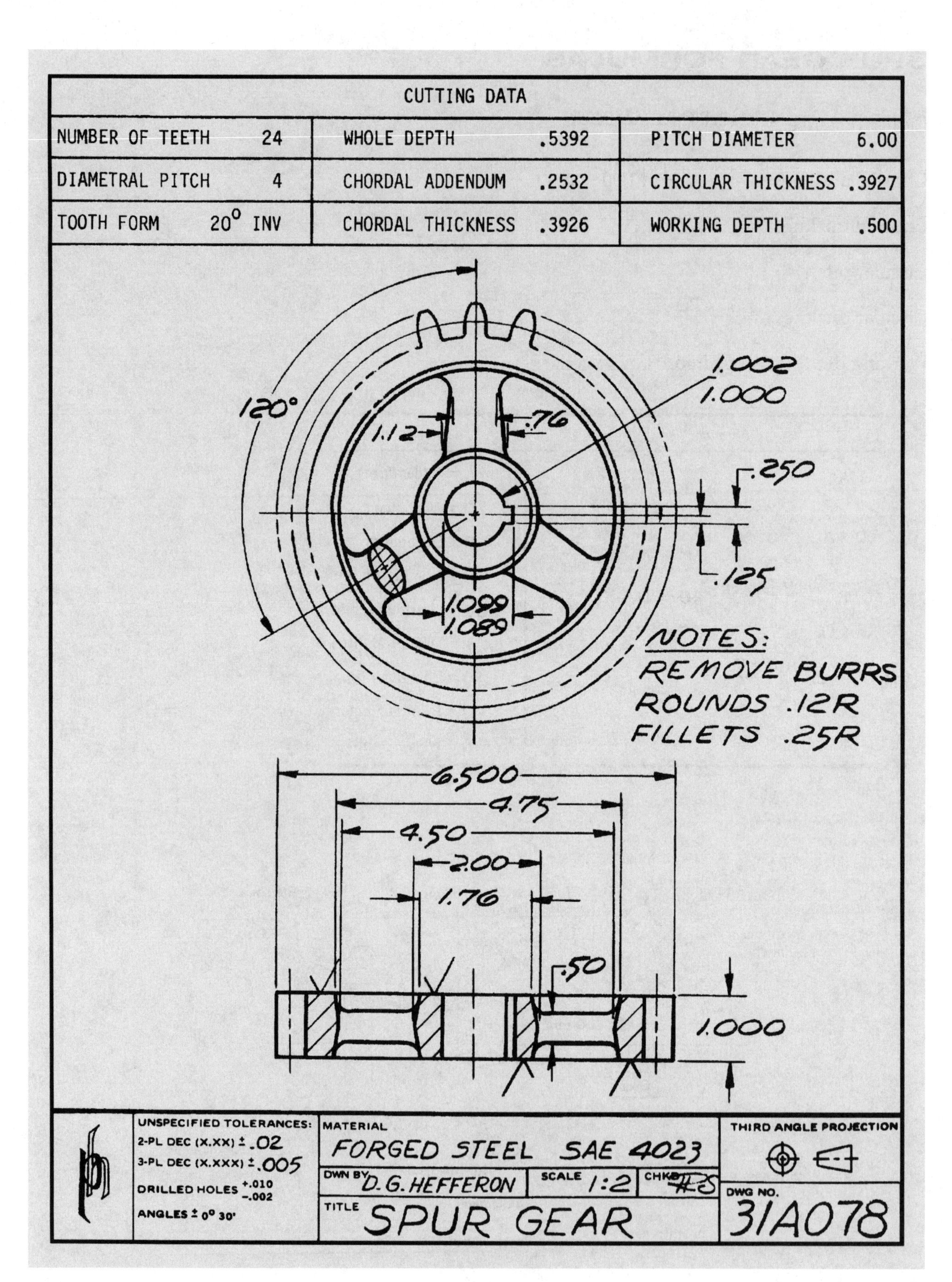
CUTTING DATA
NUMBER OF TEETH 24
DIAMETRAL PITCH 4
TOOTH FORM 20° INV
WHOLE DEPTH .5392
CHORDAL ADDENDUM .2532
CHORDAL THICKNESS .3926
PITCH DIAMETER 6.00
CIRCULAR THICKNESS .3927
WORKING DEPTH .500
1.002
1.000
120°
1.12
.76
.250
.125
1.099
1.089
NOTES:
REMOVE BURRS
ROUNDS .12R
FILLETS .25R
6.500
4.75
4.50
2.00
1.76
.50
1.000
UNSPECIFIED TOLERANCES:
2-PL DEC (X.XX) ± .02
3-PL DEC (X.XXX) ± .005
DRILLED HOLES +.010 −.002
ANGLES ± 0° 30'
MATERIAL
FORGED STEEL SAE 4023
DWN BY D.G. HEFFERON
SCALE 1:2
CHKD
TITLE SPUR GEAR
THIRD ANGLE PROJECTION
DWG NO.
31A078

SPUR GEARS

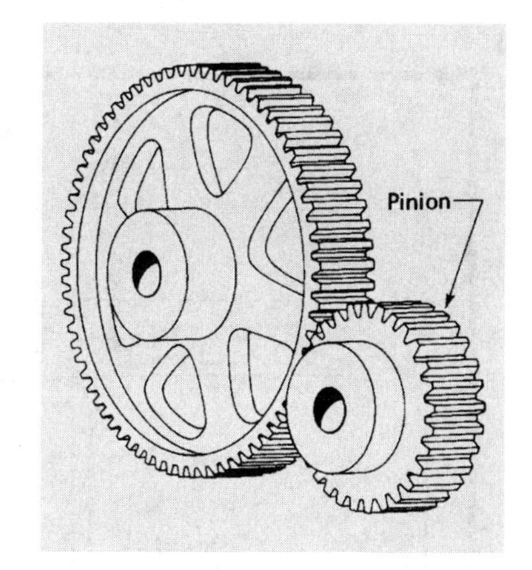

Spur gears are toothed wheels designed to transmit positive rotary motion from one shaft to another. When there is a difference in the size of gears, there will be a difference in the revolutions per minute (RPM). The smaller of two gears in mesh is known as the pinion.

Drawings of spur gears usually include only a few of the teeth, with phantom lines used to represent those omitted. All pertinent information will be listed in a table included on the drawing. Information not included on the drawing may be obtained by using the formulas listed on page 351.

INSTRUCTIONS: Refer to drawing 31A078 to answer the following questions.

1. What is the outside diameter of the spur gear? 1. ____________
2. What is the pitch diameter of the spur gear? 2. ____________
3. Are mating spur gear spacings determined by (a) their OD or (b) their PD? 3. ____________
4. What pressure angle is specified? 4. ____________
5. Calculate the addendum using the formula on page 351. 5. ____________
6. Calculate the dedendum using the proper formula. 6. ____________
7. Calculate the root diameter. 7. ____________
8. What is the minimum wall thickness between the bore and the hub? 8. ____________
9. Show the limits of the gear thickness. 9. ____________
10. What is the thickness of the spokes? 10. ____________
11. What is the width of the keyway? 11. ____________
12. What is the fillet radius? 12. ____________
13. What is the length of each spoke? 13. ____________
14. What is the approximate spoke width midway between the hub and the rim? 14. ____________
15. Would a mating gear have a diametral pitch of (a) 4, (b) 5, or (c) 8? 15. ____________

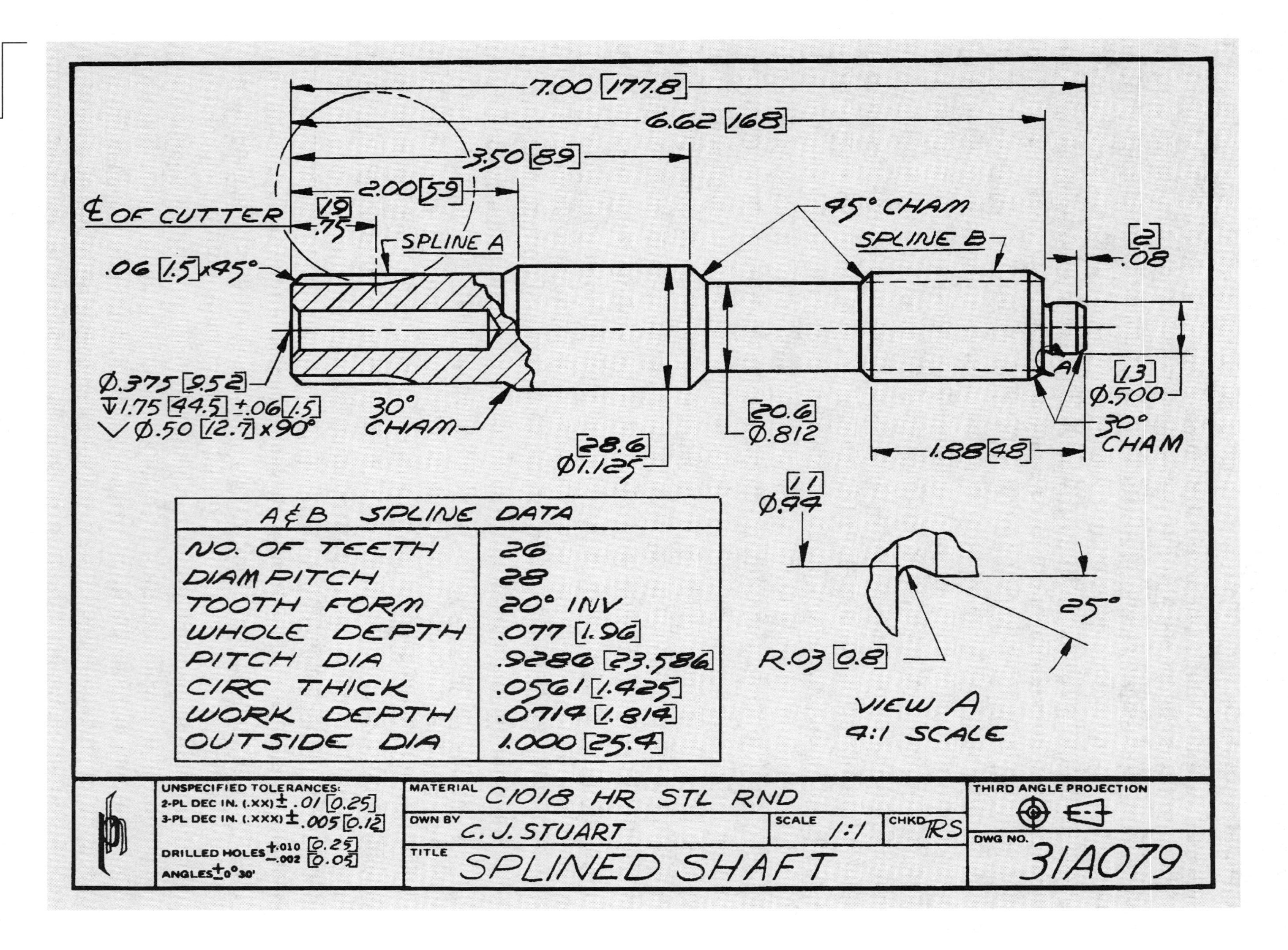

A & B SPLINE DATA	
NO. OF TEETH	26
DIAM PITCH	28
TOOTH FORM	20° INV
WHOLE DEPTH	.077 [1.96]
PITCH DIA	.9286 [23.586]
CIRC THICK	.0561 [1.425]
WORK DEPTH	.0714 [1.814]
OUTSIDE DIA	1.000 [25.4]

SPLINES

Instead of machining keyseats and keyways for keys to prevent rotation between shaft and hub, splines are often machined into both members, parallel to the axis. The teeth on a spline may have parallel sides, or they may have an involute profile. In either case, they are usually drawn in simplified form, with an accompanying table containing machining specifications.

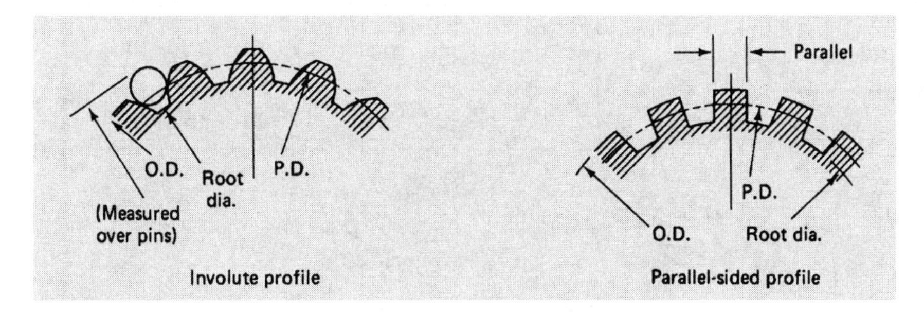

INSTRUCTIONS: Refer to drawing 31A079 to answer the following questions.

1. What method of dimensioning was used? 1. ______
2. What type of section was used on the left side of the shaft? 2. ______
3. What type of view was drawn to dimension the neck? 3. ______
4. What is the metric diameter of the neck? 4. ______
5. What is the outside diameter (inches) of splines A and B? 5. ______
6. Is the specified material (a) a carbon steel or (b) an alloy steel? 6. ______
7. How many chamfers are to be machined on the outside? 7. ______
8. What pressure angle is specified for the splines? 8. ______
9. Are the spline teeth (a) parallel-sided or (b) involute? 9. ______
10. What is the pitch diameter (inches) of the spline? 10. ______
11. What are the metric limits of the shaft length? 11. ______
12. What is the metric depth of the blind hole? 12. ______
13. What is the effective length (inches) of spline A? (Include the chamfer.) 13. ______
14. What is the length (inches) of spline B? (Include the chamfers.) 14. ______
15. What is the length (inches) of the .812 dia? 15. ______
16. What is the maximum length (inches) of the 1.125 dia? 16. ______
17. What is the minimum length (millimeters) of the .812 dia? 17. ______

FLAT ROOT EXTERNAL INVOLUTE SPLINE	
NO. OF TEETH	20
DIAMETRAL PITCH	12
PRESSURE ANGLE	20°
MINOR DIA	1.5627-1.5494

FLAT ROOT INTERNAL INVOLUTE SPLINE	
NO. OF TEETH	11
DIAMETRAL PITCH	16
PRESSURE ANGLE	20°
MINOR DIA	.6306-.6356
MAJOR DIA	.7500-.7511

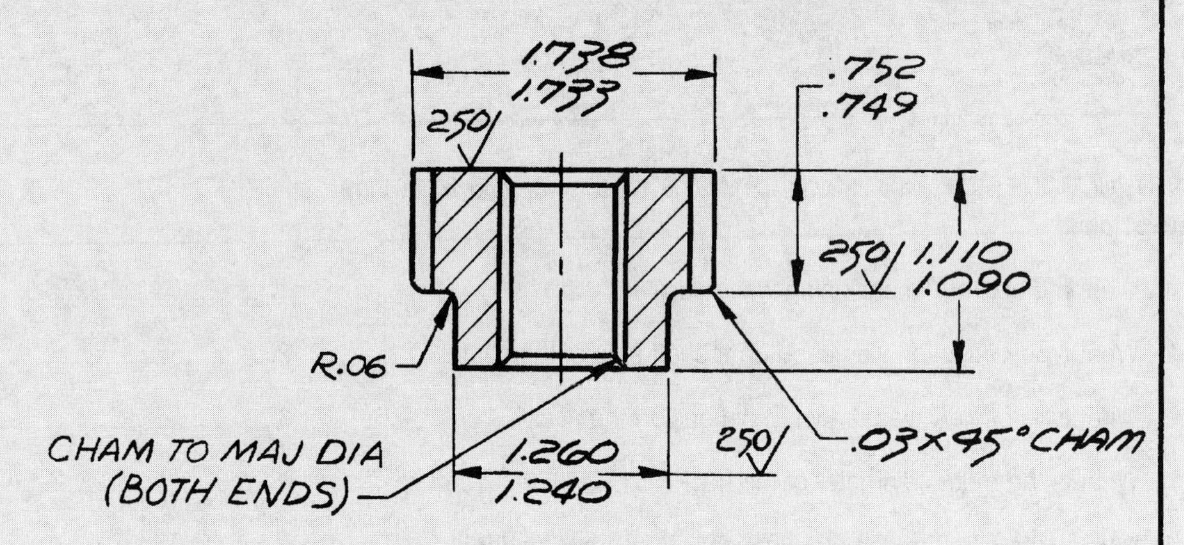

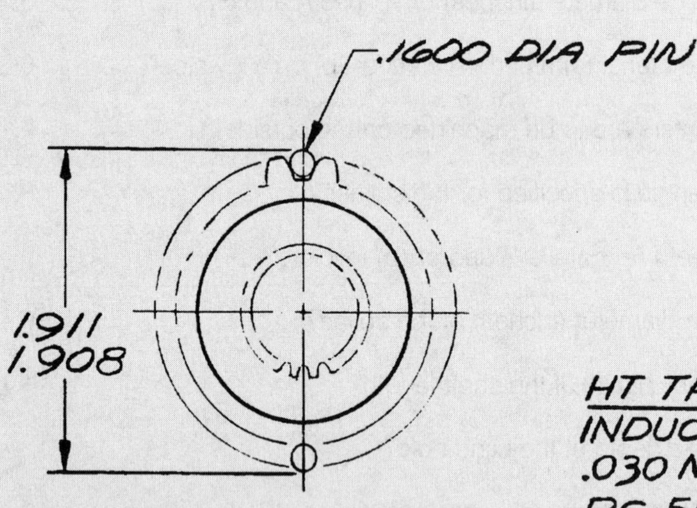

HT TR:
INDUCTION HDN
.030 MIN DP
RC 56 MIN

UNSPECIFIED TOLERANCES: 2-PL DEC (X.XX) ± .01; 3-PL DEC (X.XXX) ±; DRILLED HOLES +.010 −.002; ANGLES ± 0° 30'	MATERIAL 1.750 HR RD AISI 8620	THIRD ANGLE PROJECTION
	DWN BY G.L. DUFFY — SCALE 1:1 — CHKD RS	DWG NO. 31A080
	TITLE SPLINED HUB	

INSTRUCTIONS: Refer to drawing 31A080 to answer the following questions.

1. How many teeth does the external spline contain? 1. ______
2. How many teeth does the internal spline contain? 2. ______
3. What is the maximum OD of the part? 3. ______
4. What is the minimum ID of the part? 4. ______
5. What is the pressure angle of the teeth? 5. ______
6. Are the spline teeth (a) involute or (b) parallel-sided? 6. ______
7. What is the diametral pitch of the external spline? 7. ______
8. What is the diametral pitch of the internal spline? 8. ______
9. What is the pitch diameter of the external spline? (Calculate.) 9. ______
10. What is the pitch diameter of the internal spline? (Calculate.) 10. ______
11. What is the maximum length of the 1.250 diameter? (Disregard the radius.) 11. ______
12. What is the minimum length of the 1.250 diameter? (Disregard the radius.) 12. ______
13. What size pins are specified for measuring the external spline? 13. ______
14. How many chamfers does the part contain? 14. ______
15. What is the carbon content of the steel? 15. ______
16. What type of section is the top view? 16. ______
17. What type of lines are used to represent teeth on the front view? 17. ______
18. What is the maximum surface roughness tolerance specified? 18. ______
19. What Rockwell hardness is specified? (Include the letters.) 19. ______
20. What is the diameter of the bar stock specified? 20. ______

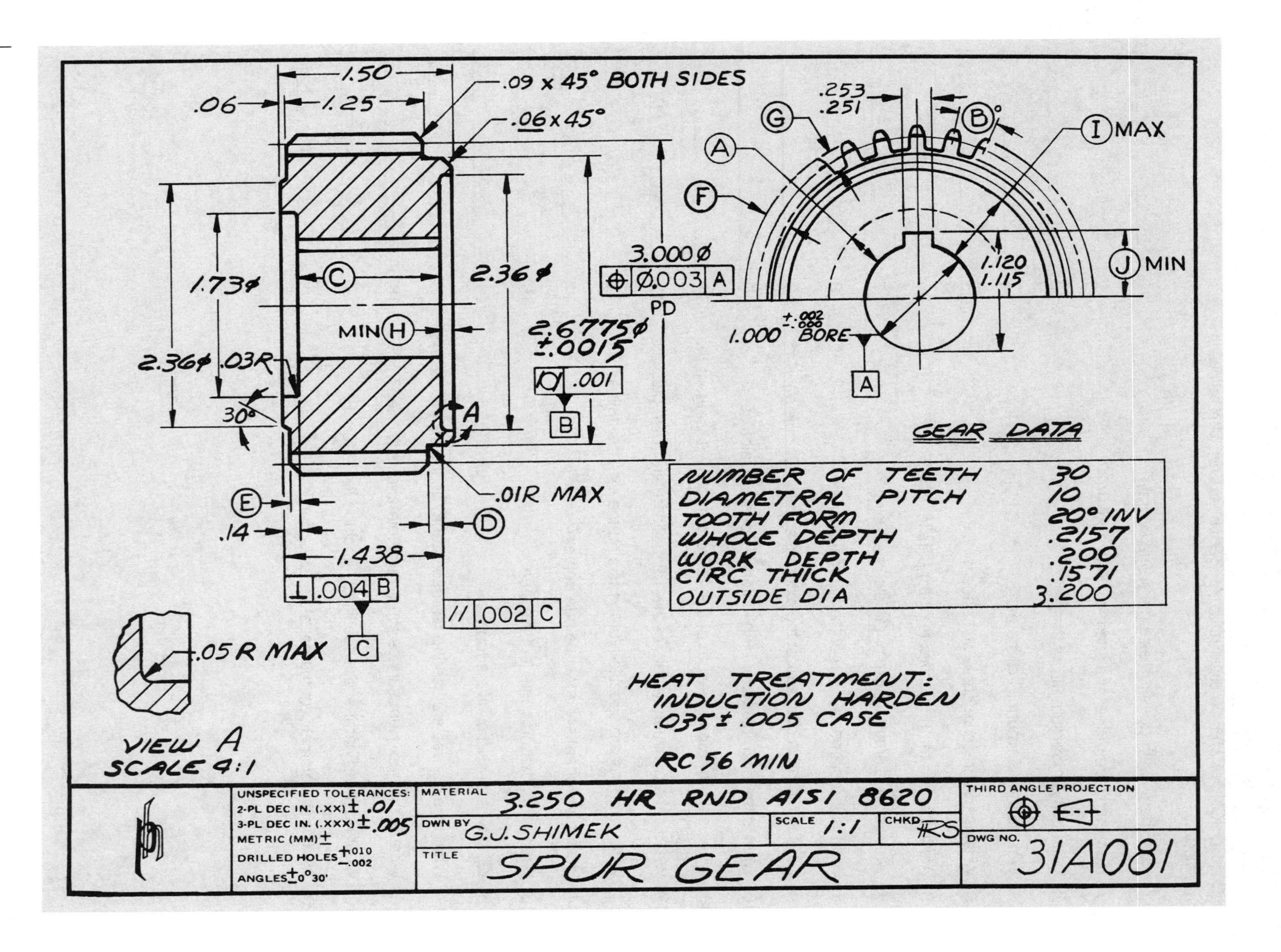

1.50
.06
1.25
.09 x 45° BOTH SIDES
.06 x 45°
1.73⌀
C
2.36⌀
MIN H
2.36⌀
.03R
30°
A
E
.14
1.438
D
.01R MAX
⊥ .004 B
C
// .002 C
3.000⌀
⊕ ⌀.003 A
PD
2.6775⌀ ±.0015
⌭ .001
B
.253 .251
G
A
F
B°
I MAX
J MIN
1.120 1.115
1.000 +.002 -.000 BORE
A
GEAR DATA
NUMBER OF TEETH 30
DIAMETRAL PITCH 10
TOOTH FORM 20° INV
WHOLE DEPTH .2157
WORK DEPTH .200
CIRC THICK .1571
OUTSIDE DIA 3.200
.05 R MAX
VIEW A
SCALE 4:1
HEAT TREATMENT:
INDUCTION HARDEN
.035 ± .005 CASE
RC 56 MIN
UNSPECIFIED TOLERANCES:
2-PL DEC IN. (.XX) ± .01
3-PL DEC IN. (.XXX) ± .005
METRIC (MM) ±
DRILLED HOLES +.010 -.002
ANGLES ± 0°30'
MATERIAL 3.250 HR RND AISI 8620
DWN BY G.J.SHIMEK
SCALE 1:1
CHKD RS
THIRD ANGLE PROJECTION
TITLE SPUR GEAR
DWG NO. 31A081

INSTRUCTIONS: Refer to drawing 31A081 to answer the following questions.

1. What diametral pitch would be required for a mating gear? 1. ____________
2. Do the material specifications designate a carbon steel or an alloy steel? 2. ____________
3. Do the geometric tolerances apply at LMC, MMC, or RFS? 3. ____________
4. Show the pressure angle of this gear's teeth. 4. ____________
5. Interpret the geometric characteristic symbol on the pitch diameter. 5. ____________
6. Interpret the geometric characteristic symbol on datum C. 6. ____________
7. Interpret the geometric characteristic symbol on datum B. 7. ____________
8. Show the addendum dimension of this gear's teeth. 8. ____________
9. Show the MMC of datum A. 9. ____________
10. Show the MMC of datum B. 10. ____________

INSTRUCTIONS: Enter the dimensions for the following letters.

Ⓐ ____________

Ⓑ ____________

Ⓒ ____________

Ⓓ ____________

Ⓔ ____________

Ⓕ ____________

Ⓖ ____________

Ⓗ MIN: ____________

Ⓘ MAX: ____________

Ⓙ MIN: ____________

UNIT 16

WELDMENTS

Weldments are units created from individual pieces of metal that have been welded together. Almost all metals can be welded, some more easily than others. Low-carbon steel, for example, can be welded much more easily than high-carbon or certain alloyed steels. Exceptionally thin material (20 ga. or thinner) and very heavy material (1 in. or thicker) may require special procedures to weld. Weldment drawings will include many different weld symbols with which you must become familiar. First, acquaint yourself with the various types of joints and welds that follow.

Types of Joints

There are five basic types of weld joints. Named for the positions of the pieces being joined, they are: butt, tee, corner, lap, and edge. Observe the illustration of each drawn below.

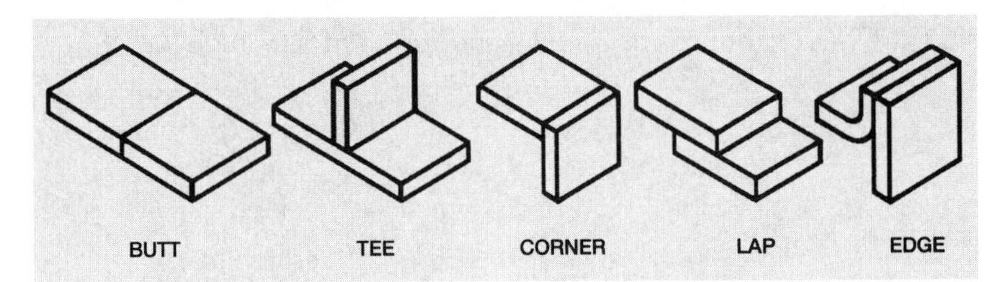

Types of Welds

There are four basic types of gas and arc welds. They are: fillet, groove, plug or slot, and back or backing weld. The common fillet weld may be used for tee, lap, or corner joints. Groove welds are named for the shape of their joints: square, bevel, V (both pieces beveled), J, and U (both pieces J-grooved). Plug or slot welds are used when fastening one piece on top of another to eliminate bolts and rivets. Back or backing welds are used on the surface of joints when no groove is used, or on the back side of a single groove weld. Shown below are the various types of gas and arc welds described above.

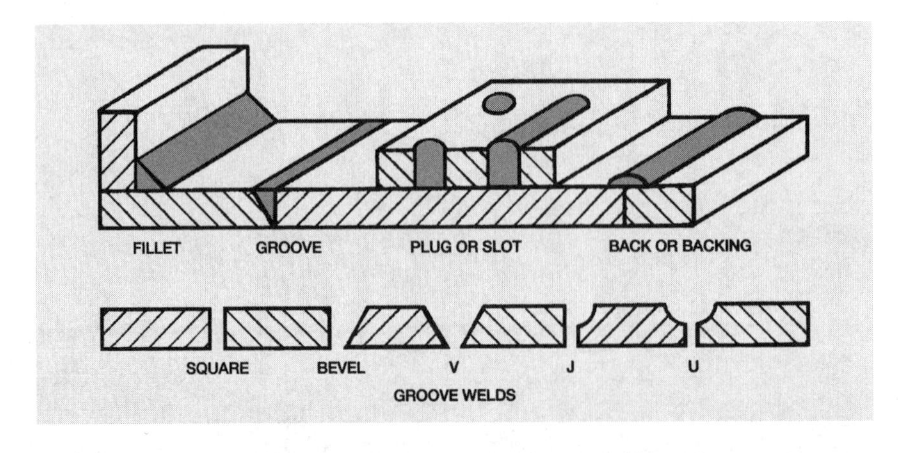

WELDING SYMBOLS

To convey the complete welding information from the designer to the welder, a series of graphical symbols has been established for use on weldment drawings. The symbols and their method of use are part of the American National Standard ANSI/AWS A2.4-98 sponsored by the American Welding Society.

Reference Line

The reference line forms the body of the welding symbol. All other elements are placed in their designated positions with respect to this line. An arrow is affixed to one end and a tail, when required, is affixed to the other end.

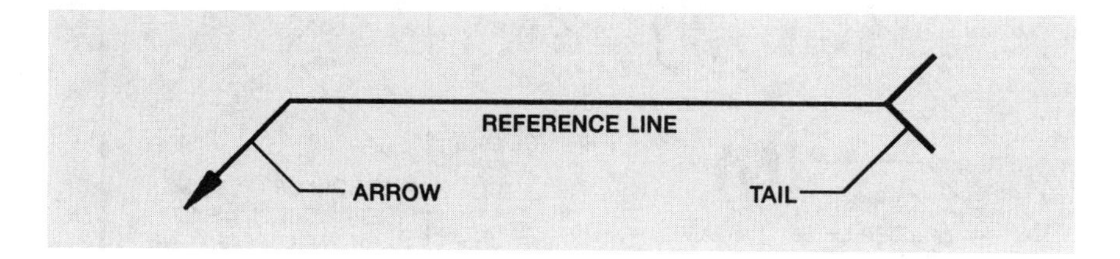

Arrow

The arrow connects the reference line to one side of the weld joint. The side touched by the arrowhead becomes known as the *arrow side,* and the opposite side is the *other side* of the joint. Bevel groove and J-groove welding symbols may include a double bend in the arrow. When this is the case the arrow is pointing to the member that is to receive the groove.

Basic Weld Symbols

The weld symbols designate the type of welding to be performed. Their position on the reference line is in the approximate center, either above, below, or on both sides of the line. Symbols below the reference line designate the weld to be on the arrow side of the joint. Symbols above the line designate the other side of the joint. Symbols on both sides of the line indicate welds on both sides of the joint. Study the basic weld symbols shown below.

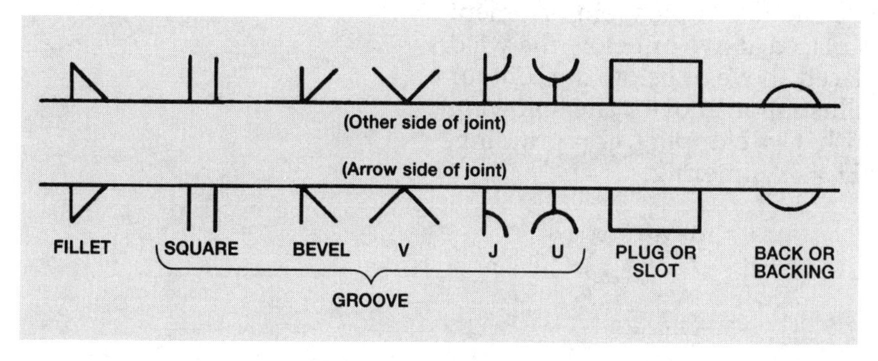

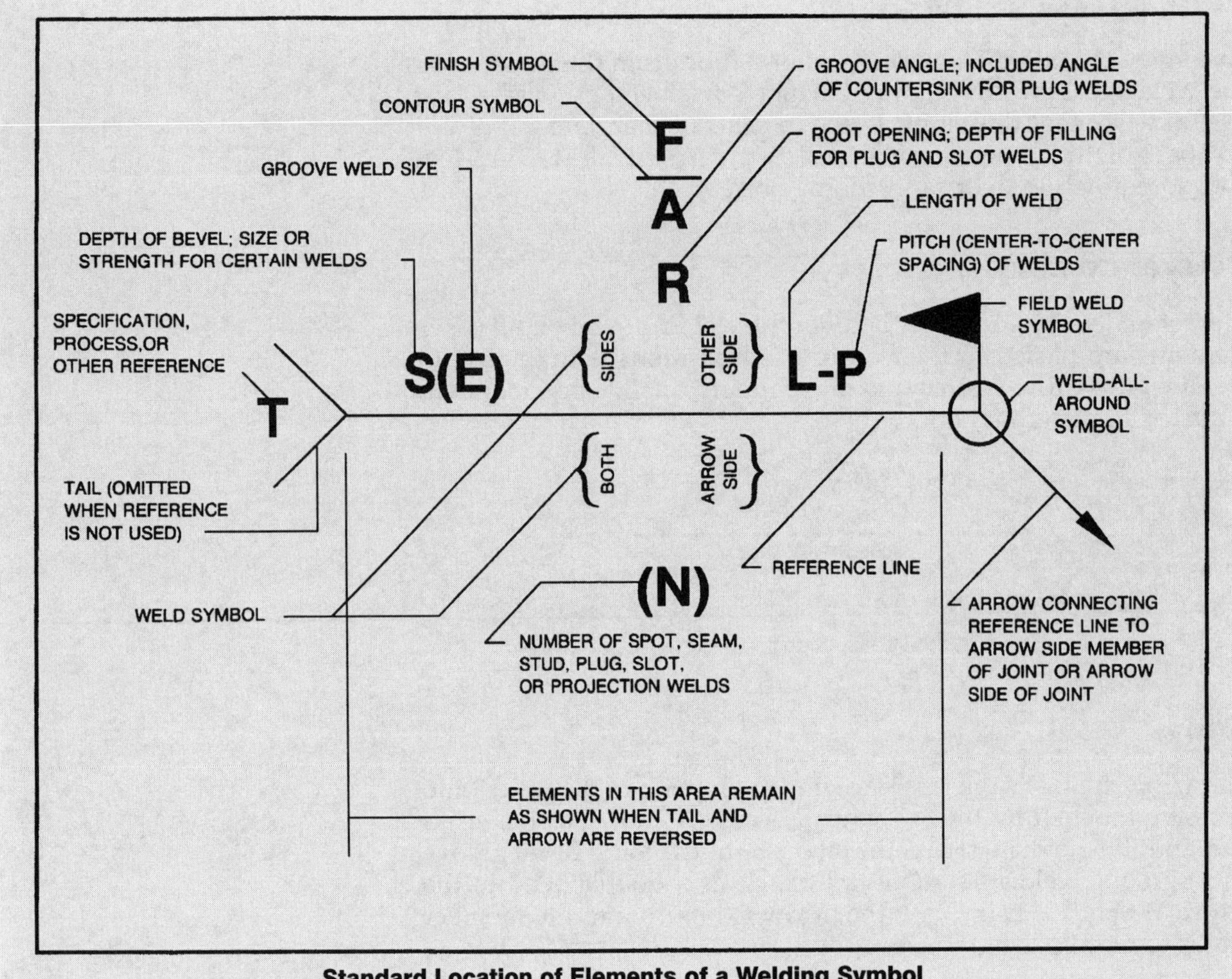

Standard Location of Elements of a Welding Symbol

(Reprinted by permission of American Welding Society)

Supplementary Symbols

Supplementary symbols convey additional information relative to the extent of the welding, where the welding is to be performed, and the contour of the weld bead. The "weld all around" and "field weld" symbols appear at the end of the reference line at the junction of the arrow. (See the example below.) The melt-thru symbol is used where 100% joint or member penetration plus reinforcement is required. The contour symbol is placed above or below the weld symbol, and the finish symbol is placed above or below the contour symbol. (See their positions in the illustration above.) The following letters are used to designate the finish: C = chipping, G = grinding, M = machining, R = rolling, and H = hammering.

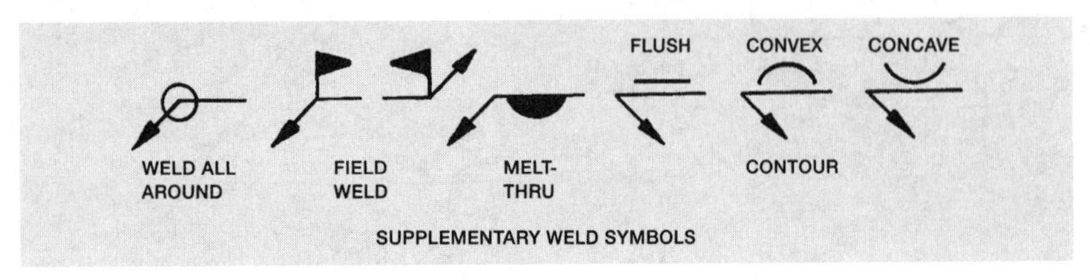

Dimensions

The size of the weld appears to the left of the basic weld symbol, and the length appears to the right of the symbol. If the length is followed by a dash, the number following the dash designates the center-to-center spacing of intermittent welds. Angles of grooves and included angles of countersinks for plug welds appear above or below the symbol. (See the illustration at the top of page 362.)

Tail

The tail appears on the end of the reference line, opposite the arrow end, whenever a specification, process, or other reference is made in the welding symbol. When no specification, process, or other reference is used with the welding symbol, the tail is omitted. Specified welding processes will appear in the tail of the welding symbol as letters. The standard letter designations for welding processes are listed alphabetically in the table below.

Letter Designations	Processes and Variations	Letter Designations	Processes and Variations
AAW	air acetylene welding	IB	induction brazing
AB	adhesive bonding	INS	iron soldering
ABW	arc braze welding	IRB	infrared brazing
AC	arc cutting	IRS	infrared soldering
AHW	atomic hydrogen welding	IS	induction soldering
AOC	oxygen arc cutting	IW	induction welding
ASP	arc spraying	LBC	laser beam cutting
AW	arc welding	LBC-A	laser beam air cutting
B	brazing	LBC-EV	laser beam evaporative cutting
BB	block brazing	LBC-IG	laser beam inert gas cutting
BMAW	bare metal arc welding	LBC-O	laser beam oxygen cutting
BW	braze welding	LBW	laser beam welding
CABW	carbon arc braze welding	LOC	oxygen lance cutting
CAC	carbon arc cutting	OAW	oxyacetylene welding
CAC-A	air carbon arc cutting	OC	oxygen cutting
CAW	carbon arc welding	OFC	oxyfuel gas cutting
CAW-G	gas carbon arc welding	OFC-A	oxyacetylene cutting
CAW-S	shielded carbon arc welding	OFC-H	oxyhydrogen cutting
CAW-T	twin carbon arc welding	OFC-N	oxynatural gas cutting
CEW	coextrusion welding	OFC-P	oxypropane cutting
CW	cold welding	OFW	oxyfuel gas welding
DB	dip brazing	OHW	oxyhydrogen welding
DFB	diffusion brazing	PAC	plasma arc cutting
DFW	diffusion welding	PAW	plasma arc welding
DS	dip soldering	PEW	percussion welding
EBC	electron beam cutting	PGW	pressure gas welding
EBW	electron beam welding	POC	metal powder cutting
EBW-HV	high vacuum electron beam welding	PSP	plasma spraying
EBW-MV	medium vacuum electron beam welding	PW	projection welding
EBW-NV	nonvacuum electron beam welding	RB	resistance brazing
EGW	electrogas welding	ROW	roll welding
ESW	electroslag welding	RS	resistance soldering
EXB	exothermic brazing	RSEW	resistance seam welding
EXBW	exothermic braze welding	RSEW-HF	high frequency seam welding
EXW	explosion welding	RSEW-I	induction seam welding
FB	furnace brazing	RSW	resistance spot welding
FCAW	flux cored arc welding	RW	resistance welding
FCAW-G	gas shielded flux cored arc welding	S	soldering
FCAW-S	self-shielded flux cored arc welding	SAW	submerged arc welding
FLB	flow brazing	SAW-S	series submerged arc welding
FLOW	flow welding	SMAC	shielded metal arc cutting
FLSP	flame spraying	SMAW	shielded metal arc welding
FOC	flux cutting	SSW	solid-state welding
FOW	forge welding	SW	stud arc welding
FRW	friction welding	TB	torch brazing
FS	furnace soldering	TC	thermal cutting
FW	flash welding	TCAB	twin carbon arc brazing
GMAC	gas metal arc cutting	THSP	thermal spraying
GMAW	gas metal arc welding	TS	torch soldering
GMAW-P	pulsed gas metal arc welding	TW	thermit welding
GMAW-S	short circuit gas metal arc welding	USW	ultrasonic welding
GTAC	gas tungsten arc cutting	UW	upset welding
GTAW	gas tungsten arc welding	UW-HF	high frequency upset welding
GTAW-P	pulsed gas tungsten arc welding	UW-I	induction upset welding
HPW	hot pressure welding	WS	wave soldering

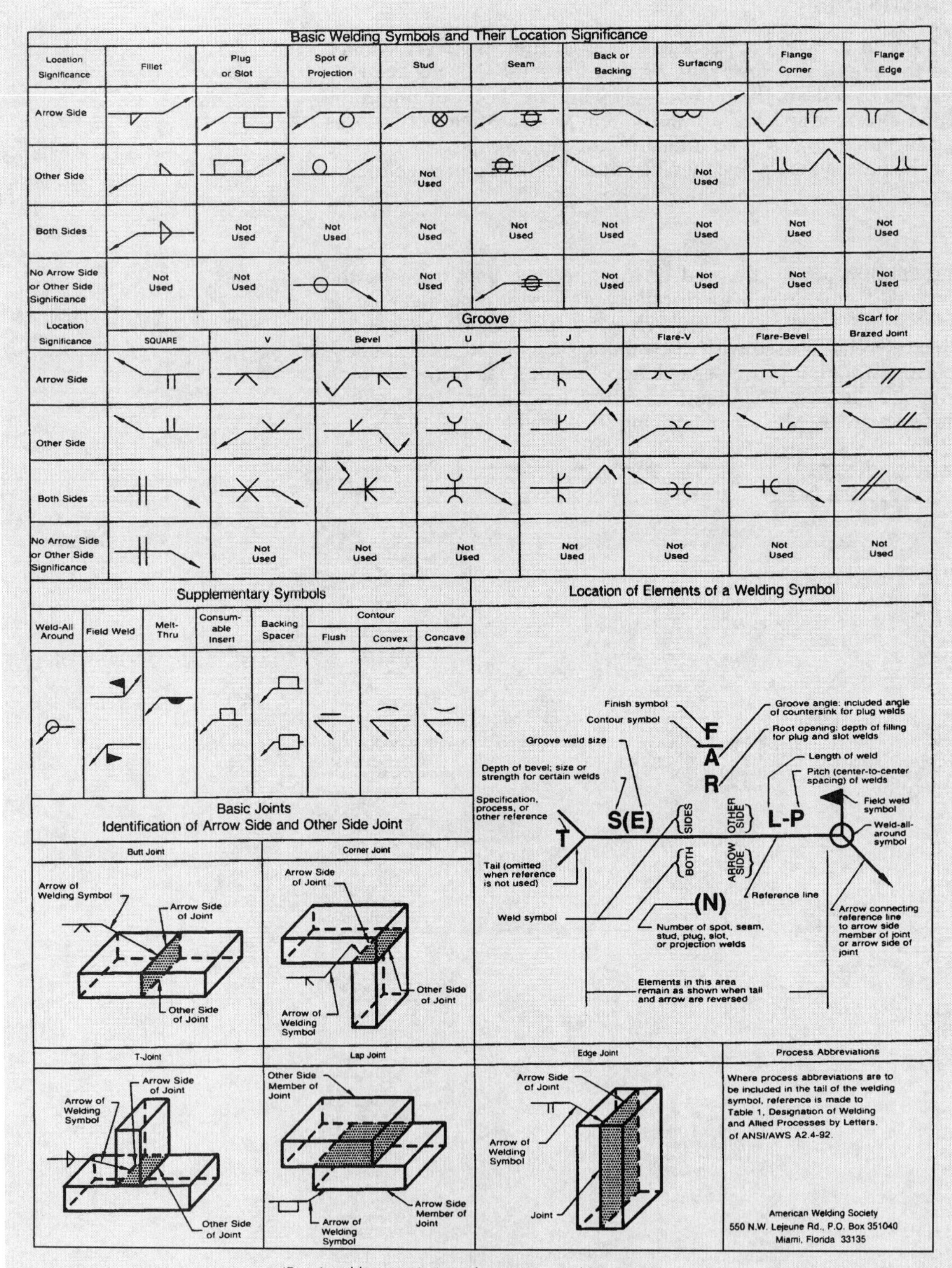

(Reprinted by permission of American Welding Society)

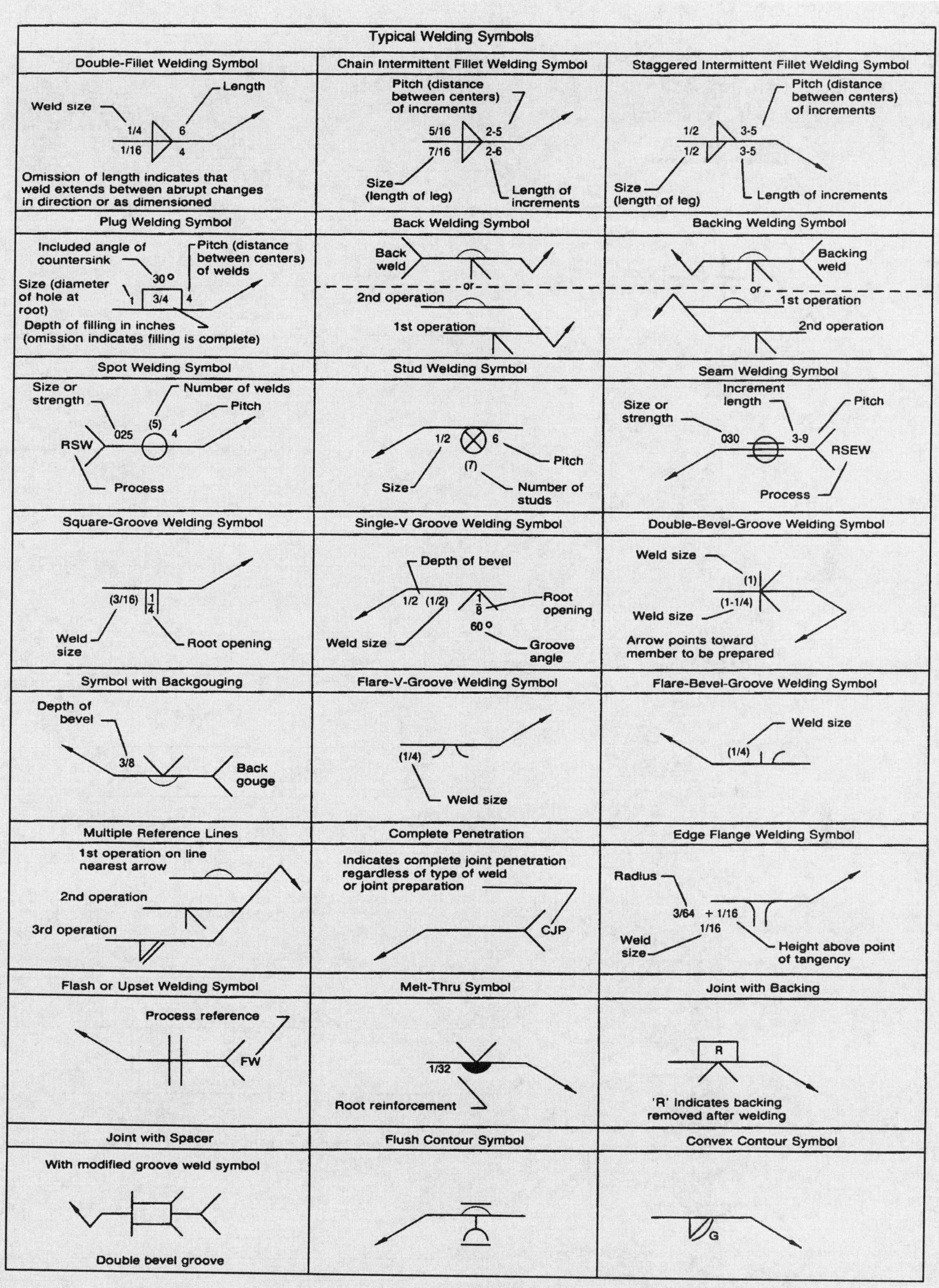

(Reprinted by permission of American Welding Society)

Welding Symbol Quiz 1

INSTRUCTIONS: List the weld name, joint side, and all other supplementary information for each of the following symbols. Refer to pages 364 and 365 if necessary.

1.

2.

3.

4. M M

5. .50

6. TB

7. 1 4 30°

8. 2-5

9.

10.

1. ______________________________

2. ______________________________

3. ______________________________

4. ______________________________

5. ______________________________

6. ______________________________

7. ______________________________

8. ______________________________

9. ______________________________

10. ______________________________

Welding Symbol Exercise

INSTRUCTIONS: Add the proper weld symbols to the reference lines on the drawing below to designate:

Ⓐ Fillet weld both sides.

Ⓑ Fillet weld all around joint.

Ⓒ V-groove weld arrow side, grind flush. Fillet weld opposite side.

Ⓓ J-groove weld both sides, then fillet weld both sides.

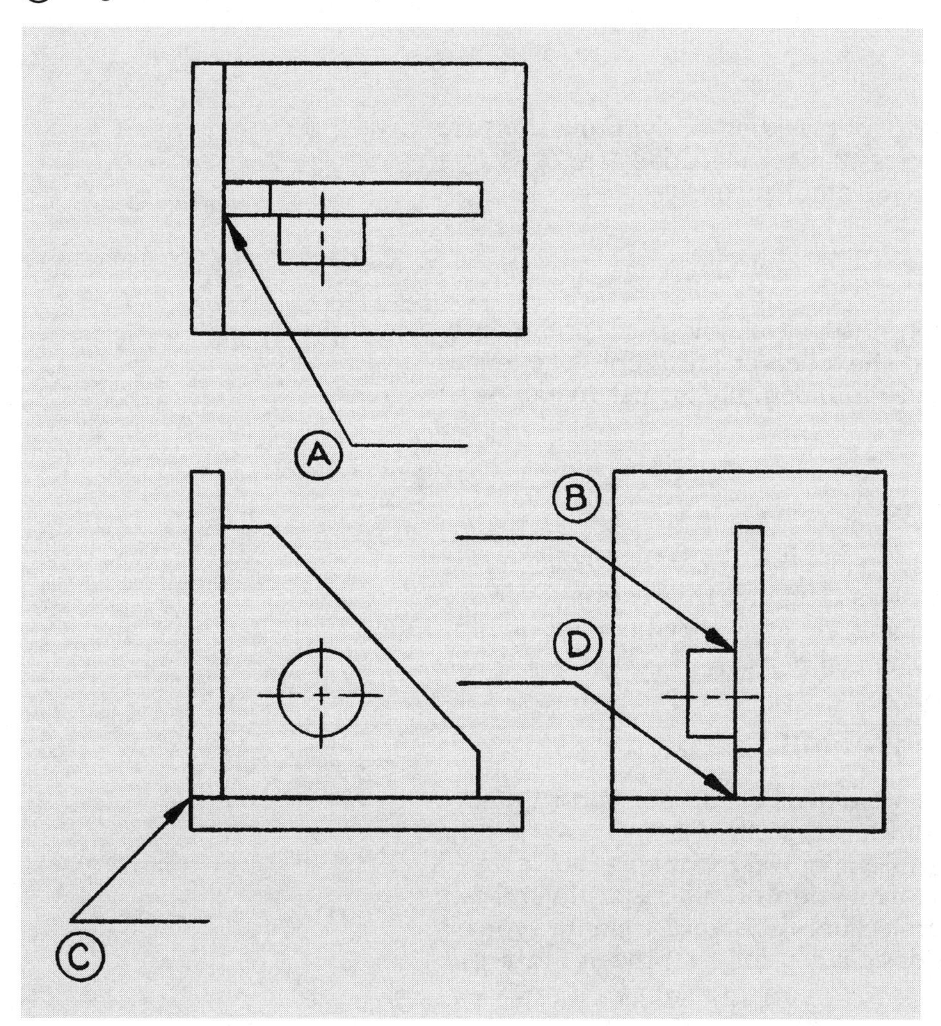

RESISTANCE WELDS

Resistance welding includes spot, projection, seam, and flash or upset welds. The process involves heating the joint with high amperage electricity and applying pressure to complete the weld.

Spot Welding

Spot welding is used mostly for fusing two or more sheets of metal together. It consists of applying pressure by means of electrodes, and then passing high current at low voltage through the sheets from one electrode to the other.

Projection Welding

Projection welding is similar to spot welding except projections are embossed into one of the pieces. Larger electrodes are used that may accommodate several welds simultaneously.

Seam Welding

Seam welding, an adaptation of spot welding, uses continuously rotating rollers for electrodes. The rollers create a series of overlapping spots as the electrodes are automatically turned on and off as they revolve.

Flash or Upset Welding

Flash or upset welding is used to join the ends of bars, rods, strips, tubing, etc. The process involves clamping the two pieces in fixtures facing each other, bringing the ends together, and holding them under pressure until fusion takes place.

Resistance Welding Symbols

Spot, seam, flash, and upset weld symbols do not normally have arrow-side or other-side significance, so they are centered on the reference line. However, projection weld symbols (abbreviated RPW) do designate either arrow-side or other-side to indicate which piece receives the projection. Resistance welding symbols include the standard letter designation for the process in the tail. (See the examples below.)

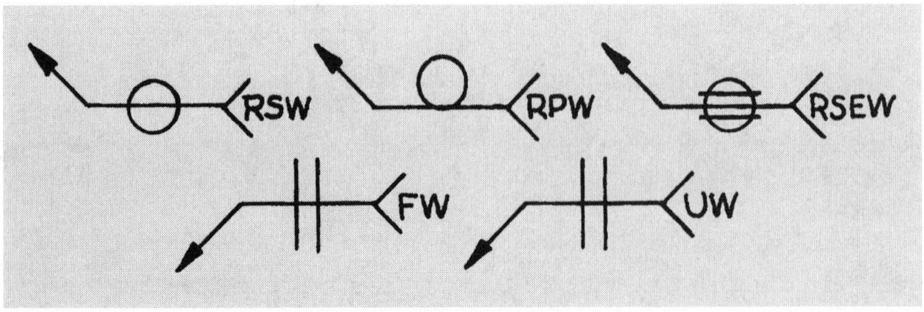

RSW RPW RSEW

FW UW

Welding Symbol Quiz 2

INSTRUCTIONS: List all of the information shown in each of the following resistance welding symbols. Refer to the table on page 363 for standard letter designations.

1. RPW

2. FW

3. .25 RSEW

4. 700 RSW

5. 500 2.00 (3) RPW

6. .50 6.00 RSW

7. 1000 5-10 RSEW

8. G G UW

9. (4) .25 3 RSW

10. (6) 3/8 4 RPW

1. ______________________________

____

2. ______________________________

____

3. ______________________________

____

4. ______________________________

____

5. ______________________________

____

6. ______________________________

____

7. ______________________________

____

8. ______________________________

____

9. ______________________________

____

10. ______________________________

____

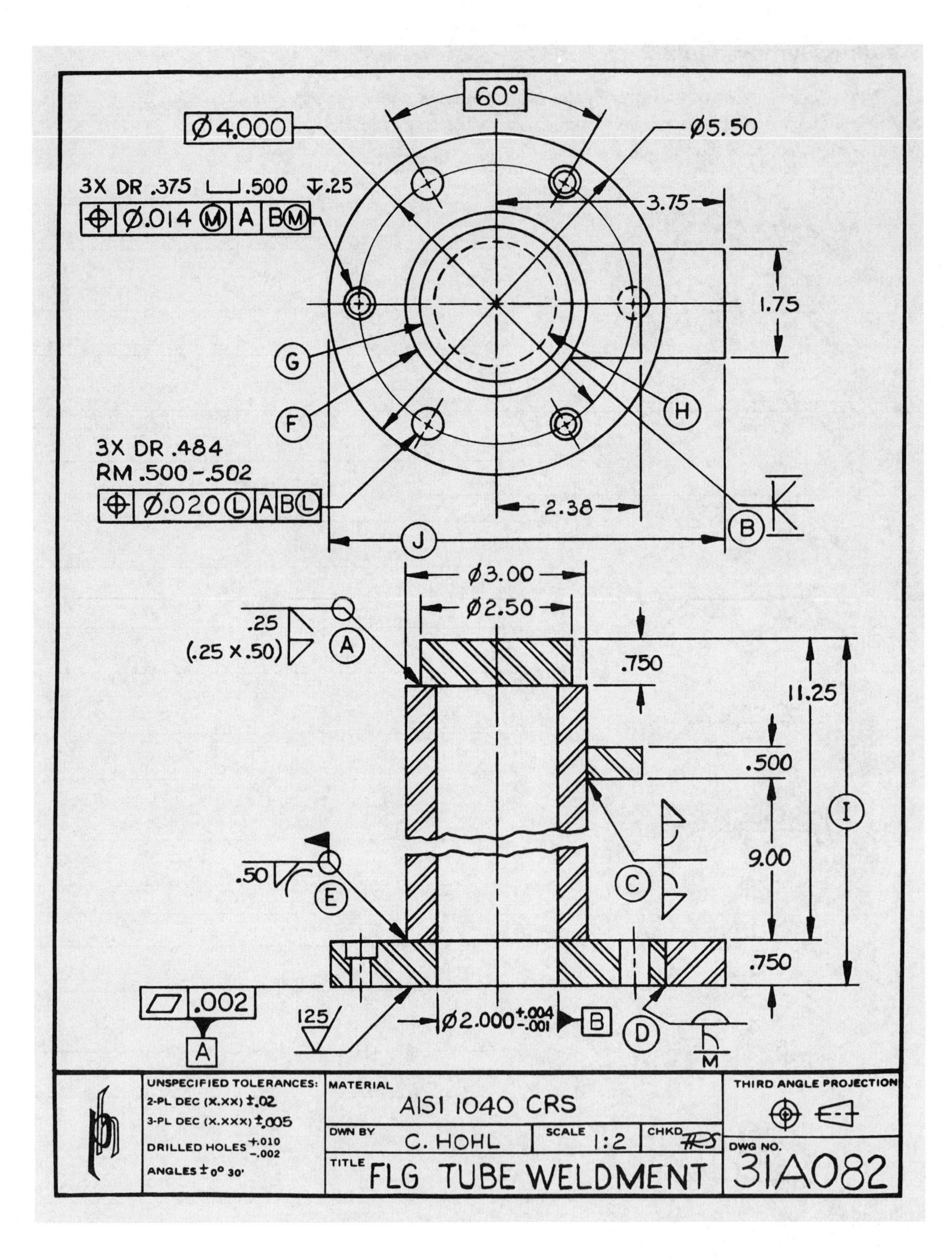
60°
Ø4.000
Ø5.50
3X DR .375 ⌴.500 ↧.25
⌖ Ø.014 Ⓜ A BⓂ
3.75
1.75
G
F
H
3X DR .484
RM .500-.502
⌖ Ø.020 Ⓛ A BⓁ
2.38
B
J
Ø3.00
Ø2.50
.25
(.25 X .50)
A
.750
11.25
.500
I
C
9.00
.50
E
.750
.002
A
125
Ø2.000 +.004 -.001
B
D
M
UNSPECIFIED TOLERANCES:
2-PL DEC (X.XX) ±.02
3-PL DEC (X.XXX) ±.005
DRILLED HOLES +.010 -.002
ANGLES ± 0° 30'
MATERIAL
AISI 1040 CRS
DWN BY
C. HOHL
SCALE
1:2
CHKD
TITLE
FLG TUBE WELDMENT
THIRD ANGLE PROJECTION
DWG NO.
31A082

INSTRUCTIONS: Refer to drawing 31A082 to answer the following questions.

1. Describe the complete specifications from the welding symbol located at Ⓐ. 1.

2. Describe the complete specifications from the welding symbol located at Ⓑ. 2.

3. Describe the complete specifications from the welding symbol located at Ⓒ. 3.

4. Describe the complete specifications from the welding symbol located at Ⓓ. 4.

5. Describe the complete specifications from the welding symbol located at Ⓔ. 5.

6. What is the MMC of diameter Ⓕ? 6. ____________

7. What is the LMC of diameter Ⓖ? 7. ____________

8. What is the MMC of diameter Ⓗ? 8. ____________

9. What is the maximum overall height dimension Ⓘ? 9. ____________

10. What is the maximum overall width dimension Ⓙ? 10. ____________

11. How many individual pieces are required to make one complete weldment? 11. ____________

12. What size is the locational tolerance zone if the three reamed holes measure exactly .500? 12. ____________

13. What size is the locational tolerance zone if the three counterbored holes measure exactly .375? 13. ____________

14. What is the geometric tolerance on datum A if the base thickness measures exactly .748? 14. ____________

15. Calculate the minimum possible material between the side of a reamed hole and the flange OD. (Consider the geometric tolerance.) 15. ____________

Welding Symbol Quiz 3

INSTRUCTIONS: Enter the letter from alongside the welding symbol that designates the following information. The same letter may be used more than once.

1. ________ Weld all around
2. ________ Field weld
3. ________ Back or backing weld
4. ________ Melt-thru
5. ________ Weld other side only
6. ________ V-groove weld
7. ________ J-groove weld
8. ________ Bevel groove weld
9. ________ Plug or slot weld
10. ________ Resistance spot weld
11. ________ Surfacing
12. ________ Flare-V weld
13. ________ Upset weld
14. ________ Flush contour
15. ________ Resistance seam weld
16. ________ Flare-bevel weld
17. ________ Concave contour
18. ________ Root opening
19. ________ Square groove weld
20. ________ Edge-flange weld

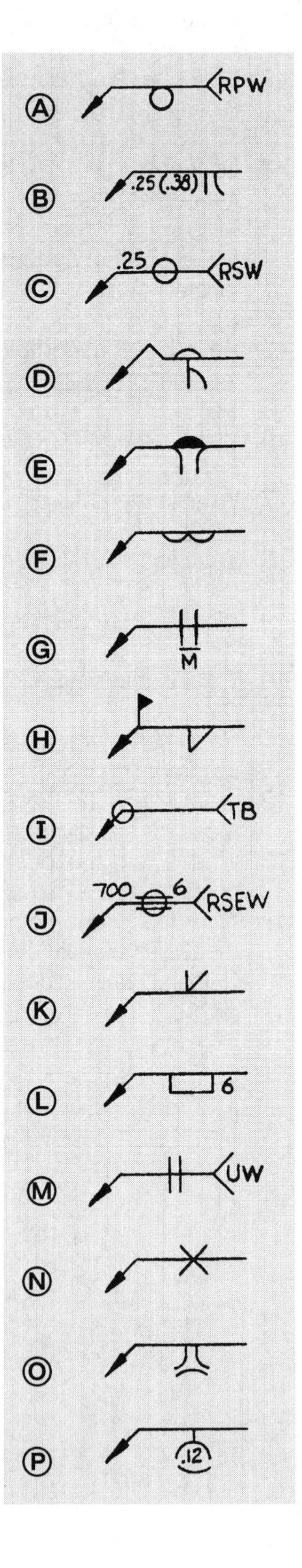

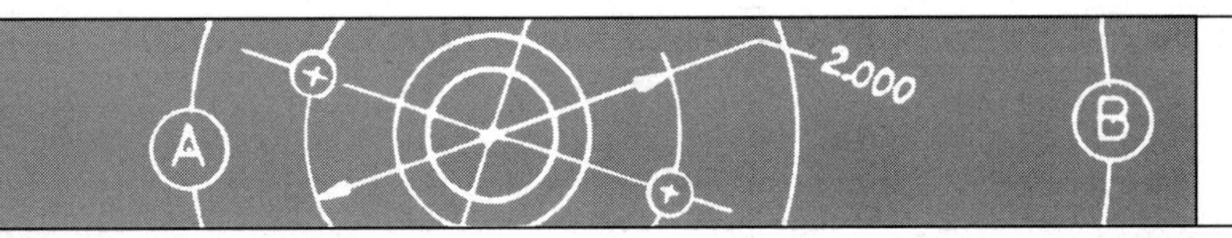

APPENDIX

MATH APPENDIX

These formulas are used to determine right angle trigonometry, measurement of dovetails, measurement of vees, determining chords, and measurement over pins.

Formulas for Right Triangle Problem Solutions

To Find Sides or Angles or Areas Use the Following Formulas

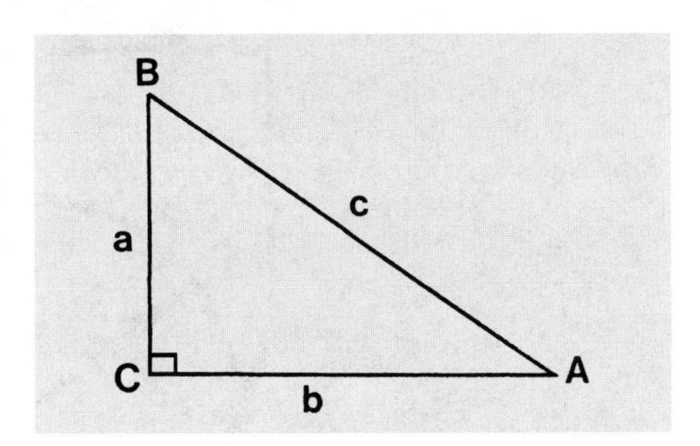

Given: ....**To Find**..........................

SIDES & ANGLES	SIDE	ANGLE	AREA
a and b	$c = \sqrt{a^2 + b^2}$	$\tan \angle A = a/b$ $\cot \angle A = b/a$ $\tan \angle B = b/a$ $\cot \angle B = a/b$	$\frac{a \times b}{2}$
a and c	$b = \sqrt{c^2 - a^2}$	$\sin \angle A = a/c$ $\csc \angle A = c/a$ $\cos \angle B = a/c$ $\sec \angle B = c/a$	$\frac{a \times \sqrt{c^2 - a^2}}{2}$
b and c	$a = \sqrt{c^2 - b^2}$	$\cos \angle A = b/c$ $\sec \angle A = c/b$ $\sin \angle B = b/c$ $\csc \angle B = c/b$	$\frac{b \times \sqrt{c^2 - b^2}}{2}$
a and $\angle A$	$b = \frac{a}{\tan \angle A}$ $b = a \times \cot \angle A$ $c = \frac{a}{\sin \angle A}$ $c = a \times \sec \angle A$	$\angle B = 90° - \angle A$	$\frac{a^2}{2 \times \tan \angle A}$
a and $\angle B$	$b = a \times \tan \angle B$ $c = \frac{a}{\cos \angle B}$ $c = a \times \sec \angle B$	$\angle A = 90° - \angle B$	$\frac{a^2 \times \tan \angle B}{2}$
b and $\angle B$	$a = b \times \tan \angle A$ $c = \frac{b}{\cos \angle A}$ $c = b \times \csc \angle B$	$\angle A = 90° - \angle B$	$\frac{b^2}{2 \times \tan \angle B}$
c and $\angle A$	$a = c \times \sin \angle A$ $b = c \times \cos \angle A$	$\angle B = 90° - \angle A$	$c^2 \times \sin \angle A \times \cos \angle A$
c and $\angle B$	$a = c \times \cos \angle B$ $b = c \times \sin \angle B$	$\angle A = 90° - \angle B$	$c^2 \times \sin \angle B \times \cos \angle B$

TO VERIFY IF A VEE HAS BEEN MACHINED TO A FINISHED SIZE

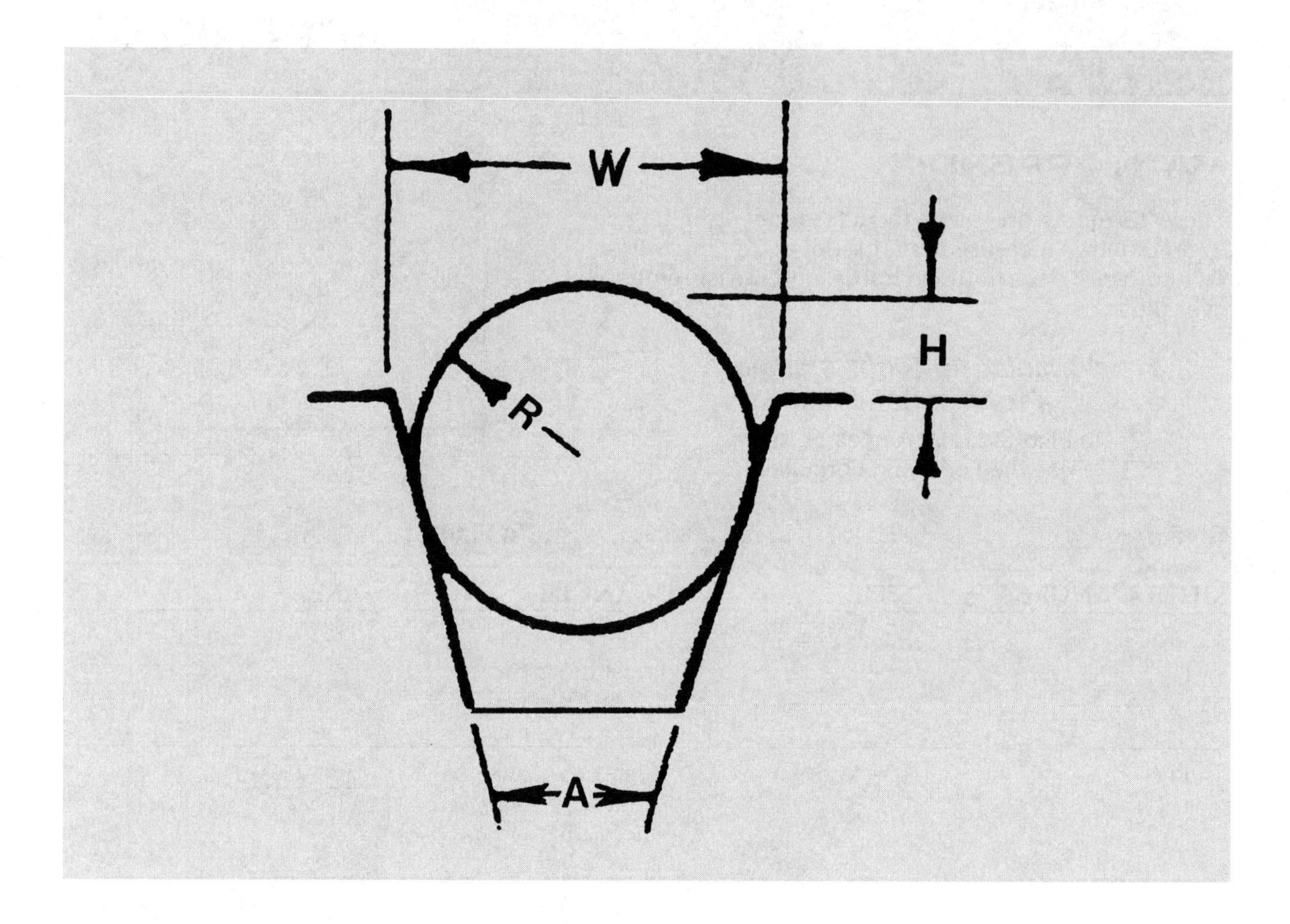

A vee that has been machined in a block cannot be measured directly. However, by placing a pin of a known diameter in the vee, the distance from the top surface where the vee is located to the top of the known diameter pin can be calculated. Then this calculated distance is compared to the actual distance on the block.

The formula for measuring a vee is:

$$H = R\left(\csc \angle\frac{A}{2} + 1\right) - \frac{W}{2}\cot \angle\frac{A}{2}$$

Where

H = Measurement from the top of a pin to the top surface where the vee is located

R = Radius of pin used to measure the vee

A = Included angle of the vee

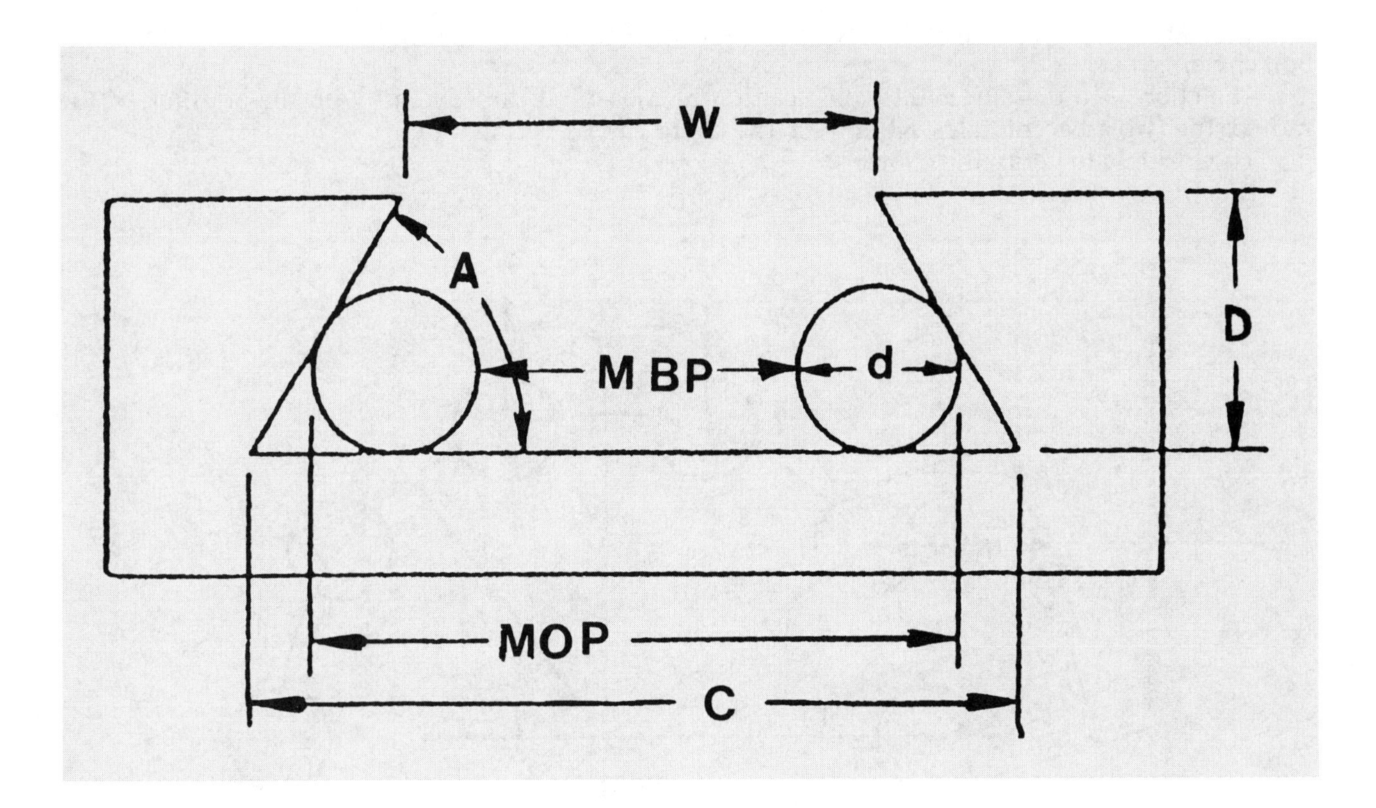

W = Opening at the top of the dovetail
C = Distance at the bottom of the dovetail
D = Depth of the dovetail
d = Diameter of precision pins used to measure dovetail
MOP = Measurement over precision pins
MBP = Measure between precision pins

$\text{MOP} = \text{W} - \text{n} + 2\text{D} \cot \angle\text{A}$ or $\text{MOP} = \text{C} - \text{n}$

$$\text{n} = \text{d}(\cot \angle\frac{\text{A}}{2} - 1)$$

MBP = MOP − diameter of 2 pins (2d)
MBP = MOP − (2d)

TO DETERMINE THE LENGTH OF A CHORD

A chord is the center-to-center distance between equally spaced adjacent holes located on the same bolt circle.

The chord is a measurement which cannot be measured directly, but with the addition of the radii of the two adjacent holes, R1 and R2, the chord can be verified.

The formula to verify the chord is:

$L = DC$

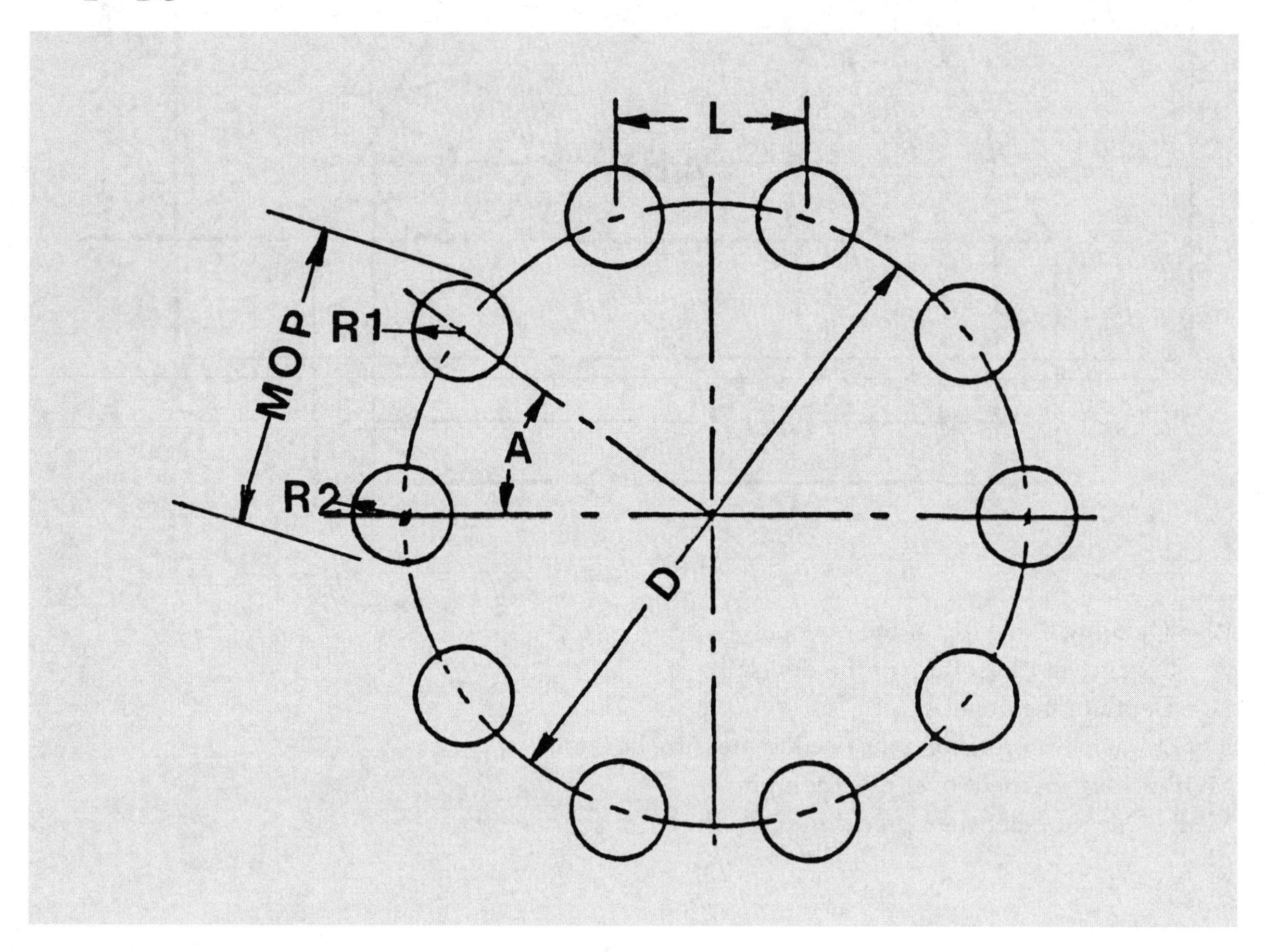

L = Length of chord
D = Diameter of bolt circle
C = A constant equal to the sin of $\frac{1}{2}$ the central angle
A = Central angle between adjacent holes
R = Radius of adjacent hole
MOP = Measurement Over Pins = L + R1 + R2

# of holes	C (constant)	# of holes	C (constant)	# of holes	C (constant)
3	.86603	7	.43388	11	.28173
4	.70711	8	.38268	12	.25882
5	.58779	9	.34202	13	.23932
6	.50000	10	.30902	14	.22252

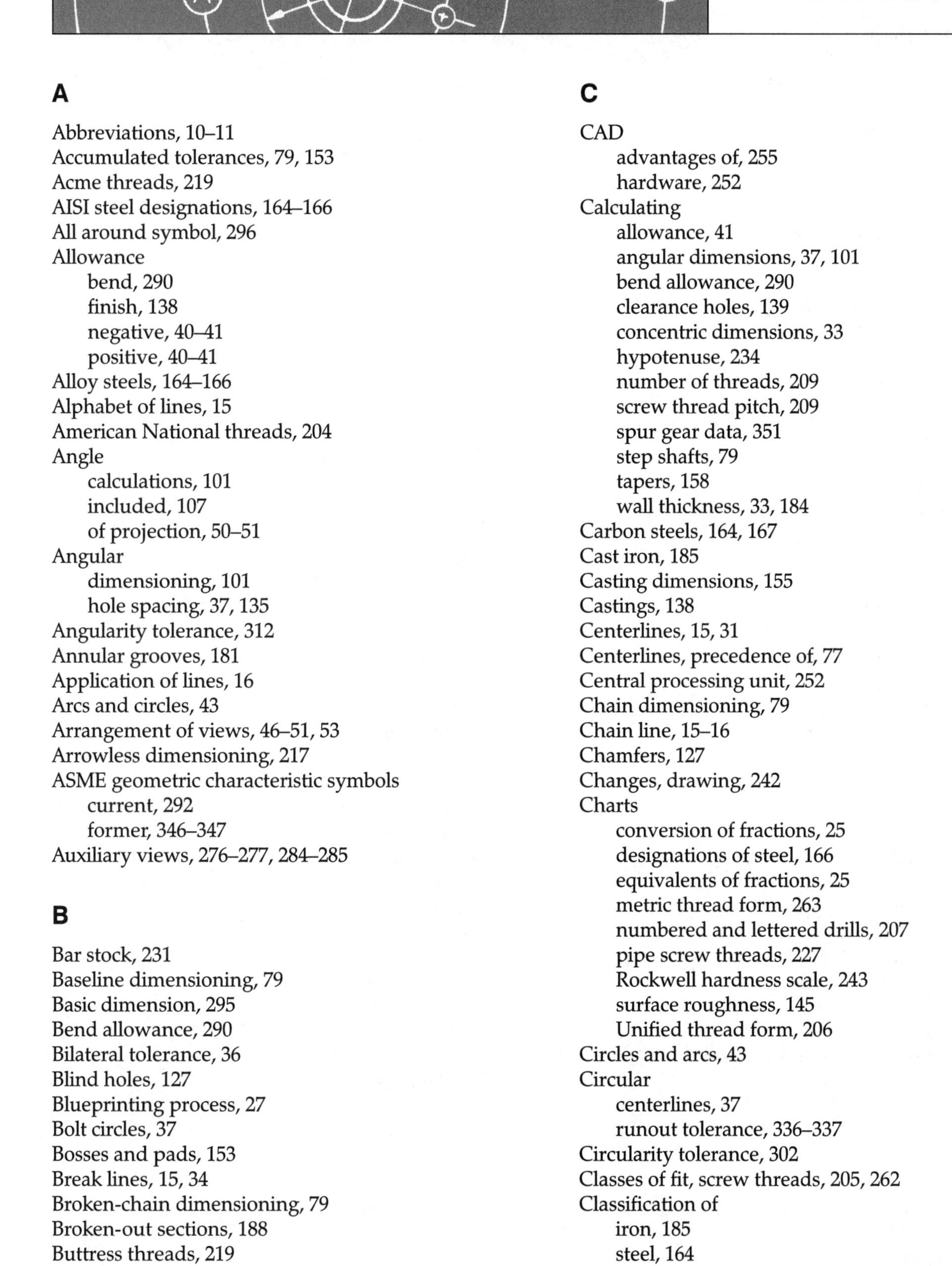

INDEX

A

B

C

D

J

K

L

M

N

O

P

R

S

T

U

V

W

Z